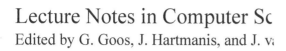

Lecture Notes in Computer Sc
Edited by G. Goos, J. Hartmanis, and J. va

T0230280

Springer

Berlin
Heidelberg
New York
Barcelona
Hong Kong
London
Milan
Paris
Tokyo

Gilles Bertrand Atsushi Imiya
Reinhard Klette (Eds.)

Digital and Image Geometry

Advanced Lectures

Springer

Series Editors

Gerhard Goos, Karlsruhe University, Germany
Juris Hartmanis, Cornell University, NY, USA
Jan van Leeuwen, Utrecht University, The Netherlands

Volume Editors

Gilles Bertrand
Groupe ESIEE, Cité Descartes
2, Bld. Blaise-Pascal, 93162 Noisy-le-Grand, France
E-mail: bertrang@esiee.fr

Atsushi Imiya
Chiba University, Inst. of Media and Information Technology
Media Technology Division
1-33 Yayoi-cho, Inage-ku, 263-8522 Chiba, Japan
E-mail: imiya@media.imit.chiba-u.ac.jp

Reinhard Klette
The University of Auckland, CITR Tamaki, Tamaki Campus
Morrin Road, Glen Innes, Auckland 1005, New Zealand
E-mail: r.klette@auckland.ac.nz

Cataloging-in-Publication Data applied for

Die Deutsche Bibliothek - CIP-Einheitsaufnahme

Digital and image geometry : advanced lectures / Gilles Bertrand ... (ed.).
- Berlin ; Heidelberg ; New York ; Barcelona ; Hong Kong ; London ; Milan ;
Paris ; Tokyo : Springer, 2001
  (Lecture notes in computer science ; Vol. 2243)
  ISBN 3-540-43079-2

CR Subject Classification (1998):I.3, I.4, G.2, G.4, G.1

ISSN 0302-9743
ISBN 3-540-43079-2 Springer-Verlag Berlin Heidelberg New York

Springer-Verlag Berlin Heidelberg New York
a member of BertelsmannSpringer Science+Business Media GmbH

http://www.springer.de

© Springer-Verlag Berlin Heidelberg 2001
Printed in Germany

Typesetting: Camera-ready by author, data conversion by DA-TeX Gerd Blumenstein
Printed on acid-free paper   SPIN 10845884   06/3142   5 4 3 2 1 0

Preface

This edited volume addresses problems in digital image geometry and topology, a special discipline of discrete mathematics. Images or discrete objects, to be analyzed based on digital image data, need to be represented, analyzed, transformed, recovered etc. These problems have stimulated many interesting developments in theoretical fundamentals of image analysis, and this volume contains reviewing articles as well as more specialized contributions on theoretical (basic) or more applied aspects of

- *digital topology*: axiomatic foundation of digital topology (E. Domínguez and A.R. Francés), programming, algorithmic, and data structure issues on cell complexes (U. Köthe and V. Kovalevsky), cellular representations of 3-manifolds (S. Matveev), and fixed points in discrete mappings (R. Tsaur and M.B. Smyth),
- *representation of images and objects*: description of a spatial modeling tool (E. Andrès, R. Breton, and P. Lienhardt), reviews on hierarchical image representations (L. Brun and W. Kropatsch) and simplicial multi-complexes for object modeling (E. Danovaro, L. De Floriani, P. Magillo, and E. Puppo), polyhedral 3D surface representations based on combinatorial topology (Y. Kenmochi and A. Imiya), and encodings of digital partitions (J. Žunić),
- *digital geometry*: discrete object reconstruction (A. Alpers, P. Gritzmann, and L. Thorens), Hough transform (P. Bhattacharya, A. Rosenfeld, and I. Weiss), a review on digital lines and digital convexity (U. Eckhardt), digital curvature (A. Imiya), Hausdorff digitization (C. Ronse and M. Tajine), and cellular convexity (J. Webster),
- *multigrid convergence*: length measurements of curves in 3D digital space via minimum-length polygons (T. Bülow, R. Klette) and via digital straight segments (D. Coeurjolly, I. Debled-Renesson, and O. Teytaud), a review on multigrid-convergent feature measurement (R. Klette), 2D curve approximation for length estimation (L. Noakes, R. Kozera, and R. Klette), and use of the relative convex hull for surface area estimation (F. Sloboda and B. Zaťko),
- *shape similarity and simplification*: rotation-invariant similarity of convex polyhedra (J.B.T.M. Roerdink and H. Bekker), surface skeletons of 3D objects (S. Svensson), a review on 3D skeletonization (J-I. Toriwaki and K. Mori), iterated morphological operations (J. Van Horebeek and E. Tapia-Rodriguez), and optimal grouping of collinear line segments (P. Veelaert).

These are the five parts of the volume, each containing five or six chapters, reflecting major developments in recent digital geometry and topology.

The volume presents extended and updated versions of 27 talks given at the winterschool "Digital and Image Geometry" (December 18 - December 22, 2000, Schloss Dagstuhl, Germany). The editors thank all the reviewers for their detailed responses and the authors for efficient collaboration in ensuring a high-quality publication. Reviewers, besides the editors, were:

E. Andrès	V. Kovalevsky
T. Asano	R. Kozera
J. Barron	W. Kropatsch
L. Brun	A. Kuba
T. Bülow	S. Matveev
D. Coeurjolly	K. Mori
M. Couprie	L. Noakes
A. Daurat	I. Nyström
I. Debled-Rennesson	L. Perroton
L. De Floriani	J.P. Réveillès
E. Domínguez	J.B.T.M. Roerdink
U. Eckhardt	C. Ronse
A.R. Francés	F. Sloboda
R. Geoghegan	P. Soille
Y. Kenmochi	M. Tajine
N. Kiryati	T. Wada
J. Klitter	J. Webster
U. Köthe	

The editors thank Michel Couprie for his crucial support in many tasks related to the editing of this volume, such as communication with authors and reviewers, and compilation of the final text.

September 2001   Gilles Bertrand, Atsushi Imiya, Reinhard Klette

Table of Contents

Part IV   Multigrid Convergence

Part V   Shape Similarity and Simplification

Part I

Topology

An Axiomatic Approach to Digital Topology *

Eladio Domínguez and Angel R. Francés

Dpt. de Informática e Ingeniería de Sistemas. Facultad de Ciencias
Universidad de Zaragoza, E-50009 – Zaragoza, Spain
ccia@posta.unizar.es

Abstract. In a series of papers the authors have developed an approach to Digital Topology, which is based on a multilevel architecture. One of the foundations of this approach is an axiomatic definition of the notion of digital space. In this paper we relate this approach with several other approaches to Digital Topology appeared in literature through a deep analysis of the axioms involved in the definition of digital space.

Keywords: digital space, lighting function

1 Introduction

The main purpose of Digital Topology is to study the topological properties of digital images. Several different approaches to Digital Topology have been proposed in literature. Let us recall Rosenfeld's graph–based approach [ ] (see also [ ], [ ]), Khalimsky's approach based in a particular topological space [ ], and Kovalevsky's approach based on abstract cell complexes [ ].

Digital images are discrete objects in nature, but they are usually representing continuous objects or, at least, they are perceived as continuous objects by observers. In a series of papers ([ ]–[ ]), the authors have developed a framework for Digital Topology that explicitly takes this fact into account and, in addition, tries to unify the previous approaches. The two foundations of our approach are a multilevel architecture and an axiomatic definition of the notion of digital space. The multilevel architecture bridges the gap between the discrete world of digital objects and the Euclidean world of their continuous interpretations. On the other hand, a digital space fixes, among all the possible continuous interpretations, just one for each digital object. This continuous interpretation of a digital object is represented at each level of the architecture using a different model; in particular, the corresponding model at the continuous level is an Euclidean polyhedron, called the *continuous analogue* of the object.

Our approach has allowed us to introduce, in a very natural way, topological notions which are generalizing the corresponding notions given in other approaches. For example, we say that a digital object is connected (a digital manifold) if its continuous analogue is a connected polyhedron (a combinatorial

* This work has been partially supported by the projects DGES PB96-1374 and DGES TIC2000-1368-C03-01.

G. Bertrand et al. (Eds.): Digital and Image Geometry, LNCS 2243, pp. 3–16, 2001.

manifold, respectively); and we have showed that, for suitable digital spaces, this notion of connectivity coincides with the (α, β)-connectivity defined on the grid $\mathbb{Z}^3$, for $\alpha, \beta \in \{6, 18, 26\}$ and $(\alpha, \beta) \neq (6, 6)$, within the graph–theoretical approach. Moreover, the digital 2-manifolds (i.e., digital surfaces) in those digital spaces are the (α, β)-surfaces ([ ]). And similarly, our notion of digital manifold generalizes also the strong 26-surfaces defined by Bertrand and Malgouyres [ ]. In addition, the bridge provided by the multilevel architecture gives us a general method to obtain results in Digital Topology by translating the corresponding results from continuous topology. We have obtained using this method general digital versions of the Jordan–Brouwer ([ ]) and Index Theorems ([ ]) for our digital manifolds, which are generalizing the well–known result for (26,6)- and (6,26)-surfaces due to Morgenthaler and Rosenfeld [ ]. Recently, we have also introduced a notion of digital fundamental group for arbitrary digital spaces and obtained a digital version of the Seifert–Van Kampen Theorem [ ].

As it was quoted above, one of the foundations of our approach is the notion of digital space. A digital space consists of a polyhedral complex K and a function that associates each digital object with a set of cells of K from which it is derived the continuous analogue of the object. That function must satisfy a certain set of axioms intended for limiting the set of possible continuous analogues to those which are not contradictory with the natural perception of objects.

The main goal of this paper is to analyze these axioms and, along the stages of this analysis, to relate our framework with some other approaches appeared in literature.

2   Our Approach to Digital Topology

In this section we briefly summarize the basic notions of our approach to Digital Topology as introduced in [ ], as well as the notation that will be used in this paper.

We represent the spatial layout of the image elements (that we will call *pixels* in 2D, *voxels* in 3D and *xels* in nD) of a digital image by a *device model*, which is a homogeneously n-dimensional locally finite polyhedral complex K. Only the n-cells in a device model K are representing xels, while the other lower dimensional cells in K are used to describe how the xels could be linked to each other. In this way, the objects displayed in digital images are subsets of the set $\text{cell}_n(K)$ of n-cells in a device model K; and, thus, we say that a subset $O \subseteq \text{cell}_n(K)$ is a *digital object in K*.

By a homogeneously n-dimensional locally finite polyhedral complex we mean a set K of polytopes, in some Euclidean space $\mathbb{R}^d$, provided with the natural reflexive, antisimetric and transitive binary relationship "to be face of", that in addition satisfies the four following properties:

1. If $\sigma \in K$ and τ is a face of σ then $\tau \in K$.
2. If $\sigma, \tau \in K$ then $\sigma \cap \tau$ is a face of both σ and τ.

3. For each point x in the underlying polyhedron $|K| = \cup\{\sigma; \sigma \in K\}$ of K, there exists a neighbourhood of x which intersects only a finite number of polytopes in K; in particular, each polytope of K is face of a finite number of other polytopes in K.
4. Each polytope $\sigma \in K$ is face of some n-dimensional polytope in K.

These complexes are particular cases of cellular complexes, as they are usually defined in polyhedral topology. So, in this paper, polytopes in K will be simply referred to as *cells*, and K itself will be called a *polyhedral complex*. Next paragraph recalls some elementary notions from polyhedral topology used in this paper. We refer to [ ] for further notions on this subject.

Given a polyhedral complex K and two cells $\gamma, \sigma \in K$, we shall write $\gamma \leq \sigma$ if γ is a face of σ, and $\gamma < \sigma$ if in addition $\gamma \neq \sigma$. A centroid-map on K is a map $c: K \to |K|$ such that $c(\sigma)$ belongs to the interior of σ; that is, $c(\sigma) \in \mathring{\sigma} = \sigma - \partial\sigma$, where $\partial\sigma = \cup\{\gamma; \gamma < \sigma\}$ stands for the boundary of σ.

Remark 1. Most examples in this paper are given on the device model R^n, termed the *standard cubical decomposition* of the Euclidean n-space $\mathbb{R}^n$. The device model R^n is the complex determined by the collection of unit n-cubes in $\mathbb{R}^n$ whose edges are parallel to the coordinate axes and whose centers are in the set $\mathbb{Z}^n$. The centroid-map we will consider in R^n associates each cube σ with its barycenter $c(\sigma)$, which is a point in the set $\mathcal{Z}^n$. Here, $\mathcal{Z} = \frac{1}{2}\mathbb{Z}$ stands for the set of points $\{z \in \mathbb{R}; z = y/2, y \in \mathbb{Z}\}$. In particular, if $\dim \sigma = n$ then $c(\sigma) \in \mathbb{Z}^n$, where $\dim \sigma$ denotes the dimension of σ; and, thus, every digital object O in R^n can be identified with a subset of points in $\mathcal{Z}^n$.

The topological properties of a digital image depend on the "continuous interpretation" that an observer makes of the object displayed in that image. The selection of such a continuous interpretation for each digital object is represented by a lighting function defined on the device model K. In the definition of lighting functions we use the following notions, which are illustrated in Fig. 1 for an object in the device model R^2.

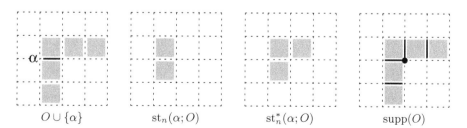

$$O \cup \{\alpha\} \qquad \mathrm{st}_n(\alpha; O) \qquad \mathrm{st}_n^*(\alpha; O) \qquad \mathrm{supp}(O)$$

Fig. 1. The support of an object O and two types of digital neighbourhoods in O for a cell α. The cells in O together with the bold edges and dots are the elements in $\mathrm{supp}(O)$

The first two notions formalize two types of "digital neighbourhoods" of a cell $\alpha \in K$ in a given digital object $O \subseteq \mathrm{cell}_n(K)$. Indeed, the *star of α in O* is the set $\mathrm{st}_n(\alpha; O) = \{\sigma \in O; \alpha \leq \sigma\}$ of n-cells (xels) in O having α as a face. Similarly, the *extended star of α in O* is the set $\mathrm{st}_n^*(\alpha; O) = \{\sigma \in O; \alpha \cap \sigma \neq \emptyset\}$ of n-cells (xels) in O intersecting α.

The third notion is the *support* of a digital object O which is defined as the set $\mathrm{supp}(O)$ of cells of K (not necessarily xels) that are the intersection of n-cells (xels) in O. Namely, $\alpha \in \mathrm{supp}(O)$ if and only if $\alpha = \cap\{\sigma; \sigma \in \mathrm{st}_n(\alpha; O)\}$. In particular, if α is a pixel in O then $\alpha \in \mathrm{supp}(O)$. Notice also that, among all the lower dimensional cells of K, only those in $\mathrm{supp}(O)$ are joining pixels in O.

To ease the writing, we use the following notation: $\mathrm{supp}(K) = \mathrm{supp}(\mathrm{cell}_n(K))$, $\mathrm{st}_n(\alpha; K) = \mathrm{st}_n(\alpha; \mathrm{cell}_n(K))$ and $\mathrm{st}_n^*(\alpha; K) = \mathrm{st}_n^*(\alpha; \mathrm{cell}_n(K))$. Finally, we shall write $\mathcal{P}(A)$ for the family of all subsets of a given set A.

Definition 1. *Given a device model K, a lighting function on K is a map $f : \mathcal{P}(\mathrm{cell}_n(K)) \times K \to \{0,1\}$ satisfying the following five axioms for all $O \in \mathcal{P}(\mathrm{cell}_n(K))$ and $\alpha \in K$:*

(1) object axiom: if $\dim \alpha = \dim K$, then $f(O, \alpha) = 1$ if and only if $\alpha \in O$;
(2) support axiom: if $\alpha \notin \mathrm{supp}(O)$ then $f(O, \alpha) = 0$;
(3) weak monotone axiom: $f(O, \alpha) \leq f(\mathrm{cell}_n(K), \alpha)$;
(4) weak local axiom: $f(O, \alpha) = f(\mathrm{st}_n^*(\alpha; O), \alpha)$; and,
(5) complement's connectivity axiom: if $O' \subseteq O \subseteq \mathrm{cell}_n(K)$ and $\alpha \in K$ are such that $\mathrm{st}_n(\alpha; O) = \mathrm{st}_n(\alpha; O')$, $f(O', \alpha) = 0$ and $f(O, \alpha) = 1$, then: (a) the set of cells $\alpha(O'; O) = \{\beta < \alpha; f(O', \beta) = 0, f(O, \beta) = 1\}$ is not empty; (b) the set $\cup\{\beta; \beta \in \alpha(O'; O)\}$ is connected in $\partial\alpha$; and, (c) if $O \subseteq \overline{O} \subseteq \mathrm{cell}_n(K)$, then $f(\overline{O}, \beta) = 1$ for every $\beta \in \alpha(O'; O)$.

If $f(O, \alpha) = 1$ we say that f lights the cell α for the object O; and the set $Lb(O) = \{\alpha \in K; f(O, \alpha) = 1\}$ consisting of all cells in K that f lights for O is said to be the light body of O.

So, according to the remarks above, we introduce the following

Definition 2. *A digital space is a pair (K, f), where K is a device model and f is a lighting function on K. That is, a model for the spatial layout of xels and a continuous interpretation for every possible digital image drawn using these xels.*

A lighting function may also satisfy some of the two following axioms:

(a) strong monotone axiom: $f(O_1, \alpha) \leq f(O_2, \alpha)$ for any pair of digital objects $O_1 \subseteq O_2 \subseteq \mathrm{cell}_n(K)$ and any cell $\alpha \in K$; and,
(b) strong local axiom: $f(O, \alpha) = f(\mathrm{st}_n(\alpha; O), \alpha)$ for every $O \subseteq \mathrm{cell}_n(K)$ and $\alpha \in K$.

Definition 3. *A lighting function f is said to be* strongly monotone *(strongly local) if it satisfies the strong monotone (respectively, strong local) axiom; otherwise, f is called* weakly monotone *(weakly local, respect.).*

The term lighting function was introduced for the first time in [ ] to refer what we call here a strongly local lighting function. While, in [ ], the general notion of lighting function given in Definition 1 was termed weak lighting function. The original set of axioms in [ ] and [ ] involved the following property

(I) for all $O \subseteq \mathrm{cell}_n(K)$ and $\alpha \in K$, $f(O, \alpha) = 1$ whenever $\alpha \in O$

instead of axiom (1) in Def. 1. Notice, however, that the conjunction of axioms (1)+(2) is equivalent to (I)+(2). Observe also that the strong local and strong monotone axioms imply the corresponding weak axioms; and, in addition, the strong local axiom also implies axiom (5) in Definition 1. Apart from these relations, it is not hard to find examples showing that axioms (1)-(5) in Def. 1 and the strong monotone and local axioms are independent; see Section 4.

We will also use the following notion in this paper.

Definition 4. *Let K be a device model. A* light body map *on K is a mapping $Lb : \mathcal{P}(\mathrm{cell}_n(K)) \to \mathcal{P}(K)$ that assigns a subset of cells of K to each digital object $O \subseteq \mathrm{cell}_n(K)$.*

Remark 2. Notice that, each lighting function on a given device model K defines a light body map on K. Moreover, the axioms given in Def. 1 can be rewritten in terms of the light body of objects as follows

(1') *Object axiom:* $\mathrm{Lb}(O) \cap \mathrm{cell}_n(K) = O$.
(2') *Support axiom:* $\mathrm{Lb}(O) \subseteq \mathrm{supp}(O)$.
(3') *Weak monotone axiom:* $\mathrm{Lb}(O) \subseteq \mathrm{Lb}(\mathrm{cell}_n(K))$.
(4') *Weak local axiom:* $\alpha \in \mathrm{Lb}(O)$ if and only if $\alpha \in \mathrm{Lb}(\mathrm{st}_n^*(\alpha; O))$.
(5') *Complement's connectivity axiom:* If $O' \subseteq O$ and $\alpha \in \mathrm{Lb}(O) - \mathrm{Lb}(O')$ is such that $\mathrm{st}_n(\alpha; O) = \mathrm{st}_n(\alpha; O')$, then $\alpha(O'; O) = \partial\alpha \cap (\mathrm{Lb}(O) - \mathrm{Lb}(O')) \neq \emptyset$, the set $\cup\{\mathring{\beta}; \beta \in \alpha(O'; O)\}$ is connected in $\partial\alpha$, and $\alpha(O'; O) \subseteq \mathrm{Lb}(\overline{O})$ for any digital object $\overline{O} \supseteq O$.
(a') *Strong monotone axiom:* if $O_1 \subseteq O_2$ then $\mathrm{Lb}(O_1) \subseteq \mathrm{Lb}(O_2)$.
(b') *Strong local axiom:* $\alpha \in \mathrm{Lb}(O)$ if and only if $\alpha \in \mathrm{Lb}(\mathrm{st}_n(\alpha; O))$.

Henceforth, we will identify a lighting function with the corresponding light body map without further comment.

Given a digital space (K, f), any digital object O in K can be naturally associated with the subspace $|\mathrm{Lb}(O)| = \cup\{\alpha; \alpha \in \mathrm{Lb}(O)\}$ of the underlying polyhedron $|K|$. Nevertheless, in general, this topological space is not suitable to represent the continuous interpretation of O provided by f. For instance, a digital object $O = \{\sigma\}$ consisting of a single n-cell $\sigma \in K$ is usually perceived as a point; while, for any lighting function, $|\mathrm{Lb}(O)| = \sigma$ is an n-dimensional ball. Anyway, a topological space, called continuous analogue, more suitable to represent such a continuous interpretation, can be derived from the light body of an object in our approach.

These continuous analogues define the fifth level of the multilevel architecture associated with each digital space. For simplicity, we introduce here only the two extreme levels of this architecture for a given digital object O in a digital space (K, f); see [ ] for a complete description of these architecture. For this, we use an arbitrary but fixed centroid–map $c : K \to |K|$ on the device model K.

The *device level* of a digital object O in (K, f) is the polyedral subcomplex $K(O) = \{\alpha \in K; \alpha \leq \sigma \in O\}$ induced by O in K. Notice that the device level of an object is fixed for any lighting function on K; that is, the discrete object displayed in a digital image does not depend on the continuous interpretation we make of it. On the other hand, the *continuous level* of O may be different for each lighting function f on K. Namely, it is defined as the underlying polyhedron $|\mathcal{A}_O|$ of the simplicial complex $\mathcal{A}_O$, whose k-simplexes are $\langle c(\alpha_0), c(\alpha_1), \ldots, c(\alpha_k) \rangle$ where $\alpha_0 < \alpha_1 < \cdots < \alpha_k$ are cells in K such that $f(O, \alpha_i) = 1$ (or, equivalently, $\alpha_i \in \mathrm{Lb}(O)$), for $0 \leq i \leq k$. The simplicial complex $\mathcal{A}_O$ is called the *simplicial analogue* of O, and the polyhedron $|\mathcal{A}_O|$ is said to be the *continuous analogue* of O.

Since continuous analogues are representing the continuous interpretation that an observer makes of digital objects, our architecture allows us to introduce digital notions in terms of the corresponding continuous ones. For example, we will say that an object O is connected if its continuous analogue $|\mathcal{A}_O|$ is a connected polyhedron; and, O is also naturally said to be a digital manifold if $|\mathcal{A}_O|$ is a combinatorial manifold. Doing that, it arises the problem of characterizing these notions in digital terms (see [ ] for connectivity, [ ],[ ] for digital surfaces and [ ] for the notion of digital fundamental group of a digital object).

3   Kovalevsky's Digital Spaces

In this section we show the relationship between the cell complex approach of Kovalevsky [ ] and our approach.

Kovalevsky uses a n-dimensional *abstract cell complex* K to represent the spatial layout of xels in a digital image, where the n-cells in K are the xels. Recall that any abstract cell complex has a well defined topological structure. In particular, for a given complex K, the following combinatorial notion of connectivity is equivalent to the usual definition in general topology: a subset $A \subseteq K$ is said to be *connected* if for each pair of cells $\alpha, \beta \in A$ there is a finite sequence $(\alpha_0, \alpha_1, \ldots, \alpha_k)$ in A such that $\alpha_0 = \alpha$, $\alpha_k = \beta$ and either α_i is a face of α_{i-1} or α_{i-1} is a face of α_i, for $1 \leq i \leq k$.

In order to avoid connectivity paradoxes ([ ]), Kovalevsky points out that a digital image should not be defined by specifying only the set of xels belonging to the object O displayed in the image. Instead, a subset $A(O) \subseteq K$ containing (possibly) some lower dimensional cells of K must be associated with that object (naturally the only n-cells in $A(O)$ correspond to xels in O). Under Kovalevsky's point of view any topological property (in particular, connectivity) of the object O is analyzed through the properties of the set $A(O)$. But, in our opinion, some type of uniform methods to associate every digital object O with such a set $A(O)$ is needed if our goal is to perform a formal development of the topological analysis of digital images. In fact, the idea of these uniform methods

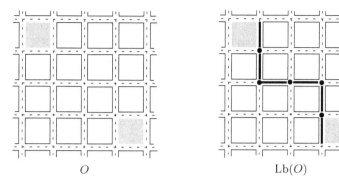

O $Lb(O)$

Fig. 2. Kovalevsky's spaces may not represent accurately our usual perception of objects

is underlying the notion of "face membership rule" introduced by Kovalevsky in [ ]. From this point of view, the theoretical development of Digital Topology take as basic structure the pair consisting of an abstract cell complex and a face membership rule.

From the remarks above, Kovalevsky's approach can be interpreted in our approach as follows.

Definition 5. *A Kovalevsky's space is a pair* (K, Lb) *where* K *is a polyhedral complex and* $Lb : \mathcal{P}(cell_n(K)) \rightarrow \mathcal{P}(K)$ *is a map satisfying*

(1') $Lb(O) \cap cell_n(K) = O$, *for any digital object* O *(see Def. 1 and Remark 2).*

Notice, however, that this notion of digital space is too much general. For example, let us consider a digital object $O \subseteq cell_2(R^2)$ consisting of two pixels (2-cells) σ, τ such that $\sigma \cap \tau = \emptyset$ (see Fig. 2). It is possible to find quite a lot Kovalevsky's digital spaces (R^2, Lb) in such a way the set $Lb(O)$ contains a sequence $(\alpha_0, \alpha_1, \ldots, \alpha_k)$ with $\alpha_0 = \sigma$, $\alpha_k = \tau$ and $\alpha_i < \alpha_{i-1}$ or $\alpha_{i-1} < \alpha_i$, $1 \leq i \leq k$. Since $Lb(O)$ is connected, O should be considered as a connected digital object in this space; but this is in contradiction with our usual perception of two isolated pixels.

In this way, it seems advisable to restrict the category of Kovalevsky's digital spaces to another category in which it is possible to find properties of general application and, moreover, whose digital spaces are not contradictory with our usual perception of digital images.

4   Analyzing the Axioms of Lighting Functions

In this section we will consider light body maps; or, equivalently, lighting functions which does not necessarily satisfy all axioms given in Def. 1. Notice, however, that for a given digital object O, we can use the set $Lb(O)$ to replicate the construction of a continuous analogue $|\mathcal{A}_O|$ for O. But doing that, the continuous analogue of a digital object may be contradictory with our usual perception of that object. Actually, each axiom in Def. 1 prevents some of such anomalous analogues as we will next justify.

4.1 Object Axiom

To justify this axiom, we shall use a quite natural argument related to the connectivity of a digital object.

As Kovalevsky, the notion of connectivity we will consider for digital objects is naturally induced by the face relationship among the cells of the device model.

Definition 6. *Let O be a digital object in a device model K and Lb a light body map defined on K. Two n-cells (xels) $\sigma, \tau \in O$ are said to be connected by a path in O if there exists a sequence $(\alpha_0, \alpha_1, \ldots, \alpha_k)$ in $Lb(O)$ such that $\alpha_0 = \sigma$, $\alpha_k = \tau$, and either $\alpha_{i-1} < \alpha_i$ or $\alpha_i < \alpha_{i-1}$, for $1 \leq i \leq k$. A subset $C \subseteq O$ is a digital component of O if any pair of n-cells $\sigma, \tau \in C$ are connected by a path in O and, in addition, no n-cell $\sigma \in C$ is connected by a path in O to a n-cell in $O - C$. Finally, a digital object is said to be connected if it has only one component.*

Notice that the cell α_1, in the previous definition, is necessarily a face of the n-cell σ. So that, for each n-cell $\sigma \in O$, the map Lb determines what faces of σ are allowed to connect σ with any other n-cell in O.

However, this natural notion of connectivity for a digital object O may not agree exactly with the connectivity of its continuous analogue $|\mathcal{A}_O|$; that is, with our continuous interpretation of the connectivity of O. On one hand, if $\sigma \notin Lb(O)$ for a xel $\sigma \in O$, the previous definition yields that $C = \{\sigma\}$ is a digital component of O; but this component is not represented in the continuous analogue since $c(\sigma) \notin |\mathcal{A}_O|$. This problem disappears if the light body map Lb satisfies the following property:

(I') for all $O \subseteq \mathrm{cell}_n(K)$, $O \subseteq Lb(O)$.

On the other hand, if a cell $\alpha \in Lb(O) - O$ is not connected by a path to some xel in O, then there is a component in $|\mathcal{A}_O|$ which does not correspond to any digital component of O. A sufficient condition to prevent this fact is the following:

(A) for any cell $\alpha \in Lb(O)$ there is a n-cell $\sigma \in O$ such that $\alpha \leq \sigma$.

In particular, none of the n-cells in the background $\mathrm{cell}_n(K) - O$ of O should belong to $Lb(O)$, since these cells are not faces of n-cells in O; otherwise, a path connecting two xels in O could intersect the background of O.

For a light body map satisfying these two properties it is immediate to show that there is a one–to–one correspondence between the digital components of any digital object O and the connected components of its continuous analogue $|\mathcal{A}_O|$; that is, the connectivity of digital objects, as it is defined in Def. 6, agrees

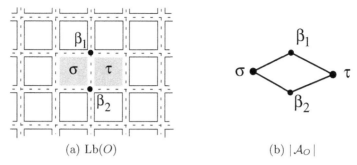

(a) Lb(O)   (b) $|\mathcal{A}_O|$

Fig. 3. In case a lighting function does not satisfy the support axiom, the continuous interpretation provided by the continuous analogue $|\mathcal{A}_O|$ of an object O may be contrary to our usual perception of that object

with the connectivity of their continuous analogues. Moreover, properties (I') and (A) together imply the *object axiom*: for any digital object $O \subseteq \mathrm{cell}_n(K)$, $\mathrm{Lb}(O) \cap \mathrm{cell}_n(K) = O$.

4.2 Support Axiom

Notice that, the object axiom and property (A) above are limiting the set of cells that may be lighted for a given digital object. In fact, these two conditions suffice to ensure that the two pixels in a digital object $O = \{\sigma, \tau\}$, with $\sigma \cap \tau = \emptyset$, are never connected by a path of lower dimensional cells "crossing" the background of O, as it could be the case in Kovalevsky's digital spaces (see Fig. 2).

However, the object axiom and property (A) do not suffice to avoid the following unusual continuous interpretation of a digital object.

Let $O = \{\sigma, \tau\}$ be a digital object in the device model R^2 with $\dim \sigma \cap \tau = 1$, and let Lb be a light body map defined on R^2 such that $\mathrm{Lb}(O) = \{\sigma, \tau, \beta_1, \beta_2\}$, where β_1, β_2 are the two vertices of the edge $\sigma \cap \tau$ (see Fig. 3(a)). Notice that Lb can be defined for the rest of objects in R^2 in such a way it satisfies both the object axiom and property (A) above. However, the corresponding continuous analogue $|\mathcal{A}_O|$ depicted in Fig. 3(b) is a continuous closed curve; but this continuous interpretation of O is contrary to our usual perception of that object.

This anomalous example vanishes if, in addition to the n-cells σ and τ, the light body $\mathrm{Lb}(O)$ is allowed to contain only the cell $\sigma \cap \tau$; that is, the only cell in $\mathrm{supp}(O)$. So, we require in general the *support axiom*: $\mathrm{Lb}(O) \subseteq \mathrm{supp}(O)$ for any object $O \subseteq \mathrm{cell}_n(K)$.

4.3 Monotone Axioms

Monotone axioms, as well as local axioms analyzed in next section, are needed to ensure a certain compatibility between the continuous analogues of different digital objects. To show this, let us consider the digital object $O_1 \subseteq \mathrm{cell}_2(R^2)$

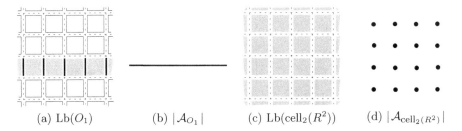

(a) Lb(O_1)   (b) $|\mathcal{A}_{O_1}|$ (c) Lb(cell$_2$(R^2))   (d) $|\mathcal{A}_{\text{cell}_2(R^2)}|$

Fig. 4. A discrete digital space may contain connected digital objects consisting of two or more pixels in case the lighting function does not satisfy the weak monotone axiom

shown in Fig. 4(a), and let Lb : $\mathcal{P}(\text{cell}_2(R^2)) \rightarrow \mathcal{P}(R^2)$ be the light body map on the device model R^2 given by Lb(O_1) = supp(O_1) and Lb(O) = O for any other object $O \neq O_1$. Notice that Lb satisfies the object and support axioms above. The continuous analogues $|\mathcal{A}_{O_1}|$ and $|\mathcal{A}_{\text{cell}_2(R^2)}|$ of the objects O_1 and cell$_2$(R^2) (i.e., the object consisting of all the pixels in the device model R^2), derived from this light body map, are those depicted in Figs. 4(b) and (d), respectively. Observe that $|\mathcal{A}_{O_1}|$ is a connected straight line, while the polyhedron $|\mathcal{A}_{\text{cell}_2(R^2)}|$ is a set of discrete points. But this continuous interpretation is contrary to our usual perception, since the only connected objects contained in a discrete ambient space are singletons.

A sufficient condition to avoid this type of examples is the *weak monotone axiom*:

For any object $O \subseteq \text{cell}_n(K)$ in a device model K, Lb(O) $\subseteq$ Lb(cell$_n(K)$). caption

It is immediate to show that if a light body map Lb satisfies this axiom, then the continuous analogue of any object O is always a subpolyhedron of the continuous analogue of the space cell$_n(K)$; that is, $|\mathcal{A}_O| \subseteq |\mathcal{A}_{\text{cell}_n(K)}|$. But, for two given digital objects $O_1 \subseteq O_2$, we cannot derive in general that $|\mathcal{A}_{O_1}|$ is a subpolyhedron of $|\mathcal{A}_{O_2}|$. A necessary condition to show this fact is the *strong monotone axiom*:

Lb(O_1) $\subseteq$ Lb(O_2), for any pair of objects $O_1 \subseteq O_2$.

Within the graph–theoretical approach to Digital Topology, Kong and Roscoe ([ ]) define the notion of (α, β)-surface on $\mathbb{Z}^3$, for $(\alpha, \beta) \in \{(6, 26), (26, 6), (6, 18),$ $(18, 6)\}$, generalizing the (6,26)- and (26,6)-surfaces of Morgenthaler and Rosenfeld [ ]. In [ ] we prove that the strong monotone axiom suffice to represent, in our framework, the continuous interpretation associated with these (α, β)-surfaces, for $(\alpha, \beta) \in \{(6, 26), (26, 6), (6, 18)\}$. More precisely, we show that there are digital spaces $(R^3, f_{\alpha\beta})$, where $f_{\alpha\beta}$ is a strongly local and strongly monotone lighting function, such that the set of digital surfaces in these spaces coincides with the set of (α, β)-surfaces. Recall that a digital surface in a digital space (K, f) is a digital object S whose continuous analogue $|\mathcal{A}_S|$ is a surface. However, it cannot be found a strongly monotone lighting function $f_{18,6}$ on R^3 such that every $(18, 6)$-surface is a digital surface in $(R^3, f_{18,6})$. Also in [ ] we show that such a digital space exists in the category of strongly local but weakly monotone lighting functions. Up to now, we have not found a digital space whose digital surfaces coincide exactly with the set of $(18, 6)$-surfaces. We conjecture, however, that this space could be found in the category of lighting functions given in Def. 1.

4.4   Local Axioms

The argument to justify these axioms is related to the physical limitations of the human perception. Indeed, our usual perception of objects is always local. So that, the continuous interpretation we can make of some piece of a digital object should be independent of any other piece of that object which is not near enough to the first one. As a consequence, it seems reasonable to ask lighting functions to satisfy some type of local property. Notice that, in case the lighting function satisfies the object axiom, such a property is not needed for the set of n-cells in the device model since, in that case, any n-cell α is lighted for an object O if and only if α belongs to O. If, in addition, the lighting function also satisfies the support axiom, then that local condition is only necessary for lower dimensional cells in the support of an object O, since $f(O, \alpha) = 0$ for any cell $\alpha \notin \mathrm{supp}(O)$.

The natural method to formalize such a local condition is to relate the lighting of a lower dimensional cell $\alpha \in K$ for a given digital object O to a certain digital neighbourhood of α in O. There are two natural possibilities for this notion of digital neighbourhood:

the star of α in O : $\mathrm{st}_n(\alpha; O) = \{\sigma \in O; \alpha \leq \sigma\}$; and
the extended star of α in O : $\mathrm{st}_n^*(\alpha; O) = \{\sigma \in O; \sigma \cap \alpha \neq \emptyset\}$.

Notice that the star of α in O coincides with the set $\mathrm{st}(\alpha; K) \cap O$, where $\mathrm{st}(\alpha; K) = \{\beta \in K; \text{ there is } \gamma \in K \text{ with } \alpha \leq \gamma \text{ and } \beta \leq \gamma\}$ is related to the smallest open neighbourhood of α in the topology of K when it is reagarded as an abstract cell complex; see [ ]. However, the polyhedron $|\mathrm{st}(\alpha; K)|$ is not a neighbourhood of α in the Euclidean space $|K|$. On the other hand, $\mathrm{st}_n^*(\alpha; O) = N(\alpha; K) \cap O$, where $N(\alpha; K) = \{\beta \in K; \beta \leq \gamma \in K, \gamma \cap \alpha \neq \emptyset\}$ is the smallest subcomplex of K which is also a neighbourhood of α in $|K|$, see [ ]. Depending on which one of these topologies we want to represent, we should use one of the two following axioms for a map $f : \mathcal{P}(\mathrm{cell}_n(K)) \times \mathcal{P}(K) \to \{0, 1\}$:

Strong local axiom : $f(O, \alpha) = f(\mathrm{st}_n(\alpha; O), \alpha)$
Weak local axiom : $f(O, \alpha) = f(\mathrm{st}_n^*(\alpha; O), \alpha)$

As it was quoted in the previous section, the category of strongly local and weakly monotone digital spaces suffices to represent the continuous interpretation of digital objects associated with the notion of (α, β)-surface. However, we

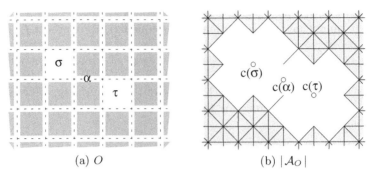

(a) O (b) $|\mathcal{A}_O|$

Fig. 5. The complement's connectivity axiom is required to obtain a right representation of the connectivity of the background of any object

show in [ ] that, for any digital space (R^3, f) in this category, the set of digital surfaces in (R^3, f) does not contain the set of all strong 26-surfaces ([ ]). On the other hand, such a digital space can be found in the category of weakly local digital spaces ([ ]).

4.5   Complement's Connectivity Axiom

As it is usual, we consider that the object displayed in a binary digital picture consists of the pixels with value 1, while the background of that picture is the set of the pixels with value 0. It is well known that the perception of a set of pixels, and so its topological properties, may be different if we consider that set as the object displayed in a picture or as the background in the complementary picture. This leads us to think that, for a given digital object O in a device model K, the properties of the background $\mathrm{cell}_n(K) - O$ of O should not be analyzed by using the continuous analogue $|\mathcal{A}_{\mathrm{cell}_n(K)-O}|$, since this polyhedron is representing the continuous interpretation of the set of pixels $\mathrm{cell}_n(K)-O$ when it is considered as the digital object displayed in a digital picture. On the contrary, the continuous interpretation of the background of the object O is better represented by the complement $|\mathcal{A}_{\mathrm{cell}_n(K)}| - |\mathcal{A}_O|$ of its continuous analogue. In particular, the connectivity of this topological space should describe the connectivity of the set of pixels $\mathrm{cell}_n(K) - O$ when it is understood as the background of O.

In [ ] we introduce a "digital" notion for the connectivity of the background of an object O at the device level of a digital space (K, f). And, moreover, we show that in case the map f is strongly local, weakly monotone and if, in addition, it satisfies both the object and support axioms, then there is a one–to–one correspondence between the digital components of the background $\mathrm{cell}_n(K) - O$ and the connected components of the topological space $|\mathcal{A}_{\mathrm{cell}_n(K)}| - |\mathcal{A}_O|$. However, next example shows that, in general, this correspondence does not hold for a map f satisfying only the weak local axiom.

Let us consider the map $f : \mathcal{P}(\mathrm{cell}_2(R^2)) \times \mathcal{P}(R^2) \rightarrow \{0, 1\}$ defined on the device model R^2 by $f(O, \alpha) = 1$ if and only if $\mathrm{st}_2^*(\alpha; R^2) \subseteq O$. Notice that f

satisfies all axioms in Def. 1 but the complement's connectivity axiom. Now, let $O = R^2 - \{\sigma, \tau\}$ be the digital object consisting of all the 2-cells in R^2 but two of them σ, τ as it is shown in Fig. 5(a). It is easily checked that, for the map f, the continuous analogue of the object $\mathrm{cell}_2(R^2)$ is the whole Euclidean plane, while $|\mathcal{A}_O|$ is the polyhedron depicted in Fig. 5(b). From these facts, we get that $|\mathcal{A}_{\mathrm{cell}_2(R^2)}| - |\mathcal{A}_O|$ is a connected space, which does not correspond with our usual perception of the background $\{\sigma, \tau\}$ of O consisting of two isolated pixels.

This paradoxical example can never occur for a lighting function f; that is, a function satifying the object, support, weak monotone and weak local axioms and, in addition, the *complement's connectivity axiom* in Def. 1:

> For objects $O' \subseteq O \subseteq \mathrm{cell}_n(K)$ and a cell $\alpha \in K$ such that $\mathrm{st}_n(\alpha; O) = \mathrm{st}_n(\alpha; O')$, $f(O', \alpha) = 0$ and $f(O, \alpha) = 1$, then: (a) the set of cells $\alpha(O'; O) = \{\beta < \alpha; f(O', \beta) = 0, f(O, \beta) = 1\}$ is not empty; (b) the set $\cup\{\overset{\circ}{\beta}; \beta \in \alpha(O'; O)\}$ is connected in $\partial\alpha$; and, (c) if $O \subseteq \overline{O} \subseteq \mathrm{cell}_n(K)$, then $f(\overline{O}, \beta) = 1$ for every $\beta \in \alpha(O'; O)$.

Actually, if O is an object in a digital space (K, f) such that f satisfies all the axioms in Def. 1, then there is a one–to–one correspondence between the digital components of $\mathrm{cell}_n(K) - O$ and the connected components of $|\mathcal{A}_{\mathrm{cell}_n(K)}| - |\mathcal{A}_O|$, as in the case of the strongly local lighting functions (see [ ]). Finally, notice that the strong local axiom implies the complement's connectivity axiom, since $f(O, \alpha) = f(\mathrm{st}_n(\alpha; O), \alpha) = 1$ and $f(O', \alpha) = f(\mathrm{st}_n(\alpha; O'), \alpha) = 0$ yield that $\mathrm{st}_n(\alpha; O) \neq \mathrm{st}_n(\alpha; O')$, and so no cell α can satisfy all the requirements in the complement's connectivity axiom.

5   Conclusions

Along the previous section we have given natural reasons justifying the necessity of several axioms to limit a general notion of digital space to a category in which certain paradoxical interpretations of digital objects cannot be done. We have pointed out various relations among these axioms. And we have also recalled some of the results reached using them, which are showing that our approach to digital topology is expressive enough to represent the continuous interpretation associated to several notions of digital surface introduced within the graph–theoretical approach to digital topology. Actually, these digital surfaces are particular cases, for suitable digital spaces, of a general notion of digital manifold that naturally arises in our approach.

As the starting point for the quoted justification, we have observed that Kovalevsky's approach to digital topology can be understood in our framework as the category of the digital spaces satisfying only the object axiom.

Finally, it is worth to point out that the topological spaces on the grid $\mathbb{Z}^n$ proposed by Khalimsky [ ] as a model for digital pictures are isomorphic to the conceptual level of suitable digital spaces defined on the device model R^n (see [ ]).

References

1. R. Ayala, E. Domínguez, A. R. Francés, A. Quintero. Digital Lighting Functions. *Lecture Notes in Computer Science.* **1347** (1997) 139–150. 3, 7, 8, 12, 13, 14

2. R. Ayala, E. Domínguez, A. R. Francés, A. Quintero. Weak Lighting Functions and Strong 26-surfaces. To appear in *Theoretical Computer Science.* 1, 7, 8, 14, 15

3. R. Ayala, E. Domínguez, A. R. Francés, A. Quintero. A Digital Index Theorem. *Proc. of the 7th Int. Workshop on Combinatorial Image Analysis IWCIA'00.* (2000) 89-101. 4

4. R. Ayala, E. Domíguez, A. R. Francés and A. Quintero, Homotopy in Digital Spaces. To Appear in *Discrete and Applied Mathematics.* 3, 4, 8

5. R. Malgouyres, G. Bertrand. Complete Local Characterization of Strong 26-Surfaces: Continuous Analog for Strong 26-Surfaces. *Int. J. Pattern Recog. Art. Intell.* **13**(4) (1999) 465-484. 4, 14

6. E. Khalimsky, R. Kopperman, P. R. Meyer. Computer Graphics and Connected Topologies on Finite Ordered Sets. *Topol. Appl.* **36** (1990) 1–17. 3, 15

7. T. Y. Kong, A. Rosenfeld. Digital Topology: Introduction and Survey. *Comput. Vision Graph. Image Process.* **48** (1989) 357–393. 3, 4, 8

8. T. Y. Kong, A. W. Roscoe, A. Rosenfeld. Concepts of Digital Topology. *Topol. Appl.* **46** (1992) 219–262. 3

9. T. Y. Kong, A. W. Roscoe. Continuous Analogs of Axiomatized Digital Surfaces. *Comput. Vision Graph. Image Process.* **29** (1985) 60–86. 12

10. V. A. Kovalevsky. Finite Topologies as Applied to Image Analysis. *Comput. Vision Graph. Image Process.* **46** (1989) 141–161. 3, 8, 9, 13

11. D. G. Morgenthaler, A. Rosenfeld. Surfaces in three–dimensional Digital Images. *Inform. Control.* **51** (1981) 227-247. 4, 12

12. A. Rosefeld. Digital Topology. *Amer. Math. Mothly* **86** (1979) 621–630. 3

13. C. P. Rourke, and B. J. Sanderson. *Introduction to Piecewise-Linear Topology.* Ergebnisse der Math. **69**, Springer 1972. 5, 13

Generic Programming Techniques that Make Planar Cell Complexes Easy to Use

Ullrich Köthe

University of Hamburg, Cognitive Systems Group
koethe@informatik.uni-hamburg.de

Abstract: Cell complexes are potentially very useful in many fields, including image segmentation, numerical analysis, and computer graphics. However, in practice they are not used as widely as they could be. This is partly due to the difficulties in actually implementing algorithms on top of cell complexes. We propose to use generic programming to design cell complex data structures that are easy to use, efficient, and flexible. The implementation of the new design is demonstrated for a number of common cell complex types and an example algorithm.

1   Introduction

Cell complexes and related structures (such as combinatorial maps) are fundamental concepts of digital topology. They are potentially very useful in many application areas. For example, they can be used in image analysis to represent a segmented image, in computational geometry and computer graphics to model 2- and 3-dimensional objects, and in numerical analysis to build finite element meshes. Yet, in practice they are not used as widely as one would expect. I believe that this is partly due to the difficulties their actual implementation and application poses to the programmer. In this paper I will present a solution to this problem that was originally developed in the context of image analysis. It is interesting to note that independent investigations in other fields (e.g. computational geometry and numerical analysis) arrived at very similar solutions [2,5]. Although these fields have quite different goals, the generic concepts to be described here seem to be equally helpful.

In the field of image analysis, image segmentation is the most important application for cell complexes. On the basis of segmented images, questions like "What are the connected regions?", "Where do regions meet?", "Where are the edges?" etc. will be answered. It is well known that naive representations of segmentation results, such as binary images or edge images, have problems in answering such questions consistently and without topological contradictions.

The most infamous problem is the so called *connectivity paradox* which occurs when one defines connected regions by determining connected components in a binary image. No matter whether 8- or 4-neighborhood is used while constructing the connected components, there are configurations that violate Jordan's curve theorem.

G. Bertrand et al. (Eds.): Digital and Image Geometry, LNCS 2243, pp. 17-37, 2001.
© Springer-Verlag Berlin Heidelberg 2001

That is, in this representation it is not always assured that a closed curve partitions the plane into exactly two distinct connected components (see e.g. [8]).

Kovalevsky [8] has shown that problems like this can be overcome by using topological cell complexes (formal definition below). In particular, planar cell complexes (i.e., 2-complexes that can be embedded in the plane) are appropriate for image analysis. On a theoretical level, researchers generally agree on the usefulness of cell complex representations, but in practice they are quite rarely used. Instead, most segmentation methods deal with topological problems heuristically, if at all. I'll demonstrate the difficulties in practical application of cell complexes by means of two typical examples: the cell complexes defined in the TargetJr image analysis framework [13] and the combinatorial maps defined by Dufourd and Puitg [3]. These examples show that current software either requires a lot of manual work to build and manipulate a cell complex or are very slow in execution.

The first example are the topological classes of TargetJr [13]. TargetJr is a very popular basic framework for image analysis research and, to a degree, represents the state-of-the-art in object-oriented image analysis software. It is written in C++ and thus allows for very fast algorithms and a moderate degree of flexibility. In the area of topology, it provides basic building blocks for 2-dimensional cell complexes, most notably: Vertex, Edge, Face, OneChain. The former three represent the different cell types, while the latter encodes a sequence of connected edges, which e.g. represent the boundary of a face. In principle, it is possible to construct any planar cell complex by using these primitives. But in practice this is very hard, because the user is responsible for ensuring consistency of the resulting complex – there is no class CellComplex that would help the programmer in constructing a topological structure. In the words of the TargetJr documentation:

> "For operations that must reuse the vertices [...] the programmer is responsible for maintaining their consistency with the topological structure as a whole." [13], http://www.targetjr.org/manuals/Topology/node18.html

Handling is further complicated by the lack of separation between the combinatorial and geometrical structures of the complex – all topology classes also carry geometric information (e.g. coordinates in case of vertices). Furthermore, when the cell complex has finally been constructed, navigation on it is also quite hard. For example, there is no obvious way of getting the ordered cycle of edges around each vertex. Essentially, TargetJr thus leaves the most difficult tasks to the user.

A totally different approach to the construction of cell complexes is described by Dufourd and Puitg [3], my second example. They use the notions of *dart* and *orbit* to define *quasi-maps*, which contain planar maps (planar cell complexes) as special cases (formal definitions below). This theory is formally specified and implemented in a functional programming languages, which leads to a very elegant and easy to use solution. For example, the orbits can be retrieved by simple functions, so that navigation on the map is very easy. Since all required constraints are automatically enforced during map construction, inconsistent structures cannot be build. This takes a lot of responsibility from the user.

However, this approach also has some disadvantages. First, it is very difficult to integrate it with other image analysis software: most image analysis software isn't written in a functional language, and it is difficult to connect software components

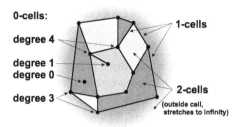

0-cells:

degree 4

degree 1
degree 0

degree 3

1-cells

2-cells
(outside cell,
stretches to infinity)

Fig. 1. Example for a planar cell complex. Some cells are marked with their types. The degree (= number of 1-cells bounded) is given for some 0-cells

written in different paradigms and languages. Second, according to the definitions in [3] even the simplest operations (getting the successor of a dart in an orbit) take *linear time* in the size of the cell complex. This leads to unacceptably slow algorithms when cell complexes of realistic size (thousands of faces) need to be processed. It is not clear to me whether a more efficient implementation (with constant complexity in the basic operations) is possible in a functional environment.

This paper proposes to use *generic programming* in order to overcome the problems mentioned. It describes possibilities to define operations on cell complexes in a way that is simultaneously efficient, flexible and easy to use. These definitions will take on the form of *abstract interfaces*. It will also be shown how these interfaces can be implemented on top of different topological data structures.

2   Cell Complexes and Related Topological Structures

Discrete topological structures can be defined in multiple ways, giving rise to different data structures. However, from the *viewpoint of algorithms* that are to operate on top of those data structures, the differences are not very significant: many topological data structures provide essentially very similar functionality to algorithms. This is the basic justification for our attempt to define a uniform abstract interface.[1]

Let us briefly review some common definitions of digital topology (detailed treatment is beyond the scope of this paper). The best known is the *cellular complex* [8]:

Definition 1: A cell complex is a triple (Z, *dim*, B) where Z is a set of *cells*, *dim* is a function that associates a non-negative integer *dimension* to each cell, and $B \subset Z \times Z$ is the *bounding relation* that describes which cells bound other cells. A cell may bound only cells of larger dimension, and the bounding relation must be transitive. If the largest dimension is k, we speak of a *k-complex*.

Two cells are called *incident* if one bounds the other. The bounding relation is used to define open sets and thus induces a topology on the cell complex: a set is called *open* if, whenever a cell z belongs to the set, all cells bounded by z do also belong to the set. In the context of image segmentation we are particularly interested in cell complexes with maximum dimension 2 which can be embedded in the plane. We will call

---

[1] The correspondences between the considered models will be formally studied elsewhere.

these *planar cell complexes*. Figure 1 shows a planar cell complex. We will use the terms *node*, *edge*, and *face* synonymously to 0-, 1-, and 2-cells respectively.

Another popular definition is that of a *combinatorial map* [3]. This definition will prove especially useful for our interface definition and implementation:

Definition 2: A *combinatorial map* is a triple (D, σ, α) where D is a set of *darts* (also known as *half-edges*), σ is a permutation of the darts, and α is an involution of the darts.

In this context, a *permutation* is a mapping that associates to each dart a unique predecessor and a unique successor. By following successor chains, we can identify the *cycles* or *orbits* of the permutation. An *involution* is a permutation where each orbit contains exactly two elements. It can be shown that the mapping $\varphi = \sigma^{-1}\alpha$ is also a permutation. The correspondence between combinatorial maps and cell complexes is established as follows: we associate each orbit in σ with a 0-cell (node), each orbit in α with a 1-cell (edge; an edge is thus a pair of half-edges), and each orbit in φ with a 2-cell (face) of a cell complex. A thus defined cell bounds another one (of higher dimension), if their corresponding orbits have a dart in common.

In the context of rectangular rasters, Khalimsky's topology is very useful [6]:

Definition 3: The Khalimsky topology of the 1-dimensional discrete line is obtained by interpreting every other point as an open or closed set respectively. The topology of a 2-dimensional rectangular raster is constructed as the product topology of a horizontal and vertical topological line.

To interpret the topological line in the terminology of cell complexes we associate dimension 1 to the open points and dimension 0 to the closed points. The dimensions in the topological raster are obtained as the sum of the vertical and horizontal dimensions. Each cell bounds the cells in its 8-neighborhood that have lower dimension. Thus, a 0-cell bounds all its 8-neighbors, and a 1-cell bounds two 2-cells. Figure 2 (left) shows a raster defined like this.

Finally, when it comes to iterative segmentation and irregular pyramids, the *block complex* is a very useful notion. The idea is to build a new cell complex on top of an existing one by merging cells into larger *block cells*, or *blocks* for short. Blocks are defined by the property that they either contain a single 0-cell, or are homeomorphic to an open *k-sphere*. (A block is homeomorphic to an open k-sphere if it is isomorphic

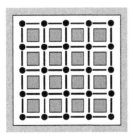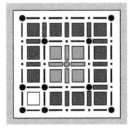

Fig. 2. left: Khalimsky topology on a square raster (0-cells: black balls, 1-cells: black lines, 2-cells: gray squares); right: block complex defined on top of a Khalimsky raster (0-blocks: large black balls, 1-blocks: black lines and small balls, 2-blocks: gray squares, rectangles and balls)

to a simply connected open set in some k-complex – see [9] for details.) A block's dimension equals the highest dimension of any of its cells.

Definition 4: A *block* complex is a complete partition of a given cell complex into blocks. The bounding relation of the blocks is induced by the bounding relation of the contained cells: a block bounds another one iff it contains a cell that bounds a cell in the other block.

Note that the induced bounding relation must meet the requirements of definition 1, i.e. blocks may only bound blocks of smaller dimension, and the relation must be transitive. This poses some restrictions on valid partitions. In particular, all 1-blocks must be sequences of adjacent 0- and 1-cells, and all junctions between 1-blocks must be 0-blocks. See figure 2 right for an example.

3 Generic Programming

Generic programming is a new programming methodology developed by A. Stepanov and D. Musser [12]. It is based on the observation that in traditional programming much time is wasted by continuously adapting algorithms to ever changing data structures. This becomes a significant problem in application areas where algorithms play a crucial role. Here, algorithms should be considered as independent building blocks rather than being dependent on specific data structures. Therefore, generic programming puts its main emphasis on *separating algorithms from data structures by means of abstract interfaces* between them.[2] Since algorithms indeed play a central role in image analysis and computational topology, generic programming suggests itself as a suitable software design method in these fields.

The algorithm centered approach of generic programming consists of 5 steps:

1. Analyze the requirements of algorithms in the intended field. Ensure that each algorithm imposes *minimal* requirements on the underlying data structures.
2. Organize the set of requirements into *concepts*. A concept contains a minimal set of functionality that is typically needed together.
3. Define abstract interfaces for all concepts. Map the definitions onto appropriate generic constructs of the target programming language (e.g. templates in C++).
4. Implement the interfaces on top of suitable data structures. A data structure may implement several concepts, and each concept may be realized by several data structures.
5. Implement algorithms in terms of abstract concepts (interfaces). Thus, algorithms can run on top of *any* data structure that implements the concepts.

The currently best example for this approach is the design of the *Standard Template Library* [1] which forms a major part of the C++ standard library. This library deals with fundamental data structures (list, arrays, trees etc.) and algorithms (sorting, searching etc.). On the basis of the steps sketched above, Stepanov and Musser arrived at the *iterator* concept as their fundamental interface abstraction. Because of its importance for our own discussion, we will describe iterators in some detail.

---

[2] This extents the traditional theory of abstract data types and object-oriented programming, where abstract interfaces have only been defined for data structures and objects.

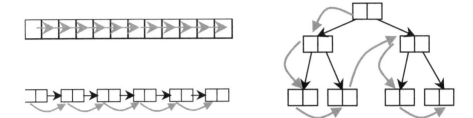

Fig. 3. Effect of a call `++iterator` on different data structures. top left: array – the iterator is moved to the next memory location; bottom left: linked list – the iterator follows the link to the next item; right: tree – the iterator realizes a pre-order traversal (black arrows denote links in the data structure, gray ones show the path of the iterator)

3.1   Iterators

Algorithm analysis revealed that many algorithms need to sequentially access the individual data items in a container. Although there exists a wealth of different sequences (such as arrays, deques, and lists), most algorithms use them in very simple ways. For example, they access one data item, process it, and then proceed to the next or previous item. It was Stepanov and Musser's key insight that the commonalties can be reinforced and the differences be encapsulated by means of *iterators*. In the example, a bi-directional iterator would be required which has the following capabilities[3]:

1. It points to a specific item in a specific container, and allows reading and possibly writing the data of that item. In C++, this is realized by the dereferencing operator: `item = *iterator`
2. It can be moved to the next and previous items. In C++, this is done using the increment and decrement operators: `++iterator` and `--iterator`
3. Two iterators can be compared to find out whether they point to the same item. This is realized by the quality operator: `iterator == end_iterator`.
   (This is used to determine the range of iteration: each data structure provides a pair of iterators that point to the start and one-past-the-end of the sequence respectively. The former iterator is incremented until it compares equal to the latter one, which signals the end of the sequence.)

Key to the efficiency of algorithms using these iterators is the additional requirement the all four operations must execute in constant time. Iterators meeting these requirements can be implemented for many different data structures, as shown in fig. 3. In case of an array, the iterator points to a memory location, and the increment operation moves it to the next memory location. In case of a linked list, a pointer to the next item is stored together with each item, the increment operator follows this link. For a tree, one can implement different iterators that realize pre-, post-, and in-order traversals respectively, by following the edges of the tree in the appropriate order.

Algorithms are then defined so that they use only iterators, not the underlying data structures. By implementing algorithms as C++ *templates*, it is possible to leave the

---

[3] In addition, Stepanov and Musser define input, output, forward, and random access iterators. These will not be needed in this paper.

iterators' types open until the algorithm is actually applied to a concrete data structure. Only then will the compiler replace the formal (place-holder) type of the iterator with an actual type. For example, the signature of the `copy()`-algorithm which copies the items from the source sequence to the target sequence, looks like this:

```
template <class SourceIterator, class TargetIterator>
void copy(SourceIterator src_begin, SourceIterator src_end,
          TargetIterator target_begin);
```

For further information about the design of the STL and generic programming in general, [1] is recommended.

In case of cyclic data structures, the concept of an iterator (which has a first and last element) is replaced by the concept of a *circulator* [5]. A circulator behaves like an iterator, with the exception that there is no special last element where further incrementing is forbidden. Moreover, a circulator needs an additional function `iterator.isSingular()` which returns `true` iff the cyclic sequence is empty. The circulator concept will be important in the context of cell complexes as their orbits form cyclic sequences.

4   Generic Concepts for Cell Complexes

The development of generic concepts for cell complexes must start with an analysis of algorithm requirements. This has been done by several researchers, e.g. [2, 5, 7]. To illustrate this process I will present a simple segmentation algorithm. A more extensive discussion of generic topological algorithms is beyond the scope of this paper and will be given elsewhere.

The algorithm "Connected Components Segmentation" groups 2-cells together according to a predicate are_similar(2-cell, 2-cell) that is applied to all adjacent 2-cells (2-cells are adjacent if they are bounded by a common 1-cell). The exact nature of the predicate is application dependent and thus left unspecified here. A unique label is assigned to the 2-cells in each connected region of "similar" cells. After labeling all 2-cells according to similarity, labels are assigned to 0- and 1-cells according to the following rule: if a 0- or 1-cell bounds only 2-cells with the same label, it is assigned this same label, otherwise it is assigned a special BORDER_LABEL. It should be noted that this algorithm essentially creates block complexes according to definition 4, with the exception that 2-blocks need not be simply connected but may have holes.

In Table 1 we give the pseudo-code of the algorithm on the left and corresponding requirements on the right. Although this simple algorithm does not exploit the entire functionality of a cell complex, it illustrates nicely how sets of abstract requirements are formulated: the algorithm can be applied to any data structure that supports the operations listed on the right. In order to maximize the scope of the algorithm these operations should represent *minimal* requirements. For example, the algorithm doesn't care in which order the 1-cells incident to a 0-cell are listed. So it would be wrong to require those cells to be listed in a particular ordered (say, in orbit order), because this would needlessly restrict the applicability of the algorithm.

Table 1. Connected components segmentation algorithm

Pseudo-code	Data structure requirements
// find connected 2-cells of single component	
define recurse_component(2-cell c):	
for each adjacent 2-cell ac **of** c **do**:	// list adjacent 2-cells of given 2-cell
if label(ac) != NO_LABEL:	// read integer attribute of 2-cell
continue *// ac already processed*	
else if are_similar(c, ac):	// evaluate predicate
label(ac) = label(c)	// assign integer attribute to 2-cell
recurse_component(ac)	
// find all connected components and	
// label cells accordingly	
define cc_segmentation(cell_complex):	
for each cell c **in** cell_complex **do**:	// find all cells in a cell complex
label(c) = NO_LABEL	// associate initial attribute with cells
maxlabel = NO_LABEL + 1	
// phase 1: 2-cell labeling	
for each 2-cell c2 **in** cell_complex **do**:	// list all 2-cells of a cell complex
if label(c2) != NO_LABEL:	// read integer attribute of cell
continue *// c2 already processed*	
else:	
label(c2) = maxlabel	// assign integer attribute to cell
maxlabel = maxlabel + 1	
recurse_component(c2)	
// phase 2: 1-cell labeling	
for each 1-cell c1 **in** cell_complex **do**:	// list all 1-cells of a cell complex
label1= label(firstBounded2Cell(c1))	// access the two 2-cells bounded ...
label2= label(secondBounded2Cell(c1))	// ... by the current 1-cell
if leftLabel == rightLabel:	
label(c1) = leftLabel	// assign integer attribute to cell
else:	
label(c1) = BORDER_LABEL	// assign integer attribute to cell
// phase 3: 0-cell labeling	
for each 0-cell c0 **in** cell_complex **do**:	// list all 0-cells of a cell complex
for each incident 1-cell ic **of** c0 **do**:	// list 1-cells bounded by current
if label(ic) == BORDER_LABEL:	0-cell
label(c0) = BORDER_LABEL	// assign integer attribute to cell
break	
else:	
label(c0) = label(ic)	// assign integer attribute to cell

Of course, it is easy to keep requirements abstract in pseudo-code. But traditional programming techniques, especially procedural programming, do not allow to carry over the abstraction into executable code: they force the programmer to fix a particular data structure so that the algorithm's potential flexibility cannot be realized in practice. This is not the case in generic programming and its realization by C++ templates – algorithms are independent of concrete data structures so that requirements can be translated into executable code *without losing abstraction*.

Analysis of topological algorithms revealed the following recurring requirements:

Combinatorial Requirements:

Analysis of the combinatorial structure
- list all k-cells of a cell complex
- list all incident cells for a given cell (in any order or in the cyclic order defined by definition 2)

Modification of the combinatorial structure
- elementary transformations between cell complexes

Data Association Requirements:

Assignment of data to cells and group of cells
- temporary data (e.g. label of the connected component each cell belongs to)
- geometric data (coordinates, model parameters)
- application data (requirements vary extremely from application to application, e.g. regions statistics, object interpretations, textures etc.)

It should be noted that this analysis suggests a strict separation between topological and geometrical properties. For a given topological structure, there are many different ways to represent its geometry – for example by using polygons, implicit equations, or sets of pixels/voxels. Separation of concerns therefore requires to treat topology and geometry as independent dimensions of abstraction. We will not explicitly deal with geometry in this paper, but concentrate on topology. However, it should be pointed out that representation of geometry is a chief application of *data association with cells* as described in the next subsection.

4.1   Data Association with the Cells

There are three basic possibilities to associate data with the cells of a cell complex:

Store Data in the Cell Data Structure: In addition to the information that encodes the combinatorial structure of the cell complex, application data can be stored directly in the cells. This data is retrieved by member functions of the cell objects. Templates are a great help in designing a cell complex that way.

Store Data outside the Cells: This technique requires to store a unique ID with each cell. This ID is used to retrieve additional application data from external data structures such as hash tables. This technique is more flexible than the first because the cell data structures need not be modified to add a new attribute, but it may be slower and requires more care from the user.

Calculate Data on Demand: Some data are redundant and need not be stored at all. If needed, these data can be calculated from other data that was stored. For example, the length of an edge can often be computed as the distance between the start and end node of the edge.

Now we want to hide these possibilities behind an interface, so that algorithms need not know how the data are stored. This is important, because otherwise all algorithms would have to be adapted when data must be stored differently. A suitable technique to achieve this is the so called *data accessor* [10].

A data accessor is a helper object that retrieves an attribute associated with a data item, given a handle to the item. It is similar to a member function in that it hides how the attribute is stored, but it is more flexible because it is independent of a particular data structure. As an example, let's consider data accessors for the length of an edge according to each of the three possibilities above:

```
struct LengthStoredInCellAccessor {
    float get(EdgeHandle edge) {
        // call member function to retrieve length attribute
        return edge.length();
    }
};
struct LengthStoredExternallyAccessor {
    Hashtable edge_length;
    float get(EdgeHandle edge) {
        // use edge ID to retrieve length from hash table
        return edge_length[edge.ID()];
    }
};
struct LengthCalculatedAccessor {
    float get(EdgeHandle edge) {
        // find differenc vector between start and end node
        Vector diff = edge.startNode().coordinate() -
                      edge.endNode().coordinate();
        // calculate length as magnitude of difference vector
        return diff.magnitude();
    }
};
```

The important point about these accessors is that they are used uniformly by algorithms: algorithms uniformly call `lengthAccessor.get(edgeHandle)`, without knowing which accessor is currently used. Iterators and circulators will play the role of cell handles, as they always refer to a "current" element. See [10] for more details.

5   Generic Concepts for Analyzing the Combinatorial Structure

Concepts for analyzing the combinatorial structure are the most interesting part of the cell complex interface. We will define this interface on the basis of the combinatorial map's darts and orbits. This doesn't imply that the interface is not applicable to other topological structures, it just simplifies the definitions.

Iterators and circulators will form the central parts of the interface. First, we need iterators that simply list all cells of a given dimension:

k-Cell Iterator: A k-cell iterator (k=0, 1, 2) iterates over all k-cells in a cell complex, in an arbitrary but fixed order.

Second, we need iterators for the different orbits that determine the combinatorial structure. Since orbits are cyclic, we actually need circulators here:

Orbit Circulator: An orbit circulator lists the darts in one of the orbits (σ, α, φ). It is bi-directional, i.e. it can step to the successor and predecessor of a given dart.

Fig. 4. Incrementing a φ-circulator which goes around the interior face of the graph. left: initial position; center: intermediate position after α-iteration of embedded circulator; right: final position after σ-iteration of embedded circulator

For all iterators and circulators, we will use the usual `++i` and `--i` notation to step forward and backward. The only exception is the α-circulator: since each α-orbit contains only two darts, we use the call `i.jumpToOpposite()` to switch between the two elements of an α-orbit, i.e. between the two ends of an edge. Circulators can thus provide all three operations simultaneously, so that σ- and φ-circulators automatically play the role of α-circulators as well. This will simplify many algorithms.

Now we are already able to give a completely generic implementation of the φ-circulator which encodes the orbit of darts around a face: since the φ-permutation can be composed from the other two, we only need a σ-circulator (which always is an α-circulator as well) to build it. Thus, we store a σ-circulator within the φ-circulator, and all φ-operations are implemented in terms of the appropriate σ and α-operations:[4]

```
template <class SigmaCirculator>
class PhiCirculator {
    SigmaCircultor sigma;
  public:
    // pass the embedded σ-circulator
    PhiCirculator(SigmaCirculator contained)
    : sigma(contained)
    {}
    // implement the φ-increment as σ⁻¹α of the embedded circulator
    void operator++() {
        contained.jumpToOpposite(); // one α iteration
        --contained;                // one backward σ iteration
    }
    // implement the φ-decrement as α⁻¹σ of the embedded circulator
    void operator++() {
        ++contained;                // one σ iteration
        contained.jumpToOpposite(); // one α⁻¹ iteration
    }
    ... // more functions
};
```

This code works regardless of what type the contained circulator has and what kind of cell complex we are using. Figure 4 illustrates a single φ-increment.

---

[4] We use C++ for the implementation examples on two reasons. First, at present there is no language-independent abstract notation for generic components. Second, it demonstrates that the concepts presented can be and actually have been implemented in practice.

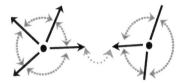

Fig. 5. Half-edges and their linking into orbits

5.1   Implementation of the σ-Circulator for a Half-Edge Data Structure

Now, we want to explain the implementation of the σ-circulator for different kinds of cell complexes. We start with a half-edge data structure, because this is the most direct implementation of a combinatorial map. The central concept of the half-edge data structure is the half-edge or dart, which explicitly stores pointers to its successors and predecessors in both the σ- and α-orbits. A typical implementation of a half-edge looks like this (see also figure 5):

```
struct HalfEdge {
    HalfEdge * sigma_next, * sigma_previous, * alpha_other;
    ... // more data not relevant here
};
```

Since the half-edge stores all necessary navigation information, the σ-circulator can simply store a pointer to the current half-edge. Upon a call to the increment operator it will look up the appropriate neighboring half-edge and replace the stored half-edge with this one. An initial half-edge is passed to the circulator in the constructor:

```
class HalfEdgeSigmaCirculator {
    HalfEdge * current;
  public:
    HalfEdgeSigmaCirculator(HalfEdge * initial)
    : current(initial)
    {}
    void operator++() {
        // goto successor in σ orbit
        current = current->sigma_next;
    };
    void operator--() {
        // goto predecessor in σ orbit
        current = current->sigma_previous;
    };
    void jumpToOpposite() {
        // goto successor (= predecessor) in α orbit
        current = current->alpha_other;
    };
    ... // more functions
};
```

This circulator gives us access to the *complete* combinatorial structure of the map: any dart, any orbit, and thus any cell can be reached by an appropriate operation sequence.

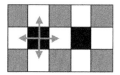 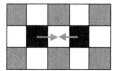

Fig. 6. left: four possible directions of the KhalimskySigmaCirculator in the σ-orbit; right: two possible directions of the circulator in the α-orbit

5.2   Implementation of σ-Circulator for Khalimsky's Topology

Since Khalimsky's topology is defined on a rectangular grid, we can use the grid coordinates to uniquely identify each cell. Note that coordinates are only used in the implementation and not exposed through the circulator interface. We assume that points whose coordinates are both odd are considered 0-cells. Then, points with two even coordinates are 2-cells, and the remaining points 1-cells. A dart is an oriented 1-cell. Hence, we add the attribute "direction" to distinguish the two darts of a 1-cell. Since there are four possible directions in a rectangular grid, we will denote directions by the integers 0...3 to encode "right", "up", "left", and "down" respectively.

Thus, a dart is uniquely determined by its direction and the coordinate of its start node. The σ-orbit around that node is obtained by incrementing the direction modulo 4. Similarly, the opposite dart in the α-orbit is found by moving the dart's coordinate two pixels in its direction (namely to the end node of the corresponding 1-cell) and reversing the direction. Figure 6 shows the possible positions of the circulator in the σ- and α-orbits. This leads to the following code for the σ-circulator:

```
class KhalimskySigmaCirculator {
    int x, y, direction;
  public:
    // init with a coordinate and a direction
    KhalimskySigmaCirculator(int ix, int iy, int idir)
    : x(ix), y(iy), direction(idir)
    {}
    void operator++() {
        // next direction counter-clockwise
        direction = (direction + 1) % 4;
    }
    void operator--() {
        // next direction clockwise
        direction = (direction + 3) % 4;
    }
    void jumpToOpposite() {
        static int xStep[] = { 2, 0, -2, 0};
        static int yStep[] = { 0, -2, 0, 2};

        x += xStep[direction];          // go to the other end of
        y += yStep[direction];          // the current 1-cell
        direction = (direction + 2) % 4; // reverse direction
    }
    ... // more functions
};
```

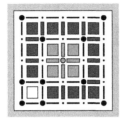 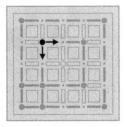

Fig. 7. left: a block complex (cf. figure 2 right); right: possible positions of the block complex σ-circulator around the black 0-cell – only two 1-cells (black) belong to a 1-block, the other two (gray) are ignored because they belong to a 2-block

5.3   Implementation of the σ-Circulator for a Block Complex

Finally, we want to demonstrate how the σ-circulator can be implemented for a block complex. A block complex differs from the two cell complex implementations described so far by the fact that blocks are made up of several cells of an underlying cell complex. Therefore, we can use the circulators of the underlying cell complex to implement the block complex's circulators. Since the circulators have identical interfaces for all cell complex types, the block complex circulators can be written completely generically, independent of the type of the underlying cell complex.

In order to analyze the structure of a block complex, we need some means to tell which kind of block each cell belongs to. Therefore, we require a data accessor that returns the dimension of the block for each 0- and 1-cell (2-cells always belong to 2-blocks). For each underlying cell complex, this accessor is the only externally visible class that must be implemented in order to create a block complex:

```
struct BlockDimensionAccessor {
    int dimensionOfEdge(SigmaCirculator) {
        ... // suitable implementation for underlying cell complex
    }
    int dimensionOfNode(SigmaCirculator) {
        ... // suitable implementation for underlying cell complex
    }
};
```

The second function will return the dimension for the *start node* of the corresponding dart. Then the σ-orbit of a given 0-block is identical to the σ-orbit of the underlying 0-cell, except that all 1-cells belonging to a 2-block are ignored (see figure 7).

The navigation in an α-orbit is slightly more difficult: given one end of a 1-block, we must find the 0-cell at the other end of this 1-block. This is essentially a *contour following* algorithm: check if the *end* node of the current dart is a 0-block. If yes, we have found the other end of the 1-block. Otherwise jump to the next dart in the 1-block and repeat (see figure 8). Therefore, the complexity of an α-increment in a block complex is proportional to the length of the path that constitutes the current 1-block. The following circulator implementation realizes this algorithm (in `jumpToOpposite()`), as well as the selection of darts for the σ-orbit (in `operator++()` and `operator--()`):

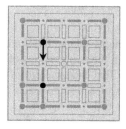 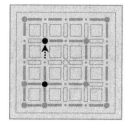 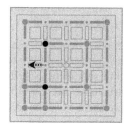

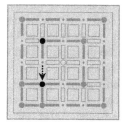 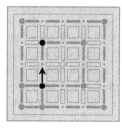

Fig. 8. Contour following when `jumpToOpposite()` is called on a block complex circulator (top left: initial position, left to right: intermediate positions, bottom right: final position)

```
template <class SigmaCirculator, class BlockDimensions>
class BlockComplexSigmaCirculator {
    SigmaCirculator embeddedSigma;
    BlockDimensions blockDimension;
  public:
    // embed a circulator for the underlying cell cimplex and a
    // block dimension accessor
    BlockComplexSigmaCirculator(SigmaCirculator embedded,
                                BlockDimensions dims)
    : embeddedSigma(embedded), blockDimension(dims)
    {}
    void operator++() {
        do {
            // increment embedded circulator ...
            ++embeddedSigma;
        // ... but ignore 1-cells that are not in 1-blocks
        } while(blockDimension.dimensionOfEdge(embeddedSigma) != 1);
    }
    void jumpToOpposite() {
        // contour following:
        // goto the other end of the current 1-cell
        embeddedSigma.jumpToOpposite();
        // until we find the other end of the 1-block ...
        while(blockDimension.dimensionOfNode(embeddedSigma) != 0) {
            // ... follow the cell chain in the 1-block
            operator++();
            embeddedSigma.jumpToOpposite();
        }
    }
    ... // more functions
};
```

5.4   Application: Generic Implementation of Connected Components Segmentation

We can now translate the pseudo-code for connected components segmentation given in table **1** into actual C++ code. The correspondences between the two versions of the algorithm should be easy to see. In order to keep the algorithm independent of the data structures, the C++ function is declared as a template of a `CellComplex` data structure and a store for additional `CellData` (which holds the cells' labels here).

The abstract requirements on the data structure are translated into generic function calls as follows: The functions `fi = c.beginFaces()` and `cc.endFaces()` specify the range of an iterator that lists all 2-cells in the cell complex `cc`. `*fi` returns a unique handle that identifies the current 2-cell. Analogous functionality is provided for 1- and 0-cells. `pi = fi.phiCirculator()` returns a circulator for the φ-orbit associated with the given 2-cell. `pi.face()` gives again a handle to that 2-cell. `pi.opposite()` retrieves the other dart in `pi`'s α-orbit and creates a φ-circulator associated with that dart. Consequently, `pi.opposite().face()` is a 2-cell adjacent to the one given by `pi.face()`. `ei.firstBoundedFace()` and `ei.secondBoundedFace()` give the two 2-cells bounded by the current 1-cell (where `ei` is an 1-cell iterator and the assignment of "first" and "second" is arbitrary). Finally, `si = ni.sigmaCirculator()` creates a circulator for the σ-orbit of the current 0-cell referenced by node iterator `ni`, and `si.edge()` returns the handle to the edge the current dart of σ-circulator `si` is part of. It shall again be emphasized that these function calls work uniformly across all kinds of topological data structures defined in this paper.

```
template <class CellComplex, class CellData>
void
connectedComponentsSegmentation(CellComplex const & cc,CellData & cd)
{
  // initialise the cells' labels to NO_LABEL
  // 2-cells
  CellComplex::FaceIterator fi = cc.beginFaces();
  for(; fi != cc.endFaces(); ++fi) cd.label(*fi) = NO_LABEL;
  // 1-cells
  CellComplex::EdgeIterator ei = cc.beginEdges();
  for(; ei != cc.endEdges(); ++ei) cd.label(*ei) = NO_LABEL;
  // 0-cells
  CellComplex::NodeIterator ni = cc.beginNodes();
  for(; ni != cc.endNodes(); ++ni) cd.label(*ni) = NO_LABEL;

  // counter for the connected components
  int current_label = NO_LABEL + 1;
  // find all connected components of 2-cells
  for(fi = cc.beginFaces(); fi != cc.endFaces(); ++fi)
  {
    if(cd.label(*fi) != NO_LABEL)
      // 2-cells has already been processed in a prior call to
      // recurseConnectedComponent() => ignore
      continue;
    cd.label(*fi) = current_label++; // start a new component
    // get the φ-circulator for the given 2-cell
    CellComplex::PhiCirculator pi = fi.phiCirculator();
    // recurse to find 2-cells belonging into current component
    recurseConnectedComponent(pi, cd);
  }
```

```
  // label the 1-cells according to the labels of the 2-cells they
bound
  for(ei = cc.beginEdges(); ei != cc.endEdges(); ++ei)
  {
    int label1 = cd.label(ei.firstBoundedFace());
    int label2 = cd.label(ei.secondBoundedFace());
    if(label1 == label2)
      // if the two 2-cells belong to the same component, the 1-cell
      // also belongs to that component
      cd.label(*ei) = label1;
    else
      // otherwise, the 1-cell is a border cell
      cd.label(*ei) = BORDER_LABEL;
  }

  // label the 0-cells according to the labels of the 1-cells they
bound
  for(ni = cc.beginNodes(); ni != cc.endNodes(); ++ni)
  {
    // get circulator for σ-orbit of current 0-cell
    CellComplex::SigmaCirculator si = ni.sigmaCirculator();

    int label = cd.label(si.edge());    // remember label of first
1-cell
    do {
      if(cd.label(si.edge()) == BORDER_LABEL)
      {
        // if any of the 1-cells bound by the current 0-cell is
        // a border cell the 0-cell is also a border cell
        label = BORDER_LABEL; break;
      }
    } while(++si != ni.sigmaCirculator());
    cd.label(*ni) = label;  // assign label to 0-cell
  }
}

// recursively mark all 2-cells of a component with the same label
template <class PhiCirculator, class CellComplex, class CellData>
void recurseConnectedComponent(PhiCirculator pi, CellData & cd)
{
  int current_label = cd.label(pi.face()); // remember current label
  do {
    // get φ-circulator for adjacent 2-cell
    PhiCirculator opposite_pi = pi.opposite();

    if(cd.label(opposite_pi.face())== NO_LABEL &&
      are_similar(pi.face(), opposite_pi.face()))
    {
      // if the adjacent 2-cell is similar and not yet assigned,
      // label it and recurse (are_similar() is not discussed here)
      cd.label(opposite_pi.face()) = current_label;
      recurseConnectedComponent(opposite_pi, cd)
    }
  }
  while(++pi != fi.phiCirculator());
}
```

The above algorithm is very efficient: since all iterator/circulator functions execute in constant time (except in a block complex), the overall algorithm has complexity

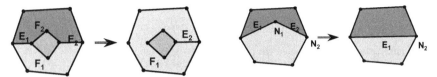

Fig. 9. left: merging of two 2-cells by removing a common 1-cell; right: merging of two 1-cells by removing the common 0-cell

$O(N\sigma + E + F\varphi)$, where N, E, and F denote the number of 0-, 1-, and 2-cells, and σ and φ are the average lengths of the σ- and φ-orbits respectively. This is much more efficient than the approach of Dufourd and Puitg [3], where the functions to access the next dart in an orbit take linear time in the size of the cell complex (i.e. $O(N + E + F)$), which makes the overall algorithm roughly $O((N + E + F)^2)$.

5.5   Elementary Transformations between Cell Complexes

The question of a uniform interface for cell complex *transformations* (insertion and removal of cells) is much more difficult to answer than that for the operations that merely navigate on an existing cell complex. This has several reasons:

1. Different applications have quite different requirements. For example, in numeric analysis (FEM) and computer graphics, mesh refinement is a key operation. It is often done by replacing a cell with a predefined refinement template, so that regularity is preserved. In contrast, in image analysis we are more interested in merging neighboring cells in order to build irregular image pyramids.

2. Different cell complex implementations support different sets of operations. For example, a half-edge data structure can essentially support any conceivable transformation. In contrast, a Khalimsky topology cannot be transformed at all. Block complexes are between the extremes: we can always merge blocks, but we can split only blocks that consist of more than one cell.

Therefore, I will only briefly show which elementary transform have emerged as particularly useful in the field of image analysis. The term "elementary" here refers to transformations that only involve few cells, so that only a very localized area of the cell complex is changed and the operation can be implemented very efficiently. More complex transformations can be build on top of the elementary ones.

The elementary transformations are often called *Euler operators* [11] because, on a planar cell complex, they must respect Euler's equation $v - e + f = 2$ (with v, e, and f denoting the number of nodes, edges, and faces respectively). In principle, two Euler operators and their inverses are sufficient to span the space of planar cell complexes (if we don't allow "holes" in faces – otherwise we will need three):

Merge Two 2-Cells: If two 2-cells are bound by a common 1-cell, we may remove that 1-cell and merge the 2-cells.

Merge Two 1-Cells: If a 0-cell bounds exactly two distinct 1-cells, we may remove the 0-cell and merge the 1-cells.

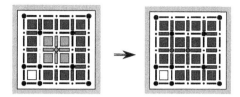

Fig. 10. merging of two 2-blocks in a block complex: when the 1-block marked gray is removed, the cells belonging to the surviving 2-block have to relabeled (in dark gray)

Figure 9 shows examples for either operation. By applying them in a suitable order, one can reduce any cell complex to a single node located in the infinite (exterior) face. The reverse operations are used to refine a cell complex. Complex operations are composed from the elementary ones. Sometimes this leads to very inefficient implementations of complex operations. Then one may add further specialized operations, but for many applications this is not necessary.

It is important to note that a direct implementation of the elementary operations is only possible if we can actually remove cells from the cell complex data structure. In case of a block complex, this is not the case. But we can still implement the Euler operators: instead of removing cells, we assign new block dimensions and labels to the cells according to the modified block structure, as figure 10 illustrates.

It should be noted that the raw Euler operators do not update the non-topological data associated to the cells involved (e.g. geometrical or statistical information). If such data are present, it is necessary to augment the Euler operators with additional code that handles the update. For example, when two 2-cells are merged, the average gray level of the combined 2-cell must be calculated from the statistics of the original cells. This can be done by wrapping the raw Euler operators into new operators (with the same interface) that carry out additional operations before and/or after calling the original Euler operator. This technique is also known as the *decorator pattern* [4]. Due to the flexibility of generic programming, templated algorithms need not be modified in order to support this: since the augmented operators still conform to the uniform interface definition, they can transparently be called instead of the raw operators, without the algorithms knowing the difference.

6   Conclusions

This article has shown that generic concepts can make the use of planar cell complexes in everyday programming much more convenient: interface concepts such as iterators, circulators, and data accessors shield algorithms from the particulars of the underlying data structure. The separation between algorithms and data structures is very important, because most of the know-how of image analysis and other fields using cell complexes lies in algorithms rather than data structures. This means that it is usually quite difficult and error prone to change existing algorithm implementations whenever the underlying data structures change. It is much easier to write some mediating iterators and accessors and leave the algorithms untouched. The resulting solutions combine the speed of C++ with the simplicity of the functional approach of Dufourd and Puitg [3].

Connected components segmentation is a good case in point. But the advantages of generic programming become even more apparent when we apply a segmentation algorithm repeatedly and in turn with Euler operations in order to generate an irregular segmentation pyramid: Initially, we interpret the entire image as a Khalimsky plane, where the pixels are considered 2-cells and the appropriate 0- and 1-cells are inserted between them. Cells are labeled by means of a segmentation algorithm to create the first level of the segmentation pyramid (usually, this is an oversegmentation). The components (blocks) at this level are then merged into single cells by a suitable sequence of Euler operators. This process is repeated to get ever coarser pyramid levels, until the final segmentation is satisfying. In order to optimize algorithm performance, it is a good idea to use a block complex on top of the Khalimsky plane for the first few levels, and switch to a half-edge data structure when regions become larger. By using the interfaces described here, we can implement segmentation and merging algorithms independently of the data structures used. Choosing the appropriate kind of cell complex at every pyramid level thus becomes very easy.

Of course, there is still much to be done. In particular, more cell complex algorithms should be implemented according to the generic paradigm to realize the advantages claimed here and gain new insights for further improvement. The interaction between topology and geometry should be investigated more deeply. A wider choice of specialized Euler operators and extension to 3D would also be desirable. Nevertheless, the approach is already working and shows promising results, with good performance (a few seconds) in segmenting cell complexes with thousands of cells.

References

1. M. Austern: *"Generic Programming and the STL"*, Reading: Addison-Wesley, 1998
2. G. Berti: *"Generic Software Components for Scientific Computing"*, PhD thesis, Fakultät für Mathematik, Naturwissenschaften und Informatik, Brandenburgische Technische Universität Cottbus, 2000
3. J.-F. Dufourd, F. Puitg: *"Functional specification and prototyping with oriented combinatorial maps"*, Computational Geometry 16 (2000) 129-156
4. E. Gamma, R. Helm, R. Johnson, J. Vlissides: *"Design Patterns"*, Addison-Wesley, 1994
5. L. Kettner: *"Designing a Data Structure for Polyhedral Surfaces"*, Proc. 14th ACM Symp. on Computational Geometry, New York: ACM Press, 1998
6. E. Khalimsky, R. Kopperman, P. Meyer: *"Computer Graphics and Connected Topologies on Finite Ordered Sets"*, J. Topology and its Applications, vol. 36, pp. 1-27, 1990
7. U. Köthe: *"Generische Programmierung für die Bildverarbeitung"*, PhD thesis, Computer Science Department, University of Hamburg, 2000 (in German)
8. V. Kovalevsky: *"Finite Topology as Applied to Image Analysis"*, Computer Vision, Graphics, and Image Processing, 46(2), pp. 141-161, 1989

9. V. Kovalevsky: *"Computergestützte Untersuchung topologischer Eigenschaften mehrdimensionaler Räume"*, Preprints CS-03-00, Computer Science Department, University of Rostock, 2000 (in German)
10. D. Kühl, K. Weihe: *"Data Access Templates"*, C++ Report Magazine, July/August 1997
11. M. Mäntylä: *"An Introduction to Solid Modeling"*, Computer Science Press, 1988
12. D. Musser , A. Stepanov: *"Algorithm-Oriented Generic Libraries"*, Software – Practice and Experience, 24(7), 623-642, 1994
13. *"TargetJr Documentation"*, 1996-2000 (http://www.targetjr.org/)

Algorithms and Data Structures for Computer Topology

Vladimir Kovalevsky

Institute of Computer Graphics, University of Rostock
Albert-Einstein-Str. 21, 18051 Rostock, Germany
kovalev@tfh-berlin.de

Abstract. The paper presents an introduction to computer topology with applications to image processing and computer graphics. Basic topological notions such as connectivity, frontier, manifolds, surfaces, combinatorial homeomorphism etc. are recalled and adapted for locally finite topological spaces. The paper describes data structures for explicitly representing classical topological spaces in computers and presents some algorithms for computing topological features of sets. Among them are: boundary tracing (n=2,3), filling of interiors (n=2,3,4), labeling of components, computing of skeletons and others.

1 Introduction: Topology and Computers

Topology plays an important role in computer graphics and image analysis. Connectedness, boundaries and inclusion of regions are topological features which are important for both rendering images and analyzing their contents. Computing these features is one of the tasks of the *computer topology*. We use this term rather than "computational topology" since our approach is analogous to that of digital geometry rather than to that of computational geometry: we are using models of topological spaces explicitly representing each element of a finite topological space as an element of the computer memory, defined by its integer coordinates. The other possible approach would be to think about the Euclidean space, to define objects by equations and inequalities in real coordinates and to approximate real coordinates on a computer by floating point variables.

Computer topology may be of interest both for computer scientists who attempt to apply topological knowledge for analyzing digitized images, and for mathematicians who may use computers to solve complicated topological problems. Thus, for example, essential progress in investigating three-dimensional manifolds has been reached by means of computers (see e.g. [14]).

Topological ideas are becoming increasingly important in modern theoretical physics where attempts to develop a unique theory of gravitation and quantum mechanics have led to the Topological Quantum Field Theory (see e.g. [2,13]), in which topology of multi-dimensional spaces plays a crucial role. This is one more possible application field for computer topology. Thus, computer topology is important both for applications in computer imagery and in basic research in mathematics and physics.

G. Bertrand et al. (Eds.): Digital and Image Geometry, LNCS 2243, pp. 38-58, 2001.
© Springer-Verlag Berlin Heidelberg 2001

We describe here among others the new data structure called the 3D cell list which allows to economically encode a segmented 3D image or a 3-manifold and gives the possibility to access topological information without searching. Then we describe algorithms on boundary tracing in 2D and 3D, filling of interiors of boundaries in nD (successfully tested up to $n=4$), component labeling, computing skeletons and computing basic topological relations in nD, $n \leq 4$. All algorithms are based on the topology of abstract cell complexes and are simpler and more economical with respect to time and memory space than traditional algorithms based on grid point models.

2   Basic Notions

In this section we recall some notions and definitions from the classical topology, which are necessary for reading the subsequent sections.

2.1   Topology of Abstract Complexes

There exist topological spaces in which any space element possesses the smallest neighborhood [1]. If the smallest neighborhood is finite then the space is called *locally finite*. Among locally finite spaces abstract cell complexes are those especially well suited for computer applications, as explained below. Abstract complexes are known since 1908 [16]. They are called "abstract" because their elements, the abstract cells, need not to be considered as subsets of the Euclidean space. A historical review may be found in [7, 17].

Definition AC: An *abstract cell complex* (AC complex) $C=(E, B, dim)$ is a set E of abstract elements (cells) provided with an antisymmetric, irreflexive, and transitive binary relation $B \subset E \times E$ called the *bounding relation*, and with a dimension function $dim: E \rightarrow I$ from E into the set I of non-negative integers such that $dim(e') < dim(e'')$ for all pairs $(e', e'') \in B$.

The maximum dimension of the cells of an AC complex is called its dimension. We shall mainly consider complexes of dimensions 2 and 3. Their cells with dimension 0 (0-cells) are called *points*, cells of dimension 1 (1-cells) are called *cracks* (edges), cells of dimension 2 (2-cells) are called *pixels* (faces) and that of dimension 3 are the *voxels*.

If $(e', e'') \in B$ then it is usual to write $e'<e''$ or to say that the cell e' *bounds* the cell e''. Two cells e' and e'' of an AC complex C are called *incident with each other in C* iff either $e'=e''$, or e' bounds e'', or e'' bounds e'. In AC complexes no cell is a subset of another cell, as it is the case in simplicial and Euclidean complexes. Exactly this property of AC complexes make it possible to define a topology on the set of abstract cells independently from any Hausdorff space.

The topology of AC complexes with applications to computer imagery has been described in [9]. We recall now a few most important definitions. In what follows we say "complex" for "AC complex".

Definition SC: A *subcomplex* $S = (E', B', dim')$ of a given complex $C = (E, B, dim)$ is a complex whose set E' is a subset of E and the relation B' is an intersection of B with $E' \times E'$. The dimension dim' is equal to dim for all cells of E'.

Since a subcomplex is uniquely defined by the subset E' it is possible to apply Boolean operations as union, intersection and complement to complexes. We will often say "subset" while meaning "subcomplex".

The *connectivity* in complexes is the *transitive hull of the incidence relation*. It can be shown that the connectivity thus defined corresponds to classical connectivity.

Definition OP: A subset OS of cells of a subcomplex S of a complex C is called *open in* S if it contains all cells of S bounded by cells of OS. An n-cell c^n of an n-dimensional complex C^n is an open subset of C^n since c^n bounds no cells of C^n.

Definition SON: The smallest subset of a set S which contains a given cell c of S and is open in S is called the *smallest (open) neighborhood* of c relative to S and is denoted by $\text{SON}(c, S)$.

The word "open" in "smallest open neighborhood" may be dropped since the smallest neighborhood is always open, however, we prefer to retain the notation "SON" since it has been used in many publications by the author.

Definition CL: The smallest subset of a set S which contains a given cell c of S and is closed in S is called the *closure* of c relative to S and is denoted by $\text{Cl}(c, S)$.

Definition FR: The *frontier* $\text{Fr}(S, C)$ of a subcomplex S of a complex C *relative to* C is the subcomplex of C containing all cells c of C whose $\text{SON}(c, C)$ contains both cells of S as well as cells of the complement $C-S$.

Illustrations to AC complexes, SONs and closures of cells of different dimensions may be found in [9, 10, 12].

Definition OF: The *open frontier* $\text{Of}(S, C)$ of a subcomplex S of a complex C *relative to* C is the subcomplex of C containing all cells c of C whose closure $\text{Cl}(c, C)$ contains both cells of S as well as cells of the complement $C-S$.

Definition BD: The boundary ∂S of an n-dimensional subcomplex S of a complex C is the union of the closures of all $(n-1)$-cells of C each of which bounds exactly one n-cell of S.

Definition TL: A connected one-dimensional complex whose each cell, except two of them, is incident with exactly two other cells, is called a *topological line*.

It is easily seen that it is possible to assign integer numbers to the cells of a topological line in such a way that a cell incident with the cell having the number k has the number $k-1$ or $k+1$. These numbers are called the *topological coordinates* of the cells [10].

Definition CR: A Cartesian (direct) product C^n of n topological lines is called an n-dimensional *Cartesian* complex [8].

The set of cells of C^n is the Cartesian product of n sets of cells of the topological lines which are the *coordinate axes* of the n-dimensional space C^n. They will be denoted by A_i, $i=1,2,...,n$. A cell of C^n is an n-tuple $(a_1, a_2,..., a_n)$ of cells a_i of the corresponding axes: $a_i \in A_i$. The bounding relation of C^n is defined as follows: the n-tuple $(a_1, a_2,..., a_n)$ is bounding another distinct n-tuple $(b_1, b_2,..., b_n)$ iff for all $i=1,2,...n$ the cell a_i is incident with b_i in A_i and $dim(a_i) \leq dim(b_i)$ in A_i.

The dimension of a product cell is defined as the sum of dimensions of the factor cells in their one-dimensional spaces. Topological coordinates of a product cell are defined by the vector whose components are the coordinates of the factor cells in their axes.

Fig. 1a shows four cells in a two-dimensional Cartesian complex: P is a 0-cell (point), C_1 and C_2 are 1-cells (a horizontal and a vertical crack), F is a 2-cell (pixel).

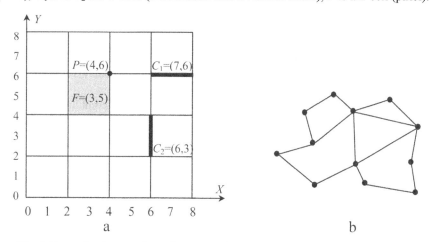

Fig. 1. Example of a two-dimensional Cartesian (a) and non-Cartesian (b) complexes

If we assign even numbers to the 0-cells and odd ones to the 1-cells of the axes then the dimension of a cell in a Cartesian complex is equal to the number of its odd coordinates.

2.2   Combinatorial Homeomorphism

The notion of the homeomorphism of two sets is a fundamental notion of topology: two sets are called homeomorphic or topologically equivalent if one of them can be mapped onto the other by a one-to-one continuous function while the inverse map is also continuous. There is another classical way to define the homeomorphism, which way is directly applicable to complexes and may be extended to other locally finite spaces. It is called the *combinatorial homeomorphism* and is based on the notion of *elementary subdivisions*.

The concept of an AC complex is too general: it is e.g. possible to define an AC complex where a 1-cell is bounded by more than two 0-cells. To avoid such situations elementary subdivisions have been defined in classical topology (see e.g. a modern survey in [17]) on the base of the topology of the Euclidean space. We give in what follows an independent, purely combinatorial definition. It is a recursive definition: we make the necessary definitions primarily for 1-cells and then for cells of still greater dimensions.

A 1-cell which is bounded by exactly two 0-cells is called *proper*.

1. An *elementary subdivision* of a proper 1-cell c^1, which is bounded by the 0-cells c_1^0 and c_2^0, replaces the complex $C'=(c_1^0 < c^1 > c_2^0)$ by the 1-complex C'' with two 1-cells c_1^1, c_2^1 and a new 0-cell c_3^0: $C''=(c_1^0 < c_1^1 > c_3^0 < c_2^1 > c_2^0)$. One or both of the 0-cells c_1^0 and c_2^0 can be missing.

2. The following Definitions should be used recursively: first for $m=1$, then for $m=2$ etc.

A m-complex arising through N ($N{\geq}0$) elementary subdivisions of a single proper m-cell is called an *open combinatorial m-ball*. When $m=1$ then it is a sequence of pairwise incident 1- and 0-cells, starting and ending with a 1-cell. A single 1-cell is also an open combinatorial 1-ball.

The boundary of an open m-ball is called a *combinatorial $(m-1)$-sphere*. When $m=1$ then it consists of exactly two 0-cells. The closure of an m-ball is called the *closed m-ball*. The union of two closed m-balls with identical boundaries is called a *combinatorial m-sphere*.

An m-cell c^m, $m>1$ is called *proper* if its boundary ∂c^m is an combinatorial $(m-1)$-sphere.

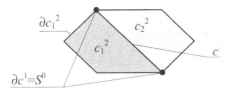

Fig. 2. Example of the elementary subdivision of a 2-cell

An *elementary subdivision* in an n-complex replaces a proper m-cell c^m, $1<m{\leq}n$, with two proper m-cells c_1^m, c_2^m and one new proper $(m-1)$-cell $c^{(m-1)}$ bounding both m-cells c_1^m and c_2^m while the boundary $\partial c^{(m-1)}$ is an $(m-2)$-sphere $S^{(m-2)}{\subset}\partial c^2$, $\partial(c_1^m{\cup}c^{(m-1)}{\cup}c_2^m)=\partial c^m$ and $c^{(m-1)}{\notin}\partial c^m$.

An AC complex is called *proper* if all its cells are proper.

Two proper AC complexes are called *combinatorially homeomorphic* if they possess isomorphic subdivisions. Fig. 3 shows an example.

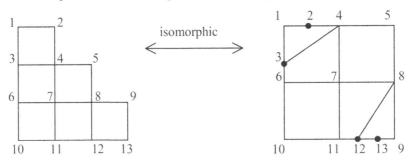

Fig. 3. A subdivision of a digitized square, which is isomorphic to a digitized triangle; black circles are new points introduced during the subdivision

2.3   Manifolds and Surfaces

Among the variety of Hausdorff spaces there are spaces possessing certain relatively simple and important topological properties. They are called manifolds. It is known that manifolds of dimension not greater than three may be triangulated. This means that there exists a simplicial complex homeomorphic to a given manifold.

An AC complex which is combinatorially homeomorphic to the triangulation of a manifold may represent the topological properties of a manifold (of dimension up to three) in the same way as the triangulation does. This fact opens the possibility to model manifolds in computers for investigating them.

Definition MA: An n-dimensional *combinatorial manifold* (n-manifold) without boundary is an n-dimensional complex M in which the boundary of the SON(P, M) of each 0-cell P is homeomorphic to an ($n-1$)-sphere. In a *manifold with boundary* the SON(P, M) of some 0-cell P may have a boundary homeomorphic to a "half-sphere", i.e. to an ($n-1$)-ball.

Surfaces in a 3D space are frontiers of 3D subsets of the space. Under rather general conditions surfaces are 2-manifolds. Conditions under which the frontier of a 3D subset is a 2-manifold as well as surfaces which are no manifolds but rather "quasi-manifolds" are considered in [11].

3   Data Structures

In this Section we consider some well-known and also some new data structures useful for representing topological information in two- and three-dimensional digitized images.

3.1   The Standard Raster

Two- and three-dimensional images are usually stored on a computer in arrays of the corresponding dimension. Each element of the array contains either a gray value, or a color, or a density. This data structure is not designed for topological calculations, nevertheless, it is possible to perform topological calculations without changing the data structure. For example, it is possible to trace and encode the boundary of a region in a two-dimensional image in spite of the apparent difficulty that the boundary consists of 0- and 1-cells, however, the raster contains only pixels which *must* be interpreted as 2-cells. The reason is that a pixel is mostly a carrier of an optical feature which is proportional to certain elementary area. Thus pixels must correspond to elementary areas which are the 2-cells rather than 0- or 1-cells whose area is zero. On the same reason voxels must correspond to 3-cells.

The tracing in the standard raster is possible because the concept of an AC complex is the *way of thinking* about topological properties of digitized images rather than a way of encoding them. Let us explain this idea for the case of tracing boundaries.

We think of a two-dimensional (2D) image as of a 2D Cartesian complex containing cells of dimensions form 0 to 2. The 2-cells (pixels) have integer coordinates which, unlike to topological coordinates of a pixel being always odd (compare Section 1.1 and Fig. 1), may take in the standard raster *any integer values*, both odd or even. Pixels are explicitly represented in the raster, whereas the 0- and 1-cells are present implicitly.

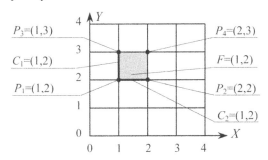

Fig. 4. Non-topological coordinates of cells of lower dimensions

Coordinate Assignment Rule: Each pixel F of a 2D image gets one 0-cell assigned to it as its "own" cell. This is the 0-cell lying in the corner of F which is the nearest to the origin of the coordinates (P_1 in Fig. 4). Also two 1-cells incident with F and with P are declared to be own cells of F (C_1 and C_2 in Fig. 4). Thus each pixel gets three own cells of lower dimensions. *All own cells of F get the same coordinates as F.*

In the three-dimensional case each voxel gets seven own cells of lower dimensions which are arranged similarly. These seven cells get the same coordinates as the corresponding voxel.

Unfortunately, some cells in the boundary of the raster remain without an "owner". In most applications this is of no importance. Otherwise the raster must be correspondingly enlarged.

According to the above rule, it is not difficult to calculate the coordinates of all pixels incident with a given point and to get the gray values of the pixels from the array. Depending on these gray values the tracing point P moves to its next position. Details of the tracing algorithm are described in Section 3.1.

The majority of low level topological problems in image processing may be solved in a similar way, i.e. without representing cells of lower dimension as elements of some multidimensional array. A typical exception is the problem of filling the interior of a region defined by its boundary. The solution is simpler when using two array elements per pixel: one for the pixel itself and one more for its own vertical crack (1-cell). The solution consists in reading the description of the boundary (e.g. its crack code), labeling all vertical cracks of the boundary in the array, counting the labeled cracks in each row (starting with 0), and filling the pixels between the crack with an even count $2 \cdot i$ and the next crack (with the count $2 \cdot i + 1$). The details of this algorithm are described in Section 3.3.

Even more complicated topological problems may be solved by means of the standard raster. For example, when tracing surfaces (Section 3.2) or producing skeletons (Section 3.5) simple pixels must be recognized. A pixel is simple relative to a given region if the intersection of its boundary with the boundary of the region is connected. It is easier to correctly recognize all simple pixels if *cells of all dimensions* of the region are labeled. To perform this in a standard raster, it is possible to assign a bit of a raster element representing the pixel F to each own cell of F. For example, suppose that one byte of a two-dimensional array is assigned to each pixel of the image shown in Fig. 4. Consider the pixel F with coordinates (1, 2) and the byte assigned to it. The bit 0 of the byte may be assigned to the 0-cell P_1, the bit 1 to the 1-cell C_1, the bit 2 to the 1-cell C_2. The remaining bits may be assigned to F itself. Similar assignments are also possible in the 3D case.

As we see, there is no necessity to allocate memory space for each cell of a complex, which would demand four times more memory space than that needed for pixels only, or eight times more than that needed for the voxels in the 3D case.

The most data structures commonly used in the processing of 2D images may be used together with the standard raster. These are primarily the run length code and the crack code. The latter differs from the widely used Freeman code in that it contains only four directions rather than eight of the Freeman code. This is due to the properties of a 2D Cartesian complex whose oriented 1-cells have exactly four different directions.

3.2   The Topological Raster

Complexes used in topological investigations by means of a computer often have a relatively small number of cells. The direct access to cells of all dimensions and the possibility to use more than two different labels for cells of lower dimensions is then more important than the possibility to save memory space. This is the case, e.g. when investigating 3-maniflds represented as boundaries of subsets in a four-dimensional space while the space is represented as a four-dimensional array. In such cases a topological raster is more suitable than a standard one.

In a topological raster each coordinate axis is a topological line (see Definition TL in Section 1.1). The 0-cells of the axis have even coordinates, the 1-cells have odd coordinates. The dimension and the orientation (if defined) of any cell may be calculated from its topological coordinates which in this case coincide with the indices of the corresponding array element. The dimension of any cell is the *number* of its odd coordinates, the orientation is specified by indicating *which* of the coordinates are odd. For example, the cell C_1 in Fig. 1 above has *one* odd coordinate and this is its X-coordinate. Thus it is a *one*-dimensional cell oriented along the X-axis. The 2-cell F has two and the 0-cell P has no odd coordinates. In a three-dimensional complex the orientation of the 2-cells may be specified in a similar way: if the ith coordinate of a 2-cell F is the only even one then the normal to F is parallel to the ith coordinate axis.

We shall show in Section 3 how topological relations between cells, like the bounding or incidence relations, may be calculated from their topological coordinates.

3.3   Data Structures Using Lists of Space Elements

A data structure designed to *efficiently* represent topological information must satisfy the following two demands:

1. The structure must contain *complete* topological information sufficient to get knowledge about topological relations among the parts of the image or of a 3D scene *without a search*. To the topological relations belong primarily the incidence and the adjacency relations (two distinct subsets are adjacent if there is a space element incident with both of them).
2. The structure must be able to correctly represent non-proper complexes which are often used in topological investigations because they contain much less elements than the corresponding proper complexes.

The notions of proper and non-proper complexes have been introduced by the author in [12]. We give here only an example and the necessary short explanation. The surface of a torus may be represented as a complex consisting of one 0-cell, two 1-cells and one 2-cell (Fig. 5a). This representation has the advantage of being very simple.

However, if one would try to interpret this representation as an AC complex, difficulties would occur since e.g. the AC complexes corresponding to Fig. 5a and Fig. 5b are the same: the same sets of four cells, the same bounding relation and the same dimensions of the cells. The difference between these two complexes is that each of the 1-cells L_1 and L_2 in Fig. 5a bounds the 2-cell *two times*, on both sides. This may be seen, if one considers the embedding of the complex in a Euclidean space: a neighborhood of a point on the 1-cell contains two half-disks each of which lies in one and the same 2-cell. However, there is no possibility to describe this relation in the language of complexes.

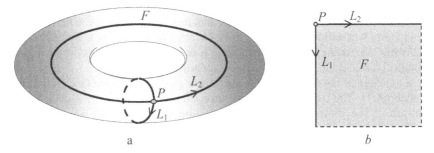

a   b

Fig. 5. Representations of the surface of a torus (a) and of a simple complex (b)

Since one of our aims is to consider a purely combinatorial approach with no relation to a Euclidean space we consider the possibility to overcome this difficulty by introducing the notion of an incidence structure [12] as explained below.

Thus when considering Fig. 5a as a representation of a complex then it is not a proper one (see Section 1.2): though each k-cell with $k>0$ is homeomorphic to an open k-ball the boundaries of the cells are not homeomorphic to $(k-1)$-spheres.

Data structures known from the literature do not fulfil the above demands 1 and 2. The classical incidence matrix (see e.g. [15]) enables one to encode any proper cell complex. It contains complete topological information. However, it is not suitable to encode non-proper complexes, as explained above. Besides that, it is not economical: it contains in the case of an n-dimensional complex

$$\sum_{k=1}^{n} N_{k-1} \cdot N_k$$

elements where N_k is the number k-dimensional cells. This number is in practically relevant cases too large. Because of this reasons data structures using "linear" rather than "quadratic" lists of space elements are preferable.

Most 2D data structures of this kind can be hardly generalized for the 3D case. So the structures using the notion of "half-edges", e.g. the FTG [5], or the n-G-map [3] would need in the 3D case the introduction of "half-faces". In this case each edge would occur in so many copies as the number of faces bounded by it. The structure would be no more a graph as this is the case for the FTG: a complete FTG structure would be needed for each 3D region, which is not economical. No suggestion for a 3D version of the FTG structure is known to the author.

In computer graphics and geometric modeling 3D list data structures are known since many years. One of the most popular is the "boundary representation" [4]. This structure enables one to easily trace the boundary of a 2D face of a body. However, to find which bodies in a 3D scene are adjacent to each other demands an exhaustive search through the descriptions of *all vertices of all bodies in the scene*. Even simpler questions, as e.g. which edges are incident with a given vertex, demand an exhaustive search to be answered. This is true for all 3D data structures known to the author.

As far as we know, the possibility to represent non-proper complexes was not discussed in the literature before the author's publication [12].

3.4   The Two-Dimensional Cell List

A 2D data structure satisfying the above mentioned demands, called the *cell list*, has been suggested by the author [9]. The peculiarity of the cell list is that the topological information, namely that of the incidence, is *explicitly* represented in it: it is possible to directly get the information about the boundaries of regions and the endpoints of lines. Information about adjacencies is available with a *restricted local search* since "adjacent" means "incident with an incident element".

The data structure of the cell list is based on the topological notion of a block complex [15] which we have adapted to AC complexes [9].

Definition BC: Consider a partition M of a complex A into subsets S_i^k. Subsets with $k=0$ are 0-cells of A; each of the subsets with $k>0$ is combinatorially homeomorphic to an open k-dimensional ball. There are a bounding relation BR and a dimension function Dim defined on M in the natural way. The triple

$B(A)=(M, BR, Dim)$

is called a *block complex of A*, the subsets S_i^k are called *k-dimensional blocks* or *k-blocks*.

Examples of two-dimensional block complexes and cell lists may be found in [9,10].

3.5   The Three-Dimensional Cell List

We call the subcomplex composed of all cells incident with a given *proper* cell c the *incidence structure* of c. In a 2D space the incidence structure of a cell consists either of a cyclic sequence or of two pairs of cells. The cyclic sequences are B-isomorphic to one-dimensional complexes (B-isomorphism [11] is a one-to-one map retaining the bounding relation but not the dimensions of cells). They can be represented in the computer as chained lists. In the 3D case the incidence structures of 0- and 3-blocks are B-isomorphic to two-dimensional complexes.

The author has shown [12] that in the case when the space is a 3-manifold the incidence structures are B-isomorphic to two-dimensional spheres. A finite 2-sphere is isomorphic to the surface of a convex polyhedron and therefore may be represented as a list of polygonal faces. This is the theoretical base of the 3D cell list. This data structure is appropriate to describe topological features of 3-manifolds and of 3D scenes containing *many bodies* which may have *common faces, edges or vertices*.

Let us consider a simple example of a topological 3D cell list with only two bodies where five faces of each cube are considered as a single face.

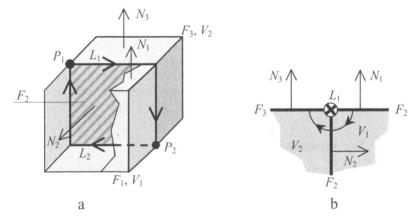

| a | b |

Fig. 6. A simple 3D block complex (a) and its cross section perpendicular to L_1 (b)

The 3D cell list of the 3D image of Fig. 6 is shown in the following tables.

List of branch points (0-blocks)

Label	N_{SON}	Lines
P_1	2	$-L_1, +L_2$
P_2	2	$+L_1, -L_2$

The partial list of the 0-blocks indicates for each 0-block P_i the number N_{SON} of all 1-blocks (lines) incident with P_i and their labels. The negative sign of a line's label in the row of P_i indicates that this line goes away from P_i. We have skipped here the geometric information, i.e. the coordinates.

In the list of 1-blocks (lines) N_{SON} denotes the number of blocks in the SON of the line L_i, i.e. the number of blocks bounded by L_i. The pointer Z_k points to the chained list containing the indices of these blocks as shown in the last column. The order of the sequence corresponds to a right-handed rotation around L_i. A negative sign before a label of a face indicates that its normal is oriented against the direction of rotation.

List of lines (1-blocks)

Label	Starting point	End Point	N_{SON}	Pointer	Chained list
L_1	P_1	P_2	5	Z_1	$-F_1 \to V_1 \to -F_2 \to V_2 \to +F_3$
L_2	P_2	P_1	5	Z_2	$-F_1 \to V_1 \to -F_2 \to V_2 \to +F_3$

List of faces (2-blocks)

Label	+Vol	−Vol	N_{Cl}	Point	Chained list
F_1	−	V_1	4	Z_3	$P_1 \to -L_2 \to P_2 \to -L_1 \to P_1$
F_2	V_1	V_2	4	Z_4	$P_1 \to -L_2 \to P_2 \to -L_1 \to P_1$
F_2	−	V_2	4	Z_5	$P_1 \to +L_1 \to P_2 \to +L_2 \to P_1$

The list of faces contains for each current face F_i the labels of two volumes bounded by F_i. The volume denoted by "+Vol" lies in the direction of the normal of F_i. N_{Cl} denotes the number of blocks in $Cl(F_i)$ which is shown as the chained list in the last column. The order of the sequence corresponds to a right-handed rotation around the normal of F_i. A negative sign before a label of a line indicates that the line is oriented against the direction of the rotation.

List of volumes (3-blocks)

Label	N_{Cl}	Faces
V_1	2	$+F_1, -F_2$
V_2	2	$+F_2, +F_3$

The partial list of volumes contains for each volume V_i the number N_{Cl} of the incident faces which are listed in the last column. The negative sign before the label of a face in the row of V_i indicates that the normal to the face is pointing away from V_i.

4   Algorithms

We describe here some algorithms for computing topological features of subsets in 2D and 3D digitized images. Since the programming languages of the C-family are now more popular than that of the PASCAL-family, we use here a pseudo-code which resembles the C-language.

4.1   Boundary Tracing in 2D Images

Boundary tracing becomes extremely simple when thinking of a 2D image as of a 2D complex. The main idea of the algorithm consists in the following: at each boundary point find the next boundary crack and make a step along the crack to the next boundary point. Repeat this procedure until the starting point is reached again. Starting points of all boundary components must be found during an exhaustive search through the whole image. The following subroutine `Trace()` is called each time when a not yet visited boundary point of a region is found.

To avoid calling `Trace()` more than once for one region, vertical cracks must be labeled (e.g. in a bit of `Image[]`) as "already visited". `Trace()` follows the boundary of one foreground region while starting and stopping at the given point (x, y). Points and cracks are present only implicitly as explained above in Section 2.1. `Trace()` starts always in the direction of the positive Y-axis. After each move along a boundary crack C the values of *only two pixels* R and L of SON(P) of the end point P of C must be tested since the values of the other two pixels of SON(P) have been already tested during the previous move. For a detailed description of this algorithm see [10].

The pseudo-code of `Trace()`

`Image[NX, NY]` is a 2D array (standard raster) whose elements contain gray values or colors. The variables `P, R, L` and the elements of the arrays `right[4]`, `left[4]` and `step[4]` are structures each representing a 2D vector with integer coordinates, e.g. `P.X` and `P.Y`. The operation "+" stands for vector addition. Text after // is a comment.

```
void Trace(int x, int y, char image[])
{ P.X=x; P.Y=y; direction=1;
  do
  { R=P+right[direction];  // the "right" pixel
    L=P+left[direction];   // the "left" pixel
    if (image[R]==foreground)
      direction=(direction+1) MOD 4; // right turn
    else
      if (image[L]==background)
        direction=(direction+3) MOD 4; // left turn
    P=P+step[direction]; //a move in the new direction
  } while( P.X!=x || P.Y!=y);
} // end Trace
```

4.2   Tracing of Surfaces in 3D

To trace surfaces of bodies in a 3D standard raster the method by Gordon and Udupa [6] uses the 2D technique in 2D slices. The code of a single closed surface is disconnected, i.e. it consists of isolated codes of the slices; 33% of the pixels are visited twice, which is not economical, and a body whose parts have only common cracks but no common faces are considered as disconnected.

A more efficient method producing a single connected sequence of code elements for each closed surface is that of [11]. According to the method the program chooses an arbitrary pixel of the surface S as the starting pixel and labels its closure. Then it traces the open frontier Of(L, S) (see Section 1.1) of the set L of labeled cells, encodes the pixels of Of(L, S) (1 byte per pixel), and labels the closures of simple pixels (Section 2.1). This ensures that L remains homeomorphic to a closed 2-ball. It has been proved that if the surface S is homeomorphic to a sphere then the traced sequence is a Hamilton path: each pixel is visited exactly once. Otherwise there remain a few non-simple pixels which are visited twice. Their code elements are attached to the end of the sequence of simple pixels. Thus the code sequence is always connected.

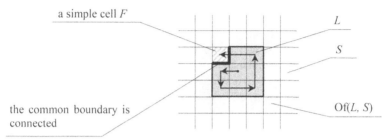

a simple cell F

L

S

the common boundary is connected

Of(L, S)

Fig. 7. The moves at the beginning of the tracing

A verbal description of the algorithm follows. A detailed description and the related proofs may be found in [11].

4.2.1   The Algorithm

Notations: S is the surface to be traced. It must be a 2-quasi-manifold. $L{\subset}S$ is the subset of labeled cells; it is homeomorphic to a closed 2-ball. The "rest sequence" is the set of non-simple pixels at the stage when all simple pixels of S are already labeled. The rest sequence is empty if the genus of S is zero.

1. Take any pixel of S as the starting pixel F_0, label its closure and save its coordinates as the starting coordinates of the code. This is the seed of L. Denote any one crack of the boundary of Fr(F_0,S) as C_{old} and find the pixel F of S which is incident with C_{old} and adjacent to F_0. Set F_{old} equal to F_0 and the logical variable $REST$ to FALSE. $REST$ indicates that the tracing of the rest sequence is running.
2. (Start of the main loop) Find the crack C_{new} as the first unlabeled crack of Fr(F,S) encountered during the scanning of Fr(F,S) clockwise while starting with the end point of C_{old} which is in Fr(L,S). If there is no such crack and F is labeled stop the algorithm: the encoding of S is finished.
3. If F is simple label its closure.
4. Put the direction of the movement from F_{old} to C_{old} and that of the movement from C_{old} to F into the next byte of the code. If the pixel F is non-simple set the corresponding bit in the code (to recognize codes of non simple pixels in the ultimate sequence).
5. If $REST$ is TRUE check, whether F is equal to F_{stop} and C_{new} is equal to C_{stop}. (These variables were defined in item 6 of the previous loop). If this is the case stop the algorithm and analyze the rest sequence to specify the genus of S as explained in [11]. Delete multiple occurrences of pixels from the rest sequence.

6. If F is simple set *REST* equal to FALSE; else set F_{stop} equal to F, C_{stop} equal to C_{new} and *REST* equal to TRUE.

7. Set F_{old} equal to F. Find the pixel F_{new} of S incident with C_{new} and adjacent to F. Set F equal to F_{new} and C_{old} equal to C_{new}. Go to item 2.

End of the algorithm.

4.3   Filling the Interiors of Boundaries in Multi-dimensional Images

To test whether an n-cell P of an n-space lies in the interior of a given closed boundary it is necessary to count the intersections of the boundary with a ray from P to any point outside the space; however, it is difficult to distinguish between intersection and tangency (Fig. 8 a and b). The solution becomes easy if the boundary is given as one or many $(n-1)$-dimensional manifolds in a Cartesian AC complex and the "ray" is a sequence of alternating n- and $(n-1)$-cells all lying in one row of the raster (Fig. 8c).

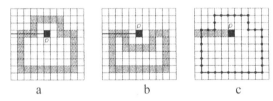

a   b   c

Fig. 8. Intersection (a) and tangency (b) are difficult to distinguish in "thick" boundaries; this is easy at boundaries in complexes (c)

In a 2D image the boundary must be a closed sequence of cracks and points (Fig. 8c). Then intersections are only possible at vertical cracks and the problem of distinguishing between intersections and points of tangency does not occur. The method has been successfully implemented for dimensions n=2, 3, 4.

4.3.1   The Pseudo-Code

Denote by F the current n-cell of the n-dimensional standard raster. Choose a coordinate axis A of the Cartesian space (e.g. $A=X$ in the 2D case). Denote by C(F) the own $(n-1)$-cell of F, whose normal is parallel to A (e.g. the vertical crack of F in the 2D case). Label all $(n-1)$-cells of the given boundary whose normal is parallel to A. In the 2D case when $A=X$ these are the vertical cracks of the given boundary.

```
for each row R parallel to A do
  { BOOLEAN fill=FALSE;
    for each n-cell F in the row R do
    { if C(F) is labeled then fill=1-fill; // inverting fill
      if fill is TRUE then F=foreground;
      else               F=background;
    }
  }
```

4.4   Component Labeling in an n-Dimensional Space

We consider here the simplest case of a 2D binary image in a standard raster while the algorithm is applicable also to multi-valued and multi-dimensional images in a topological raster. It is expedient to consider a multi-dimensional image as a one-dimensional array *Image*[*N*]. For example, in the 2D case the pixel with coordinates (x, y) may be accessed as *Image*[$y \cdot NX + x$] where *NX* is the number of pixels in a row.

In a standard raster a function must be given which specifies which raster elements are adjacent to each other and thus are connected if they have the same color. In our simple 2D example we use the well-known "8-adjacency" of the foreground pixels and the "4-adjacency" of the background pixels. In the general case the adjacency of the *n*-cells of an *n*-dimensional complex must be specified by rules specifying the membership of cells of lower dimensions [9] since an adjacency of *n*-cell depending on their "color" is not applicable for multi-valued images.

In a topological raster the connectivity of two cells is defined by their incidence which in turn is defined by their topological coordinates (see below Condition 3.6.3 in Section 3.6).

4.4.1   The Algorithm

Given is a binary array Image[] of N elements and the functions NumberNeighb(color) and Neighb(i,k): the first one returns the number of adjacent pixels depending on the color of a given pixel; the second one returns the index of the kth neighbor of the ith pixel. As the result of the labeling each pixel gets additionally (in another array Label[]) the label of the connected component which it belongs to.

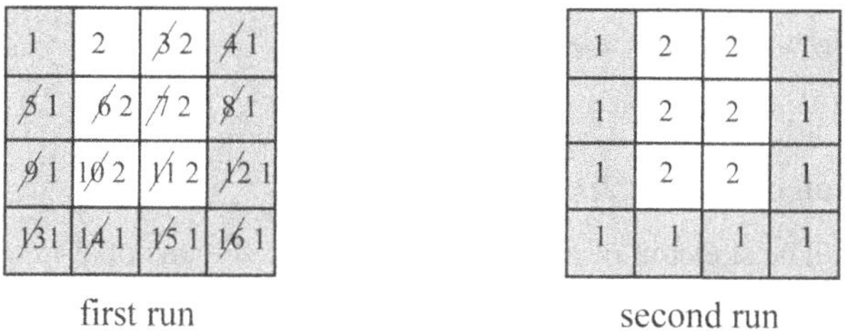

first run   second run

Fig. 9. Illustration to the algorithm of component labeling

4.4.2   The Pseudo-Code

Allocate the array Label[N] of the same size as Image[N]. Each element of "Label" must have at least $\log_2 N$ bits, where N is the number of elements in Image. In the first loop each element of Label gets its own index as its value:

```
for (i=1; i<N; i++) Label[i]=i;
for (i=1; i < N; i++)
{ color=Image[i];
    for (j=0; j<NumberNeighb(color); j++)
    { k=Neighb(i, j); //the index of the jth neighbor of i
      if (Image[k]==color) SetEquivalent(i,k,Label);
    }
} // end of the first run
SecondRun(Label,N); // end of the algorithm
```

The subroutine SetEquivalent() makes the preparation for labeling the pixels having the indices i and k as belonging to one and the same component. For this purpose one of the pixels gets the index of the "root" of the other pixel. The function

`Root()` returns the last value in the sequence of indices where the first index `k` is that of the given pixel, the next one is the value of `Label[k]` etc. until `Label[k]` becomes equal to `k`. The subroutine `SecondRun()` replaces the value of `Label[k]` by the value of a component counter or by the root of `k` depending on whether `Label[k]` is equal to `k` or not.

4.4.3   Pseudo-Codes of the Subroutines

```
subroutine SetEquivalent(i,k,Label)
{ if (Root(i,Label)<Root(k,Label))
                  Label[Root(k,Label)]=Root(i,Label);
  else Label[Root(i,Label)]=Root(k,Label);
} // end of SetEquivalent

int Root(k, Label)
{ do
  { if (Label[k]==k) return k;
    k=Label[k];
  } while(1);
} // end of Root

subroutine SecondRun(Label,N)
{ count=1;
  for (i=0; i<N; i++)
  { value=Label[i];
    if (value==i)
    { Label[i]=count;
      count=count+1;
    }
    else Label[i]=Label[value];
  }
} // end of SecondRun
```

4.5   Skeleton of a Set in 2D

Definition SK: The skeleton of a given set T in a two-dimensional image I is a subset $S \subset T$ with the properties:

a) S has the same number of connected components as T;
b) The number of connected components of $I{-}S$ is the same as that of $I{-}T$;
c) Certain singularities of T are retained in S.

Singularities may be defined e.g. as the "end points" in a 2D image or "borders of layers" in a 3D image etc.

A well-known difficulty in calculating skeletons is that it is impossible to remove all simple pixels simultaneously without violating the above conditions. However, representing an image as a complex C makes it possible to calculate the skeleton by a procedure which may be either sequential or parallel. It is based on the notion of the *open frontier* (s. Section 1.1 above). The procedure consist in removing simple non-singular cells of T alternatively from the frontier Fr(T, C) and from the open frontier Of(T, C). A cell c of the frontier Fr(T, C) (respectively of Of(T, C)) is *simple* if the intersection of SON(c)$-\{c\}$ (respectively Cl(c)$-\{c\}$) with both T and its complement $C{-}T$ is connected. We present a simple version for a 2D topological raster.

4.5.1   The Algorithm

Let $C[NX, NY]$ be a 2D array with topological coordinates. The subset T is given by labeling cells of all dimensions of T: $C[x, y] > 0$ iff the cell $(x, y) \in T$. To delete a cell means to set its label $C[x, y]$ to zero. A 0- or 2-cell c is considered as *singular* iff it is incident with exactly one labeled cell other than c.

To calculate the skeleton of T run the following loop:

    do   { Scan C and delete all simple and non-singular cells of $T \cap Fr(T, C)$;
                CountClose = number of cells deleted during this scan;
                Scan C and delete all simple and non-singular cells of $T \cap Of(T, C)$;
                CountOpen = number of cells deleted during this scan;
    } while (CountClose+CountOpen > 0);
    // end of Algorithm

Fig. 10 shows an example.

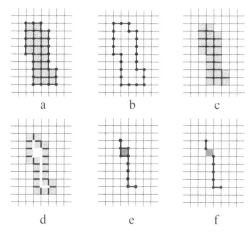

Fig. 10. a) a given 2D subcomplex T; b) its frontier Fr; c) the set $T–Fr$: the simple cells of the frontier deleted; d) the open frontier Of of the set $T–Fr$; e) the set $T–Fr–Of$: the simple cells of the open frontier deleted;   f) the skeleton

The result may be, if desired, easily transformed either to a sequence of pixels or to 1-complex containing only points and cracks.

4.6   Algorithms for Topological Investigations

Topological computations are particularly simple in a Cartesian complex with topological coordinates. We present in the following sections some basic algorithms.

4.6.1   Computing the Dimension of a Cell in an *n*-Dimensional Space

If $X=(X_1, X_2, ..., X_n)$ is a cell of an *n*-dimensional Cartesian complex then

$$Dimension(X) = \sum_{i=1}^{n} X_i \, MOD \, 2.$$

4.6.2   Condition of Bounding in an n-Complex: the Cell A Bounds the Cell B

A_i is the ith coordinate of the cell A; dim(A_i) is the dimension of A_i in the ith coordinate axis (dim(A_i) is either 0 or 1).

The condition: $\forall i = 1...n;$ dim$(A_i) \leq$ dim$(B_i) \wedge MaxDif = 1;$

where dim$(A_i) = A_i$ MOD 2; and $MaxDif = \max |A_i - B_i|;$ $i=1...n.$

4.6.3   Condition of Incidence: The Cell A Is Incident with the Cell B

A bounds B OR B bounds A OR $A = B;$

4.6.4   Computing the SON of a k-Cell A in an n-Dimensional Space

To explain the idea we first show as an example all cells of the SON of a 1-cell A in a 3D space. Let $k=1$ and $A=(2, 3, 6)$. A has *two even coordinates*. It is possible to change *one or both of them* by ± 1 to get a cell bounded by A.

The number N of cells bounded by A: $N = 2 \cdot C_2^1 + 2^2 \cdot C_2^2 = 2 \cdot 2 + 4 \cdot 1 = 8;$ where

C_k^i denotes the number of combinations of i elements from k.

The cells bounded by A are:

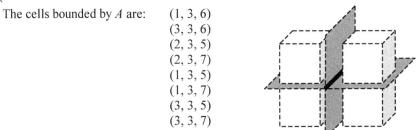

(1, 3, 6)
(3, 3, 6)
(2, 3, 5)
(2, 3, 7)
(1, 3, 5)
(1, 3, 7)
(3, 3, 5)
(3, 3, 7)

Fig. 11. The SON of a 1-cell in a 3D space

Now we present the pseudo-code of a function computing the SON of a cell "Cell" in a 4-dimensional space. The coordinates of all cells $C \in$ SON(*Cell*) will be saved in the array SON[ ][4].

4.6.5   The Pseudo-Code

```
void SaveSON(int Cell[4], int SON[ ][4])
{ int C[4], step[4];
 //"step" contains the increments of the coordinates: 1 for even and 0 for odd ones
 for (k=0; k<4; k++)
 { if ((Cell[k] MOD 2)==0) step[k]=1;
   else step[k]=0;
 }
 i=0;
 // four nested loops; C runs through all cells bounded by Cell:
 for (C[3]=Cell[3]-step[3]; C[3]≤Cell[3]+step[3];C[3]++)
 for (C[2]=Cell[2]-step[2]; C[2]≤Cell[2]+step[2];C[2]++)
 for (C[1]=Cell[1]-step[1]; C[1]≤Cell[1]+step[1];C[1]++)
 for (C[0]=Cell[0]-step[0]; C[0]≤Cell[0]+step[0];C[0]++)
 { for (k=0; k<4; k++) SON[i][k]=C[k];
   i=i+1;
 }
} // end of SaveSON
```

A similar algorithm for computing the closure of a cell in a 4-dimensional space:

```
void SaveClosure(int Cell[4], int Cl[ ][4])
{ int C[4], step[4];
//"step" contains the increments of the coordinates: 0 for even and 1 for odd ones
for (k=0; k<4; k++)
{ if ((Cell[k] MOD 2)==1) step[k]=1;
else step[k]=0;
}
i=0;
// four nested loops; C runs through all cells bounding Cell:
for (C[3]=Cell[3]−step[3]; C[3]≤Cell[3]+step[3];C[3]++)
for (C[2]=Cell[2]−step[2]; C[2]≤Cell[2]+step[2];C[2]++)
for (C[1]=Cell[1]−step[1]; C[1]≤Cell[1]+step[1];C[1]++)
for (C[0]=Cell[0]−step[0]; C[0]≤Cell[0]+step[0];C[0]++)
{ for (k=0; k<4; k++) Cl[i][k]=C[k];
i=i+1;
}
} // end of SaveClosure
```

5   Conclusion

We have demonstrated that abstract cell complexes may be successfully used for modeling locally finite topological spaces satisfying the classical axioms on a computer and for solving topological problems. We have suggested a new data structure, the 3D cell list, for encoding 3D segmented images or 3-manifolds in such a way that topological relations between subsets may be immediately extracted from the structure without searching. Another important property of the cell list is the possibility to consistently describe the so-called non-proper complexes having the advantage of being very simple although not representable by classical means since in a non-proper complex a cell may *multiply* bound another cell.

We have presented descriptions and/or pseudo-codes of seven basic algorithms for computing topological features of subsets of an abstract cell complex represented on a computer as an n-dimensional array, $n≤4$. Some of the described algorithms are the necessary tools when implementing high-level topological algorithms, namely:

Automatic calculation the cell list of a segmented (labeled) n-dimensional image (n=2 or 3);

Handle decomposition of a 3-manifold in a 4D space, which is useful for comparing two manifolds with each other.

Identification of faces of a polyhedron represented by a cell list and producing the cell list of a 3-manifold with the aim of comparing two manifolds with each other.

Modeling of linked spheres in an n-dimensional space (n=3, 4, 5) with the aim to experimentally investigate their topological properties.

The letter algorithms have been developed and successfully tested by the author, however, they cannot be described here because of paper size limitations.

An important open problem is, whether the described approach, which yields very simple descriptions of 3-manifolds, may contribute to the problem of their classification.

References

1. Alexandroff, P.: Diskrete Räume. Mat. Sbornik. 2 (1937) 501-518
2. Barett, J.: Quantum Gravity as Topological Quantum Field Theory. Jour. Math. Phys. 36 (1995) 6161-6179
3. Bertrand, Y., Fiorio, Ch., Pennaneach, Y.: Border Map: A Topological Representation for nD Image Analysis. In: Bertrand, G., Couprie, M., Perroton, L. (eds.): Discrete Geometry for Computer Imagery. Lecture Notes in Computer Science, Vol. 1568, Springer-Verlag, Berlin Heidelberg New York (1999) 242-257.
4. Encarnacao, J., Strassler, W., Klein, R.: Computer Graphics. R. Oldenbourg Verlag, Munich (1997)
5. Fiorio, Ch.: A Topologically Consistent Representation for Image Analysis: The Frontier Topological Graph. In: Miguet, S., Montanvert, A., Ubéda, S. (eds.): Discrete Geometry for Computer Imagery. Lecture Notes in Computer Science, Vol. 1176, Springer-Verlag, Berlin Heidelberg New York (1996) 151-162
6. Gordon, D., Udupa, J.K.: Fast Surface Tracking in Three-Dimensional Binary Images. Computer Vision, Graphics and Image Processing 45 (1989) 196-214,
7. Klette, R.: Cell Complexes through Time. University of Auckland, CITR-TR-60, June 2000.
8. Kovalevsky, V.: On the Topology of Digital Spaces. In: Proceedings of the Seminar "Digital Image Processing",.Technical University of Dresden (1986) 1-16
9. Kovalevsky, V.: Finite Topology as Applied to Image Analysis. Computer Vision, Graphics and Image Processing 45 (1989) 141-161
10. Kovalevsky, V.: Finite Topology and Image Analysis. In: Hawkes, P. (ed.): Advances in Electronics and Electron Physics, Vol. 84. Academic Press (1992) 197-259
11. Kovalevsky, V.: A Topological Method of Surface Representation. In: Bertrand, G., Couprie, M., Perroton, L. (eds.): Discrete Geometry for Computer Imagery. Lecture Notes in Computer Science, Vol. 1568. Springer-Verlag, Berlin Heidelberg New York (1999), 118-135
12. Kovalevsky, V.: A New Means for Investigating 3-Manifolds. In: Borgefors, G., Nyström, I., Sanniti di Baja, G. (eds.): Discrete Geometry for Computer Imagery. Lecture Notes in Computer Science, Vol. 1953. Springer-Verlag, Berlin Heidelberg New York (2000) 57-68
13. Lawrence, R.: Triangulation, Categories and Extended Field Theories. In: Baadhio, R., Kauffman, L. (eds.): Quantum Topology. World Scientific, Singapore (1993) 191-208
14. Matveev, S.: Computer Classification of 3-Manifolds. Russian Journal of Mathematical Physics 7 (2000) 319-329
15. Rinow, W.: Textbook of Topology. VEB Deutscher Verlag der Wissenschaft, Berlin (1975)
16. Steinitz, E.:Beitraege zur Analysis Situs, Sitzungsbericht Berliner Mathematischer Gesellschaft, Vol.7. (1908) 29-49
17. Stillwell, J.: Classical Topology and Combinatorial Group Theory. Springer-Verlag, Berlin Heidelberg New York (1995)

Computer Presentation of 3-Manifolds

Sergei Matveev

Chelyabinsk State University
Chelyabinsk, 454021 Russia
matveev@csu.ru

Abstract. Our goal is to describe an economic way of presenting 3-manifolds numerically. The idea consists in replacing 3-manifolds by cell complexes (their special spines) and encoding the spines by strings of integers. The encoding is natural, i.e., it allows one to operate with manifolds without decoding. We describe an application of the encoding to computer enumeration of 3-manifolds and give the resulting table. A brief introduction into the theory of quantum invariants of 3-manifolds is also given. The invariants were used by the enumeration for automatic casting out of duplicates. Separately, we investigate 3-dimensional submanifolds of R^3. Any such submanifold can be presented by a 3-dimensional binary picture. We give a criterion for a 3-dimensional binary picture to determine a 3-manifold.

1 Cell Complexes

In topology of manifolds the notion of a cell complex is usually considered in a more general sense than in computer topology(see [ ], [ ]). So we briefly recall it, restricting ourselves to the 2-dimensional case. We prefer to do that inductively.

1. A 0-dimensional cell complex X^0 is a finite collection of points called *vertices*.
2. An 1-dimensional cell complex X^1 is obtained from a 0-dimensional complex X^0 by attaching several 1-dimensional cells (i.e., arcs). The endpoints of the arcs are attached to the vertices of X^0, and the arcs must have no other common points. It is easy to see that a 1-dimensional cell complex is nothing more than a graph (loops and multiple edges are allowed).
3. A 2-dimensional cell complex X^2 is obtained from a 1-dimensional complex X^1 by attaching several 2-dimensional cells. In other words, we take a collection $\{D_1, \ldots, D_n\}$ of disjoint 2-dimensional discs and attach each disc D_i to X^1 via an *attaching map* $\varphi_i \colon \partial D_i \to X^1$. It is convenient to assume that the inverse image $\varphi^{-1}(e)$ of each open 1-cell e of X^1 consists of open connected subarcs of ∂D_i such that each of them is mapped onto e homeomorphically. In other words, we require that the boundary curve $l_i = \varphi_i(\partial D_i)$ of every 2-cell passes along the edges monotonically, without returns inside them. X^1 is called the *1-dimensional skeleton* of X^2.

We point out that in general the incidence relation between cells does not determine X^2 uniquely. It is necessary to know in which direction the boundary

G. Bertrand et al. (Eds.): Digital and Image Geometry, LNCS 2243, pp. 59–74, 2001.
© Springer-Verlag Berlin Heidelberg 2001

curve of each 2-cell passes along each edge as well as a cyclic order in which l_i passes along different edges. For example, consider two cell complexes obtained from a rectangle by identifying its vertices to one point and by different identifications of the sides (see Fig. 1). Each of them has one vertex, two loop edges, and one 2-cell attached to each edge exactly twice. Nevertheless, the complexes are different, since one of them is a torus while the other is a 2-sphere with three identified points.

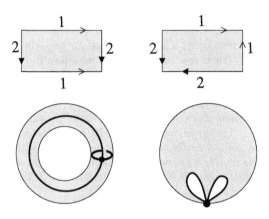

Fig. 1. Two complexes with the same incidence relation

On the other hand, information on the graph X^1, on the orientations of edges, and on the order in which the boundary curves of the 2-cells of X^2 pass along the oriented edges is quite sufficient for reconstructing X^2. The corresponding data can be written numerically as follows. First, we enumerate the vertices and the edges, and write down all the edges as a sequence of pairs $(i_1, j_1), \ldots, (i_m, j_m)$. Here i_k, j_k are the numbers assigned to the vertices joined by the edge number k. Simultaneously, we orient each edge by an arrow directed from the vertex no. i_k to the vertex no. j_k. Second, for each 2-cell we write down a string $(p_1\, p_2\, \ldots\, p_n)$ of non-zero integers ($\pm$ edges' numbers) which show how the boundary curve of the 2-cell passes along the edges. The sign of p_i shows us the direction in which the curve passes along the edge. For example, the complexes in Fig. 1 can be described by the following data:

1. Edges: $(1, 1), (1, 1)$ (in both cases);
2. 2-Cells: $(1\, 2\, -1\, -2)$ in the first case and $(1\, -1\, 2\, -2)$ in the second.

Obviously, the total number of vertices, the sequence of pairs, and the collection of the strings determine X^2. It is less obvious that if the link of each vertex is connected (i.e., if each vertex of X^2 has a cone neighborhood with a connected base), then X^2 is completely determined by the strings only. Indeed, to recover X^2, it suffices to do the following:

1. Realize the strings by boundaries of disjoint polygons with oriented and numbered edges;
2. Identify the edges that have the same numbers via orientation preserving homeomorphisms.

Then the 2-dimensional cell complex thus obtained is homeomorphic to X^2. Later on we will use this observation for encoding cell complexes by strings only.

2   Spines of 3-Manifolds

A compact topological space M is called a *3-manifold*, if every point $x \in M$ has a ball or half-ball neighborhood such that x corresponds to the center. The union of all points of M which admit no ball neighborhood is called the *boundary* of M and denoted by ∂M. The boundary of any 3-manifold M is a compact surface. If the boundary is empty, M is called *closed*.

Definition. Let P be a subpolyhedron of a 3-manifold M with nonempty boundary. Then P is called a *spine* of M, if the difference $M \setminus P$ is homeomorphic to the direct product of ∂M and a half-open interval. By a spine of a closed 3-manifold M we mean a spine of the punctured manifold $M_0 = M \setminus V$, where V is an open 3-ball in M.

The direct product structure allows us to squeeze (or collapse) M continuously onto P. Usually one considers spines that cannot be collapsed onto smaller subpolyhedra. That happens when no triangulation of a spine has principal simplices with free faces (then we cannot start the collapse). For example, a natural spine of the 3-ball B^3 is a point. Nevertheless, B^3 has many other spines, the most famous is the *Dunce hat D* obtained from a triangle by identifying its edges as shown in Fig. 2. There is an evident collapsing of B^3 onto the lateral surface of a cone contained in it, which determines a collapsing of B^3 onto the Dunce hat.

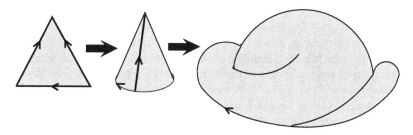

Fig. 2. Dunce hat

3   Special Spines

A spine of a 3-manifold carries much information about the manifold, in particular, all its homotopy characteristics such as the Euler characteristic and the fundamental group. Nevertheless, different 3-manifolds can have homeomorphic spines. For example, the punctured solid torus and the punctured manifold $S^2 \times S^1$ are distinct, but can be collapsed onto a wedge of a 2-sphere and a circle. To avoid that, we restrict our attention to so-called special spines having a very nice local structure.

Definition. A 2-dimensional polyhedron is called *special* if it can be presented as a cell complex such that every vertex has a neighborhood E homeomorphic to the open cone over the complete graph Δ on four vertices.

We will call E a *butterfly*. It is a strange butterfly indeed. Its *body* consists of four segments having a common endpoint, and it has six *wings* (see Fig. 3, where E is shown in three different forms). A spine of a 3-manifold is called *special*, if it is a special polyhedron.

The 1-dimensional skeleton of a special polyhedron is a regular graph of valence four. The edges consist of triple points, and the vertices can be viewed as crossing points of triple lines. One can regard special spines as *generic spines* or spines *in general position*.

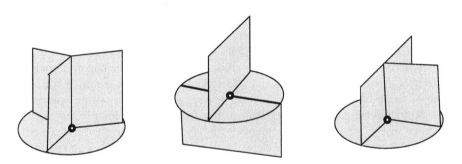

Fig. 3. Equivalent ways of looking at vertices

Theorem 1 ([ ]) *Any 3-manifold possesses a special spine.*

This is not surprising, since in squeezing a manifold to its spine we have a lot of freedom, which is sufficient for avoiding non-general position singularities. Two examples of special spines of the 3-ball (*Bing's House with Two Rooms* and the *Abalone*) are shown in Fig. 4. Notice that the Dunce hat is not special, since its unique vertex has no neighborhood of the required type.

Let us demonstrate how B^3 can be collapsed onto Bing's House. First, we squeeze B^3 onto the cube shown in Fig. 4. Then we penetrate through the upper tube into the lower room and exhaust the interior of the room. At last, we do the same with the upper room.

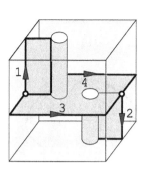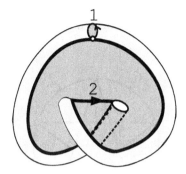

Fig. 4. Bing's House and the Abalone

To collapse the 3-ball onto the Abalone, we collapse it onto the Abalone with a filled tube. Then, starting from the ends of the tube, we push in the 3-dimensional material of the tube until we get a meridional disc.

Theorem 2 ([ ], [ ]) *If two 3-manifolds possess homeomorphic special spines, then they are homeomorphic (we assume that either both manifolds are closed or both have nonempty boundaries).*

In other words, every 3-manifold can be reconstructed from any its special spine in a unique way. The reason why the theorem is true consists in a simple local structure of special spines. Combining this fact with the above numerical description of cell complexes, we get a very economic encoding of 3-manifolds by strings of integers. For example, Bing's House and the Abalone can be presented respectively by three and two strings: (1) (2) (1 3 -2 -3 4 2 -4 -1 4 -3) and (-1) (1 2 2 1 -2).

The dodecahedron space (Poincaré homology sphere) can be presented by the following six strings describing its minimal special spine:

(1 3 -9 -2 -6)   (3 5 -6 -4 -8)   (5 2 -8 -1 -10)
(2 4 -10 -3 -7)   (4 1 -7 -5 -9)   (6 7 8 9 10).

In Fig. 5 (left) we show the 1-dimensional skeleton of the spine. The spine can be obtained from the skeleton by attaching six 2-cells along the curves corresponding to the above six strings. There is a convenient way to present the spine graphically by drawing a regular neighborhood of the skeleton in the spine, see Fig. 5 (right). The neighborhood is obtained from the 1-dimensional skeleton by attaching six annuli. The remaining part of the spine consists of six discs, each contained in the interior of the corresponding 2-cell. The spine can be obtained from the neighborhood by attaching these discs to all six closed curves in its boundary.

The Poincaré sphere is the simplest closed 3-manifold (different from the standard sphere S^3) whose homology groups coincide with the ones of S^3. In particular, its first homology group is trivial. The Poincaré sphere can be obtained from the regular dodecahedron by identifying its pentagonal faces such that each face is glued with the opposite one along a homeomorphism between them induced by a translation and rotation by $2\pi/10$, see Fig. 6. The images of faces of the dodecahedron under gluing form the spine described above.

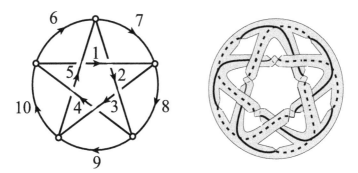

Fig. 5. A special spine of the dodecahedron space and its 1-dimensional skeleton

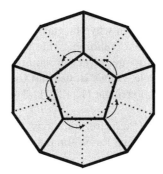

Fig. 6. A face identification schema for the regular dodecahedron producing the Poincaré sphere

Just as the Poincaré sphere, any special spine admits a similar graphical presentation by a neighborhood of its singular graph. Such presentations for Bing's House with two rooms and the Abalone are shown in Fig. 7. Fig. 8 (left) presents another special spine of the 3-ball, a mutant of Bing's House. It can be called a *double Abalone*, see Fig. 8 (right).

4   Moves on Special Spines

Thanks to large freedom in constructing, every 3-manifold possesses many different special spines. How can one describe them all? The question is important,

Fig. 7. Graphical presentations of Bing's House and the Abalone

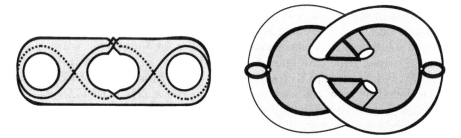

Fig. 8. Double Abalone, a mutant of the Bing House

since it is related to comparing different numerical codes of the same 3-manifold. Let us describe a move T on special spines which is sufficient for the purpose.

Consider two typical fragments E_1, E_2 of special spines. E_1 is a regular neighborhood of an edge in a special spine. It consists of a "cap" and a "cup" joined by a segment, with three attached "wings" (see Fig. 9, to the left). Since E_1 contains two vertices, it can be considered as a "double butterfly". E_2 is the union of the lateral surface of a cylinder, a middle disc, and three wings (see Fig. 9, to the right). Note that the fragments have the same natural boundary consisting of two circles joined by three arcs.

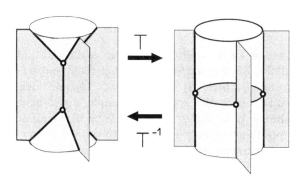

Fig. 9. The T-move

Definition. The elementary move T on a special spine P consists in removing a fragment $E_1 \subset P$ and replacing it by E_2.

Notice that T increases by one the number of vertices of a spine, while the inverse move T^{-1} decreases it. Also note that we start and finish with special spines that have more than one vertex each.

Theorem 3 ([ ], [ ]) *Let P, Q be two special polyhedra with more than one vertices each. Then P, Q are spines of the same 3-manifold if and only if one can take P to Q by a finite sequence of moves $T^{\pm 1}$.*

The proof of the theorem is nontrivial and based on well-known fact that any two triangulations of a 3-manifold have a common star subdivision. It is worth noticing that, from an algorithmic point of view, Theorem 3 does not give us a solution of the homeomorphism problem for 3-manifolds, which is still open. Of coarse, given two 3-manifolds M_1, M_2, we can construct their special spines P_1, P_2 and attempt to transform P_1 to P_2 by a sequence of moves $T^{\pm 1}$. Having found at least one such sequence, we can conclude that M_1, M_2 are homeomorphic. Nevertheless, this method never tells us that M_1, M_2 are distinct, since we are unable to test all possible sequences of moves (there are infinitely many of them).

We remark that moves $T^{\pm 1}$ can be easily realized on the level of numerical codes of the spines. For example, to realize the move T, we do the following:

1. Take the number i presenting a given edge and find in the strings all three triples $\pm i_1 \ \pm i \ \pm i_2$ consisting of $\pm i$ and two its neighbors. These triples present the wings of E_1.
2. Find in the strings all six neighboring pairs of the type $\pm i_1 \ \pm i_2$, where $i_1, i_2 \neq \pm i$ are contained in the above triples. The pairs present the six triangles forming the cup and the cap of E_1.
3. Insert a new string of three integers corresponding to the edges of the new 2-cell which appears under T, and replace the triples for the wings and the pairs for the triangles of E_1 by the pairs for the wings and the triples for the quadrilaterals of E_2.

5   Tabulation of 3-Manifolds

Recall that the set of all 3-manifolds possesses a natural additive structure: given two 3-manifolds M_1, M_2, one can construct their *connected sum* $M_1 \# M_2$ by removing two open 3-balls $B_1 \subset M_1, B_2 \subset M_2$ and identifying the two boundary spheres arising in this way. For example, any 3-manifold M admits trivial decompositions $M = M \# S^3$ and $M = S^3 \# M$. Manifolds admitting no nontrivial decompositions are called *prime* (with an exception of S^3, which is a neutral element). It is known that any orientable 3-manifold can be presented as a connected sum of prime factors and that the factors are determined in a unique way. Therefore the tabulation of 3-manifolds can be reduced to the tabulation of prime ones.

This very much resembles the situation with natural numbers. Indeed, everybody knows that any natural number can be decomposed into a product of prime numbers, and the decomposition is unique. Prime manifolds are thus analogous to prime numbers and were named after them.

We define the *complexity* $c(M)$ of a prime 3-manifold $M \neq S^2 \times S^1$ to be the minimal number of vertices of its special spines. According to this definition, the complexity of every 3-manifold is greater than 0 (since any special polyhedron must have at least one vertex). On the other hand, there are exactly three prime 3-manifolds whose natural minimal spines are in some sense simpler than any special polyhedron: the projective space RP^3, $S^2 \times S^1$, and the lens space $L_{3,1}$ (the quotient space of S^3 by a linear action of the cyclic group of order 3). Their natural minimal spines are a projective plane, a 2-sphere with an attached arc, and a *triple hat* (the 2-dimensional cell complex presented by the string (1 1 1)). The natural minimal spine of S^3 (a point) is also simpler than any other polyhedron. As an exception, we set the complexity of these four manifold to be zero.

It turns out that this complexity is naturally related to all the known methods of presenting manifolds and adequately describes complexity of manifolds in the informal sense of the expression. In particular, if a 3-manifold M admits a triangulation containing n tetrahedra, then $c(M) \leq n$.

Another nice property of the notion is that there are only finitely many closed prime manifold of a fixed complexity. To see that, let us show that for any integer k the number of special polyhedra with k vertices is finite. First we enumerate all regular graphs of degree 4 with k vertices. Clearly, there is only a finite number of them. Given a regular graph, we replace each vertex v by a butterfly E_v (a copy of the standard vertex singularity) such that the four-segment body of E_v coincides with the union of initial segments of edges emanating from v, see Fig. 10. For each edge e of G, we glue together the

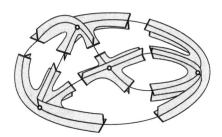

Fig. 10. Constructing spines from butterflies

butterflies corresponding to its endpoints: three wings of the first butterfly are glued to the three wings of the second. It can be done in six different ways. We get a 2-dimensional polyhedron whose natural boundary consists of circles. Attaching 2-cells to the circles, we obtain a special polyhedron. It is clear that all special polyhedra with k vertices can be obtained in this way. Not all of them are spines of 3-manifolds. Nevertheless, the resulting finite set of special polyhedra contains all special spines with k vertices.

It follows that for any k there are only finitely many closed prime 3-manifolds of complexity k. So it is natural to tabulate 3-manifolds in order of increasing complexity. The following table had been obtained by means of a computer [, ,]. $N(k)$ is the number of closed orientable prime 3-manifolds of complexity k.

k	0	1	2	3	4	5	6	7	8	9
$N(k)$	3	2	4	7	14	31	74	175	436	1160

Let us comment the table.

1) As we have mentioned above, prime manifolds of complexity 0 are RP^3, $S^2 \times S^1$, and $L_{3,1}$ (S^3 is usually considered as a manifold which is not prime).

2) All manifolds of complexity ≤ 5 are Seifert manifolds (see [ ]) and have finite fundamental groups. They are quotient spaces of S^3 by linear actions of different finite groups.

3) The table contains only one nontrivial homology sphere (the Poincaré sphere, which has complexity 5).

4) All six flat (i.e., having Euclidean geometric structure) 3-manifolds have complexity 6; the torus $S^1 \times S^1 \times S^1$ is one of them. Also, the list of manifolds of complexity 6 contains representatives of all the four known series of elliptic 3-manifolds [ ].

5) All manifolds of complexity ≤ 8 are graph-manifolds of Waldhausen [ ]. The first two manifolds admitting a complete hyperbolic structure appear on the level of complexity 9. One of them has the smallest hyperbolic volume among all known closed orientable hyperbolic 3-manifolds [ ]. Hyperbolic manifolds with the four next values of the volume have complexity 10.

The computer program enumerating 3-manifolds works in the following way. It first looks through all the regular graphs of degree 4 with a given number of vertices and, for each graph, lists all the possible gluings together of butterflies corresponding to its vertices. Then we attach 2-cells to the boundary circles of the resulting 2-dimensional polyhedra. Each special polyhedron P thus obtained is tested in the following respects :

1. Can it be thickened, i.e., is it a spine of a 3-manifold M?
2. Is M closed and orientable?
3. Is M prime?
4. Is P a minimal spine of M?
5. Is M different from all the manifolds obtained on the previous steps?

Having obtained a negative answer to at least one of the questions, the computer refuses to consider the proposed choice of pasting the butterflies and goes to the next version. In the converse case, the result is stored.

Let us comment how the computer performs the tests. Answers to the first two questions can be easily obtained algorithmically. There is also an algorithm answering the third question, but it is enormously huge and cannot be implemented in practice. No complete algorithms for answering the last two questions are known, even theoretically. So to answer questions 3-5 we are compelled to content ourselves with partial algorithms.

To answer the questions 3,4, we try to apply to P different combinations of simplification moves that preserve M, see [ ]. One of the moves is T^{-1}. Other moves consist in adding to P one 2-cell, removing another one, and collapsing the resulting spine as long as possible. Any simplification move transforms P either to a simpler special spine, or to a collection of simpler special spines joined by arcs. So spines admitting simplification moves are not interesting: either they are non-minimal (in the first case) or they determine reducible manifolds (in the second one).

To answer the last question, we apply two algorithms. The first one tries to transform P by successive application of moves $T^{\pm 1}$ to a special spine obtained earlier and thus prove that M is not new. The second one compares M with all the manifolds obtained on the previous steps by calculating different 3-manifold invariants, in particular, by calculating homology groups and Turaev-Viro invariants (see the next section). An important experimental observation is that up to the level of complexity 10 the joint work of these algorithms always gives a definitive answer: we either prove that M is a new manifold (then we include it into the table) or prove that it is a duplicate (then we reject it).

6   The Turaev-Viro Invariants

How can one prove that two given 3-manifolds are not homeomorphic? Usually this can be done by means of various invariants, i.e., numerical or algebraic characteristics of manifolds. We compute the invariants of the manifolds and compare them. If they are distinct, then so are the manifolds; the converse is not always true. We describe here the so-called *state sum*, or *quantum*, invariants introduced by V. Turaev and O. Viro. Deep notions of statistical mechanics and quantum physics lie behind them ([ ]). The construction of the invariants proceeds in a number of steps.

Step 1. Choose an integer $N > 1$.

Step 2. Consider the set integers $\mathcal{C} = \{0, 1, \ldots, N - 1\}$. These integers will called *colors*. Let us associate to each color i a real number w_i called its *weight*. We will consider w_i as variables whose values are to be determined.

Step 3. Consider the set of all butterflies with wings colored by the colors from the palette $\mathcal{C}$. Two colored butterflies are said to be equivalent, if there is a color preserving homeomorphism between them. The set is finite, since the butterfly has 6 wings, and thus we have $\leq N^6$ colorings.

Step 4. To each colored butterfly we assign a real number called the *weight* of the butterfly. There arises the following problem: how to denote colored butterflies and their weights? Note that a colored butterfly is uniquely determined by the triple of colors i, j, k of wings adjacent to an edge, and the triple k, l, m of colors of the opposite wings (two wings are *opposite*, if their intersection is a vertex rather than an edge, see Fig. 11). For typographical convenience, and

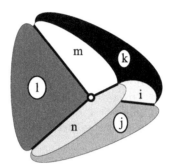

Fig. 11. Colored butterfly

following some earlier conventions, such a butterfly will be denoted by (2×3)-matrix $\begin{pmatrix} i & j & k \\ l & m & n \end{pmatrix}$. The weight of this butterfly is denoted by $\begin{vmatrix} i & j & k \\ l & m & n \end{vmatrix}$ and called a *quantum* $(q - 6j)$-*symbol*, for reasons we will not go into here. We will think of the symbols as variables whose values are also to be determined.

Step 5. Let P be a special polyhedron, $V(P)$ the set of its vertices, and $C(P)$ the set of its 2-cells. A *coloring* of P is a map $\xi : C(P) \to \mathcal{C}$. This means that the color $\xi(c)$ is used for a 2-cell c. Denote by $Col(P)$ the set of all colorings of P. It consists of $N^{\#C(P)}$ elements, where N is the number of colors and $\#C(P)$ is the number of 2-cells in P. To each coloring $\xi \in Col(P)$ we associate its *weight* $w(\xi)$ according to the rule

$$w(\xi) = \prod_{v \in V(P)} \begin{vmatrix} i & j & k \\ l & m & n \end{vmatrix}_v \prod_{c \in C(P)} w_{\xi(c)}. \tag{1}$$

Note that the coloring ξ determines a coloring of a neighborhood of every vertex $v \in V(P)$. It means that in a neighborhood of v we see a colored butterfly $\begin{pmatrix} i & j & k \\ l & m & n \end{pmatrix}_v$ with the $(q-6j)$-symbol $\begin{vmatrix} i & j & k \\ l & m & n \end{vmatrix}_v$. Every 2-cell c of P is colored by the color $\xi(c)$ having weight $w_{\xi(c)}$. Therefore, the right-hand side of the formula is the product of all symbols and weights of used colors (with multiplicity).

Definition. The *weight* of a special spine P of P is given by the formula

$$w(P) = \sum_{\xi \in Col(P)} w(\xi).$$

Step 6. Of course, the weight of a special polyhedron depends heavily on the specific values of color weights and symbols. So we run into the following problem: under what conditions is the weight of a spine an invariant of the manifold, i. e., does not depend on the choice of a specific spine? To answer, let us try to subject the variables to constraints which would insure that the weight of a special polyhedron is invariant with respect to T-moves. In order to do that, let us write down the following system of equations:

$$\begin{vmatrix} i & j & k \\ l & m & n \end{vmatrix} \begin{vmatrix} i & j & k \\ l' & m' & n' \end{vmatrix} = \sum_z w_z \begin{vmatrix} i & m & n \\ z & n' & m' \end{vmatrix} \begin{vmatrix} j & l & n \\ z & n' & l' \end{vmatrix} \begin{vmatrix} k & l & m \\ z & m' & l' \end{vmatrix}, \qquad (2)$$

where $i, j, k, l, m, n, l', m', n', z$ run over the palette $\mathcal{C}$.

The geometric meaning of these equations is clear from Fig. 12. Let a special polyhedron P_2 be obtained from a special polyhedron P_1 by one T-move. Our aim is to choose values of weights of the colors and symbols of the colored butterflies so that for any P_1, P_2 the equality $w(P_1) = w(P_2)$ holds. The left-hand side of each equation equals the contribution of the fragment E_1 to the weight $w(P_1)$, and the right-hand side is equal to the contribution of E_2 to $w(P_2)$. We take into account only those weights and symbols that occur in both $w(P_1), w(P_2)$. They correspond to five butterflies (two on the left and three on the right) and a unique 2-cell colored by z, which occurs on the right-hand side but not on the left one.

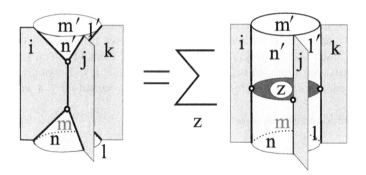

Fig. 12. Graphical interpretation of the equation

It follows from Theorem 3 that if the weights of colors and symbols of butterflies form a solution of the system, then the weight of a special spine is an invariant of the manifold.

One can easily see that the number of equations is much greater than the number of variables. Nevertheless, V. Turaev and O. Viro found solutions for arbitrary N. Saying "found" I mean that literally: they found them in a paper devoted to a different subject (see [ ])!

Explicit expressions for the solutions are too cumbersome to include them here. We restrict ourselves to the simplest case $N = 2$, when only two colors are used, and describe one of three independent solutions which are possible in this case. Let $\varepsilon = (1+\sqrt{5})/2$ or $\varepsilon = (1-\sqrt{5})/2)$. Then the solution is given by the formulas $w_0 = 1, w_1 = \varepsilon$, $\begin{vmatrix} 0\ 0\ 0 \\ 0\ 0\ 0 \end{vmatrix} = 1$, $\begin{vmatrix} 0\ 0\ 0 \\ 1\ 1\ 1 \end{vmatrix} = \varepsilon^{-1/2}$, $\begin{vmatrix} 0\ 1\ 1 \\ 0\ 1\ 1 \end{vmatrix} = \begin{vmatrix} 0\ 1\ 1 \\ 1\ 1\ 1 \end{vmatrix} = \varepsilon^{-1}$, and $\begin{vmatrix} 1\ 1\ 1 \\ 1\ 1\ 1 \end{vmatrix} = \varepsilon^{-2}$.

From the theoretical point of view, the calculation of invariants causes no problems, since the weights and symbols are known. In order to perform it, one must exhaust all colorings, calculate their weights by multiplying the weights of colors and the symbols of colored butterflies, and sum up all the numbers thus obtained. Computer programs based on this approach work well as long as both the number of colors and the number of vertices of the spine are about ten or less.

7   Submanifolds of R^3

In this section we describe the class of compact 3-manifolds which can be embedded into R^3. The class seems to be interesting, since any such manifold M can be realized as a union of unit cubes of the standard cubic lattice in R^3. One can think of the cubes forming the union as being black, and of all other lattice cubes as white. Thus we get a presentation of M by a 3-dimensional binary picture.

There is a simple criterion to decide if a given 3-dimensional binary picture presents a 3-manifold. We say that two lattice cubes are related by a *jump*, if they share a common 2-dimensional face and have the same color. We will distinguish black and white jumps according to the color of the cubes.

Criterion. *A 3-dimensional binary picture represents a 3-dimensional manifold if and only if any two one-colored cubes can be joined by ≤ 3 jumps of the same color.*

The forbidden situations are shown in Fig. 13:

1. There are two black cubes that have no common edges, but have a common vertex incident to six white cubes.
2. There are two black cubes that have no common 2-dimensional face, but have a common edge incident to two white cubes.
3. There are two white cubes that have no common edges, but have a common vertex incident to six black cubes.

The evident duality between conditions 1 and 3 and the self-duality of condition 2 reflects the fact that M is a manifold if and only if so is the background. It would be interesting to find similar conditions for n-dimensional binary pictures presenting n-dimensional manifolds.

Let us investigate 3-manifolds in R^3 from a purely topological point of view. To describe a general way for constructing them, we introduce two basic moves. Let M be a compact 3-dimensional submanifold of R^3.

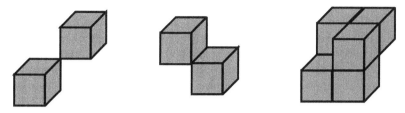

Fig. 13. Three situations that force a non-manifold binary picture

Move A. We attach to M a solid tube $D^2 \times [0, 1]$ that intersects M along two discs $D^2 \times \{0, 1\} \subset \partial M$;

Move B. We cut off from M a solid tube $D^2 \times [0, 1] \subset M$ that intersects ∂M along two discs $D^2 \times \{0, 1\}$. In other words, we drill in M a tubular hole.

Move C. (Puncturing.) We cut off a 3-ball from the interior of M.

Note that solid tubes and tubular holes may be knotted and linked, and run inside each other, see Fig 14.

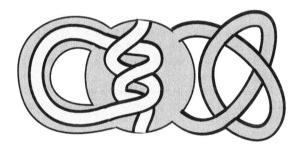

Fig. 14. 3-Ball with two solid tubes and a tubular hole

Theorem 4 *Any compact 3-dimensional submanifold of R^3 can be obtained from a 3-ball by a finite sequence of Moves A-C. If the boundary ∂M of M is connected, then M can be obtained from a 3-ball by Moves A, B only.*

The proof is based on the well-known fact that any connected closed surface $F \neq S^2$ in R^3 is compressible in the following sense: there exists an embedded disc $D \subset R^3$ such that $D \cap F = \partial F$ and the curve ∂D does not bound a disc in F. D is called a compressing disc for F, since cutting F along ∂D and attaching two parallel copies of D to the two boundary circles of the resulting surface transforms F to a simpler surface F'. Doing that as long as possible, we get a collection of disjoint 2-spheres.

Now, if we have a submanifold M of R^3, we can apply this procedure to the boundary ∂M of M. Each time when the compressing disc is in M, we cut off a solid tube, i.e., perform the inverse of Move A. If the compressing disc lies in the outside of M, we fill a tubular hole, i.e., perform the inverse of Move B.

To finish the argument, we simply go in the inverse direction: instead of cutting tubes and filling holes, we drill holes and attach tubes.

The paper is based on my talk given at the winterschool on Digital and Image Geometry at Dagstuhl, December 2000. I thank the organizers for the possibility to participate in that very interesting school. The paper is written under financial support of RFBR (grant N 99-01-00813), INTAS (project 97-808), and the program "Universities of Russia".

References

1. Casler, B. G.: An embedding theorem for connected 3-manifolds with boundary. Proc. Amer. Math. Soc. **16**(1965), 559-566. 62, 63
2. Fomenko, A. T., Matveev, S. V.: Isoenergetic surfaces of integrable Hamiltonian systems, enumeration of 3-manifolds in order of increasing complexity and volume computation of closed hyperbolic 3-manifolds. Usp. Math. Nauk **43**(1988), 5-22 (Russian; English transl. in Russ. Math. Surv. 43 (1988), 3-24. 68
3. Kirillov, A. N., Reshetikhin, N. Yu: Representations of the algebra $U_q(sl_2)$, q-orthogonal polynomials, and invariants for links. Infinite dimensional Lie algebras and groups. Adv. Ser. in Math. Phys. (V. G. Kac, ed.) **7**(1988), 285-339. 71
4. Klette, R.: Cell complexes through time. Proc. Vision Geometry IX, San Diego, **SPIE-4117**(2000), 134-145. 59
5. Kovalevsky, V. A.: Finite topology as applied to image analysis. Computer Vision, Graphics, and Image Processing **46**(1989), 141-161. 59
6. Martelly, B., Petronio, C.: 3-Manifolds having Complexity at Most 9. Geometry and Topology, 2001 (to appear). 68
7. Matveev, S. V.: Transformations of special spines and the Zeeman conjecture. Izv. AN SSSR **51**(1987), N 5, 1104-1115 (Russian; English transl. in Math. USSR Izv. **31**(1988), N 2, 423-434). 63, 66
8. Matveev, S. V.: Complexity theory of 3-manifolds. Acta Appl. Math. **19**(1990), N 2, 101-130. 68
9. Matveev, S. V.: Computer Recognition of Three Manifolds. Experimental Mathematics. **7**(1998), N 2, 153-161. 68, 69
10. Milnor, J.: Groups which act on S^n without fixed points. Amer. J., Math., **79** (1967), 623-630. 68
11. Piergallini, R.: Standard moves for standard polyhedra and spines, III Convegno Nazionale di Topologia Trieste, 9-12 Giugno **1986**(1988), 391-414. 66
12. Seifert, Y.:Topologie dreidimensionalen gefaserter Räume, Acta Math. **60**(1933), 147-238. 68
13. Turaev, V. G., Viro, O. Y.: State sum invariants of 3-manifolds and quantum $6j$-symbol. Topology **31**(1992), N 4, 865-902. 69
14. Waldhausen, F.: Eine Klasse von 3-dimensionalen Mannigfaltigkeiten I, II, Invent. Math., **3**(1967), 308-333; Invent. Math., **4**(1967), 87-117. 68

"Continuous" Multifunctions in Discrete Spaces with Applications to Fixed Point Theory

Rueiher Tsaur and Michael B. Smyth *

Department of Computing, Imperial College
London SW7 2BZ

Abstract. In this paper, a sound notion of "continuous" multifunctions for discrete spaces is introduced. It turns out that a multifunction is continuous (= both upper and lower semi-continuous) if and only if it is "strong", or equivalently it takes (original) neighbours into (induced) neighbours w.r.t. Hausdorff metric. Generalizing various previous results, we show that any finite contractible graph has the almost fixed point property (afpp) for strong multifunctions. We then turn to the fixed clique property, or fcp (a desirable strengthening of afpp). We are able to establish this property only under much more explicitly geometric assumptions than in the case of afpp. Specifically, we have to restrict to convex-valued multifunctions, over spaces admitting a suitable notion of convexity; and, for the best results, to the standard n-dimensional digital space with $(3^n - 1)$-adjacency. In this way we obtain what may be considered (with reservations, discussed in the Conclusion) as an analogue of the classical Kakutani fixed point theorem.

1 Introduction

In [ ], a natural analog of the ϵ-δ definition of continuous functions for $2n$-adjacent n-dimensional digital pictures has been studied and investigated. It was also shown that any finite $(3^n - 1)$-adjacent n-dimensional digital picture has the almost fixed point property. In [ ], Boxer continued Rosenfeld's investigation for $2n$-adjacent n-dimensional digital pictures and examined several important classes of continuous functions including homeomorphisms, retractions and homotopies.

In this paper we shall investigate some like topics, but in the more general setting of many-valued functions. As indicated in [ ], there is a special need for multifunctions on discrete spaces in approximating ordinary continuous functions on classical (topological) spaces. To be more specific, it is not hard to see that multifunctions appear more naturally than single-valued functions in discrete (digital) structures. The point may perhaps be made clear via a simple example:

---

* The second author has been partially supported by EPSRC grant on "Digital Topology and Geometry: an Axiomatic Approach with Applications to GIS and Spatial Reasoning".

consider a continuous function f from unit interval I to I defined by

$$f(x) = \begin{cases} 2x, & 0 \le x \le 0.5 \\ 2 - 2x, & 0.5 \le x \le 1. \end{cases}$$

It is easy to see that during the approximation of f, the discrete representations $\dot{f}$ and $\ddot{f}$ of f may become multifunctions, i.e. $\dot{f} : K_4 \rightarrow K_4$, $00 \mapsto \{00, 01, 10\}$, $01 \mapsto \{01, 10, 11\}$, $10 \mapsto \{01, 10, 11\}$ & $11 \mapsto \{00, 01, 10\}$; and $\ddot{f} : K_8 \rightarrow K_8$, $000 \mapsto \{000, 001, 010\}$, $001 \mapsto \{001, 010, 011, 100\}$, $010 \mapsto \{011, 100, 101, 110\}$, $011 \mapsto \{101, 110, 111\}$, $100 \mapsto \{101, 110, 111\}$, $101 \mapsto \{011, 100, 101, 110\}$, $110 \mapsto \{001, 010, 011, 100\}$ & $111 \mapsto \{000, 001, 010\}$. Fig. 1 shows the diagrams of $\dot{f}$ and $\ddot{f}$. Hence the discrete representations of a function may become many-valued rather than single-valued.

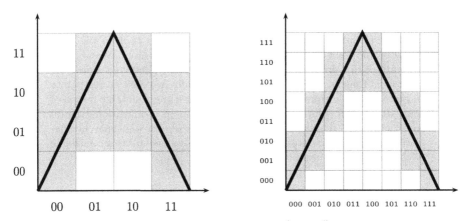

Fig. 1. The diagrams of $\dot{f}$ and $\ddot{f}$

As in the cited studies of Rosenfeld and Boxer, we take graphs (perhaps of some restricted class) as our underlying model of digital space. In looking for a notion of "well-behaved" multifunction between graphs (defined over vertices), it is apparent that there is a range of possibilities. Specifically, if f maps vertices of graph G to sets of vertices of graph H, we may ask for either a weak or strong relation to hold between $f(x)$ and $f(y)$ (assumed non-empty), when x is adjacent to y in G:

Weak there exist a vertex in $f(x)$ and a vertex in $f(y)$ which are adjacent (in H).
Strong every vertex in $f(x)$ is adjacent to some vertex in $f(y)$, and vice versa.

(There is at least one intermediate possibility as well.) In previous work [, , ,] we have shown that the weak relation is appropriate for digital approximation of ordinary continuous functions. On the other hand, the strong relation might for example be appropriate for the digital representation of smooth motion of bodies.

In this paper we focus on the "continuous" multifunctions corresponding to the strong relation. In Section 2 we give several alternative characterizations of the multifunctions in question, especially in terms of upper and lower semicontinuity in closure spaces. Section 3 contains the main result of the paper: all finite contractible graphs have the almost fixed point property (afpp) with respect to these (multi)functions. This generalizes previous afpp results for single-valued functions, including that of Rosenfeld [ ]. In the remainder of the paper we compare the afpp with another adaptation (suitable for digital spaces) of the fixed point property, namely the fixed clique property (fcp). It turns out that, in contrast with the single-valued situation, there is a sharp divergence between the afpp and the fcp for multifunctions. We are only able to secure the fcp when severe restrictions, of a "geometrical" nature, are placed on both the spaces and the (multi)functions in question.

2   Continuous Multifunctions

We briefly review the concept of strong multifunctions: by a *graph* G we mean a set X of *vertices* (or *points*), with a reflexive and symmetric relation E the *edges*, also known as a *tolerance graph* in [ ]. For graphs G_1 and G_2, a mapping $f : G_1 \rightarrow G_2$ is a *graph homomorphism* if f maps vertices of G_1 to vertices of G_2, preserving the tolerance relation. By a *multifunction* $f : G_1 \rightarrow G_2$ we understand a function that assigns to a vertex x of G_1 a non-empty subset $f(x)$ in G_2.

Definition 1. *Let* $G = (X, E)$ *be a graph. The* strong *power graph of* G, $\mathcal{P}_s(G)$, *has the non-empty finite subsets of* X *as vertices, and the edge set* E_s, *where*

$$AE_sB \Leftrightarrow \forall x \in A, \exists y \in B, xEy \ \& \ \forall y \in B, \exists x \in A, xEy.$$

Notice that the relation E_s is the conjunction of the two "weak power relations" E_0, E_1 (in the notation of Brink [ ]), defined by:

$$AE_0B \Leftrightarrow \forall x \in A, \exists y \in B, xEy$$

$$AE_1B \Leftrightarrow \forall y \in B, \exists x \in A, xEy.$$

Of course, since we have assumed E to be symmetric, we have $E_1 = E_0^{-1}$. Some remarks on the non-symmetric (digraph) case follow Proposition 3.

Definition 2. *Let* G *and* H *be graphs. The multifunction* f *from* G *to* H, $f : G \rightarrow H$, *is* strong *if regarded as a (single-valued) mapping into* $\mathcal{P}_s(H)$, *it is a graph homomorphism.*

Strong multifunctions may be simply characterized as follows:

Proposition 1. *The multifunction $f : G \to H$ is strong if and only if for any pair of vertices x, y of G, we have $h_H(f(x), f(y)) \le 1$, where h_H is the Hausdorff metric induced from the natural graph metric d_H.*   □

In order to compare this property with the usual continuity conditions for multifunctions in topology, it is useful to consider closure spaces [ ]:

Definition 3 (Čech). *A* closure space *is a pair (X, cl), where* $\mathrm{cl} : \mathcal{P}(X) \to \mathcal{P}(X)$ *is the* closure *operator which assigns to each subset $S \subseteq X$ a subset $\mathrm{cl}(S)$ called* closure *of S, satisfies*

(1) $\mathrm{cl}(\emptyset) = \emptyset$
(2) $S \subseteq \mathrm{cl}(S)$
(3) $\mathrm{cl}(S \cup T) = \mathrm{cl}(S) \cup \mathrm{cl}(T)$.

A topological space is a closure space in which the closure is idempotent. On the other hand, we shall view a graph $G = (X, E)$ as a closure space by taking cl as:

$$\mathrm{cl}(S) = \{x \in X \mid \exists y \in S, xEy\}.$$

(It is easy to characterize graphs as closure spaces satisfying certain conditions, see [ ].) Exactly as in topology, the *interior* operator of a closure space (X, cl) is defined by:

$$\mathrm{int}(S) = (\mathrm{cl}(S^c))^c.$$

Notice that in a graph (X, E) we have:

$$\mathrm{int}(S) = \{y \in X \mid N(y) \subseteq S\},$$

where the (1-)*neighbourhood* $N(y)$ is $\{x \in X \mid xEy\}$.

The basic notions of topology generally extend to closure spaces, provided we formulate them in terms of interior and closure, rather than in terms of open and closed sets. Thus we have the following definition for the standard continuity conditions on many-valued functions:

Definition 4. *Let X, Y be closure spaces, and $f : X \to Y$ a multifunction. Then f is said to be* lower semi-continuous (lsc) *if $f(\mathrm{cl}(S)) \subseteq \mathrm{cl}(f(S))$ for all $S \subseteq X$; and* upper semi-continuous (usc) *if $f^{-1}(\mathrm{int}(T)) \subseteq \mathrm{int}(f^{-1}(T))$ for all $T \subseteq Y$, where $f^{-1}(T) = \{x \in X \mid f(x) \subseteq T\}$.*

Taking graphs as closure spaces (as above), we then have:

Proposition 2. *Let $f : (X, E) \to (Y, E')$ be a graph multifunction. The following are equivalent:*

(1) *f is strong;*
(2) *f is lsc;*
(3) *f is usc.*

Proof. Notice that, in the case of a graph with edge relation R, $A \subseteq \mathrm{cl}(B)$ if and only if $\forall x \in A, \exists y \in B, xRy$, that is, if and only if AR_0B. Thus, lower semi-continuity of f reduces to the condition:

$$\forall x, y \in X : xEy \Rightarrow f(x)E_0'f(y).$$

By the symmetry of E and E', this is indeed equivalent to the statement that f is strong. Hence (1) $\Leftrightarrow$ (2).

Similarly, upper semi-continuity of f reduces to

$$\forall x, y \in X : xEy \Rightarrow f(x)E_1'f(y),$$

so that, again by symmetry of E and E', we have (1) $\Leftrightarrow$ (3). $\square$

The equivalence of upper and lower semi-continuity is a consequence of symmetry. The preceding definitions and (part of the) proof make sense also for reflexive digraphs, and it is easy to see that in that context we obtain:

Proposition 3. *A graph multifunction $f : G \rightarrow H$, where G, H are reflexive digraphs, is strong if and only if it is continuous (that is, both upper and lower semi-continuous).* $\square$

What happens in the case of non-reflexive digraphs? In that case, the corresponding "topological" structures are not closure spaces as ordinarily understood. The same kind of analysis is still possible, however: see [ ] for a discussion of "neighbourhood spaces" in relation to digraphs.

The "strong" relation E_s has been studied in the algebraic literature as the power of a (binary) relation E. It can be generalized to n-ary relations, see Brink [ ]. Multifunctions based on this and other power constructs (but for partial orders rather than relations in general) have been studied extensively in domain theory/denotational semantics: see [,] for a survey. In [,] we investigated this topic in the context of graphs and simplicial complexes.

3   Strong Almost Fixed Point Properties

Let $G = (X, E)$ be a graph and $f : G \rightarrow G$ a self-mapping multifunction. Then we say that $x \in X$ is an *almost fixed point* of f if x is adjacent to some element of $f(x)$. Notice that this condition can be written as $xEf(x)$, if we extend the edge relation E to subsets A, B of X by: $AEB \Leftrightarrow \exists a \in A, \exists b \in B, aEb$ (in effect, the "weak" relation mentioned in the Introduction). Let ϕ be a property of sets of vertices (points) of G. Then G is said to possess the *ϕ-almost fixed point property* (*ϕ-afpp*) if for any multifunction $f : G \rightarrow G$ which sends each vertex of G into a subset of G satisfying ϕ (f is said to be a *ϕ-multifunction* in this case), f has an almost fixed point.

A finite graph $G = (X, E)$ is said to be *contractible* (or *dismantlable*) if there is a linear ordering $<$ on X, say $\langle x_1, x_2, \ldots, x_m \rangle$, such that, for all $i \leq m - 1$, there exists a y differs from x_i, in the subgraph of G induced by $\{x_j \in X \mid j \geq i\}$, satisfying $N(x_i) \subseteq N(y)$ (in this case, it is said that y *dominates* x_i, or that x_i is *dominated* by y). Informally, the idea is that the graph may be contracted (or retracted) step by step to a vertex. For the "continuous" treatment in terms of homotopy, establishing the connection with topology, see [, ,].

Theorem 1. *Every finite contractible graph has the almost fixed point property for strong multifunctions (strong afpp).*

Proof. Let $G = (X, E)$ be a finite contractible graph and $f : G \to G$ a self-mapping strong multifunction. Since G is contractible, there exist $x, y \in X$ such that $N(x) \subseteq N(y)$ (**Condition 1**), and the induced subgraph $G' = (X \setminus x, E')$, $E' = E \cap (X \setminus x \times X \setminus x)$ is contractible (**Condition 2**).

We prove by induction on $\#X$ that f has an almost fixed point. *Initial step:* It is clear that G has the strong afpp when $\#X = 1$ or 2. *Inductive step:* We assume that G has the strong afpp when $\#X \leq m \in \mathbb{N}$, and we claim that G has the strong afpp when $\#X = m+1$. Define the multifunction $g : G' \to G'$ by

$$g(u) = \begin{cases} f(u), & x \notin f(u) \\ (f(u) \setminus x) \cup \{y\}, & x \in f(u). \end{cases}$$

We claim that g is a strong multifunction. Indeed, let $a, b \in X \setminus x$ such that $aE'b$. We have aEb and hence $f(a)E_s f(b)$. Now $g(a), g(b)$ are obtained from $f(a), f(b)$ by replacing any occurrences of x by y; since $N(x) \subseteq N(y)$, it is clear that such a replacement cannot cause E'_s to fail. That is, $f(a)E_s f(b) \Rightarrow g(a)E'_s g(b)$. Thus g is strong.

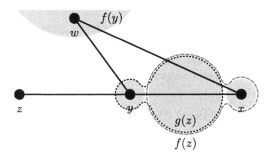

Fig. 2. $x, y, z, w, g(z), f(z)$ and $f(y)$

Therefore by our inductive assumption and **Condition 2**, there exists a $z \in X \setminus x$ such that z is an almost fixed point of g, i.e. $zE'g(z)$. Then (since $f(z)$ differs from $g(z)$ by, at most, having x in place of y) z must be an almost fixed point also of f, except in the following case: y is the only vertex of $g(z)$ adjacent to $z, x \in f(z)$, but $y \notin f(z)$, see Fig. 2. Since yEz, $f(y)E_s f(z)$. So $f(y)$ contains a vertex, say w, adjacent to x, hence adjacent to y ($\because N(x) \subseteq N(y)$). Thus y is an almost fixed point of f. $\square$

Corollary 1. *Every finite contractible graph has the afpp for single-valued graph homomorphisms.* □

In Euclidean geometry, the "fixed point property" is linked with Euclidean convexity structure by the *Kakutani fixed point theorem*: in 1941, Kakutani generalized Brouwer's fixed point theorem for closed n-balls to multifunctions by showing that any closed n-ball has the *fixed point property* for usc closed convex-valued multifunctions [,].

At first sight, this seems to have no bearing on what we are doing here. However, in the next section we shall go on from the afpp to the *fixed clique* property (fcp). We will find that the fcp fails badly for general continuous multifunctions, and that the Kakutani theory gives useful guidance on the conditions that need to be imposed to get positive results.

4   Strong Fixed Clique Properties

If S is a set and $D = \{c_1, \ldots, c_k\}$ is a collection of distinct non-empty subsets of S whose union is S (a cover of S), we say $G = (X, E)$ is the *intersection graph* of D if $X = D$ with c_i and c_j adjacent whenever $i \neq j$ and $c_i \cap c_j \neq \emptyset$ [ ].

By the *induced graph* of an n-dimensional digital space (cellular or grid point space), we mean the appropriate *neighbourhood graph* in the case of a grid, and the intersection graph of the closed cells of highest dimension in the case of a cell complex (for information on digital topology/geometry and neighbourhood graphs, we refer [,]).

It is clear that the induced graph of a finite n-dimensional regular orthogonal grid (with $2n$-adjacency) does not have the afpp. Therefore, for his almost fixed point theorem in [ ], Rosenfeld allowed also "diagonal" neighbours (giving $(3^n - 1)$-adjacency). In terms of our preferred underlying model, namely the regular (finite) n-dimensional cell complex (forming a cube), this amounts to working with the intersection graph of the closed n-cells, which we denote by K^n.

Definition 5. *Let $G = (X, E)$ be a graph, and $f : G \to G$ a ϕ-multifunction. Then a clique $C \subseteq X$ is said to be* fixed *by f if $C \subseteq f(C)$. The graph G is said to possess the ϕ-fixed clique property (ϕ-fcp) if for any ϕ-multifunction $f : G \to G$, there exists a clique C of G such that C is fixed by f.*

The notion of ϕ-fcp clearly generalizes the notion of fcp for single-valued graph homomorphisms as in [, , ,]. Furthermore, any singleton in the graph is a clique:

Proposition 4. *Let G be a graph. Then*

(1) *G has the ϕ-fcp $\Rightarrow$ G has the fcp for single-valued graph homomorphisms;*
(2) *G has the ϕ-fcp $\Rightarrow$ G has the ϕ-afpp.* □

Rosenfeld's almost fixed point theorem can be generalized by

Theorem 2. *K^n has the fcp for single-valued graph homomorphisms.*

Proof. Since the lattice points (with usual coordinates) of K^n can be listed (ordered) by lexicographic ordering $\le$, and it is clear that $(K^n, \le)$ satisfies the hypothesis of contractible graphs, we have: K^n is contractible. The result then follows from the fact that every finite contractible graph has the fcp for single-valued graph homomorphisms in [,]. □

The following examples suggest that, in contrast with the single-valued situation, some consideration of convexity will be needed in studying the ϕ-fcp:

Example 1. As in Fig. 3, we define $f : W_4 \to W_4$ by $a \mapsto \{c, d\}$, $b \mapsto \{a, d\}$, $c \mapsto \{a, b\}$, $d \mapsto \{b, c\}$ and $e \mapsto \{a, b, c, d\}$. Then clearly f has no fixed clique.

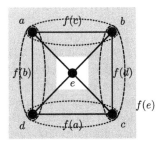

Fig. 3. W_4 has no strong fixed clique property

Example 1 shows by a counter-example that finite contractible graphs have no strong fcp for multifunctions. In fact, more is conveyed by it, that is, Helly graphs [ ] do not have the strong fcp for multifunctions. Note that there is a "hole" in the image of e, $f(e)$.

Let $G = (X, E)$ be a graph, and A a subset of X. Then A is said to be a *geodesic convex* set of G if it contains the set of vertices of any *geodesic*, i.e. shortest path joining any two vertices in A.

Example 2. As in Fig. 4, it is easy to see that $g : K_3^2 \to K_3^2$ is a strong geodesic convexity multifunction, where K_3^2 is K^2 with length 2, i.e., $\#K^1 = 3$. Clearly, g does not fix any clique of K_3^2.

The last example shows that $K^n, n \ge 2$, does not have the fcp for strong geodesic convexity multifunctions. To obtain positive results, we shall need to consider other notions of convexity than this geodesic convexity. The following is a more-or-less standard notion of convexity structure in a graph:

A *graph convexity* [,] on a connected graph $G = (X, E)$ is a pair $(X, \mathcal{C})$, where $\mathcal{C} \subseteq \mathcal{P}(X)$ is an algebraic closure system of "convex" subsets of X, such that

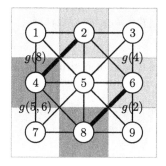

x	$g(x)$
$1, 3, 7, 9$	$\{5\}$
2	$\{6, 8\}$
4	$\{2, 6\}$
$5, 6$	$\{4, 8\}$
8	$\{2, 4\}$

Fig. 4. g has no fixed clique in K_3^2

(1) $\emptyset, X \in \mathcal{C}$
(2) $\bigcap D \in \mathcal{C}$ for any $D \subseteq \mathcal{C}$
(3) $\bigcup T \in \mathcal{C}$ for any chain $T \subseteq \mathcal{C}$
(4) C induces a connected subgraph of G for any $C \in \mathcal{C}$.

Suppose now that G is finite (and connected). In this case condition (3) and the algebraicity property are superfluous. Let M be the collection of all m-neighbourhoods, $m \in \mathbb{N}$, of all vertices of G; and let $\blacksquare(G)$ be the closure under intersection of M. Then we have:

Lemma 1. $\blacksquare(K^n)$ *is a graph convexity.*   □

Lemma 1 can be extended to Helly graphs [ ]. We may call $\blacksquare(G)$, the *neighbourhood convexity* of the graph G. For K^n, it is clear that $\blacksquare(K^n)$ contains $\emptyset$, K^n, any singleton, any vertical, horizontal line and any rectangle of K^n: in effect, the convex sets are just the rectangular blocks.

Proposition 5. *Let $f : K^n \to K^n$ be a self-mapping strong multifunction mapping each vertex to a neighbourhood convex set of K^n. Then there exists a single-valued graph homomorphism $\dot{f} : K^n \to K^n$ such that, for any vertex x of K^n, $\dot{f}(x) \in f(x)$.*

$\dot{f}$ may be said to be a *continuous selection* of f. For definition of continuous selection on topological spaces, we refer [ ].

Proof (of Proposition 5). If $K^n = (X, E)$, define $\dot{f} : K^n \to K^n$ by

$$\dot{f}(x) = \inf(f(x))$$

for any $x \in X$. That is, $\dot{f}(x) = (\inf(\mathrm{Pr}_1(f(x))), \ldots, \inf(\mathrm{Pr}_n(f(x))))$, where Pr_j is the *projection* of K^n onto its jth factor.

Clearly, for any $x \in X$, $\dot{f}(x)$ is unique and non-empty (hence well-defined), satisfying $\dot{f}(x) \in f(x)$. We show that $\dot{f}$ is a graph homomorphism. Indeed, for any neighbourhood convex sets A and B, if AE_sB, then $(\inf(A))E(\inf(B))$: If

$w \in A$ (in particular, if $w = \inf(A)$), then $\exists z \in B$ such that $\mathrm{Pr}_j(z) - \mathrm{Pr}_j(w) \leq 1$, for all $j \in \mathbb{N}, 1 \leq j \leq n$. Hence $\inf(\mathrm{Pr}_j(B)) - \inf(\mathrm{Pr}_j(A)) \leq 1$. By symmetry, $|\inf(\mathrm{Pr}_j(A)) - \inf(\mathrm{Pr}_j(B))| \leq 1$. □

Proposition 5 may be seen as a special case of digital *Michael's continuous selection theorem* for the *n*-dimensional unit cube I^n (see also [ ]).

Corollary 2. *K^n has the fcp for strong multifunctions which map each vertex to a neighbourhood convex set of K^n.*

Proof. By Proposition 5 and Theorem 2. □

5   A Kakutani Analogon?

The "block" convexity considered in the last section has some good mathematical properties, but is obviously too restrictive for many applications. In this section we focus on the digital space K^n, and by taking advantage of the geometry of K^n we shall be able to obtain (positive) results for much more liberal and realistic notions of convexity than before.

Let $\beta(K^n)$ be the *boundary* of K^n, for example, $\beta(K^2) = K^1 \times \{1, \#K^1\} \cup \{1, \#K^1\} \times K^1$; and $\varepsilon(K^n) = K^n \setminus \beta(K^n)$ the *core* of K^n. See Fig. 5 for example of $\beta(K^2)$ and $\beta(\varepsilon(K^2))$. For any $x = (x_1, \ldots, x_n) \in \beta(K^n)$, there exists an unique $y = (y_1, \ldots, y_n) \in \beta(\varepsilon(K^n))$, namely $y_i = x_i$ if $1 < i < \#K^1$, $y_i = x_i + 1$ if $i = 1$ and $y_i = x_i - 1$ if $i = \#K^1$, such that y dominates x, i.e. $N(x) \subseteq N(y)$. (In the case that $\#(K^1) = 1$, then everything is trivial.) Therefore the mapping $r : K^n \to \varepsilon(K^n), x \mapsto x$ if $x \notin \beta(K^n), x \mapsto y$ if $x \in \beta(K^n)$, is a retraction; we shall call it a *border retraction* (of K^n).

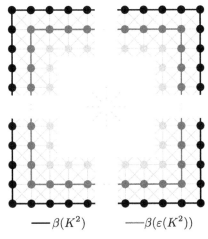

$$\text{---}\beta(K^2) \qquad \text{---}\beta(\varepsilon(K^2))$$

Fig. 5. $\beta(K^2)$ and $\beta(\varepsilon(K^2))$

We shall work with a class RC of "convex" subsets of K^n, required to satisfy the following properties:

(1) RC is closed under border retractions.
(2) Every $A \in$ RC is $2n$-connected. Moreover, if $x, y \in A \in$ RC, then x and y can be connected by a $2n$-adjacency path π in A, such that every vertex (cell) $z = (z_1, \ldots, z_n)$ of π is "coordinatewise" between x and y: that is, $z_1 \in [x_1, y_1], \ldots, z_n \in [x_n, y_n]$ (equivalently, π is contained in the smallest rectangular box which has x, y as vertices).

For example, if vertices x, y are diagonally adjacent, then $\{x, y\}$ is not RC-convex (because not $2n$-connected). Again, in the 2-dimensional case, if x and y lie in the same horizontal or vertical line, and $x, y \in A \in$ RC, then every vertex between x and y is also in A.

A suitable choice for RC may be arrived at as follows. Recall that for a given subset S in the Euclidean space $\mathbb{R}^n$, the *outer Jordan digitization* of S (w.r.t. K^n), denoted by $J^+(S)$, is the collection of cells (considered as closed) having a non-empty intersection with S [ ].

Definition 6. *Let A be a subset of K^n. Then A is a digital convex (DC) set in K^n if there exists an Euclidean convex set $S \subseteq \mathbb{R}^n$ such that A is the intersection graph of $J^+(S)$.*

Digital convex sets are not closed under taking border retractions, for example, in Fig. 6, the set A is a digital convex set, but $r(A)$ is not. Clearly, any digital convex set has the property (2). Moreover, this property will be preserved (starting with a digital convex set) under taking border retractions. Thus, we may take for RC the class of those sets S such that S can be obtained from a digital convex set by a finite sequence of border retractions.

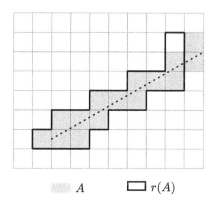

A $\quad\quad$ $\square\, r(A)$

Fig. 6. A is a digital convex set but $r(A)$ is not

In the following, we assume the collection RC (of the "convex" sets) to be fixed.

Theorem 3. *K^n has the fcp for strong multifunctions which map each vertex to a RC-set of K^n.*

The following definition and lemma are useful for the proof of Theorem 3:

Definition 7. *The fixed clique C (fixed by f) is said to be irreducible if there is no proper subset C' of C satisfying $C' \subseteq f(C')$.*

Lemma 2. *Let $G = (X, E)$ be a graph and $f : G \rightarrow G$ a self-mapping multifunction. If $C \subseteq X$ is an irreducible fixed clique of f, then for any $x \in C$ there exists one and only one $y \in C$ such that $x \in f(y)$.*

Proof. Define a directed graph $H = (C, R)$ by

$$uRv \equiv v \in f(u)$$

for all $u, v \in C$. Clearly any cycle in H determines a fixed set (clique) of f; since C is irreducible, every cycle is *Hamiltonian*. It is then easy to see that each vertex of H has an unique predecessor. Indeed C consists of a single cycle, and the function $\hat{f} : C \rightarrow C$ obtained by restricting and corestricting f to C is a bijection. $\qquad\square$

Proof (of Theorem 3). The following basic fact is used in the remainder of the proof (E is the edge relation of K^n):

($\bullet$) If A, B are any two subsets of K^n such that AE_sB, then $r(A)E_sr(B)$ (because the composition of two strong multifunctions is a strong multifunction, see [,]; and every single-valued graph homomorphism is a strong multifunction).

Let $f : K^n \rightarrow K^n$ be any self-mapping strong RC-multifunction, that is, a strong multifunction which maps each vertex of K^n to a RC-set of K^n. We show that f has a fixed clique by using induction on the length of K^n, denoted by $\ell(K^n)$. *Initial step:* It is clear that K^n has the fcp for strong RC-multifunctions when $\ell(K^n) = 0$ or 1. *Inductive step:* We assume that K^n has the fcp for strong RC-multifunctions when $\ell(K^n) \leq m - 1 \in \mathbb{N}$, and we claim that K^n has the fcp for strong RC-multifunctions when $\ell(K^n) = m$. Define the multifunction $g : \varepsilon(K^n) \rightarrow \varepsilon(K^n)$ by

$$g(x) = (r \circ f)(x).$$

By the "basic fact" ($\bullet$) of K^n, it is easy to see that g is a strong RC-multifunction. Therefore by our inductive assumption, g has a fixed clique $C \subseteq \varepsilon(K^n)$. W.L.O.G., we may assume that C is irreducible. We consider the case that $C \cap \beta(\varepsilon(K^n)) \neq \emptyset$ (otherwise, we are done):

Case 1. $\#C = 1$: If $C = \{x\}$, we consider the case that $x \notin f(x)$ only. Clearly, $x \in \beta(\varepsilon(K^n))$, therefore there is a dominated vertex $y \in \beta(K^n)$ of x, such that $y \in f(x)$. Since f is strong, and xEy, there exists a $z \in f(y)$ such that zEy (and hence zEx). Again, since $\{x, y, z\}$ is a clique, and f is strong, we claim that there exists a $w \in f(z)$ such that $\{x, y, z, w\}$ is a clique. Indeed, since f is strong, there exist $w_1, w_2 \in f(z)$ such that $w_1 E y$ (hence $w_1 E x$) and $w_2 E z$. Then by the property of RC-sets, there exists a $w \in f(z)$ such that $\{x, y, z, w\}$ is a clique. Finally, since K^n is finite, then by repeating this procedure, there must exist a fixed clique (containing x or not) of f.

Case 2. $C \subseteq \beta(\varepsilon(K^n))$ with $\#C \neq 1$: This can be dealt with by an extension of the argument for the Case 1; details of the proof are omitted here.

Case 3. Let x, y be vertices in C such that $y \in g(x)$. We only need to consider the case that $y \in \beta(\varepsilon(K^n))$. If $y \in f(x)$, then nothing needs to be shown. Hence we suppose $y \notin f(x)$, that is, there is a dominated vertex $z \in \beta(K^n)$ of y, such that $z \in f(x)$. Since f is strong, there exists some vertex $v \in C \cap \varepsilon(\varepsilon(K^n)), v \in f(u)$ for some $u \in C$, such that $vEf(x)$. Hence there exists $w \in M$, where M is the maximal clique containing C, such that $w \in f(x)$ (otherwise, this would violate the structure of $f(x)$ since $f(x)$ is a RC-set). Since K^n is finite, then by repeating the procedure described above, we conclude that there must exist a fixed clique ($\subseteq M$). $\qquad\square$

Corollary 3. *K^n has the fcp for strong multifunctions which map each vertex to a digital convex set of K^n.* $\qquad\square$

6 Conclusion

Taking our cue from Pultr [ ], we might describe Theorem 3 as an "analogon" of the Kakutani fixed point theorem. In [ ] we proposed a precise criterion for saying that a given digital theorem is a *version* of a classical (continuous) theorem. That criterion requires, in particular, that the classical theorem can be recovered by a limiting process from the digital one. This is by no means the case for Theorem 3 in relation to Kakutani's Theorem. For an adequate digital version, in the strict sense, we need to work with weak rather than strong multifunctions: see [ ]. In this paper we have chosen to study strong multifunctions; one of the advantages of this choice is that, unlike in [,], we are able to give purely combinatorial proofs for all our results.

It is our belief that, in developing digital topology/geometry, single-valued functions are unduly restrictive. They are not flexible enough: no "stretching" is possible with them. (Poston [ ] writes of "chain mail geometry".) This circumstance, we conjecture, explains why relatively little has been done with functions. We hope that multifunctions will be developed and used more in the future.

References

1. Berge, C.: Espaces Topologiques: Fonctions Multivoques. Dunod (1959) 81
2. Border, K. C.: Fixed Point Theorems with Applications to Economics and Game Theory. Cambridge University Press (1985) 81, 83, 84
3. Boxer, L.: Digitally continuous functions. Pattern Recognition Lett. **15** (1994) 833–839 75

4. Brink, C.: Power structures. Algebra Universalis **30** (1993) 177–216. 77, 79
5. Čech, E.: Topological Spaces. John Wiley (1966) 78
6. Duchet, P., Meyniel, H.: Ensemble convexes dans les graphes I, théorèmes de Helly et de Radon pour graphes et surfaces. European J. Combin. **4** (1983) 127–132 82
7. Duchet, P.: Convex sets in graphs II, minimal path convexity. J. Combin. Theory Ser. B **44** (1988) 307–316 82
8. Harary, F.: Graph Theory. Addison-Wesley (1969) 81
9. Klette, R.: Digital geometry- the birth of a new discipline (preprint) 81, 85
10. Kong, T. Y., Rosenfeld, A.: Digital topology: introduction and survey. Computer Vision, Graphics and Image Processing **48** (1989) 357–393 81
11. Nowakowski, R., Rival, I.: Fixed-edge theorem for graphs with loops. J. Graph Theory **3** (1979) 339–350 81
12. Poston, T.: Fuzzy geometry. Ph.D. Thesis. University of Warwick (1971) 77, 80, 81, 82, 87
13. Pultr, A.: An analogon of the fixed-point theorem and its application for graphs. Comment. Math. Univ. Carolin. (1963) 80, 81, 82, 87
14. Quilliot, A.: Thèse de 3° cycle. Université de Paris VI (1978)
15. Quilliot, A.: On the Helly property working as a compactness criterion on graphs. J. combin. Theory Ser. A **40** (1985) 186–193 81, 82
16. Quilliot, A.: Une extension du problème du point fixe pour des graphes simples. Discrete Math. **111** (1993) 435–445 80
17. Rosenfeld, A.: 'Continuous' functions on digital pictures. Pattern Recognition Lett. **4** (1986) 177–184 75, 77, 81
18. Smyth, M. B.: Semi-metrics, closure spaces and digital topology. Theoret. Comput. Sci. **151** (1995) 257–276 78, 79
19. Smyth, M. B., Webster, J.: Finite approximation of functions using inverse sequences of graphs. In: Edalat, A., Jourdan, S. and McCusker, G., eds., Advances in Theory and Formal Methods in Computing: Proceedings of the Third Imperial College Workshop. Imperial College Press (1996) 76
20. Smyth, M. B., Tsaur, R.: A digital version of the Kakutani fixed point theorem for convex-valued multifunctions (to appear in Electron. Notes Theor. Comput. Sci.) 76, 79, 86, 87
21. Smyth, M. B., Tsaur, R.: AFPP vs FPP: the link between almost fixed point properties of discrete structures and fixed point properties of spaces (submitted for publication) 75, 76, 79, 86, 87
22. Stoltenberg-Hansen, V., Lindström, I., Griffor, E. R.: Mathematical Theory of Domains. Cambridge Tracts in Theoretical Computer Science **22**. Cambridge University Press (1994) 79
23. Tsaur, R., Smyth, M. B.: Convexity and fixed point properties in graphs: a study of many-valued functions in Helly and dismantlable graphs (in preparation) 83
24. Webster, J.: Topology and measure theory in the digital setting: on the approximation of spaces by inverse sequences of graphs. Ph.D. Thesis. Imperial College (1997) 76

Part II

Representation

SpaMod: Design of a Spatial Modeling Tool

Eric Andres, Rodolphe Breton, and Pascal Lienhardt

Université de Poitiers (France), Laboratoire IRCOM-SIC
SP2MI, Boulevard Marie et Pierre Curie
BP 30179, 86962 Futuroscope-Chasseneuil cedex - France
{Andres,Breton,Lienhardt}@sic.sp2mi.univ-poitiers.fr

Abstract. The aim of this paper is to present the design of a spatial modeling tool, called SpaMod, that is currently developed in Poitiers (France). SpaMod will allow users to represent and manipulate both discrete and continuous representations of geometrical objects. It is a topology based geometric modeling tool with three types of embeddings: the classical Euclidean embedding, the discrete matrix embedding (voxel or inter-pixel) and the discrete analytical embedding (discrete objects defined by inequations). In order for such a tool to fulfill its role, all three embeddings have to be available together. Conversions between embeddings is thus an important however, in 3D, still partially open question.

Keywords: Modeler, Discrete, Reconstruction, n-G-map

1 Introduction

With the present paper we undertake an investigation aimed at proposing a software platform, called *SpaMod* that stands for Spatial Modeling Tool, that will allow us to represent and manipulate both discrete and continuous representations of geometrical objects. For fourty years software plateforms have been developed to create and manipulate computer models that aim at reproducing real world objects and simulating real world phenomena. They are unavoidable in any industrial manufacturing process. The objects that are manufactured nowadays are almost all designed with the help of some modeling tools, may it be a screw or an airplane. However, most available softwares are only able to handle either discrete (medical imaging, image processing libraries, etc.) or continuous (CAD tools for example) objects. There is an important advantage at disposing of both representations as one can mix synthetic and acquired data sets. With SpaMod, we want to be able to create new discrete or continuous objects, to import discrete objects obtained from direct acquisition (numerical photography, magnetic resonance imaging, pet scan, etc.) or import continuous objects obtained from classical modelers. Most importantly, we want to represent objects both in discrete and in continuous forms and be able to update all the representations in case of any modification.

We have chosen to include discrete geometry into classical Euclidean 3D modeling methods, and more precisely topology-based geometric modeling methods

G. Bertrand et al. (Eds.): Digital and Image Geometry, LNCS 2243, pp. 91–108, 2001.
© Springer-Verlag Berlin Heidelberg 2001

often used for Boundary Representation. Another type of modeling method has been chosen by Kaufman [ ] that has designed a Constructive Solid Geometry (CSG) type tool. The principal difference between Kaufmans' approach and ours' consists in the definition of the kernel. He considers only discrete object as collections of elementary elements (voxels) and performs boolean operations on these sets. In our approach, we consider an object as a topological structure embedded in a discrete or/and a continuous space. The considered operations are two fold: topological operations that modify the topological structure of the objects and embedding operations that modify their geometrical shapes. This allows us multiple different embeddings (especially both discrete and continuous) for a same topological entity.

The topological layer we chose for SpaMod is based on a combinatorial model called *chain of maps* [ ]. This topological model allows a very large class of topological objects (manifold, *non-manifold*, oriented, non oriented, bounded, unbounded, etc.) as SpaMod is not, for the moment, oriented towards a specific application. The topological model can be restricted afterwards relatively easily.

Three types of embeddings are proposed in SpaMod. Each of which can be declined in several different forms. These three embedding types are the classical Euclidean embedding, the classical discrete matrix embedding and the new discrete analytical embedding. An analytical discrete object is a discrete object defined by inequations rather than as a set of discrete points. The purpose of this last embedding type was at first just to enable the representation of objects such as, for instance, discrete 3D planes of Reveillès [ ]. This allows the integration of recognition algorithms such as the ones of I. Debled-Renesson [ ], L. Papier [,] or J. Vittone [ ]. It appears now that this embedding type is of fundamental importance because it represents a compact way of representing discrete data and it is an alternative (to the Marching Cube [ ] for instance), much more compact, way of reconstructing Euclidean data sets out of discrete data.

The software SpaMod we are designing is far from being finished and there are many problems that remain open. In this paper we explain the architecture of SpaMod. We address the fundamental choices we made and explain the steps that have to be achieved to reach the goals we have set up. We show that the open problems that this project has brought up are fundamental to the community and that solving them will address applications that go far beyond this modeler. This modeler is also intended at being a development environment for the computer graphics and, more precisely, for the discrete geometry community for implementing and testing algorithms.

In section 2 we present the kernel of SpaMod. We detail the two layers of the kernel: the topological layer and the embeddings layer. We first explain the topological model, the discrete analytical model and the discrete matrix model. Then, we end this section by explaining how the different embeddings are linked to the topological model. In section 3, we present the conversions between the different embeddings proposed in SpaMod. Then, we will eventually conclude in section 4 with some open problems and what remains to be done to achieve this project.

2   SpaMod's Kernel

The central scheme of SpaMod is organized in two layers: the topological layer and the embedding layer. The embedding layer is itself subdivided into three different types of embedding: the Euclidian embedding, the discrete analytical embedding and the discrete matrix embedding (Fig. 1). In figure 1, the parts in grey represent a classical Euclidean topology based geometrical modeling tool. As we have already explained, our aim is not to create a discrete-only modeling tool but to propose a modeler that is able to handle different types of embedding. The modeler is designed in such a way that the discrete embeddings can be added as extensions on already existing modeling softwares. Let us examine each component of the diagram.

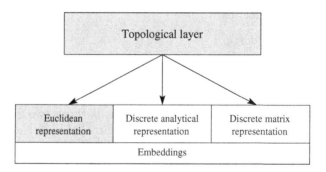

Fig. 1. Topological layer and embeddings layer

2.1   Topological Layer

The topological layer is a central element of the modelers' kernel as it handles the topology of the objects. The different embeddings are attached onto the topological structure. The way this attachement is organized is presented in section 2.3. This layer is based on a combinatorial model that describes the objects' subdivision, i.e. the set of i-cells, for $i = 0, 1, 2, 3$ (vertex, edge, face, volume) and the border relationships between these cells. Different topological models have been proposed in [, ,]. In order to be able to propose a wide variety of objects, we have selected the chain of maps model [ ]. This model is able to handle orientable and non-orientable, bounded and non-bounded, manifold and non-manifold topological objects.

Although the structure used is cellular, it is based on simplicial structures (simplicial sets) [ ]. However we use cellular structures since some operations on a simplicial structure do not always lead to another simplicial structure, or at least, it would be very time-consuming. The choice of the model of chains of maps rests on several studies. For instance, Brisson compared this model with the CW-complex one [ ] and Lienhardt compared the data structures used here with the ones used in Boundary Representation [ ].

SpaMod is for the moment not dedicated to specific applications and the topological model that we chose is therefore as large as possible to avoid limitations. The topological model can however easily be tailored down for specific uses as it is written in object-oriented C++ programming language.

The *n-chain of maps* [ ] explicitely describes each i-cell of the subdivision by using an *i-dimensional generalized map*. We are going to present briefly the basics of our topological model by providing the definition of an *n-dimensional generalized map* (or *n-G-map*) (Def. 1) and the notion of orbits (Def. 2). We will then explain the model we use.

Definition 1. (n-G-map). *An n-G-map, $n \in \mathbb{N}$, is defined by an $(n+2)$-tuple $G = (D, \alpha_0, \ldots, \alpha_n)$, such that:*

- *D is a finite set of darts,*
- *$\alpha_0, \ldots, \alpha_n$ are permutations on D, such that:*
    - *for $0 \leq i \leq n$, α_i is an involution[1],*
    - *for $0 \leq i < i + 2 \leq j \leq n$, $\alpha_i \circ \alpha_j$ is an involution.*

Figure 2 illustrates an example of a 2-G-map. It shows two 2-simplices (or 2-cells), where each 2-simplex is composed of six darts linked by α_0 and α_1. Note that 2-dimensional darts, i.e. darts of 2-G-maps, will be called *2-darts*. A 2-dart can also be linked by α_2 when two 2-cells share a common edge (Fig. 2).

Definition 2. (Orbit). *Let $d \in D$ and Φ a subset of the permutations $\{\alpha_0, \ldots, \alpha_n\}$. Let us assume that $\langle \Phi \rangle$ is the subgroup of permutations generated by Φ. Then, the orbit of a dart d is $\langle \Phi \rangle(d) = \{\phi(d), \phi \in \langle \Phi \rangle\}$. If Φ is empty, then $\langle \Phi \rangle(d) = \{d\}$.*

In other words, an orbit is a set of darts representing a topological entity. Thus, orbits can be considered as the link between topological cells and their geometrical   embeddings.   Precisely,   an i-cell   is   an   orbit   for $\{\alpha_k : 0 \leq k \leq n, k \neq i\}$. For instance (Fig. 2), to describe the vertex V_1, i.e. 0-cell, whose topological representation is encircled with the dashed ellipse, we have to consider one of the darts in this ellipse, and apply α_1, α_2 and $\alpha_2 \circ \alpha_1$ to it. You can notice that the starting dart does not matter.

When observing the topological view of figure 2, we can easily associate each orbit with their corresponding geometrical entity, as summerized in table 1.

Although the n-G-map model is powerful, we use another model in SpaMod: the *n-chain of maps*. Among all its assets, this model allows the modelisation

---

[1] α is an involution $\Leftrightarrow \alpha \circ \alpha = Id$

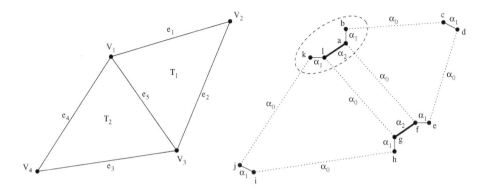

Fig. 2. On the left, a classical view of two 2D triangles T_1 and T_2 sharing a common edge. On the right, their topological modelisation. The set of darts is $D = \{a, b, c, d, e, f, g, h, i, j, k, l\}$

Table 1. Associations orbit/geometrical entity

Geometrical entity	Orbit
vertex V_1	$\langle \alpha_1, \alpha_2 \rangle (k) = (k, l, a, b)$
vertex V_2	$\langle \alpha_1, \alpha_2 \rangle (c) = (c, d)$
vertex V_3	$\langle \alpha_1, \alpha_2 \rangle (e) = (e, f, g, h)$
vertex V_4	$\langle \alpha_1, \alpha_2 \rangle (i) = (i, j)$
edge e_1	$\langle \alpha_0, \alpha_2 \rangle (b) = (b, c)$
edge e_2	$\langle \alpha_0, \alpha_2 \rangle (d) = (d, e)$
edge e_3	$\langle \alpha_0, \alpha_2 \rangle (h) = (h, i)$
edge e_4	$\langle \alpha_0, \alpha_2 \rangle (j) = (j, k)$
edge e_5	$\langle \alpha_0, \alpha_2 \rangle (a) = (a, f, g, l)$
triangle T_1	$\langle \alpha_0, \alpha_1 \rangle (a) = (a, b, c, d, e, f)$
triangle T_2	$\langle \alpha_0, \alpha_1 \rangle (g) = (g, h, i, j, k, l)$

of non-manifold objects. For instance, a tetrahedron and three edges sharing a common vertex (Fig. 8) is non-manifold and cannot be represented with the n-G-map model. We won't recall the definition of this model as it will be too long, but an interested reader will find all the details in [ ].

To conclude, we would like to stress on our choice of these models for SpaMod. The main reason is that they are defined for n-dimensional objects of any kind and therefore, they allow high modeling variety. They are also coherent as they are based on a unique elementary object: the dart. Moreover, the constraints make sure that performing any operation on an object will result in a correctly modeled object.

2.2   Embeddings Layer

The embedding layer decribes the embedding of each topological subdivision into a geometrical space, i.e. the spatial position and form of the object. Due to the specification of SpaMod, we propose three different types of embeddings to work with (Fig. 1). We present those three types embeddings in this section.

Euclidean Embedding   The first embedding model used is the classical Euclidean model. An Euclidean object is represented by real number coordinate points for each vertex of the topological model. The edges, faces and volumes of the modeled objects are interpolated from their vertices. The topological layer and the Euclidean embedding form a classical Euclidean topology based modeling tool with all the classical operations that exist in such softwares (in grey on figure 1). We need this type of embeddings in order to be able to import Euclidian models (VRML objects for instance) for later discretization or as result of reconstruction algorithms (such as Marching Cube ones). SpaMod is therefore not a discrete-only modeling tool but a classical topology-based geometrical modeling tool with two extra types of embeddings that we will describe just next. For the moment we have limited ourself to a polygonal Euclidean embedding because, to the best authors' knowledge, no convincing general discrete counterpart models for Beziers, splines, . . . curves and surfaces have been proposed so far.

Discrete Analytical Embedding   The aim of this representation form is to represent a discrete object by an analytical description, i.e. a set of analytical inequations. This type of representation is by definition independent of the number of discrete points composing a discrete object. This type of embedding is new and central to our purpose. The aim is to be able to handle discrete analytical objects such as discrete lines and planes of Reveillès [ ], circles and spheres of Andres [ ], supercover [,], standard [ ], naive [ ], . . . models. Many papers have dealt these last decades with 2D discrete analytical straight line [, ,] and 3D plane segment [, , , , , ,] recognition algorithms. Our aim is to propose a platform where such algorithms can be developed, implemented, tested and finally used in industrial applications. We have implemented two discrete analytical embedding models so far: the supercover and the standard model. These two models are both discrete analytical models and geometrically coherent.

Definition 3. *A discrete model is said to be geometrically coherent if the discretization of an Euclidean object A depends only on A.*

This means, for instance, that the discretization of a 3D straight line segment D is the same if D is just an isolated 3D straight line segment, the edge of a 3D triangle, or even the edge of a 3D tetrahedron. This notion is not verified by most discrete models such as, for example, the discrete primitives proposed by Cohen and Kaufman [ ].

In SpaMod we have implemented two different discrete analytical models so far; the supercover model and the standard model. Let us present briefly the supercover model.

Definition 4. *The supercover* $\mathbb{S}(E) \subset \mathbb{Z}^n$ *of a general Euclidean object* $E \subset \mathbb{R}^n$ *can be defined in several equivalent manners:*

$$\mathbb{S}(E) = \{p \in \mathbb{Z}^n \,|\, \mathbb{V}(p) \cap E \neq \varnothing\} = \left\{p \in \mathbb{Z}^n \,\middle|\, d_\infty(p, E) \leq \frac{1}{2}\right\}$$
$$= (\mathbb{V}(0) \otimes E) \cap \mathbb{Z}^n$$

where $\mathbb{V}(p)$ *is a voxel centered on* p, d_∞ *is the block distance,* $\otimes$ *represents the Minkowsky sum.*

We will not detail the supercover model. This would go beyond the scope of this paper. The reader can find extended details on this model in [,]. Let us just state that, intuitively, the supercover of a Euclidean object E is composed of all the discrete points whose voxels are in contact with E. The supercover model is by definition geometrically coherent since the supercover of an object is independent of it being part or not of another geometrical structure. The supercover of a vertex (resp. edge) of a triangle is, by definition, the supercover of a point (resp. straight line segment).

The supercover is also an analytical discrete model. It has been shown that the supercover of any semi-algebraic set can be described analytically [ ]. Let us just examine, as an example, the supercover of the triangle T_1 of figure 2 (see Fig. 4 in section 2.3). The supercover of the triangle T_1 is given by seven inequations (in this example, the equations number 17, 5, 6, 20, 11, 22 and 4) and this independently of its size. If one or two edges would have been parallel to an axis then only six or even five inequations would have been necessary to describe the supercover of T_1. In SpaMod, the representation of the supercover of a convex set corresponds simply to a set of inequations. A concave object is defined as unions of convex shaped objects (i.e. as union of simplices).

One drawback with the supercover model is that it has sometimes *bubbles*. For instance, the supercover of the 2D Euclidean point $\left(\frac{1}{2}, \frac{1}{2}\right)$ is the set of points $\{(0,0), (0,1), (1,0), (1,1)\}$. The fact that a Euclidean point has four discrete points as embedding can raise some problems in some algorithms, especially those aimed at recognition. To avoid this problem we have introduced the *standard discrete analytical model* [ ]. The standard model is obtained from the supercover model by changing some inequations of the supercover representation to strict inequations. When this is done in a correct way (see [ ] for more details) then we obtain objects without bubbles. The standard model of an Euclidean point is always an unique discrete point, the standard model of a 2D line is always 4-connected without simple points, a standard 3D line and 3D-plane is always 6-connected without simple points. This is proved in [ ]. This means that the standard model is especially well adapted to the inter-pixel K^2 discrete space.

Discrete Matrix Embedding The model we choose for a discrete matrix embedding is of course the same as for the discrete analytical embedding. Both embedding types represent the same object. The classical way of representing a discrete object is to enumerate its points. This set of discrete points is contained in a finite matrix that represents a finite part of space. The discrete matrix in SpaMod embedding is proposed in two forms:

- The first form is the classical voxel form. A discrete object is formed of voxels corresponding each to a discrete point in a matrix.
- The second form is the *Khalimsky-Kovaleski* (K^2) representation form [ ], also called *inter-pixel* or *abstract cell complexe* representation form. A voxel is not represented as such, instead the voxels' vertices, edges, faces and its volume are each represented explicitly. The K^2 discrete space can also easily be stored in a matrix. The K^2 discrete space representation has many interesting properties and will probably, in authors' opinion, become prevalent over the classical voxel representation in applications. Several questions of course need to be addressed such as, for example: Do we just represent the inter-pixel border of an object or its complete cell subdivision?

Example To conclude this section, let us consider the topological triangles T_1 and T_2 of figure 2. The Euclidean embedding will simply consist in the four vertex coordinates $(0,0), (3,8), (9,1)$ and $(12,10)$. The discrete matrix embedding for the supercover model will consist of all the (dark and light) grey points in the following matrix (Fig. 3). With the standard model, the three grey points marked with a circle need a closer look. The two points which are in the outline of T_1 and T_2 do not belong to the triangles as they do not belong to the standard straight line segments $(0,0) - (9,1)$. But the marked point that is on the common edge belongs to the triangles as it belongs to the interior of T_2.

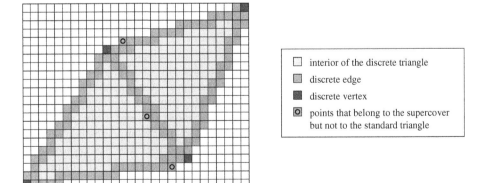

Fig. 3. Discrete matrix embedding

2.3   Link between the Topological Layer and the Discrete Analytical Embedding

We are now going to explain on an example how the link between the topological layer and the embedding layer is established. Then we will explain where the discrete analytical information is stored.

As we have seen in the previous section, an analytical object is defined by a set of inequations that describe its geometrical embedding.

It would have been a great loss of space to store those inequations in each dart of the corresponding orbit. That is why this information is only held by one dart. To obtain the geometrical embedding of an i-cell, we have to randomly pick up a dart of that cell and search for this special dart all over its orbit. A dart can contain the information related to several objects, i.e. it can hold the inequations of a vertex and the ones of an edge if it belongs to the two orbits.

Let us focus on our current example, the two 2D triangles T_1 and T_2 of figure 2. They obviously share the edge e_5. Figure 2 shows the classical Euclidean embedded view and the topological representation of T_1 and T_2. Figure 4 illustrates the analytical view of those triangles, and shows the 26 inequations related to that representation.

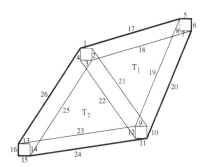

Fig. 4. An analytical representation of T_1 and T_2 with inequations numbered from 1 to 26. The outline is thickened

The supercover analytical representation of these triangles is defined by the set of inequations given in table 2. Figure 3 illustrates all the discrete points verifying the inequations of table 2.

The eight bold inequations in table 2 define the global object (thickened outline of figure 4) made of T_1 and T_2. We need however all the other inequations too. For instance, if we want to perform an operation on T_1 alone, we need Eq_4, Eq_{11} and Eq_{22}.

The information is not duplicated. Each simplex only holds the inequations that are peculiar to it, and holds pointers to the others of lower dimensions. Table 3 illustrates that.

Table 2. The 26 inequations that describe T_1 and T_2

(Eq_1)	$y - 17/2 \leq 0$	(Eq_2)	$x - 7/2 \leq 0$
(Eq_3)	$-y + 15/2 \leq 0$	(Eq_4)	$-x + 5/2 \leq 0$
$\mathbf{(Eq_5)}$	$\mathbf{y - 21/2 \leq 0}$	$\mathbf{(Eq_6)}$	$\mathbf{x - 25/2 \leq 0}$
(Eq_7)	$-y + 19/2 \leq 0$	(Eq_8)	$-x + 23/2 \leq 0$
(Eq_9)	$y - 3/2 \leq 0$	(Eq_{10})	$x - 19/2 \leq 0$
(Eq_{11})	$-y + 1/2 \leq 0$	(Eq_{12})	$-x + 17/2 \leq 0$
(Eq_{13})	$y - 1/2 \leq 0$	(Eq_{14})	$x - 1/2 \leq 0$
$\mathbf{(Eq_{15})}$	$\mathbf{-y - 1/2 \leq 0}$	$\mathbf{(Eq_{16})}$	$\mathbf{-x - 1/2 \leq 0}$
(Eq_{17})	$-2x + 9y - 143/2 \leq 0$	(Eq_{18})	$2x - 9y + 121/2 \leq 0$
(Eq_{19})	$-9x + 3y - 84 \leq 0$	$\mathbf{(Eq_{20})}$	$\mathbf{9x - 3y + 72 \leq 0}$
(Eq_{21})	$7x + 6y - 151/2 \leq 0$	(Eq_{22})	$-7x - 6y + 125/2 \leq 0$
(Eq_{23})	$-x + 9y - 5 \leq 0$	$\mathbf{(Eq_{24})}$	$\mathbf{x - 9y - 5 \leq 0}$
(Eq_{25})	$8x - 3y - 11/2 \leq 0$	$\mathbf{(Eq_{26})}$	$\mathbf{-8x + 3y - 11/2 \leq 0}$

In table 3, Eq_i $(1 \leq i \leq 26)$ is the i-th inequation (given in table 2) and $\uparrow Eq_i$ $(1 \leq i \leq 26)$ denotes a pointer (in a programming view) to the i-th inequation. For instance, the discrete analytical embedding of triangle T_1 is held by dart a and is composed of 7 pointers to the actual inequations. The discrete analytical embedding of edge e_5 is also held by dart a (for example) and is composed of 2 inequations (the supercover line) Eq_{21} and Eq_{22}, and links to the 4 other inequations (belonging to the edge).

Table 3. One possibility to store the inequations

Geometrical entity	Dart	Inequations held
vertex V_1	k	Eq_1, Eq_2, Eq_3, Eq_4
vertex V_2	c	Eq_5, Eq_6, Eq_7, Eq_8
vertex V_3	e	$Eq_9, Eq_{10}, Eq_{11}, Eq_{12}$
vertex V_4	i	$Eq_{13}, Eq_{14}, Eq_{15}, Eq_{16}$
edge e_1	b	$Eq_{17}, Eq_{18}, \uparrow Eq_3, \uparrow Eq_4, \uparrow Eq_5, \uparrow Eq_6$
edge e_2	d	$Eq_{19}, Eq_{20}, \uparrow Eq_5, \uparrow Eq_6, \uparrow Eq_{11}, \uparrow Eq_{12}$
edge e_3	h	$Eq_{23}, Eq_{24}, \uparrow Eq_9, \uparrow Eq_{10}, \uparrow Eq_{15}, \uparrow Eq_{16}$
edge e_4	j	$Eq_{25}, Eq_{26}, \uparrow Eq_1, \uparrow Eq_2, \uparrow Eq_{15}, \uparrow Eq_{16}$
edge e_5	a	$Eq_{21}, Eq_{22}, \uparrow Eq_1, \uparrow Eq_4, \uparrow Eq_{10}, \uparrow Eq_{11}$
triangle T_1	a	$\uparrow Eq_4, \uparrow Eq_5, \uparrow Eq_6, \uparrow Eq_{11}, \uparrow Eq_{17}, \uparrow Eq_{20}, \uparrow Eq_{22}$
triangle T_2	g	$\uparrow Eq_1, \uparrow Eq_{10}, \uparrow Eq_{15}, \uparrow Eq_{16}, \uparrow Eq_{21}, \uparrow Eq_{24}, \uparrow Eq_{26}$

3   Embedding Conversions

One of the first reasons why the implementation of embedding conversions operations are necessary is the fact that we want to be able to import objects created or acquired externally. When an object is imported, we need to construct the other, remaining, embeddings and, if necessary, a topological representation of the object. We propose three types of embeddings in SpaMod, thus we distinguish six different conversions between embeddings as shown in figure 5. Let us examine these different types of conversions.

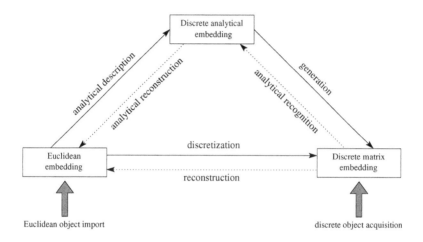

Fig. 5. Embedding conversions

3.1   From Euclidean to Discrete Analytical Embedding

To perform such a conversion is easy. We just need to compute the inequations given by a discrete analytical model, like the supercover one [, ,].

3.2   From Discrete Analytical to Discrete Matrix Embedding

Again, this conversion is not much of a problem. Indeed, from a discrete analytical representation it follows that all the points whose coordinates verify the inequations belongs to the discrete object. So, we can use a brute-force algorithm and check all the points to find those that match the set of inequations. This is of course not very efficient but it shows that this operation is not a theoretical problem. In SpaMod we generate discrete objects in the matrix representation by incremental algorithms, which are far more efficient.

3.3 From Euclidean to Discrete Matrix Embedding

For a given discrete model, this conversion is usually not a problem since all the discrete models propose algorithms to generate discrete objects. That is also what we do for the supercover and the standard model.

3.4 From Discrete Matrix to Euclidean Embedding

To convert directly a discrete object, that is a set of voxels, in an Euclidean object is possible and several methods have been proposed, such as the Marching Cubes algorithm [ ]. The problem with these methods is that the number of polygons of the resulting Euclidean embedding is proportional to the size of the original discrete matrix object. This is not a compact representation of the Euclidean object and often requires additional steps of simplification in order to reduce the number of Euclidean polygons. In this case the discrete matrix object and the Euclidean object do not represent the same object anymore. That is why we propose a new alternative two step process described below.

3.5 From Discrete Matrix to Discrete Analytical Embedding

We called this operation *analytical recognition*, as it permits to find the inequations that are verified by the voxels of the discrete matrix object. This is the first step of the "discrete to Euclidean" analytical transformation process. This operation has been partially studied by I. Debled-Renesson [ ], L. Papier [ ], J. Vittone [ ], etc. The 2D case has been sucessfully solved by I. Debled-Rennesson. The 3D case remains however an open problem. For the moment, given a set of voxels, we can compute the features of the analytical plane which they belong to. We now want to describe totally the edges and vertices of that plane piece. There are several research leads. We are trying to adapt the naive objects recognition algorithm of Vittone [ ] to our needs.

3.6 From Discrete Analytical to Euclidean Embedding

Once the inequations have been found, there are generally several Euclidean objects that fit those inequations. That is to say, for instance, several Euclidean edges can result in the same discretized edge. So the reverse operation is a serious issue. This so-called *analytical reconstruction* step, the second one of the "discrete to Euclidean" analytical process, consists in finding the Euclidean object that fits the best. For instance, given a discrete edge, we obtain its analytical representation and then we choose an Euclidean edge amongst the possible ones (Fig. 6). There is no problem when, for instance, an edge of a triangle is not attached to another one. In case we have a 3D discrete edge that is part of two discrete triangles then we need to choose an Euclidean edge that will be part of both 3D Euclidean triangles (i.e. planes). This is not easy and we have seen in the 2D case that there is not always a solution (Fig. 7). There are still many open problems here.

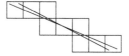

Fig. 6. A discrete edge and two Euclidean edges that match

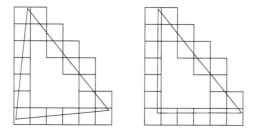

Fig. 7. A discrete triangle and two Euclidean triangles that match

4   Conclusion

In this paper we have presented the architecture of a modeling tool that works in a topological space, in the Euclidean space and in discrete space. The software, called Spamod for Spatial Modeling Tool, is currently in development in our laboratory. We have presented the different layers and explained how we implemented the different embeddings. We propose three types of embeddings in SpaMod. The Euclidean embedding, the discrete matrix embedding (discrete information is stored in a matrix) and the discrete analytical embedding (discrete objects are represented by inequations). See figure 8 for an example of each view. The discrete analytical embedding provides a compact representation form for discrete information. We have implemented two coherent discrete models that provide analytical descriptions of objects: the supercover and the standard model. This software is intended as discrete object creation and manipulation tool and algorithm development platform for the discrete geometry community. SpaMod serves also as an important research tool. The software is far from finished. We have still to solve some difficult open problems. One of these very important open problems is the analytical recognition of discrete polygons. Different research groups work already with different approaches on this problem. The problem is to describe a piece of a discrete analytical plane as a discrete 3D polygon. Solving this problem will provide a compact representation of discrete object and several important geometrical properties such as normal vectors. From the discrete polygons one can construct corresponding Euclidean polygons. This reconstruction process, once realised, offers compact Euclidean reconstructed objects out of discrete objects. This method could replace the Marching Cube method efficiently. This research does not seem to be irrealistic to achieve as it has already been solved in 2D in medical imaging application [ ].

Acknowledgements

We wish to thank Ivan Salam for his work on this field and particularly on SIGM (Sigm Is a Geometrical Modeler), the first version of the topological based discrete modeler.

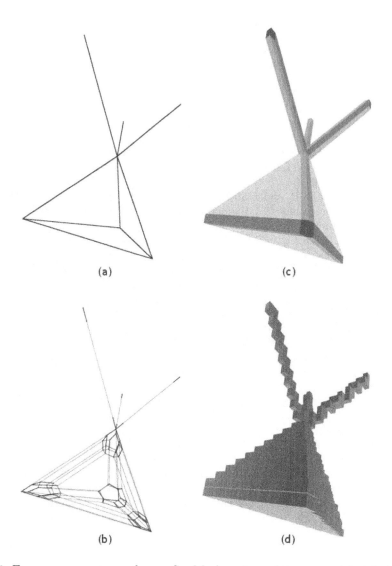

Fig. 8. Four screen captures, from a SpaMod session, of the same object (a tetrahedron and three edges): (a) the classical view, (b) the topological modelisation, (c) the discrete analytical representation, (d) the discrete matrix representation

References

1. Andres E., *Discrete Circles, Spheres and Hyperspheres*, Comp. & Graphics, Vol. 18, No. 5, 1994, pp. 695-706. 96
2. Andres E., *Modélisation analytique discrète d'objets géométriques*, Habilition, Poitiers University, December 2000. 96, 97
3. Andres E., Acharya R. and Sibata C., *Discrete Analytical Hyperplane*, Graphical Models and Image Processing, Vol. 59, No. 5, September 1997, pp. 302-309.
4. Andres E., Debled-Renesson I., Sibata C., Acharya R and Shin K., *Linear Contour Segmentation and its Application to the Computation of Stereotactic Radiosurgery Dose Distribution*, SPIE, Medical Imaging '96, New Port Beach - California (USA), SPIE 2710, Image Processing, february 1996. 104
5. Andres E., Nehlig P. and Françon J., *Supercover of Straight Lines, Planes and Triangles*, 7th International Workshop on Discrete Geometry for Computer Imagery, Montpellier, France, December 1997, Lecture Notes in Computer Science, No. 1347, pp. 243-254. 96, 97, 101
6. Andres E., Nehlig P. and Françon J., *Tunnel-Free Supercover 3D Polygons and Polyhedra*, Eurographics'97, Budapest, Hungary, 4-8 September 1997. Computer Graphics Forum, Vol. 16 (3), Conference issue, C3-C13. 101
7. Andres E., Sibata C. and Acharya R., *The Supercover 3D Polygon*, 6th International Workshop on Discrete Geometry for Computer Imagery, Lyon, France, November 1996, Lecture Notes in Computer Science, No. 1176, pp. 237-242. 101
8. Bresenham J. E., *Algorithm for Computer Control of a Digital Plotter*, A. C. M. Transaction on Graphics, Vol. 4, No. 1, 1965, pp. 25-30.
9. Brimkov V. E., Andres E. and Barneva R. P., *Object Discretization in Higher Dimensions*, 9th International Conference, Uppsala, Sweden, December 2000, Lecture Notes in Computer Science, No. 1953, pp. 210-221. 97
10. Brimkov V., Barneva R. and Nehlig P., *Optimally Thin Tunnel-Free Triangular Meshes*, International Workshop on Volume Graphics, 24-25 March 1999, Swansea, UK, pp. 103-122.
11. Brisson E., *Representing Geometric Structuresin d Dimensions: Topology and Order*, Discrete Comput. Geom., No. 9 (1993), pp. 387-426. 94
12. Debled-Rennesson I., *Étude et reconnaissance des droites et plans discrets*, Phd thesis, Louis Pasteur University, Strasbourg, France, December 1995. 92, 96, 102
13. Debled-Rennesson I. and Reveillès, *An Incremental Algorithm for Digital Plane Recognition*, 4th International Workshop on Discrete Geometry for Computer Imagery, Grenoble, France, September 1994, DGCI'94, pp. 207-222. 96
14. Elter H. and Lienhardt P., *Cellular Complexes as Structured Semi-simplicial Sets*, International Journal of Shape Modeling, Vol. 1, No. 2 (1994), pp. 191-217. 92, 93, 94, 95
15. Françon J., *Topologie de Khalimsky et Kovaleski et algorithmique graphique*, 1st Symposium on Discrete Geometry for Computer Imagery, Strasbourg, France, September 1991. 98
16. Françon J., Schramm J.-M. and Tajine M., *Recognizing Arithmetic Straight Lines and Planes*, 6th International Workshop on Discrete Geometry for Computer Imagery, Lyon, France, November 1996, Lecture Notes in Computer Science, No. 1176, pp. 141-150. 96
17. Kaufman A., Cohen D. and Yagel R., *Volume Graphics*, IEEE Computer, Vol. 27, No. 7, July 1993, pp. 51-64. 92, 96

18. Kim C. E. and Stojmenović I., *On the Recognition of Digital Planes in Three Dimensional Space*, Pattern Recognition Letters, 1991, 12, pp. 612-618. 96

19. Kovalevsky V., *Digital Geometry Based on the Topology of Abstract Cell Complexes*, 3th International Conference on Discrete Geometry for Computer Imagery, Strasbourg, France, DGCI 1993, pp. 259-284.

20. Lienhardt P. and Elter H., *Different Combinatorial Models Based on the Map Concept for the Representation of Different Types of Cellular Complexes*, Proceedings IFIP TC 5/WG II, Working Conference on Geometric Modeling in Computer Graphics, Genova, Italy, 1993. 93

21. Lienhardt P., *N-dimensional Generalized Combinatorial Maps and Cellular Quasi-manifolds*, International Journal of Computational Geometry and Applications, Vol. 4, No. 3, 1994, pp. 275-324. 93, 94

22. Lienhardt P., *Subdivisions of N-dimensional Space and N-dimensional Generalized Maps*, Proceedings 5th Annual A. C. M. Symposium on Computational Geometry, Saarbrüken, Germany, 1989, pp. 228-236.

23. Lienhardt P., *Topological Models for Boundary Representations: a Comparison with N-dimensional Generalized Maps*, Computer Aided Design 23, 1 (1991), pp. 59-82. 93, 94

24. Lorensen W. E. and Cline H. E., *Marching Cube: a High Resolution 3D Surface Construction Algorithm*, Computer Graphics 21, 4 (1987), pp. 163-169. 92, 102

25. Papier L., *Polydrisation et visualisation d'objets discrets tridimensionnels*, PhD thesis, Louis Pasteur University, Strasbourg, 2000. 102

26. Papier L. and Françon J., *Évaluation de la normale au bord d'un objet discret 3D*, C. F. A. O. and Informatique Graphique, Vol. 13, No. 2, 1998, pp. 205-226. 92

27. Papier L. and Françon J., *Polyhedrization of the Boundary of a Voxel Object*, 8th International Workshop on Discrete Geometry for Computer Imagery, Marne-la-Vallée, France, March 1999, Lecture Notes in Computer Science, No. 1568, pp. 425-434. 92

28. Reveillès J.-P., *Combinatorial pieces in digital lines and planes*, S. P. I. E. Vision Geometry IV, Vol. 2573, 1995. 96

29. Reveillès J.-P., *Géométrie discrète, calcul en nombres entiers et algorithmique*, Habilitation, Louis Pasteur University, Strasbourg, France, December 1991. 92

30. Salam I., Nehlig P. and Andres E., *Discrete Ray-Casting*, 8th International Workshop on Discrete Geometry for Computer Imagery, Marne-la-Vallée, France, March 1999, Lecture Notes in Computer Science, No. 1568, pp. 435-446.

31. Schramm J. M., *Coplanar Tricubes*, 7th International Workshop on Discrete Geometry for Computer Imagery, Montpellier, France, December 1997, Lecture Notes in Computer Science, No. 1347, pp. 87-98. 96

32. Stojmenović I. and Tošić R., *Digitization schemes and the Recognition of Digital Straight Lines, Hyperplanes and Flats in Arbitrary dimensions*, Contemporary Mathematics, 1991, 119, pp. 197-212. 96

33. Tajine M. and Ronse C, *Hausdorff Discretizations of Algebraic Sets and Diophantine Sets*, 9th International Conference, Uppsala, Sweden, December 2000, Lecture Notes in Computer Science, No. 1953, pp. 99-110. 97

34. Vialard A., *Chemins Euclidiens: un modèle de représentation des contours discrets*, Phd thesis, Bordeaux 1 University, 1996.

35. Vittone J., *Caractérisation et reconnaissance de droites et de plans en géométrie discrète*, Phd thesis, Joseph Fourier - Grenoble 1 University, December 1999. 92, 102

36. Vittone J. and Chassery J.-M., *Coexistence of Tricubes in Digital Naive Plane*, 7th International Workshop on Discrete Geometry for Computer Imagery, Montpellier, France, December 1997, Lecture Notes in Computer Science, No. 1347, pp. 99-110. 96

37. Vittone J. and Chassery J.-M., *Recognition of Digital Naive Planes and Poly-hedrization*, 9th International Conference, Uppsala, Sweden, December 2000, Lecture Notes in Computer Science, No. 1953, pp. 296-307. 96, 102

Introduction to Combinatorial Pyramids

Luc Brun[1] and Walter Kropatsch[2] [*]

[1] Laboratoire d'Études et de Recherche en Informatique(EA 2618)
Université de Reims - France
brun@leri.univ-reims.fr
[2] Institute for Computer-aided Automation, Pattern Recognition and Image
Processing Group
Vienna Univ. of Technology- Austria
krw@prip.tuwien.ac.at

Abstract. A pyramid is a stack of image representations with decreasing resolution. Many image processing algorithms run on this hierarchical structure in $\mathcal{O}(\log(n))$ parallel processing steps where n is the diameter of the input image. Graph pyramids are made of a stack of successively reduced graphs embedded in the plane. Such pyramids overcome the main limitations of their regular ancestors. The graphs used in the pyramid may be region adjacency graphs or dual graphs. This paper reviews the different hierarchical data structures and introduces a new representation named combinatorial pyramid.

1 Introduction

Regular image pyramids have been introduced in 1981/82 [ ] as a stack of images with decreasing resolutions. Such pyramids present interesting properties for image processing and analysis such as [ ]: The reduction of noise, the processing of local and global features within the same frame, the use of local processes to detect global features at low resolution and the efficiency of many computations on this structure. Regular pyramids are usually made of a very limited set of levels (typically $\log(n)$ where n is the diameter of the input image). Therefore, if each level is constructed in parallel from the level below, the whole pyramid may be built in $\mathcal{O}(\log(n))$ steps. Since 1981, regular pyramids have been widely used in image segmentation, shape analysis, surface reconstruction and motion analysis (see e.g. [ ] for a survey of these applications). However, the rigidity of regular pyramids induces several drawbacks such as the shift-dependence problem and the limited number of regions encoded at a given level of the pyramid [ ]. Irregular pyramids overcome these negative properties while keeping the main advantages of their regular ancestors [ ]. These pyramids are defined as a stack of successively reduced graphs. Each graph is built from the graph below by selecting a set of vertices named *surviving vertices* and mapping each non surviving vertex to a surviving one [ ]. Typically, such graphs represent

---

[*] Thanks to L. Lucas and M. Mokhtari for proofreading. This Work was supported by the Austrian Science Foundation under P14445-MAT.

G. Bertrand et al. (Eds.): Digital and Image Geometry, LNCS 2243, pp. 108–128, 2001.
© Springer-Verlag Berlin Heidelberg 2001

the pixel neighborhood, the region adjacency, or the semantical context of image objects. Since images are samplings of a plane, the graphs are embedded in the plane by definition. Hence they are planar and the dual graph is well defined.

Two kinds of graphs are usually used within the pyramid framework: the simple graph data structure encodes each relation between two vertices by a single edge. The dual graph data structure is based on an explicit encoding of both the initial graph and its dual.

Combinatorial map first introduced in 1960 by Edmonds [ ] may be considered as a planar graph with an explicit encoding of the neighborhood's orientation of each vertex. This graph encoding induces several advantages besides the other graph representations such as the explicit encoding of the orientation of the plane or an implicit encoding of the dual graph.

This paper presents the usual hierarchical data structures in Section 2. Then the combinatorial maps are presented together with their main properties in Section 3. A combination of the combinatorial and Irregular Pyramid frameworks is then presented in Section 4.

2   Hierarchical Data Structures

2.1   Regular Pyramids

A regular pyramid is defined as a sequence of images with exponentially reduced resolution. Each image of this sequence is called a *level* of the pyramid. The lowest level corresponds to the original image while the highest one corresponds to the weighed average of the original image. Using the neighborhood relationships defined on each image the *Reduction window* relates each pixel of the pyramid with a set of pixels defined in the level below. The pixels belonging to a reduction window are the *children* of the pixel which defines it. Such a pixel is the *father* (filled circles in Fig. 1) of the pixels belonging to its reduction window. This father-child relationship maybe extended by transitivity to any level of the pyramid. The set of children of one pixel in the base level image is named its *receptive field*. Within the regular pyramid framework the shape and the cardinal of the reduction window remains the same for all pixels of the pyramid.

Another parameter of a regular pyramid is its *Reduction factor* which encodes the ratio between the size of two successive images in the pyramid. This ratio remains constant between any two levels of a regular pyramid. A *reduction function* computes the contents of a father from the contents of the pixels in the reduction window. A regular pyramid is thus formally defined by the ratio $N \times N/q$, where $N \times N$ denotes the size of the reduction window while q denotes the reduction factor. Different types of pyramids may be distinguished according to the ratio between the size of the reduction window and the reduction factor:

- If $N \times N/q < 1$, the pyramid is named a *non-overlapping holed pyramid*. Within such pyramid, some pixels have no fathers [ ] (e.g., the center pixel in Fig. 2(a)).

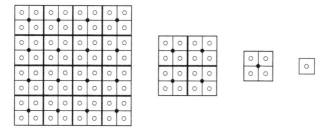

Fig. 1. A $2 \times 2/4$ regular pyramid

- If $N \times N/q = 1$, the pyramid is called a *non overlapping pyramid without hole*(see e.g. in Fig. 2(b)). Within such pyramids, each pixel in the reduction window has exactly one father [ ].
- If $N \times N/q > 1$, the pyramid is named an *Overlapping pyramid* (see e.g. in Fig. 2(c)). Each pixel of such a pyramid has several *potential parents* [ ]. If each child selects one parent, the set of children in the reduction window of each parent is re-arranged. Consequently the receptive field may take any form inside the original receptive field.

2.2 Irregular Pyramids

Beside their interesting properties mentioned in Section 1, regular pyramids have several drawbacks. Indeed, the fixed size and shape of the reduction window together with the fixed value of the reduction factor induce a poor ability of regular pyramids to adapt their structure to the data. The rigidity of regular pyramids induces several drawbacks such as the shift-dependence problem and the limited number of regions encoded at a given level of the pyramid [ ].

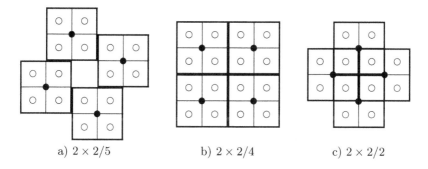

a) $2 \times 2/5$ b) $2 \times 2/4$ c) $2 \times 2/2$

Fig. 2. Three different types of regular pyramids

Irregular pyramids overcome these negative properties while keeping the main advantages of their regular ancestors [ ]. These pyramids are defined as a stack of successively reduced graphs (G_0, G_1, G_2, G_3 in Fig. 3). Each graph is built from the graph below by selecting a set of vertices named surviving vertices and mapping each non-surviving vertex to a surviving one [ ]. Therefore each non-surviving vertex is the child of a surviving one which represents all the non surviving vertices mapped to it and becomes their father. If the initial graph $G_0 = (V_0, E_0)$ is defined from the regular grid, we may associate one pixel to each vertex (in V_0) and link vertices by edges in E_0 according to the adjacency relationships defined on the regular grid(G_0 in Fig. 3).

The reduction operation was first introduced by Meer [,] as a stochastic process. Using such framework, the graph $G_{l+1} = (V_{l+1}, E_{l+1})$ defined at level $l + 1$ is deduced from the graph defined at level l by the following steps:

1. The selection of the vertices of G_{l+1} among V_l. These vertices are the surviving vertices of the decimation process.
2. A link of each non surviving vertex to a surviving one. This step defines a partition of V_l.
3. A definition of the adjacency relationships between the vertices of G_{l+1} in order to define E_{l+1}.

2.3   Selection of Surviving Vertices

In order to obtain a fixed decimation ratio between each level, Meer [ ] imposes the following constraints on the set of surviving vertices:

$$\forall v \in V_l - V_{l+1} \, \exists v' \in V_{l+1} : \, (v, v') \in E_l \qquad (1)$$
$$\forall (v, v') \in V_{l+1}^2 : \, (v, v') \notin E_l \qquad (2)$$

Constraint (1) insures that each non-surviving vertex is adjacent to at least a surviving one. Constraint (2) insures that two adjacent vertices cannot both survive. These constraints define a *maximal independent set* (MIS).

Meer [ ] defines a set of surviving vertices fulfilling the maximal independent set requirements thanks to an iterative stochastic process: The outcome of a

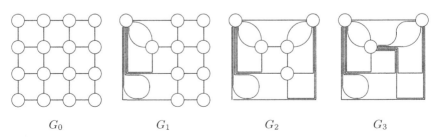

G_0 $\qquad\qquad$ G_1 $\qquad\qquad$ G_2 $\qquad\qquad$ G_3

Fig. 3. An irregular pyramid using the 4-adjacency of pixels

random variable uniformly distributed between $[0, 1]$ is associated to each vertex. Then each vertex associated to a local maximum of the random variable survives. This process is iterated on the remaining non-surviving vertices until the set of surviving vertices fulfills the maximal independent set requirements. This purely random process has been adapted by Montanvert [ ] to the connected component analysis scheme by restricting the decimation process to a set of subgraphs of the graph G_l. Note that, using such restriction, the MIS is defined separately on each subgraph. Therefore, two surviving vertices may be adjacent in G_l if they do not belong to the same subgraph. The definition of the subgraphs is equivalent to the definition of a function λ from E_l to $\{0, 1\}$. For each $(v, v') \in E_l$, $\lambda((v, v'))$ is set to 1 if v and v' belong to the same subgraph and 0 otherwise. Applied to the connected component analysis framework, the function λ is simply defined by setting $\lambda((v, v'))$ to 1 if v and v' belong to the same component(each surrounded by a closed curve in Fig. 4).

Applied to the segmentation scheme, the function λ should be adapted at each level in order to take into account the greater variability of the input data. Jolion [ ] improves the adaptability of the decimation process by using the local maxima of an interest operator instead of the local maxima of the random variable. For example, within the segmentation framework, Jolion defines the operator of interest as a decreasing function of the gray level variance computed in the neighborhood of each vertex. This operator provides a location of surviving vertices in homogeneous regions.

2.4   Parent-Child Definition

Given the set of surviving vertices, constraint (1) of a MIS insures that each non-surviving vertex is adjacent to at least one surviving vertex. Using the stochastic decimation process of Meer [ ] and Montanvert [ ], each non surviving vertex is attached to its adjacent surviving neighbor with the greatest value of the random variable. Jolion [ ] uses a contrast measure such as the gray level difference, to link each non-surviving vertex to its least contrasted surviving neighbor. Each surviving vertex is thus the father of all non-surviving vertices attached to it. The set of children of one father is called its reduction window by reference to Regular

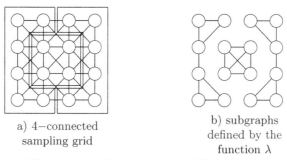

a) 4−connected
sampling grid

b) subgraphs
defined by the
function λ

Fig. 4. The connected components defined by function λ

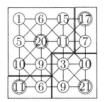

Fig. 5. Construction of a Maximal Independent Set using a random variable. The outcome of the random variable are displayed inside each vertex

pyramid notations. Fig. 5 represents the outcomes of the random variable on a graph defined from the 4×4 8-connected sampling grid. The surviving vertices induced by the outcomes of this random variable are represented by an extra circle. The boundary of each reduction window is superimposed to the figure.

2.5   Connecting Surviving Vertices

The last step of the decimation process consists in connecting surviving vertices in G_{l+1}. Meer [ ] joins two fathers by an edge if they have adjacent children (e.g. vertices labeled 15 and 7 in Fig. 6(a)). Since each child is attached directly to its father, two adjacent fathers at the reduced level are connected in the level below by paths of length less than three. Let us call these paths *connecting paths*. Using Constraint (2) of a MIS, two surviving vertices cannot be adjacent in G_l. Therefore, the length of the connecting paths is between 2 and 3. Using the restriction of the decimation process to subgraphs [,] defined by the function λ, two surviving vertices may be adjacent by an edge mapped to 0 by λ. In such cases, the length of the connecting path is equal to 1.

2.6   Reaching the Top of the Pyramid

The irregular pyramid is thus built recursively from G_0 (the original sampling grid) to the apex G_t. The apex of the pyramid cannot be reduced by the decimation process. This case occurs when G_t is reduced to a single vertex or when no

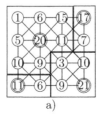
a)
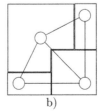
b)

Fig. 6. The reduced graph deduced from the maximal independent set (a) and the Parent-Children relationships (b)

decimation rule may be applied on the graph. For example, within the connecting component analysis scheme, the pyramid may be reduced until all connected components of the initial graph are reduced to a single vertex [ ]. Within the segmentation framework, the decimation process may also stop when the difference between the value of any adjacent vertices exceeds a threshold. Another strategy consists in decimating the initial graph until it reduces to a single vertex and to select a set of roots within the pyramid. Jolion [ ] defines a root as a vertex whose contrast measure with its parent exceeds a threshold. This strategy is also used within the regular pyramid framework [ ] and allows us to select roots at different levels of the pyramid.

2.7   The Need for Multi-edges and Self-Loops

The stochastic pyramids used by Meer [ ] and Montanvert [ ] or the adaptive pyramids defined by Jolion [ ] combine the advantages of their regular ancestors with a greater adaptability to the input data. However, both the stochastic and adaptive reduction process utilize simple graphs e.g. graphs without multiple edges and self-loops. Therefore, if the graphs of the pyramid are used to encode a partition, an edge between two vertices in a simple graph pyramid encodes disconnected boundaries between the associated regions. Moreover, the lack of self-loops does not allow to differentiate inclusions from adjacencies relationships.

Fig. 7 illustrates these inadequacies. The left and right regions of the left image share two distinct boundaries. However, a reduction process such as the one defined in the stochastic pyramid framework provides a final graph encoding these multiple boundaries by a single edge.

This last drawback is induced by the definition of the edges between the surviving vertices. Indeed, two surviving vertices having several adjacent children will be connected by a single edge. A better understanding of this drawback may be achieved by interpreting the decimation process in terms of *edge contractions* and *edge removals*. The contraction of one edge (v, v') consists to identify v and v' into a new vertex v'' and to remove the edge (v, v'). Note that, if v and v' are connected by an other edge, this last one becomes a self-loop doubly incident to v''. Since the contraction operation first identifies two distinct vertices, the

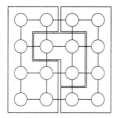 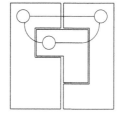

Fig. 7. Inadequacy of the stochastic or adaptive decimation process

contraction of self-loops is not defined. The removal of one edge consists to remove it from the set of edges of the graph.

Using the edge contractions and removal operations the decimation process performed by Meer [ ], Montanvert [ ] and Jolion [ ] is equivalent to the application of the following steps:

1. The selection of a set S of surviving vertices.
2. The definition of a set of edges N linking each non-surviving vertex to its father.
3. The contraction of the set of edges N.
4. The removal of all multiple edges and self-loops.

The definitions of the sets S and N vary according to the considered method. However, within the stochastic [,] or adaptive [ ] framework, the step 4 of the decimation process removes all multiple edges and self-loops between surviving vertices.

2.8   Contraction Kernels

Kropatsch [ ] encodes the step 3 of the decimation process using a *Contraction Kernel*. A Contraction Kernel is defined on a graph $G = (V, E)$ by a set of surviving vertices S, and a set of non surviving edges N (represented by bold lines in Fig. 8) such that:

– (V, N) is a spanning forest of G,
– Each tree of (V, N) is rooted by a vertex of S.

Since the set of non surviving edges N forms a forest of the initial graph, no self loop may be contracted and the contraction operation is well defined. If the graph is deduced from the initial sampling grid, each vertex on the border of this grid has to be adjacent to a special vertex encoding the background of the grid and called the *background vertex*. In order to preserve the image boundary, any edge encoding an adjacency relationship between a vertex of the graph and the background vertex must be excluded from N. Therefore, an

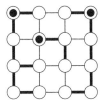

Fig. 8. A contraction Kernel (S, N) composed of three trees. Vertices belonging to S are represented with a filled circle inside($\bullet$)

initial graph G_0 deduced from the sampling grid, may at most be reduced up to a graph G_t defined by two vertices connected by a single edge. If each vertex of G_0 is associated to one pixel of the sampling grid, the two vertices encode respectively the image and its background. The edge between the two vertices corresponds to the image boundary.

The decimation of a graph by Contraction Kernels differs from the Stochastic [,] or Adaptive [ ] decimation processes on the two following points:

– First, using Contraction Kernels, the set of surviving vertices is not required to form a MIS. Therefore, two surviving vertices may be adjacent in the contracted graph. This last property avoids the use of the function λ defined by Montanvert [ ] and Jolion [ ].
– Secondly, using Contraction Kernels a non surviving vertex is not required to be directly linked to its father but may be connected to it by a branch of a tree (see Fig 8). The set of children of a surviving vertex may thus vary from a single vertex to a tree with any height.

Fig. 9 illustrates the decimation performed by the Contraction Kernel displayed in Fig. 8 by contracting successively each tree of the forest (V, N). Note that all the contractions may be performed in parallel.

The contraction of a graph reduces the number of vertices while maintaining the connections to other vertices. As a consequence, the decimation of a graph by a Contraction Kernel may induce the creation of some redundant edges. The characterization of these edges requires a better description of the topological relationships between the objects described by the graph. Since images are often projections of the reality into a plane, the image structure is planar and can be described by planar graphs. Moreover, if the initial graph is deduced from an initially planar grid (like 4-neighborhood) the initial graph and its reduced versions are planar. Given an initial planar graph G, the vertices of its dual $\overline{G}$ are located inside every face of G. The edges of $\overline{G}$ connect those dual vertices of which the corresponding faces are adjacent (lines between boxes in Fig. 10(a)). Note that the construction of the dual graph induces a one to one mapping between the edges of both graphs. Therefore, if N denotes a set of edges of

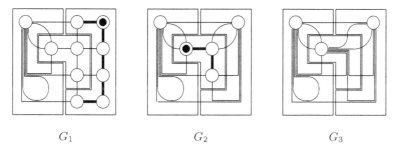

G_1 G_2 G_3

Fig. 9. The successive contractions of the trees defined in Fig. 8

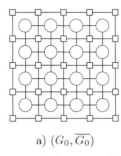 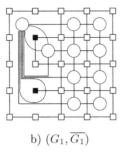

a) $(G_0, \overline{G_0})$ b) $(G_1, \overline{G_1})$

Fig. 10. The initial planar grid with the associated dual gaph(a), and characterization of redundant edges(b). Dual vertices incident to redundant dual edges are represented by filled boxes

the initial graph, we can denote by $\overline{N}$ the set of associated edges in the dual graph. A well known result of graph theory [ ] states that if G' is obtained from $G = (V, E)$ by contracting a set $N \subset E$, the dual of G' may be obtained from $\overline{G}$ by removing $\overline{N}$ in $\overline{G}$. Conversely, if G' is obtained from G by removing a set $N \subset E$, $\overline{G'}$ is obtained from $\overline{G}$ by contracting the edges $\overline{N}$ in $\overline{G}$.

The characterization of redundant edges using the dual of the contracted graph is illustrated in Fig 10. If the initial graph is deduced from the initial 4-connected sampling grid, each dual vertex encodes a corner of a pixel and each edge of the dual graph encodes a side of a pixel. The upper filled box of Fig 10(b) represents a dual vertex with a degree equal to two. The two vertices defining the face of this dual vertex correspond to two regions of the initial image sharing a boundary artificially split in two subpaths by the dual vertex. The contraction of any of the two dual edges incident to this dual vertex concatenates the two paths. The contraction of this dual edge has to be followed by the removal of the associated edge in the initial graph to maintain the duality between both graphs. In the same way, the lower filled box of Fig 10(b) represents a dual vertex with a degree one. The empty self-loop associated to this dual vertex encodes an adjacency relationship between two vertices contracted in a same vertex by the Contraction Kernel. This useless[1] adjacency relation is deleted by removing the self-loop and contracting the associated dual edge.

The *level generation* from a pair of dual graphs $(G_0, \overline{G_0})$ using a Contraction Kernel (S, N) is thus performed in two steps respectively illustrated in Fig. 11(b) and Fig. 11(c):

1. A set of edge contractions on G_0 encoded by the Contraction Kernel (S, N). The dual of the contracted graph G_1 is computed from $\overline{G_0}$ by removing the dual of the edges contained in N.
2. The removal of redundant edges encoded by a Contraction Kernel applied on the dual graph. Let us call this kernel a *Removal Kernel*. The edge con-

---

[1] A self-loop is 'useless' if it does not surround any surviving part of the graph

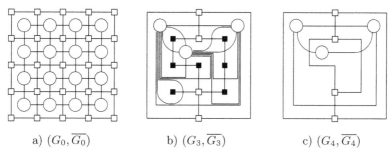

a) $(G_0, \overline{G_0})$ b) $(G_3, \overline{G_3})$ c) $(G_4, \overline{G_4})$

Fig. 11. Two steps of the graph reduction operation encoded by Contraction Kernels

tractions performed in the dual graph has to be followed by edge removals in the initial one in order to preserve the duality between the reduced graphs.

If each vertex of the original graph G_0 encodes a pixel of the sampling grid, each vertex of the contracted graphs encodes a connected region of this grid. The preservation of the meaningful edges by the two step strategy defined by Kropatsch induces a one to one mapping between the edges of the graph and the region boundaries. Experiences with connected component analysis [ ], with universal segmentation [ ], and with topological analysis of line drawings [ ] show the great potential of this concept.

3   Combinatorial Maps

Combinatorial maps and generalized combinatorial maps define a general framework which allows to encode any subdivision of nD topological spaces orientable or non-orientable with or without boundaries. The concept of maps has been first introduced by Edmonds [ ] in 1960 and later extended by several authors such as Lienhardt [ ], Tutte [ ] and Cori [ ]. This model has been applied to several fields of computer imagery such as geometrical modeling [ ], 2D segmentation [,] and graph labeling [ ]. An exhaustive comparison of combinatorial maps with other boundary representations such as cell-tuples and quad-edges is presented in [ ]. Recent trends in combinatorial maps apply this framework to the segmentation of 3D images [,] and the encoding of hierarchies [,].

The remaining of this paper will be based on 2D combinatorial maps which will be just called combinatorial maps. A Combinatorial map may be deduced from a planar graph by splitting each edge into two half edges called darts. The relation between two darts d_1 and d_2 associated to the same edge is encoded by the permutation α which maps d_1 to d_2 and vice-versa. The permutation α is thus an involution and its orbits are denoted by $\alpha^*(d)$ for a given dart d. These orbits encode the edges of the graph. Moreover, each dart is associated to a unique vertex. The sequence of darts encountered when turning around a vertex is encoded by the permutation σ. Using a counter-clockwise orientation, the orbit $\sigma^*(d)$ encodes the set of darts encountered when turning counter-clockwise around the vertex encoded by the dart d. A combinatorial map can thus be formally defined by:

Definition 1. Combinatorial Map

A combinatorial map G is the triplet $G = (\mathbf{D}, \sigma, \alpha)$, where $\mathbf{D}$ is the set of darts and σ, α are two permutations defined on $\mathbf{D}$ such that α is an involution:

$$\forall d \in \mathbf{D} \quad \alpha \circ \alpha(d) = d$$

Note that, if the darts are encoded by positive and negative integers (as in Fig 12(a)), the involution α may be implicitly encoded by:

$$\forall d \in \mathbf{D} \quad \alpha(d) = -d$$

This convention is often used for practical implementations of combinatorial maps [,] where the permutation σ is simply implemented by an array of integers (see Fig. 12 (d)).

Using the combinatorial map formalism, the dual $\overline{G}$ of a combinatorial map $G = (\mathbf{D}, \sigma, \alpha)$ is defined as $\overline{G} = (\mathbf{D}, \varphi = \sigma \circ \alpha, \alpha)$. The orbits of the permutation φ encode the set of darts encountered when turning counter-clockwise around the vertices of the dual graph (Fig 12(b)). Such orbits may also be interpreted as the sequence of darts encountered when turning around a face. Note that due to our orientation conventions, each dart of an orbit have its associated face on its right (e.g. the φ-orbit $(1, 3, 2)$ in Fig 12(a)).

A combinatorial map and its dual being deduced one from the other by a basic transformation, many properties of one graph remain true for its dual. For example, if the initial combinatorial map is connected its dual is also connected [ ]. Moreover, many particular configurations such as self-loops (see Table 1) remain particular configurations in the dual combinatorial map (see [ ] for a demonstration of these results).

Contraction kernels described in Section 2.8 may induce the creation of some redundant edges. Using the dual graph framework, such redundant edges are characterized by their associated dual edges which are incident to dual vertices with a degree lower than 3. Using the combinatorial map framework one of the dart of a redundant edge $\alpha^*(d)$ is either a self-direct-loop (see Table 1) or belongs

Table 1. Relationships between particular configurations in the original and dual combinatorial maps

	$(\mathbf{D}, \sigma, \alpha)$	$(\mathbf{D}, \varphi, \alpha)$
self-loop	$\alpha(d) \in \sigma^*(d)$	bridge
bridge	$\alpha(d) \in \varphi^*(d)$	self-loop
self-direct-loop	$(\sigma(d) = \alpha(d)$ or $\sigma(\alpha(d)) = d)$	pendant dart
pendant dart	$(\sigma(d) = d$ or $\sigma(\alpha(d)) = \alpha(d))$	self-direct-loop

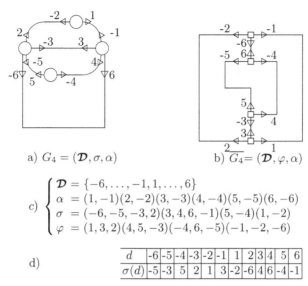

a) $G_4 = (\mathcal{D}, \sigma, \alpha)$ b) $\overline{G_4} = (\mathcal{D}, \varphi, \alpha)$

c) $\begin{cases} \mathcal{D} = \{-6, \ldots, -1, 1, \ldots, 6\} \\ \alpha = (1, -1)(2, -2)(3, -3)(4, -4)(5, -5)(6, -6) \\ \sigma = (-6, -5, -3, 2)(3, 4, 6, -1)(5, -4)(1, -2) \\ \varphi = (1, 3, 2)(4, 5, -3)(-4, 6, -5)(-1, -2, -6) \end{cases}$

d)

d	-6	-5	-4	-3	-2	-1	1	2	3	4	5	6
$\sigma(d)$	-5	-3	5	2	1	3	-2	-6	4	6	-4	-1

Fig. 12. The combinatorial maps corresponding to $\overline{G_4}$ and its dual. The lower part of this figure represents an encoding of the permutation σ by an array of integers

to an orbit of φ with a cardinal equal to 2. This last condition characterizes dual vertices with a degree 2 and may be checked by: $\varphi^2(d) = d$ or $\varphi(\sigma(d)) = \alpha(d)$.
    Combinatorial maps have thus the following interesting properties:

- The darts are ordered around each vertex and face. Note that, this information is not encoded by the simple graph data structure used by Meer [ ], Montanvert [ ] and Jolion [ ] nor explicitly available in dual graph data structures.
- The simplicity and the efficiency of the computation of the dual combinatorial map avoids an explicit encoding of the dual graph.
- The combinatorial map formalism may be extended to any dimensions [ ].

4   Combinatorial Pyramids

The aim of combinatorial pyramids is to combine the advantages of combinatorial maps with the reduction scheme defined by Kropatsch [ ]. A combinatorial pyramid will thus be defined by an initial combinatorial map successively reduced by a sequence of contraction or removal kernels defined in the combinatorial map framework. The definition of combinatorial pyramids requires thus a formal definition of the contraction and removal operations in the combinatorial map framework.
    Using combinatorial maps, the removal operation (see Definition 2) is encoded as an update of the permutation σ. The analytic form of Definition 2 is

given by proposition 1. Note that this last operation may be performed efficiently if σ is encoded by an array of integers.

Definition 2. Removal Operation

Given a combinatorial map $G = (\mathcal{D}, \sigma, \alpha)$ and a dart $d \in \mathcal{D}$. If $\alpha^*(d)$ is not a bridge, the combinatorial map $G' = G \setminus \alpha^*(d) = (\mathcal{D}', \sigma', \alpha)$ is defined by:

- *$\mathcal{D}' = \mathcal{D} \setminus \alpha^*(d)$ and*
- *σ' is defined as:*

$$\forall\, d \in \mathcal{D}' \quad \sigma'(d) = \sigma^n(d) \text{ with } n = Min\{p \in \mathbb{N}^* \ / \ \sigma^p(d) \notin \alpha^*(d)\}$$

Note that the bridges are excluded from Definition 2 in order to keep the number of connected components of the combinatorial map. Moreover, since both the dart d and its opposite $\alpha(d)$ are removed, the involution α of G' is restricted to the reduced dart set $\mathcal{D} \setminus \alpha^*(d)$. Although, the involution α of the reduced graph is mathematically different from the one of the original graph, the implicit encoding of the permutation α by the sign (see Section 3) may remain unchanged in both graphs. We thus consider that the involution α is unchanged by the removal operation.

Proposition 1. *Given a combinatorial map $G = (\mathcal{D}, \sigma, \alpha)$ and a dart $d \in \mathcal{D}$ which is neither a bridge nor a self-direct-loop, the combinatorial map $G \setminus \alpha^*(d) = (\mathcal{D} \setminus \alpha^*(d), \sigma', \alpha)$ is defined by:*

$$\begin{cases} \forall d' \in \mathcal{D} \setminus \sigma^{-1}(\alpha^*(d)) & \sigma'(d') = \sigma(d') \\ \sigma'(\sigma^{-1}(d)) = \sigma(d) \\ \sigma'(\sigma^{-1}(\alpha(d))) = \sigma(\alpha(d)) \end{cases}$$

As stated in Section 2.8, the removal of one edge on a planar graph induces the contraction of its associated edge in its dual. Conversely, the contraction of one edge on the initial graph is equivalent to the removal of the associated dual edge in the dual graph (Fig. 12). This last property allows us to define the contraction operation as a removal performed in the dual combinatorial map.

Definition 3. Contraction Operation

Given a combinatorial map $G = (\mathcal{D}, \sigma, \alpha)$ and a dart $d \in \mathcal{D}$ which is not a self-loop. The contraction of dart d creates the combinatorial map $G/\alpha^*(d)$ defined by:

$$G/\alpha^*(d) = \overline{\overline{G} \setminus \alpha^*(d)}$$

Note that this operation is well defined since d is a self-loop in G iff it is a bridge in $\overline{G}$. Thus, any sequence of removal or contraction operations will preserve the number of connected components of the initial graph. This last property is useful in the irregular pyramids framework because it attempts to simplify the initial planar map while preserving its essential topological properties. Moreover, since the involution α is not modified by the removal operation and remains the same for one combinatorial map and its dual (see Section 3) it is also unchanged by the contraction operation.

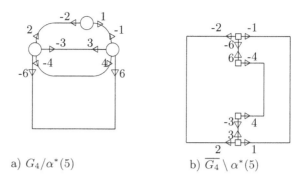

a) $G_4/\alpha^*(5)$ b) $\overline{G_4} \setminus \alpha^*(5)$

Fig. 13. Contraction of the edge $\alpha^*(5)$ from G_4 and the associated dual removal

Since the contraction operation is equivalent to a removal operation performed on the dual combinatorial map, the analytic form of Definition 3 may be deduced from the analytic form of the removal operation by substituting φ by σ in Proposition 1. This analytical form, rewritten in terms of modifications of the permutation σ is provided in Proposition 2. The proof of this result together with the description of the special cases excluded from Propositions 1 and 2 may be found in [ ].

Proposition 2. *Given a combinatorial map* $G = (\mathcal{D}, \sigma, \alpha)$ *and a dart* $d \in \mathcal{D}$ *which is neither a pendant edge nor a self loop. The combinatorial map* $G/\alpha^*(d) = (\mathcal{D} \setminus \alpha^*(d), \sigma', \alpha)$ *is defined by:*

$$\begin{cases} \forall d' \in \mathcal{D} \setminus \sigma^{-1}(\alpha^*(d)) & \sigma'(d) = \sigma(d) \\ \sigma'(\sigma^{-1}(d)) & = \sigma(\alpha(d)) \\ \sigma'(\sigma^{-1}(\alpha(d))) = \sigma(d) \end{cases}$$

Note that, since the dual graph is implicitly encoded, any modification of the initial combinatorial map will also modify its dual. Therefore, using combinatorial maps the dual combinatorial map is both implicitly encoded and updated. Moreover, the updates of the permutation σ by the contraction operation preserves the orientation of the original combinatorial map. For example, let us suppose that the empty vertices displayed in Fig 14(a) represents two adjacent regions surrounded by the regions 1,2,3,4,a,b,c,d. This last sequence encodes the order in which the adjacent regions will be encountered when turning counterclockwise around the two adjacent regions. The order of this sequence is preserved when the two regions are merged in Fig. 14(b), by the contraction of their common edge.

Definitions 2 and 3 define the value of the permutation σ after the removal or the contraction of one edge. Such definitions may be extended to the removal or the contraction of a set of edges [ ]. However, in order to avoid testing the preconditions of these operations for each edge we have to adapt the concept of contraction kernel (see section 2.8) to the combinatorial map framework.

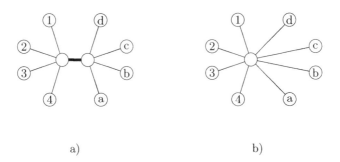

a)   b)

Fig. 14. Preservation of the orientation using combinatorial maps

Definition 4. Contraction Kernel

Given a connected combinatorial map $G = (\mathcal{D}, \sigma, \alpha)$ and a non-empty set of darts $\mathcal{B} \subset \mathcal{D}$ encoding the adjacency relationship between G and its background. The set $K \subset \mathcal{D}$ will be called a contraction kernel iff:

1. *K is a forest of G,*
2. *K does not intersect $\mathcal{B}$.*

The set $\mathcal{SD} = \mathcal{D} - K$ is called the set of surviving darts.

Note that the set of surviving darts of a contraction kernel cannot be empty since it contains at least $\mathcal{B}$. Unlike dual graph contraction kernels, contraction kernels defined within the combinatorial map framework do not specify the surviving vertices. Within such framework, the surviving vertices are implicitly defined by the orbits of the updated permutation σ.

Given an initial combinatorial map $G_0 = (\mathcal{D}, \sigma, \alpha)$ and a contraction kernel K, the contracted combinatorial map $G' = G/K = (\mathcal{SD}, \sigma', \alpha)$ may be deduced from G by the iterative contraction of each dart contained in K using proposition 2. However, this last solution does not allow a straightforward parallel implementation and induces unnecessary updates of the permutation σ. The parallel computation of the contracted combinatorial map is based on the concept of *connecting walk*:

Definition 5. Connecting Walk

Given a combinatorial map $G = (\mathcal{D}, \sigma, \alpha)$, a contraction kernel K and a dart $d \in \mathcal{SD}$, the connecting walk associated to d is equal to:

$$CW(d) = d, \varphi(d), \dots, \varphi^{n-1}(d) \text{ with } n = Min\{p \in \mathbb{N}^* \mid \varphi^p(d) \in \mathcal{SD}\}$$

Note that, only the first dart of each connecting walk survives. Each connecting walk $CW(d)$ connects the surviving darts d and $\varphi^n(d)$ by a sequence of non surviving darts.

Connecting walks may be considered as an extension of the connecting paths used by Jolion [ ] and Montanvert [ ](see Section 2.5). However, using contraction kernels, the MIS requirements are substituted by the more general concept of forest requirement. Therefore, the paths defined by Montanvert and Jolion have to be replaced by the more general concept of walks [ ]. Moreover, since the trees of the forest may have any depth, the length of a connecting walk is not bounded by 3. Finally, using combinatorial maps, a connecting walk is an ordered sequence of darts which preserves the order defined on $\mathcal{D}$ by the permutation φ. More precisely, given a surviving dart d and its connecting walk we have [ ]:

$$\forall d \in \mathcal{SD} \quad \varphi'(d) = \varphi^n(d) \tag{3}$$

Where $n = Min\{p \in \mathbb{N}^* \mid \varphi^p(d) \in \mathcal{SD}\}$ and $\varphi' = \sigma' \circ \alpha$ denotes the φ permutation of the contracted graph G/K. The permutation α being an involution, the permutation σ' may be retrieved from φ' by $\sigma' = \varphi' \circ \alpha$. We have thus:

$$\forall d \in \mathcal{SD} \quad \sigma'(d) = \varphi^n(\alpha(d)) \text{ with } n = Min\{p \in \mathbb{N}^* \mid \varphi^p(\alpha(d)) \in \mathcal{SD}\}$$

Note that, $\varphi^{n-1}(\alpha(d))$ is the last dart of $CW(\alpha(d))$. Therefore, the permutation α remains unchanged in the contracted combinatorial map while the permutation σ' maps each surviving dart d to the φ-successor of the last dart of $CW(\alpha(d))$. Therefore, the computation of the σ-successor of a dart d in the contracted combinatorial map requires to traverse $CW(\alpha(d))$. Sequential algorithm 1 computes the σ-successor of all the surviving darts in the contracted combinatorial map. Since the set of connecting walks forms a partition of $\mathcal{D}$ (see proof in [ ]), Algorithm 1 has to traverse $|\mathcal{D}|$ darts. Its complexity is thus equal to $\mathcal{O}(|\mathcal{D}|)$.

```
1   dart contracted_sigma(G = (𝒟, σ, α), K)
2   {
3       For each d ∈ 𝒮𝒟 = 𝒟 − K
4       do
5           d' = φ(α(d)) = σ(d) // Second dart of CW(α(d))
6
7           while(d' ∈ K) // computation of CW(α(d))
8               d' = φ(d')
9           σ'(d) = d'
10      done
11  }
```

Algorithm 1: *Computation of the permutation σ' of the contracted combinatorial map $G' = (\mathcal{SD}, \sigma', \alpha)$*

Using the dual graph pyramid scheme, a removal kernel is defined as a contraction kernel applied on the dual combinatorial map. Therefore, all results already obtained on contraction kernels remain valid in the dual combinatorial map for removal kernels [ ]. The connecting walks defined by removal kernels may be constructed as a sequence of σ-successors of a surviving dart:

$$\forall d \in \mathcal{SD} \; CW_R(d) = d.\sigma(d)\ldots\sigma^{n-1}(d) \text{ with } n = Min\{p \in \mathbb{N}^* \mid \sigma^p(d) \in \mathcal{SD}\}.$$

The σ-successor of any surviving dart d in the removed combinatorial map is then defined as $\sigma^n(d)$ where $\sigma^{n-1}(d)$ is the last dart of $CW_R(d)$. The updated permutation σ' may thus be computed from the original combinatorial map by a traversal of the connecting walks defined by the removal kernel. This traversal may be performed by a slightly modified version of Algorithm 1 where line 8 is replaced by the instruction $d' = \sigma(d)$.

Using the same scheme as dual graph pyramids (see section 2.8), the level generation using a contraction kernel K is performed in two steps:

1. One set of edge contractions encoded by a contraction kernel K_1.
2. The removal of redundant edges encoded by a removal kernel K_2. Such redundant edges are incident to faces with a degree one or two (see Section 2.8 and 3).

5   Conclusion

In this paper we have described different hierarchical models and some of the basic processes working on them: regular, irregular and combinatorial pyramids. All three possess the following important well known properties [ ]:

− Independence to resolution,
− global to local interactions,
− $log(image\_diameter)$ parallel complexity.

In the remaining of this section, we want to differentiate the advantages and drawbacks of the three types of pyramids discussed in the paper. The first part of Table 2 describes the properties of the basic entities represented at each level of the structure. In the second part, we consider the properties of the projection of these basic entities into the image plan.

Each level of the hierarchy encodes a set of resolution cells ("region") and their respective adjacencies. A resolution cell is explicitly encoded by a pixel or a vertex of the graph structure while it is implicitly coded by a σ-orbit of the combinatorial map. An adjacency relation between two resolution cells is implicitly encoded by a crack between two 4-adjacent pixels while it is explicitly coded as an edge of the graph or a couple of darts $(d, \alpha(d))$ in the other hierarchies. A point is defined where three or more resolution cells meet. It is implicitly defined as a pixel corner in the regular structure and as a φ-orbit in the combinatorial map while it is not encoded by the simple graph pyramids and explicitly encoded as a dual vertex in the dual graph. A combinatorial map is the only structure which explicitly codes the orientation of the resolution cells local arrangements by the permutation σ.

Table 2. Basic hierarchical entities and properties of their image embedding

Pyramids	Regular	Irregular		Combinatorial
		Simple graph	Dual graph	
Region	pixel	vertex	vertex	σ-orbit
Adjacency	crack	edge	edge	$(d, \alpha(d))$
Point	pixel corner	-	dual vertex	φ-orbit
orientation	-	-	implicit	σ, φ
Properties of image embedding				
Receptive field	regular shape	arbitrary shape	arbitrary shape	arbitrary shape
Segmentation scheme	linking	adaptive deci-mation	adaptive dual contraction	adaptive dual contraction
Region(vertex)	not connected	connected	connected	connected
Boundaries (edge)	not connected	not connected	connected	connected

The receptive field of a resolution cell defines its projection into the image plane. The regular pyramid is the only structure which restricts the form of the receptive fields to regular shapes. Using hierarchies, the segmentation of the image plane is achieved by selecting a set of roots(Section 2.6). The receptive field of each root defines a region of the image plane. The pyramid linking [ ] algorithm adapts the shape of the regular pyramid receptive fields to the image content. All other hierarchies may use the receptive fields obtained from the pyramid construction scheme. Moreover, the pyramid linking algorithm is the only segmentation scheme which may produce non connected regions (e.g. [ ]). A boundary is defined as the image embedding of an adjacency relation. This boundary may be not connected using regular or simple graph hierarchies.

Dual graph and Combinatorial pyramids are thus the only hierarchies which provide both connected regions and connected boundaries. Combinatorial pyramids combine the advantages of dual graph pyramids with an explicit orientation of the boundary segments of the embedded object thanks to the permutation σ. Moreover, using combinatorial maps, the dual graph is both implicitly encoded and updated. Finally, the combinatorial map formalism is defined in any dimensions.

References

1. E. Ahronovitz, J. Aubert, and C. Fiorio. The star-topology: a topology for image analysis. In *5th DGCI Proceedings*, pages 107–116, 1995. 118
2. Y. Bertrand, G. Damiand, and C. Fiorio. Topological map: Minimal encoding of 3d segmented images. In J. M. Jolion, W. Kropatsch, and M. Vento, editors, 3rd *Workshop on Graph-based Representations in Pattern Recognition*, pages 64–73, Ischia(Italy), May 2001. IAPR-TC15, CUEN. 118

3. Y. Bertrand and J. Dufourd. Algebraic specification of a 3D-modeler based on hypermaps. *CVGIP: Graphical Models and Image Processing*, 56(1):29–60, Jan. 1994. 118

4. M. Bister, J. Cornelis, and A. Rosenfeld. A critical view of pyramid segmentation algorithms. *Pattern Recognit Letter.*, 11(9):605–617, Sept. 1990. 108, 110, 125

5. J. P. Braquelaire and L. Brun. Image segmentation with topological maps and inter-pixel representation. *Journal of Visual Communication and Image representation*, 9(1), 1998. 118, 119

6. J. P. Braquelaire, P. Desbarats, and J. P. Domenger. 3d split and merge with 3-maps. In J. M. Jolion, W. Kropatsch, and M. Vento, editors, 3rd *Workshop on Graph-based Representations in Pattern Recognition*, pages 32–43, Ischia(Italy), May 2001. IAPR-TC15, CUEN. 118

7. L. Brun. *Segmentation d'images couleur à base Topologique*. PhD thesis, Université Bordeaux I, 351 cours de la Libération 33405 Talence, December 1996. 119

8. L. Brun and W. Kropatsch. Dual contractions of combinatorial maps. Technical Report 54, Institute of Computer Aided Design, Vienna University of Technology, lstr. 3/1832,A-1040 Vienna AUSTRIA, January 1999. 118, 119, 122

9. L. Brun and W. Kropatsch. Pyramids with combinatorial maps. Technical Report PRIP-TR-057, PRIP, TU Wien, 1999. 118, 124, 125

10. P. Burt, T.-H. Hong, and A. Rosenfeld. Segmentation and estimation of image region properties through cooperative hierarchial computation. *IEEE Transactions on Systems, Man and Cybernetics*, 11(12):802–809, December 1981. 108, 110, 114, 126

11. R. Cori. *Un code pour les graphes planaires et ses applications*. PhD thesis, Université Paris VII, 1975. 118

12. J. Edmonds. A combinatorial representation for polyhedral surfaces. *Notices American Society*, 7, 1960. 109, 118

13. F. Harary. *Graph Theory*. Addison-Wesley, 1972. 117, 124

14. J. Jolion and A. Montanvert. The adaptive pyramid : A framework for 2d image analysis. *Computer Vision, Graphics, and Image Processing*, 55(3):339–348, May 1992. 112, 113, 114, 115, 116, 120, 124

15. W. G. Kropatsch. From equivalent weighting functions to equivalent contraction kernels. In E. Wenger and L. I. Dimitrov, editors, *Digital Image Processing and Computer Graphics (DIP-97): Applications in Humanities and Natural Sciences*, volume 3346, pages 310–320. SPIE, 1998. 109, 110, 115, 120

16. W. G. Kropatsch and S. BenYacoub. Universal Segmentation with P*IRR*amids. In A. Pinz, editor, *Pattern Recognition 1996, Proc. of 20th ÖAGM Workshop*, pages 171–182. OCG-Schriftenreihe, Österr. Arbeitsgemeinschaft für Mustererkennung, R. Oldenburg, 1996. Band 90. 118

17. W. G. Kropatsch and M. Burge. Minimizing the Topological Structure of Line Images. In A. Amin, D. Dori, P. Pudil, and H. Freeman, editors, *Advances in Pattern Recognition, Joint IAPR International Workshops SSPR'98 and SPR'98*, volume Vol. 1451 of *Lecture Notes in Computer Science*, pages 149–158, Sydney, Australia, August 1998. Springer, Berlin Heidelberg, New York. 118

18. W. G. Kropatsch and H. Macho. Finding the structure of connected components using dual irregular pyramids. In *Cinquième Colloque DGCI*, pages 147–158. LLAIC1, Université d'Auvergne, ISBN 2-87663-040-0, September 1995. 118

19. P. Lienhardt. Subdivisions of n-dimensional spaces and n-dimensional generalized maps. In *Annual ACM Symposium on Computational Geometry, all*, volume 5, 1989. 120

20. P. Lienhardt. Topological models for boundary representations: a comparison with n-dimensional generalized maps. *Computer-Aided Design*, 23(1):59–82, 1991. 118

21. J. Marchandier, S. Michelin, and Y. Egels. A graph labelling approach for connected feature selection. In A. Amin, F. J. Ferri, P. Pudil, and F. J. Iñesta, editors, *Advances in Pattern Recognition, Joint IAPR International Workshops SSPR'2000 and SPR'2000*, volume Vol. 1451 of *Lecture Notes in Computer Science*, pages 287–296, Alicante, Spain, August 2000. Springer, Berlin Heidelberg, New York. 118

22. P. Meer. Stochastic image pyramids. *Computer Vision Graphics Image Processing*, 45:269–294, 1989. 111, 112, 113, 114, 115, 116, 120

23. A. Montanvert, P. Meer, and A. Rosenfeld. Hierarchical image analysis using irregular tessellations. *IEEE Transactions on Pattern Analysis and Machine Intelligence*, 13(4):307–316, APRIL 1991. 108, 111, 112, 113, 114, 115, 116, 120, 124

24. P. F. Nacken. Image segmentation by connectivity preserving relinking in hierarchical graph structures. *Pattern Recognition*, 28(6):907–920, June 1995. 126

25. A. Rosenfeld, editor. *Multiresolution Image Processing and Analysis*. Springer Verlag, Berlin, 1984. 108

26. W. Tutte. *Graph Theory*, volume 21. Addison-Wesley, encyclopedia of mathematics and its applications edition, 1984. 118

Representing Vertex-Based Simplicial Multi-complexes

Emanuele Danovaro, Leila De Floriani, Paola Magillo, and Enrico Puppo

Department of Computer and Information Sciences (DISI)
University of Genova, Via Dodecaneso 35, 16146 Genova - Italy
{danovaro,deflo,magillo,puppo}@disi.unige.it
http://www.disi.unige.it/research/Geometric_modeling/

Abstract. In this paper, we consider the problem of representing a multiresolution geometric model, called a Simplicial Multi-Complex (SMC), in a compact way. We present encoding schemes for both two- and three-dimensional SMCs built through a vertex insertion (removal) simplification strategy. We show that a good compression ratio is achieved not only with respect to a general-purpose data structure for a SMC, but also with respect to just encoding the complex at the maximum resolution.

1 Introduction

Simplicial complexes (triangle and tetrahedral meshes) are popular representations for surfaces, objects, scalar and vector fields in applications, such as Geographic Information Systems (GISs), computer graphics, virtual reality, medical imaging, finite element analysis. The availability in the applications of very large meshes has led to an intense research activity on mesh simplification and on multiresolution models.

Data simplification is a tool for reducing a mesh to a manageable size: it reduces the resolution of a mesh (i.e., the number and the density of its cells), while loosing accuracy. In many cases, however, simplification needs to be applied selectively, i.e., only in those portions of a mesh that are less relevant for the current application. Since the regions of interest for an application may vary dynamically, on-line simplification would be important. Mesh simplification, on the other hand, is an expensive task, which cannot be performed on-line on meshes of large size.

The basic idea underlying *multiresolution modeling* consists of de-coupling the simplification phase (performed off-line) from the selective refinement phase (performed on-line). A multiresolution model basically encodes the steps performed by a simplification process as a partial order, from which a virtually continuous set of meshes at different Levels Of Detail (LODs) can be extracted. In [ ], we have described a general multiresolution model, called a Multi-Complex (MC), which consists of directed acyclic graph representing a partially ordered set of mesh updates. The Multi-Complex is dimension-independent as well as independent of the specific simplification strategy used for its construction. On

G. Bertrand et al. (Eds.): Digital and Image Geometry, LNCS 2243, pp. 129–149, 2001.
© Springer-Verlag Berlin Heidelberg 2001

the other hand, the multiresolution approach tends to introduce an overhead due to the need of encoding several complexes at various LODs as well as the partial order.

The fundamental operation on a multiresolution model is the extraction of meshes at variable resolution, which satisfy user-defined criteria, and have a minimum size. Such operation is known as *selective refinement* [ ]. Most common application-dependent queries on two- and three-dimensional meshes reduce to instances of selective refinement, since they need a high resolution just in a specific part of the mesh. Some examples are contour, or isosurface extraction, windowing, intersection with a plane, or with a polygonal line [ ].

In this paper, we consider a specialization of the MC to simplicial complexes, that we call a Simplicial Multi-Complex (SMC). We describe first a general data structure for SMCs which is independent of the dimension and of the construction technique. However, such structure has a large overhead since it is about three times larger than a standard data structure representing the simplicial complex available from the SMC the maximum resolution.

Then, we present compact data structures for a specific class of SMCs in two and three-dimensions, constructed on the basis of a vertex-insertion, or vertex-removal simplification strategy. We show that such structures provide good compression ratios not only with respect to the general SMC data structure, but even with respect to encoding just the original mesh at full resolution. All data structures are analyzed with respect to both their space requirements and their efficiency in supporting the primitives needed for implementing selective refinement.

The rest of the paper is organized as follows. Section 2 introduces multiresolution modeling and the vertex-based SMC. Section 3 is concerned with general issues about encoding multiresolution models, and includes a brief overview of related work. Section 4 describes the selective refinement query. Section 5 presents a general-purpose data structure for the SMC. Section 6 describes and analyzes compact ways for encoding vertex removals in both two and three dimensions. Section 7 presents comparisons among the various data structures. Finally, Section 8 contains some concluding remarks.

2 Multiresolution Modeling and the Simplicial Multi-complex

We denote with k and d two natural numbers such that $k \leq d$. A k-*dimensional simplex* σ is the locus of points that can be expressed as the convex combination of $k + 1$ affinely independent points in $\mathbb{E}^d$, called the *vertices* of σ. Any simplex having its vertices at a subset of the vertices of σ is called a *facet* of σ.

A *simplicial complex* in $\mathbb{E}^d$ is a finite set Σ of simplices of heterogeneous dimension such that: for any simplex $\sigma \in \Sigma$, all facets of σ are in Σ; and, for any pair of simplices $\sigma_1, \sigma_2 \in \Sigma$, their intersection is either empty or a common facet of σ_1 and σ_2.

A k-dimensional simplicial complex (for short, a k-complex) is *regular* if and only if every simplex is a facet of some k-simplex. A regular k-complex is completely defined by its k-simplices. Here, we will deal only with regular complexes whose domain is a manifold. Therefore, we will identify a k-complex with the set of its k-simplices (hereafter called just simplices, whenever no ambiguity arises). The manifold assumption is consistent with the modification operator we adopt to generate multiple levels of details (i.e., vertex insertion / removal, which will be defined later), since such operator preserves the manifold property of a mesh.

The general idea behind *data simplification techniques* is to perform a sequence of atomic modifications on a given mesh by either removing details from a mesh at high resolution, or adding details to a coarse mesh. Specific simplification algorithms are characterized by using a certain type of modification operators to act on the mesh. The most popular ones are *vertex insertion, vertex removal*, and *edge collapse*.

A survey on data simplification for triangle meshes can be found in [ ]. Data simplification algorithms for tetrahedral meshes have been proposed more recently in the literature, and they are based on either edge collapse [,], or on vertex insertion [,]. Algorithms for vertex insertion (and removal) on tetrahedral meshes have been proposed in [ ].

Multiresolution models of meshes are built on top of a data simplification process. They encode the range of Levels of Detail (LODs) generated during the simplification, and arrange them by defining suitable dependency relations, which are used to guide the extraction of specific LODs.

Several two-dimensional multiresolution models have been proposed in the literature for terrains in Geographic Information Systems (GISs), and for surfaces in computer graphics and virtual reality. They can be either based on nested regular meshes [, ,], or on irregular triangular meshes [, , ,]. A fewer number of three-dimensional multiresolution representations have also been proposed, mainly for describing scalar fields in scientific visualization. Again, they can be classified into models based on regular nested meshes [, , ,], and models based on irregular simplicial complexes [ ]. In [ ], we have shown that all proposed models are special cases of a Multi-Complex (MC), a general and dimension-independent framework for multiresolution modeling based on cell complexes.

Here, we consider the *Simplicial Multi-Complex* (SMC), which specializes the MC defined in [ ] to simplicial cells. The SMC is a general multiresolution model for simplicial complexes that is completely independent on the dimension of the complex, on the dimension of the embedding space, and on the strategy used to obtain the various LODs included in the model.

The definition of a SMC is based on the concept of a *local update*: an operator that, when repeatedly applied to a complex, monotonically increases or decreases its resolution (i.e., the number and density of its simplices). In this paper, we will consider local updates which refine a complex through the insertion of a new vertex, or coarsen it through the removal of an existing vertex.

Any sequence of updates starts from an initial complex and terminates with a final complex which represent the two extrema of the Level Of Details (LODs) spanned by the update process. We call *reference complex* the complex at maximum resolution, and *base complex* the one at minimum resolution. Any coarsening sequence can be reversed into a refinement one, and vice versa. We consider a *local update* as an invertible operation that either refines or coarsens a complex locally by replacing a connected subset of its simplices with another set of simplices having a higher or a lower resolution, respectively, in such a way that the result is still a regular and manifold complex (see [ ] for conditions to guarantee these properties). An update is denoted as a pair $u \equiv (u^-, u^+)$, where u^- represents the coarsening operation, and u^+ represents its inverse refinement operation. Alternatively, u^- and u^+ can be seen as sets of simplices: u^- is the set of simplices that disappear from the mesh when refining it, and appear when coarsening it, and the opposite holds for u^+.

In case of *vertex-based updates* (vertex removal and vertex insertion), u^- is a simplicial complex filling a k-dimensional polytope without internal vertices, and u^+ is a set of k-dimensional simplices all incident in the new vertex v_u. The boundaries of u^- and u^+ are the same.

A SMC abstracts from the totally ordered sequence of updates performed during a mesh simplification process, and defines the *partial order* describing the *mutual dependencies* between pairs of updates. For the purpose of defining the partial order, we look at a simplification process in the coarse-to-fine direction. An update u_2 of a given sequence is said to *depend* on another update u_1 if and only if $u_1^+ \cap u_2^- \neq \emptyset$, i.e., if u_2^+ remove some simplex created by u_1^+.

A SMC encodes a *partial order* describing the *mutual dependencies* between updates as a Directed Acyclic Graph (DAG). In a SMC $\mathcal{M} = (U, A)$, nodes in U represent updates, while arcs in A represent dependency relations between updates: an arc $<u_1, u_2>$ is in A if and only if update u_2 depends on u_1. The creation of the base complex is also represented as a special node, which is the unique root of the DAG. Figure 1 shows an example of a two-dimensional SMC.

We recall that the construction of a vertex-based SMC relies on an algorithm for data simplification based on vertex insertion or removal. Several simplification algorithms have been proposed in the literature both in two dimensions (for triangle meshes representing either scalar fields or free-form surfaces, see [ ] for a survey), and in three dimensions (for tetrahedral meshes, see [, ,]). The various existing proposals differ in the strategy they use to drive simplification, involving issues such as bounding the geometric error, preserving a good aspect ratio of simplices, etc. However, such choices do not affect the possibility of building a vertex-based SMC, nor the design of data structures to encode it.

3   Encoding Multiresolution Models

The effectiveness of a multiresolution model is highly dependent on its storage requirements. A multiresolution data structure is characterized by the way

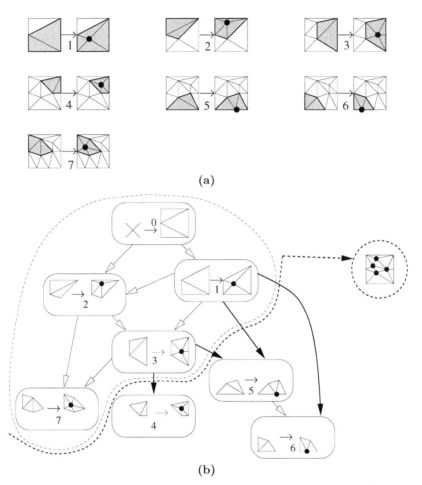

Fig. 1. (a) A sequence of vertex insertions refining a two-dimensional simplicial complex, where the mesh portion affected by each update is highlighted. (b) The corresponding SMC: the dashed line encloses a consistent subset S of nodes; the triangle mesh Σ_S is shown in the circle on the right

in which it encodes the individual *mesh updates* and the *dependency relations* among updates.

A data structure for a two-dimensional SMC, described in [ ], adopts an explicit representation of both the updates (encoded by listing all involved triangles) and the dependency links (encoded as a DAG). Such structure introduces an overhead which can be too high to be used for representing very large data sets.

More compact data structures are designed for certain subclasses of SMCs, i.e, SMCs built by using vertex insertion / removal, or edge collapse, by exploiting the topological and geometric nature of specific simplification operators. This allows defining compact encodings for the updates and sometimes for the partial order.

In the two-dimensional case, compact encodings of edge collapses has been first proposed by Hoppe [ ]. A very compact way to represent dependency links has been proposed by El-Sana and Varshney for the special case of general edge collapse [ ]. Other compact structures for dependencies have been defined based on binary trees of vertices [,], but they fail to represent the natural notion of dependency (see [ ] for a discussion).

A compact encoding of a generalized edge collapse has been proposed by Popovic and Hoppe in [ ]. This method has been developed for a quite general situation, in which the update can affect the topology, as well as the regularity of the mesh. A very compact data structure for encoding three-dimensional SMCs built based on edge collapse has been proposed in [ ].

Much less work has been performed for encoding SMCs built based on vertex insertion and vertex removal. In two dimensions, a method has been proposed by Klein and Gumhold [ ]. Based the observation that representing a vertex-removal reduces to describing the way in which the polygonal hole left by the removed vertex must be triangulated, and that such polygon has a bounded number of vertices, they encode all equivalence classes of the possible triangulations of such polygons, and, for each removed vertex, they store a reference to the appropriate equivalence class. In [ ], a compact way to encode dependencies is also described, which will be recalled in Section 5.1. To our knowledge, no proposal exists for the three-dimensional case.

In this paper, we give our proposal of compact data structures for two- and three-dimensional SMCs.

We denote with n, s, m and a the number of vertices, simplices, nodes (updates) and arcs (dependency links) in a SMC $\mathcal{M}$. We denote with $\bar{n}$ and $\bar{s}$ the number of vertices and simplices in the reference complex of $\mathcal{M}$, respectively (see Table 1). We also assume that the base complex has a very small, and, therefore, negligible number of vertices and simplices.

From Euler formula in two dimensions we have that $\bar{s} \simeq 2\bar{n}$. In three dimensions, $\bar{s}$ is $O(\bar{n}^2)$ in the worst case, but it has experimentally been found that, for real datasets, $\bar{s} \approx 6\bar{n}$ [ ].

Table 1. Parameters expressing the size of a SMC

Symbol	meaning	context
n	# vertices	whole SMC
s	# simplices	whole SMC
m	# nodes	whole SMC
a	# arcs	whole SMC
$\bar{n}$	# vertices	reference complex
$\bar{s}$	# simplices	reference complex
P	$\#u^-, \#u^+ < P$	SMC

Table 2. Relations that are assumed among the parameters of Table 1, in the two- and in the three-dimensional case

Relation	restrictions	justification
$\bar{s} \simeq 2\bar{n}$	2D	provable fact
$\bar{s} \approx 6\bar{n}$	3D	experimentals
$n = m = \bar{n}$	vertex-based	provable fact
$\#u^-, \#u^+ < 16$	2D, vertex-based	construction
$\#u^-, \#u^+ < 32$	3D, vertex-based	construction
$s \approx 6n$	2D	experimentals
$s \approx 18n$	3D	experimentals
$a \approx 3m$	2D	experimentals
$a \approx 6m$	3D	experimentals

For vertex-based SMCs, $n = m = \bar{n}$, since (disregarding the vertices of the base complex) each node adds a vertex, and all such vertices form the vertices of the reference complex.

We also assume an upper bound P on the number of simplices involved in a single update, such that $\#u^-, \#u^+ < P$. This is achieved by imposing constraints on the mesh simplification algorithm used to build the SMC. The values of $P = 16$ in two dimensions, and $P = 32$ in three dimensions are reasonable in practice. Under this assumption, also the number of arcs entering and leaving each SMC node is bounded, and thus a and s are linear in m.

It has been experimentally found that in the two-dimensional case, $s \approx 6n$ and $a \approx 3m$; while, in three dimensions, $s \approx 18n$ and $a \approx 6m$. All such relations are summarized in Table 2.

Finally, we assume that vertex coordinates can be described as 32 bit floating-point numbers. All logarithms appearing in formulas are intended in base 2.

4   Selective Refinement

The data structures proposed in this paper are evaluated not only based on their space requirements, but also with respect to their efficiency in supporting the operation of selective refinement. We remark that such structures, although called "compressed", are complete and effective structures on which algorithms for selective refinement can be run: they do not need to be decoded into a more explicit representation before use.

The extraction of a complex from a SMC reduces to selecting a subset $S = \{u_1, \ldots, u_k\}$ of the SMC nodes and performing the corresponding updates $u_1^+, \ldots, u_k^+$ on the base complex. Set S must be *consistent* with the partial order defined in the DAG: for every node $u \in S$, all the predecessors of u must be also in S. A consistent set S may contain more updates in some areas, and fewer updates elsewhere, hence defining a complex Σ_S whose resolution is variable in space.

The fundamental query on a multiresolution model, called *selective refinement*, consists of extracting a complex Σ such that the resolution of Σ satisfies some user-defined requirements (e.g., having a certain accuracy over a region of interest), and Σ contains a minimal number of simplices. Given a SMC $\mathcal{M} = (U, A)$, the resolution requirements for extracting a complex from $\mathcal{M}$ are expressed by a means of a Boolean function τ defined on the SMC nodes: $\tau(u) = \textbf{true}$ if and only if update u is necessary for achieving the desired resolution. The value of τ over an update u is application dependent.

A selective refinement query applied to a SMC $\mathcal{M} = (U, A)$, based on τ, extracts the complex Σ_S, associated with a consistent set S of $\mathcal{M}$, such that:

(i) for each simplex $\sigma \in \Sigma_S$, the update $u = (u^-, u^+)$, with $\sigma \in u^-$, is such that $\tau(u) = \textbf{false}$.
(ii) S is the consistent set with the minimum cardinality that satisfies the previous condition.

If function τ is such that the set of nodes $\{u | \tau(u) = \textbf{true}\}$ is connected, then the result of the above query is unique.

Algorithms for solving such query on a SMC $\mathcal{M} = (U, A)$ perform a traversal of the DAG describing $\mathcal{M}$ to find set S. The basic algorithm, that we call *static*, is based on *node expansion*. It starts with a set S containing just the root, and with Σ_S equal to base complex. Then, it adds nodes to S as long as condition (i) is not satisfied. Adding a node u to S locally refines Σ_S by applying update u^+, and may cause recursive insertion of ancestors of u into S. A *dynamic* variant of the algorithm starts each time from the consistent set S generated by a previous query, and performs *node contraction* prior to performing node expansion as in the static case: in this phase, all those nodes that are not necessary to satisfy condition (i) are eliminated from S.

Details on algorithms are given in [ ]. For the purpose of this paper, it is sufficient to notice that primitive operations needed to support such algorithms are concerned with traversal of dependency links in the SMC, and maintenance of a current status, i.e., a consistent set S of SMC nodes together with its associated complex Σ_S.

DAG traversal primitives retrieve the set of direct ancestors and direct descendants of a node $u \in U$:

– operation $\textbf{pred}(u)$ returns the set $\{u_i | <u_i, u> \in A\}$.
– operation $\textbf{succ}(u)$ returns the set $\{u_i | <u, u_i> \in A\}$.

Operations $\textbf{pred}$ and $\textbf{succ}$ are used to test the feasibility of an update $u \in U$ in the current status (S, Σ_S). If $u \notin S$, adding it to S produces a consistent set (i.e., update u^+ can be applied to Σ_S) if and only if $\textbf{pred}(u) \subset S$. If $u \in S$, removing it from S produces a consistent set (i.e., update u^- can be applied to Σ_S) if and only if $\textbf{succ}(u) \cap S = \emptyset$.

Finally, two operations modify the current status (S, Σ_S) by performing a feasible update u^- or u^+:

– operation $\textbf{refinement}(u, S, \Sigma_S)$ returns a new status $(S', \Sigma_{S'})$, where $S' = S \cup \{u\}$ and $\Sigma_{S'} = \Sigma_S \setminus \{u^-\} \cup \{u^+\}$.

– operation **abstraction**(u, S, Σ_S) returns a new status $(S', \Sigma_{S'})$, where $S' = S \setminus \{u\}$ and $\Sigma_{S'} = \Sigma_S \setminus \{u^+\} \cup \{u^-\}$.

It is important to implement the primitive operations with an optimal time complexity, i.e., the cost of **pred** and **succ** must be linear in the number of output nodes, and that of **refinement** and **abstraction** must be linear in $\#(u^- \cup u^+)$.

4.1 Encoding the Current Complex

In this Subsection, we introduce a data structure for simplicial complexes that is used to encode the current mesh Σ_S during selective refinement as well as the final result of selective refinement algorithms.

The simplest way to encode a simplicial complex Σ is through an *indexed data structure* which stores, for each k-simplex σ, references to the $k+1$ vertices of σ, while each vertex is stored as a tuple of d coordinates. A variation is the indexed data structure *with adjacencies*, that stores, in addition, the $k+1$ simplices adjacent to each simplex σ along a $(k-1)$-dimensional facet.

If complex Σ has $\hat{n}$ vertices and $\hat{s}$ simplices, then the space required for encoding it into this data structure is equal to $d\hat{n}$ bits for the vertex coordinates, while connectivity information requires $\hat{s}(k+1)\log\hat{n}$ bits (in the variant without adjacencies) and $\hat{s}(k+1)(\log\hat{n} + \log\hat{s})$ bits (in the variant with adjacencies). We recall that $\hat{s} \simeq 2\hat{n}$ in two dimensions $(k=2)$, and $\hat{s} \simeq 6\hat{n}$ in three dimensions $(k=3)$. It follows that connectivity information requires $6\hat{n}\log\hat{n}$ bits ($12\hat{n}\log\hat{n} + 6\hat{n}$ with adjacencies) in the two-dimensional case. In three dimensions, $24\hat{n}\log\hat{n}$ bits ($48\hat{n}\log\hat{n} + 72\hat{n}$ with adjacencies) are required.

5 A Data Structure for General Updates

Representing a SMC $\mathcal{M} = (U, A)$ requires encoding two kinds of information: the *updates* associated with DAG nodes, and *dependency links* represented by DAG arcs. In this Section, we describe a general-purpose SMC data structure suitable to represent *any* SMC. Such structure is called *explicit* because it represents an update u by explicitly listing the simplices belonging to sets u^- and u^+.

5.1 Encoding Dependencies

The dependency relation is encoded by associating with each node u the direct ancestors and the direct descendants of u, i.e., the two sets of nodes corresponding to the result of primitives **pred**(u) and **succ**(u). For each such set, references to all the updates in the set are stored as well as the cardinality of the set.

Summing up over all nodes $u \in U$, it results in $2m\log P$ bits for encoding the cardinalities of the two sets **pred**(u) and **succ**(u), and $2a\log m$ bits for referencing all the nodes in such sets. For a vertex-based SMC, the above cost reduces to:

– $8\bar{n} + 6\bar{n}\log\bar{n}$ in two dimensions ($P = 16, a \approx 3m$), and

– $10\bar{n} + 12\bar{n} \log \bar{n}$ in three dimensions ($P = 32, a \approx 6m$).

The implementation of operations **pred** and **succ** is straightforward, and their time complexity is linear in the number of the output nodes.

Klein and Gumhold [ ] propose a DAG encoding with a reduced memory cost. For each node u, they define a cyclic linked list, called a *loop*, which contains update u followed by all its direct ancestors (i.e., the loop of u contains the set of nodes defined by $\textbf{pred}(u) \cup \{u\}$). Figure 2 shows a DAG and the loops needed to encode it.

A node u appears in its own loop and in all the loops defined by its direct descendants. For each of such loops, u stores a forward pointer (implementing the linked list) plus an integer, the loop identifier, which is used to distinguish the loop each pointer belongs to. The number of loops (i.e., $\#\textbf{succ}(u) + 1$) is also stored at u.

The total number of links to describe the arcs of a SMC is thus $a + m$. Since each node has at most P direct ancestors or descendants, we need $\log P$ bits for a loop identifier and to count the loops to which a node belongs; a pointer to a node requires $\log m$ bits.

The number of bits to encode the whole DAG is $m \log P + (\log P + \log m)(a + m) = m(2 \log P + \log m) + a(\log P + \log m)$. In the special case of a vertex-based SMC, we have:

– $20\bar{n} + 4\bar{n} \log \bar{n}$ bits in two dimensions, and
– $40\bar{n} + 7\bar{n} \log \bar{n}$ bits in three dimensions.

This way of encoding the DAG supports operations **pred** with an optimal cost. On the contrary, implementing $\textbf{succ}(u)$ needs examining all the loops traversing node u, which leads to a sub-optimal time complexity in the worst case. However, such complexity is amortized in the selective refinement algorithm; therefore this data structure exhibits an efficient expected behavior even in this case.

5.2   Encoding Updates

In the general case, each node u is represented as two sets of k-simplices u^- and u^+, where simplices are given as tuples of vertices. Vertices, simplices, and nodes

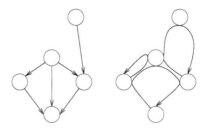

Fig. 2. A DAG (on the left) and the loops needed to encode it (on the right)

are stored in arrays. The simplices introduced by a given update are encoded in consecutive positions in the simplex array. If two nodes u_i and u_j are stored at consecutive positions in the node array, then the simplices of u_i^+ are stored next to the simplices of u_j^+ in the simplex array.

For each vertex, we store its d coordinates. For each simplex, we store references to its $k + 1$ vertices. For each node u, we store the number of simplices in u^- plus references to all simplices in u^-, and a reference to the first simplex in u^+. Given a node u, all the simplices of set u^+ are retrieved by scanning the simplex array starting at the first simplex of u^+ and stopping at the first simplex of the update following u in the node array. This requires a linear time in the cardinality of u^+.

Encoding vertices requires $32dn$ bits. The overall size of the simplex array is $(k + 1)s \log n$ bits. Summing up over all nodes $u \in U$, it results in $m \log P$ bits for encoding the number of simplices in u^-; $s \log s$ bits for storing all sets u^-; and $m \log s$ bits for referencing the first simplex in u^+.

Disregarding the cost of purely geometric information (i.e., vertex coordinates), we concentrate on the space needed to represent updates and the simplices involved in them.

In the two-dimensional case, this is equal to $4m + (s + m) \log s + 3s \log m$ bits, which, for vertex-based SMCs, is bounded from above by $25\bar{n} + 25\bar{n} \log \bar{n}$.

In the three-dimensions, we have $5m + (s + m) \log s + 4s \log m$ bits, bounded by $100\bar{n} + 91\bar{n} \log \bar{n}$ in the case of a vertex-based SMC.

Assuming that the current complex Σ_S is encoded in an indexed data structure *without* adjacencies, operation **refinement** (resp., **abstraction**) can be performed on Σ_S by simply deleting the simplices of u^- (resp., u^+), and replacing them with the simplices of u^+ (resp., u^-). This can be done in time linear in $\#(u^- \cup u^+)$, i.e., optimal. If needed, we can reconstruct all adjacency relations of the output complex in a post processing stage, in computational time $O(N \log N)$, where N is the number of $(k - 1)$-facets in the output complex.

Table 3 summarizes the memory requirements of an explicit encoding of a vertex-based SMC in two and in three dimensions.

Table 3. Summary of memory requirements for explicitly encoding a vertex-based SMC

dimension	standard dependencies	loop-based dependencies	updates
2D	$8\bar{n} + 6\bar{n} \log \bar{n}$	$20\bar{n} + 4\bar{n} \log \bar{n}$	$25\bar{n} + 25\bar{n} \log \bar{n}$
3D	$10\bar{n} + 12\bar{n} \log \bar{n}$	$40\bar{n} + 7\bar{n} \log \bar{n}$	$100\bar{n} + 91\bar{n} \log \bar{n}$

6   Compact Encoding of Vertex-Based Updates

In this Section, we present compact data structures for representing updates in the class of *vertex-based* SMCs in two and three dimensions. Such structures exploit the specific properties of such type of updates in order to reduce space complexity. The idea is to encode an implicit, procedural, description of an update, which contains sufficient information to perform operations **refinement** and **abstraction** on the current complex.

If u is a vertex insertion/removal, operation **refinement**(u, S, Σ_S) reduces to locating a k-dimensional polytope π_u on the current complex Σ_S, which bounds the set of simplices of u^-: all simplices internal to π_u must be deleted. The simplices of u^+ are created by simply joining each face of π_u to vertex v_u. The converse operation **abstraction**(u, S, Σ_S) consists of deleting all simplices incident in v_u from Σ_S (i.e., simplices that form u^+), and re-tesselating the resulting polytope π_u, without using any internal point. In general, π_u admits more than one tesselation: we must recreate the one corresponding to u^-.

A family of encoding structures for vertex-based updates can be developed that do not store all the simplices in u^- and u^+ (as in the explicit data structure), but just those information, that are sufficient to perform the following two tasks:

1. Given the current complex Σ_S and an update u, such that u^+ is feasible, recognize the simplices forming u^- among those of Σ_S.
2. Given the polytope π_u bounding u^-, build u^-.

In all data structures we present, Tasks 1 and 2 need a boundary $(k-1)$-dimensional facet γ_u of π_u as a starting point. When either update u^+ or u^- is to be performed on the current complex Σ_S, we need to identify facet γ_u on Σ_S. The way to identify it is different in case we are going to perform u^+ (i.e., $u^- \subseteq \Sigma_S$) or u^- (i.e., $u^+ \subseteq \Sigma_S$). In both cases, we rely on labelling all simplices of the current mesh u^+ with integer numbers.

In the case of a dynamic algorithm for selective refinement, a simplex $\sigma \in \Sigma_S$ can have been generated either during expansion or during contraction. Simplex σ is labelled with one bit to discriminate between the two cases, and with an integer on $\log P$ bits which uniquely identifies σ among the simplices that have been introduced in Σ_S together with σ. This label is generated in the following way. When we perform an update u^+ on Σ_S, we label the new simplices (i.e., those of u^+) in an (arbitrary) order which is always the same every time the update is performed, and such that simplex σ_{u^+} containing γ_u is the first one. When we perform an update u^-, we label the new simplices (i.e., those of u^-) in an (arbitrary) order which is always the same every time the update is performed, and such that simplex σ_{u^-} which contains facet γ_u is the first one. Figure 3 shows a two-dimensional example of simplex labelling in the two cases. In a static algorithm, we have a simplified case since contraction is never performed and, thus, all simplices are generated during expansion.

All structures that we present share a common framework that is described below. They all store the vertices of the SMC, each as a tuple of coordinates, with a cost of $d\bar{n}$ bits. Since each update u corresponds to a vertex v_u, updates

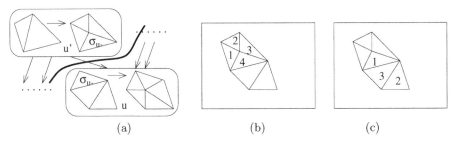

Fig. 3. (a) The current state of a selective refinement algorithm seen on the MT: the thick line separates nodes in S from nodes not in S, denoting that $u' \in S$ and $u \notin S$. (b) The current complex in case σ_{u^-} is generated by executing u'^+. (c) The current complex in case σ_{u^-} is generated by u^-

and vertices are re-numbered in such a way that a node u and its corresponding vertex v_u have the same label. Thus, the relation between u^+ and v_u is encoded at a null cost.

Moreover, for each update u, we encode information needed to retrieve γ_u when we are going to perform u^+ (i.e, when $u^- \subseteq \Sigma_S$):

- an index on $\log P$ bits that identifies the simplex $\sigma_{u^-} \in u^-$ such that γ_u is a facet of σ_{u^-};
- an index on $\log(k+1)$ bits that identifies γ_u among the $k+1$ $(k-1)$-facets of σ_{u^-};

and information to retrieve γ_u when we are going to perform u^- (i.e, when $u^+ \subseteq \Sigma_S$:

- an index on $\log P$ bits that identifies the simplex $\sigma_{u^+} \in u^+$ such that γ_u is a facet of σ_{u^+};

Note that we do not need an index to identify γ_u among the $(k-1)$-facets of σ_{u^+}, since γ_u is automatically identified as the facet of σ_{u^+} opposite to v_u.

Encoding such information requires $2 \log P + 2$ bits per update, i.e., $10\bar{n}$ bits in two dimensions and $12\bar{n}$ bits in three dimensions.

In order to retrieve σ_{u^-} from the index stored in u, we need to know the update u', among the direct ancestors of u, such that $\sigma_{u^-} \in u'^+$. Such update u' is encoded at no cost by adopting the convention that, in DAG encoding, **pred**(u) contains u' in the first position.

The index stored with an update u to identify σ_{u^-} is the same as the number that σ_{u^-} receives when σ_{u^-} is created in the current complex by performing update u'^+ (in Fig. 3b, this index is 4). This information, together with the node u' such that $\sigma_{u^-} \in u'^+$, enables us to retrieve σ_{u^-} on the current mesh, in case σ_{u^-} is in Σ_S because it has been created by performing u'^+. In case σ_{u^-} is in Σ_S because it has been created by performing u^-, then σ_{u^-} is simply retrieved as the simplex having the first label among the ones generated by u^-.

In order to retrieve σ_{u+} from the index stored in u, we need to know the update u'', among the direct descendants of u, such that $\sigma_{u+} \in u''^{-}$. Such update u'' is encoded at no cost by adopting the convention that, in DAG encoding, $\mathbf{succ}(u)$ contains u'' in the first position. If the compact DAG encoding is adopted, then the loop corresponding to u'' is the first one among the loops traversing u, after the loop of u itself.

The index stored with u to identify simplex σ_{u+} is the same as the number that σ_{u+} receives when σ_{u+} is created in the current complex by performing update u''^{-}. This information, together with the node u'' such that $\sigma_{u+} \in u''^{-}$, enables us to retrieve σ_{u+} on the current complex Σ_S, in the case in which σ_{u+} has been created by performing u''^{-}. If σ_{u+} is in the current complex Σ_S because it has been created by performing u^{+}, then σ_{u+} is simply retrieved as the simplex having the first label among the ones generated by u^{+}.

In order to perform operations **refinement** and **abstraction**, the current complex Σ_S needs to be encoded into an indexed structure *with* adjacencies (see Section 4.1).

The remaining information stored at a node u, i.e., the encoding of u^{-}, are described in the following subsections for the two- and for the three-dimensional case, respectively.

6.1   2D Vertex-Based Updates

To encode the triangulation u^{-} of a node u, a stream of two bits for each triangle can be used, following an approach proposed by Taubin et al. [ ] in the context of progressive mesh compression.

The construction of the bit stream is performed through a recursive procedure that encodes one triangle $\sigma \in u^{-}$ at a time. At each recursive call, the algorithm maintains a current triangulation T and a current edge e on its boundary, and generates the code for a triangle σ bounded by e. At the beginning, the current triangulation T is initialized as u^{-} and the current edge e as γ_u. Triangle σ is the triangle of u^{-} lying on the left side of e. The algorithm looks at the two edges of triangle σ different from e and at the two parts in which such edges partition the current triangulation in order to assign a code to σ and to decide about recursive calls to be made:

- If both parts are non-empty, then code 00 is added to the bit stream.
- If only the left part is non-empty, then code 01 is added.
- If only the right part is non-empty, then code 10 is added.
- If both parts are empty, then code 11 is added.

Then, a recursive call is performed on each of the non-empty parts, starting from the boundary edge it shares with triangle σ. On each branch, the recursion stops when a code 11 is generated. Figure 4a shows the codes generated for a sample mesh u^{-}.

Encoding an update u, such that u^{-} contains h triangles, requires $2h$ bits. The value of h does not need to be encoded since the structure of the bit stream

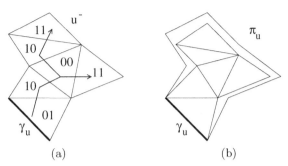

Fig. 4. (a) Codes assigned by Taubin's method to the triangles of u^-; arrows correspond to the traversal of u^-. (b) Topological reconstruction of u^-

(basically, a preorder visit of the recursion tree, where leaves are characterized by code 11) can be exploited to reconstruct its length. The total length af all bit streams is equal to $2 \sum_{u \in U} \#u^- = 2(s - \bar{s})$, which is about $8\bar{n}$ in our case (two-dimensional, vertex-based SMCs).

In order to perform task 1 in operation **refinement**, we run the same recursive procedure described above, starting from edge γ_u on the current mesh Σ_S as our working edge e. In this case, the bit stream is known, and we need to recover u^-. At each call, we consider the current working edge e, and recognize the triangle σ, adjacent to e, as belonging to u^-. Then, depending on the code found in the stream, we decide, for each edge of σ, different from e, if a recursive call must be invoked on that edge. The recursive calls progressively "consume" the bit stream, so that each call always reads the next code which has not been read by any call executed before it (either at a deeper or at an outer recursion level).

Task 2 in operation **abstraction** is performed in a very similar way. In this case, at each step, we generate a triangle σ, instead of recognizing it on the current mesh. Note that this procedure allows reconstructing just the *topology* of triangle mesh u^-: when we create a triangle σ, we do not know which vertex of π_u must become the third vertex of σ. However, when the topology of u^- has been built, it is easy to map the vertices of π_u onto it (see Fig. 4b). This simply reduces to a synchronous traversal of π_u and of the boundary of the triangulation, starting at one of the two vertices of γ_u (for which the mapping is known).

For each node u, the computational cost of performing either task 1 or task 2, with the methods described above, is linear in the number of triangles of u^-.

6.2   3D Vertex-Based Updates

In the three-dimensional case, tetrahedralization u^- is described as a *tetrahedron spanning tree* rooted at σ_{u^-}, which is encoded as a bit stream. The tetrahedron spanning tree is constructed as follows. We start from $\sigma_{u^-} \in u^-$, and traverse all

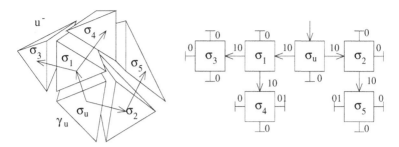

Fig. 5. The tetrahedron spanning tree

tetrahedra of u^- in a breadth-first fashion. Each triangular facet of a traversed tetrahedron is labeled in the following way:

- with 0, if the facet is a face of polyhedron π_u;
- with 10, if the facet is adjacent to a tetrahedron that belongs to u^- and has not yet been traversed;
- with 11, if the facet is adjacent to a tetrahedron that belongs to u^- and has already been traversed.

Three facets are labelled for each tetrahedron. This is also true for σ_{u^-}, since we know that one of its facets (i.e., γ_u) is on π_u. An example of a tetrahedron spanning tree with facet labelling is shown in Fig. 5.

If u^- contains h tetrahedra and has q external facets, then the stream contains $q - 1 + 2(3h - (q-1)) = 6h - q + 1$ bits . Note that the number q of external facets is equal to the number of tetrahedra in u^+. The bitstream length does not need to be stored since the structure of the bit stream can be exploited to reconstruct its length.

By summing up over all nodes, we have $6\sum_{u \in U} \#u^- - \sum_{u \in U} \#u^+ + m = 6(s - \bar{s}) - s + m$ bits, which, in the case of three-dimensional vertex-based SMCs, reduces to $6(18\bar{n} - 6\bar{n}) - 18\bar{n} + \bar{n} = 55\bar{n}$ bits.

In order to perform Task 1 in the **refinement** operation, we start from σ_{u^-}, and use the bit stream as a mask to retrieve all tetrahedra of u^- by traversing them in the same sequence as they have been traversed for creating the tetrahedron spanning tree.

In order to perform Task 2 in the **abstraction** operation, we start from γ_u and scan the bit stream to recreate the tetrahedra of the spanning tree one at a time. Each time we traverse a facet marked 0, we glue it to a face of π_u; each time we traverse a facet marked 10, we create a new tetrahedron of u^-; each time we traverse a facet marked 11, we glue two tetrahedra of u^- at that face. The other face to be glued (in cases 0 and 11) is uniquely determined because the tetrahedralization we are creating has no internal vertices.

7 Comparisons

For the purpose of comparison, we consider the space required to encode the reference complex in an indexed data structure with adjacencies (see Section 4.1). The space required by the *reference complex* with such structure is equal to $d\bar{n}$ bits for the vertex coordinates, while connectivity data require $6\bar{n} + 12\bar{n}\log\bar{n}$ bits in two dimensions, and $72\bar{n} + 48\bar{n}\log\bar{n}$ bits in three dimensions.

The storage cost for vertex coordinates is the same, i.e, $d\bar{n}$ bits, also in all SMC data structures. Encoding the partial order (i.e., the DAG) of a SMC in the standard way requires $8\bar{n} + 6\bar{n}\log\bar{n}$ bits in two dimensions, and $10\bar{n} + 12\bar{n}\log\bar{n}$ bits in three dimensions. With compact DAG encoding based on loops, this cost becomes $20\bar{n} + 4\bar{n}\log\bar{n}$ bits in two dimensions, and $40\bar{n} + 7\bar{n}\log\bar{n}$ bits in three dimensions.

The amount of space needed to encode the updates through the *explicit data structure* is equal to $25\bar{n} + 25\bar{n}\log\bar{n}$ in two dimensions, and $100\bar{n} + 91\bar{n}\log\bar{n}$ in three dimensions. Adopting the compact encoding of updates described in Section 6 reduces the cost of encoding all updates to $14\bar{n}$ and $62\bar{n}$ in the two and in the three dimensional case, respectively.

In Tables 4 and 5, we compare the storage requirements of the reference complex, of the explicit data structure, and of the data structure using compact update encodings, in two and three dimensions, respectively; for the two-dimensional case (see Table 4), we also show the cost of the data structure proposed by Klein and Gumhold in [ ]. We have considered vertex-based SMCs with 10K, 100K, 500K, 1M and 2M vertices. The costs are expressed as number of bits per vertex; geometric information (i.e., vertex coordinates) are equal in all structures and, therefore, are not considered in the tables. The number of bits shown for the data structure of Klein and Gumhold does not include the representation of equivalence classes of triangulations. In [ ], the authors report that, for 100K vertices, 120 bits per vertex are necessary, which allows thinking that, in general, the total cost of the structure is about twice the number of bits reported in the last column of Table 4.

Table 4. Comparisons of data structures in the two-dimensional case. We show the sizes of the reference complex, of explicit SMC data structure with both variants of DAG encoding (standard or compact), of the data structure for SMCs with compression of updates, with both variants of DAG encoding, and of the data structure in [ ]

$\bar{n}$	refer. complex	expl. SMC std. DAG	expl. SMC comp. DAG	comp. SMC std. DAG	comp. SMC comp. DAG	[ ]
10K	174	467	451	110	94	117
100K	210	560	538	128	106	135
500K	234	622	596	140	114	147
1M	246	653	625	146	118	153
2M	258	684	654	152	122	159

Table 5. Comparisons of data structures in three dimensions. We show the sizes of the reference complex, of explicit SMC data structure, with both variants of DAG encoding, of the data structure for SMCs with compression of updates, with both variants of DAG encoding

$\bar{n}$	refer. complex	expl. SMC std. DAG	expl. SMC comp. DAG	comp. SMC std. DAG	comp. SMC comp. DAG
10K	744	1552	1512	245	205
100K	888	1861	1806	281	226
500K	984	2067	2002	305	240
1M	1032	2170	2100	317	247
2M	1080	2273	2198	329	254

We compare the data structure adopting a compact representation of updates against an explicit one, when the DAG is encoded in the same way. We have that, in two dimensions, the compact version saves about $77\% - 78\%$ with respect to the explicit one (with standard DAG encoding), and $80\% - 82\%$ (with compact DAG encoding). In three dimensions, we have a $85\% - 86\%$ saving with standard DAG encoding, and a $87\% - 89\%$ saving with compact DAG encoding.

Now, we compare the two structures using compact update encoding with respect to the reference complex. In two dimensions, such structures save about $37\% - 42\%$ (with a standard DAG representation), and about $46\% - 53\%$ (with a compressed DAG); the memory reduction increases for larger values of $\bar{n}$. In three dimensions, the above figures are $68\% - 70\%$ (with standard DAG) and $73\% - 77\%$ (with compressed DAG).

In two dimensions, the use of a compact DAG representation gives a storage compression of about 5% (with explicit update encoding) and $15\% - 20\%$ (with implicit update encoding), where the best reduction is achieved for larger values of $\bar{n}$. In three dimensions, the compression rate is about 4% (with explicit update encoding) and $18\% - 23\%$ (with implicit update encoding).

An explicit encoding of updates allows a lighter working data structure for the current mesh during selective refinement, since maintaining adjacencies is not needed. But if adjacencies are required on the output complex, then there is an overhead to reconstruct them as a post-processing step. Implicit update encodings need to maintain adjacencies during selective refinement, but they produce an output complex with adjacencies at no additional cost.

8   Concluding Remarks

In this paper, we have considered a specific class of multiresolution simplicial complexes, built based on a vertex-removal/insertion simplification strategy in two and three dimensions. We have described compact encoding structures for such SMCs. SMCs built through vertex removal from a mesh at the maximum resolution, or through vertex insertion from a mesh at a coarse resolution are extensively used in applications for modelling surfaces and two- and three-dimensional fields.

The resulting data structures are particularly interesting since we have shown that they achieve a compression of almost 80% (in two dimensions) and 87% (in three dimensions) with respect to a general-purpose structure for encoding the SMC and a compression of almost 50% (in two dimensions) and 70% (in three dimensions) with respect to encoding the mesh at the maximum resolution.

It would be interesting to compare, in terms of performances, a vertex-based SMC described as a compressed data structure with hierarchical data structures based on regular recursive subdivisions even if these latter have a more restricted applicability (they can be used only with regularly distributed data points and not for free-form surfaces).

Further developments of the research described in this paper are investigating issues related to the study of more compact data structures for storing the DAG, the design of a vertex-based update encoding in the arbitrary k-dimensional case, and the development of out-of-core data structures for a vertex-based a SMC as well as out-of-core algorithms for selective refinement.

Acknowledgement

This work has been partially supported by the Research Training Network EC Project on *Multiresolution in Geometric Modelling* (MINGLE), under contract HPRN-CT-1999-00117, by the EC Project *Augmented Reality for Remotely Operated Vehicles based on 3D acoustical and optical sensors for underwater inspection and survey* (ARROV) under contract G3RD-CT-2000-00285, and by the project funded by the Italian Ministry of University and Scientific and Technological Research (MURST) on *Representation and Processing of Spatial Data in Geographic Information Systems*.

References

1. P. Cignoni, D. Costanza, C. Montani, C. Rocchini, and R. Scopigno. Simplification of tetrahedral volume with accurate error evaluation. In *Proceedings IEEE Visualization'00*, pages 85–92. IEEE Press, 2000. 131, 132

2. P. Cignoni, L. De Floriani, C. Montani, E. Puppo, and R. Scopigno. Multiresolution modeling and rendering of volume data based on simplicial complexes. In *Proceedings 1994 Symposium on Volume Visualization*, pages 19–26. ACM Press, October 17-18 1994. 131

3. P. Cignoni, L. De Floriani, P. Magillo, E. Puppo, and R. Scopigno. *TAn2* - visualization of large irregular volume datasets. Technical Report DISI-TR-00-07, Department of Computer and Information Science, University of Genova (Italy), 2000. (submitted for publication). 134

4. L. De Floriani, P. Magillo, and E. Puppo. Efficient implementation of multi-triangulations. In *Proceedings IEEE Visualization 98*, pages 43–50, Research Triangle Park, NC (USA), October 1998. 130, 131, 133, 136

5. L. De Floriani, P. Magillo, and E. Puppo. Multiresolution representation of shapes based on cell complexes. In L. Perroton G. Bertrand, M. Couprie, editor, *Discrete Geometry for Computer Imagery, Lecture Notes in Computer Science*, volume 1568, pages 3–18. Springer Verlag, New York, 1999. 129, 131, 132, 134

6. M. Duchaineau, M. Wolinsky, D. E. Sigeti, M. C. Miller, C. Aldrich, and M. B. Mineed-Weinstein. ROAMing terrain: Real-time optimally adapting meshes. In *Proceedings IEEE Visualization'97*, pages 81–88, 1997. 131

7. W. Evans, D. Kirkpatrick, and G. Townsend. Right triangular irregular networks. Technical Report 97-09, University of Arizona, May 1997. Algorithmica, 2001, to appear. 131

8. J. El-Sana and A. Varshney. Generalized view-dependent simplification. *Computer Graphics Forum*, 18(3):C83–C94, 1999. 131, 134

9. M. Garland. Multiresolution modeling: Survey & future opportunities. In *Eurographics '99 – State of the Art Reports*, pages 111–131, 1999. 130, 131, 132

10. R. Grosso and G. Greiner. Hierarchical meshes for volume data. In *Proceedings of the Conference on Computer Graphics International 1998 (CGI-98)*, pages 761–771, Los Alamitos, California, June 22-26 1998. IEEE Computer Society. 131

11. S. Gumhold, S. Guthe, and W. Straßer. Tetrahedral mesh compression with the cut-border machine. In *Proceedings IEEE Visualization'99*, pages 51–58. IEEE, 1999. 134

12. M. H. Gross and O. G. Staadt. Progressive tetrahedralizations. In *Proceedings IEEE Visualization'98*, pages 397–402, Research Triangle Park, NC, 1998. IEEE Comp. Soc. Press. 131, 132

13. H. Hoppe. View-dependent refinement of progressive meshes. In *ACM Computer Graphics Proceedings, Annual Conference Series, (SIGGRAPH '97)*, pages 189–198, 1997. 131, 134

14. R. Klein and S. Gumhold. Data compression of multiresolution surfaces. In *Visualization in Scientific Computing '98*, pages 13–24. Springer-Verlag, 1998. 134, 138, 145

15. M. Lee, L. De Floriani, M., and H. Samet. Constant-time neighbor finding in hierarchical meshes. In *Proceedings International Conference on Shape Modeling*, Genova (Italy), May 7-11 2001. *in print*. 131

16. D. Luebke and C. Erikson. View-dependent simplification of arbitrary polygonal environments. In *ACM Computer Graphics Proceedings, Annual Conference Series, (SIGGRAPH '97)*, pages 199–207, 1997. 131, 134

17. P. Lindstrom, D. Koller, W. Ribarsky, L. F. Hodges, N. Faust, and G. A. Turner. Real-time, continuous level of detail rendering of height fields. In *Computer Graphics Proceedings, Annual Conference Series (SIGGRAPH '96), ACM Press*, pages 109–118, New Orleans, LA, USA, Aug. 6-8 1996. 131

18. M. Ohlberger and M. Rumpf. Adaptive projection operators in multiresolution scientific visualization. *IEEE Transactions on Visualization and Computer Graphics*, 5(1):74–93, 1999. 131

19. J. Popovic and H. Hoppe. Progressive simplicial complexes. In *ACM Computer Graphics Proceedings, Annual Conference Series, (SIGGRAPH '97)*, pages 217–224, 1997. 131, 134

20. K. J. Renze and J. H. Oliver. Generalized unstructured decimation. *IEEE Computational Geometry & Applications*, 16(6):24–32, 1996. 131

21. J. R. Shewchuck. Tetrahedral mesh generation by delaunay refinement. In *Proceedings 14th Annual Symposium on Computational Geometry*, pages 86–95, Minneapolis, Minnesota, June 1998. ACM Press. 131, 132

22. G. Taubin, A. Guéziec, W. Horn, and F. Lazarus. Progressive forest split compression. In *Computer Graphics (SIGGRAPH '98 Proceedings)*, pages 123–132. ACM Press, 1998. 142

23. Y. Zhou, B. Chen, and A. Kaufman. Multiresolution tetrahedral framework for visualizing regular volume data. In *Proceedings IEEE Visualization'97*, pages 135–142. IEEE Press, 1997. 131

Discrete Polyhedrization of a Lattice Point Set

Yukiko Kenmochi[1][*] and Atsushi Imiya[2]

[1] Laboratoire A2SI, ESIEE
Cité Descartes, B. P. 99, 93162 Noisy-Le-Grand Cedex, France
[2] Institute of Media and Information Technology, Chiba University
1-33 Yayoi-cho Inage-ku 263-8522 Chiba, Japan

Abstract. We introduce a polyhedral representation of surfaces for analysis and recognition of three-dimensional digital images. Our representation is based on combinatorial topology. By using a discrete version of combinatorial topology we also present an algorithm for reconstruction of a polyhedron in a discrete space from a set of lattice points.

1 Introduction

In this paper, given a 3-dimensional digital image, we present a method to reconstruct a polyhedron for a 3-dimensional object in the image. Such polyhedrization problem is one of the fundamental problems in the fields of digital image analysis, recognition and visualization, and has been called the border tracking problem [, ,]; sometimes the words of "border" and "tracking" are replaced by other words such as "boundary" or "contour" and "following", respectively.

We state our polyhedrization problem more precisely and mathematically. Let us consider a 3-dimensional discrete space such as the set $\mathbf{Z}^3$ of all lattice points whose coordinates are all integers in the 3-dimensional Euclidean space $\mathbf{R}^3$ as the mathematical space model for a 3-dimensional digital image. Given a subset $\mathbf{V}$ of $\mathbf{Z}^3$ for a digitized 3-dimensional object, we consider construction of a polyhedron such that

(A) the vertices of the polyhedron are all lattice points which are the border points of $\mathbf{V}$ in the sense of general topology/set theory [ ];
(B) any adjacent vertices of the polyhedron are neighboring in the sense of the traditional 6-, 18- or 26-neighborhood system for digital image analysis [ ];
(C) it has the topological structures of two-dimensional surfaces in the sense of combinatorial and algebraic topology [,].

We call such a polyhedron a discrete polyhedron hereafter.

In the following sections, we first give the classical set-theoretical definition of border points of $\mathbf{V} \subset \mathbf{Z}^3$ [, ,] and see that the topological surface structures may be useful for border tracking in $\mathbf{Z}^3$. Among the various approaches for defining discrete surfaces in $\mathbf{Z}^3$, we then choose an approach based on combinatorial

---

[*] The first author has been on leave from School of Information Science, JAIST, Japan, thanks to the JSPS postdoctoral fellowships for research abroad from October 2000.

G. Bertrand et al. (Eds.): Digital and Image Geometry, LNCS 2243, pp. 150–162, 2001.
© Springer-Verlag Berlin Heidelberg 2001

and algebraic topology [,]. To construct a discrete polyhedron satisfying the conditions of (A), (B) and (C) in $\mathbf{Z}^3$, we need to build up the discrete version of combinatorial and algebraic topology. We define discrete cells of the dimension from 0 to 3 under the conditions (A) and (B) and propose an interpretation of a digitized 3-dimensional object $\mathbf{V}$ as a cellular complex consisting of discrete cells. The same concept for the image interpretation by cellular complexes is already introduced by several authors such as [,]. In [ ], for instance, only the 6-neighborhood system is considered while we consider either of 6-, 18- or 26-neighborhood system. Thus, we say that our framework is more general in the sense of neighborhood systems. In [ ], 3-dimensional discrete cells which are called "continuous analogs" are defined in a different way from ours for the different purpose which is the digital fundamental group.

For practical use, we also present an effective polyhedrization algorithm to generate a discrete polyhedron directly from a given subset $\mathbf{V}$ of $\mathbf{Z}^3$ similarly to the marching cubes algorithm [ ]. We have presented a similar algorithm in our previous work [ ], but the theory is limited only for the cases of 6- and 26-neighborhood systems. By using cellular complexes instead of simplicial complexes for a 3-dimensional image interpretation, we succeed to extend our work for all 6-, 18- and 26-neighborhood systems. Finally, we clarify the the relation between a set of boder points based on set theory and our discrete polyhedron.

2   Border Points and Surfaces in $\mathbf{Z}^3$

2.1   Border Points in $\mathbf{Z}^3$

Similarly to the definition of border points in $\mathbf{R}^3$ [ ], a set of border points of a finite subset $\mathbf{V}$ in $\mathbf{Z}^3$ is defined by using a neighborhood system. We have the m-neighborhood system in $\mathbf{Z}^3$ such as

$$\mathbf{N}_m(\boldsymbol{x}) = \{\boldsymbol{y} \in \mathbf{Z}^3 : \|\boldsymbol{x} - \boldsymbol{y}\| \leq \sqrt{t}\}, \tag{1}$$

where $t = 1, 2, 3$ for each $m = 6, 18, 26$. If $\mathbf{N}_m(\boldsymbol{x}) \subseteq \mathbf{V}$, then such a point $\boldsymbol{x}$ is called an interior point of $\mathbf{V}$ [,]. The set of interior points of $\mathbf{V}$ is called the interior of $\mathbf{V}$ and denoted by

$$Int_m(\mathbf{V}) = \{\boldsymbol{x} \in \mathbf{V} : \mathbf{N}_m(\boldsymbol{x}) \subseteq \mathbf{V}\}.$$

Clearly, $Int_m(\mathbf{V})$ depends on the value of m for $m = 6, 18, 26$. If a point $\boldsymbol{x} \in \mathbf{V}$ is not an interior point of $\mathbf{V}$, then $\boldsymbol{x}$ is called a border point of $\mathbf{V}$ [, ,] and the set of all border points is denoted by

$$Br_m(\mathbf{V}) = \mathbf{V} \setminus Int_m(\mathbf{V}), = \{\boldsymbol{x} \in \mathbf{V} : \mathbf{N}_m(\boldsymbol{x}) \cap \bar{\mathbf{V}} \neq \emptyset\}. \tag{2}$$

Let us say that $Br_m(\mathbf{V})$ is connected or m'-connected if any pair of $\boldsymbol{x}, \boldsymbol{y} \in Br_m(\mathbf{V})$ has a point sequence $\boldsymbol{x}_1 = \boldsymbol{x}, \boldsymbol{x}_2, \ldots, \boldsymbol{x}_k = \boldsymbol{y}$ such that all $\boldsymbol{x}_i \in Br_m(\mathbf{V})$

and $x_{i+1} \in N_{m'}(x_i)$. It is known that $Br_6(\mathbf{V})$ is 18- or 26-connected and $Br_{18}(\mathbf{V})$ and $Br_{26}(\mathbf{V})$ are 6-connected [ ]. Note that $m \neq m'$.

Let us consider boundary points. Roughly speaking, in $\mathbf{Z}^3$, a finite subset $\mathbf{V}$ and the complement $\bar{\mathbf{V}}$ have "different boundaries" even if there exists "the common boundary" in $\mathbf{R}^3$ [ ]. It is easy to see that $Br_m(\mathbf{V}) \cap Br_m(\bar{\mathbf{V}}) = \emptyset$ from (2). Clifford explained such "different boundaries" using an example such as a heap of white marbles on the top of which black marbles are put in [ ]. He declared that the boundary of the white part would be a layer of white marbles and the boundary of the black part would be a layer of the black marbles, that is, the two adjacent parts have different boundaries when they are divided into two parts. We admit the existence of "different boundaries" in $\mathbf{Z}^3$ as well as other work such as [,]. We focus on one of the "boundaries" which is the border $Br_m(\mathbf{V})$ of $\mathbf{V}$ and construct a discrete polyhedron such that all the vertices are in $Br_m(\mathbf{V})$. In the historical sense, such problem is related the border tracking problem presented in the next subsection.

2.2   Border Tracking in $\mathbf{Z}^3$

One of several well-known approaches for border tracking in $\mathbf{Z}^3$ is tracking the common faces between two unit cubes centered at the points p in $\mathbf{V} \subset \mathbf{Z}^3$ and q in $\bar{\mathbf{V}} = \mathbf{Z}^3 \setminus \mathbf{V}$. Such faces are represented by the ordered pair (p, q) [ ]. Since $q \in \mathbf{N}_6(p) \cap \bar{\mathbf{V}}$, the set of all p of such pair (p, q) becomes equal to $Br_6(\mathbf{V})$ of (2). In this approach, the "surface" is represented by a set of square faces so that the square faces are of cubes whose centroid are lattice points p. The topological structures of such "surfaces" are given as cellular complexes in [ ]. However, the vertices of a polyhedron are not lattice points because the faces exist between lattice points. Since we would like to take the purely discrete approach, we first consider how to to construct a topological structure for 2-dimensional surfaces only using lattice points in $\mathbf{Z}^3$, and then clarify the relation between such surfaces and the set-theoretical border points.

2.3   Definitions of Surfaces in $\mathbf{Z}^3$

The approaches for defining 2-dimensional surfaces in $\mathbf{Z}^3$ are mainly classified into the following three types:

1. the approach based on Jordan surface theorem [ ] so that a set of lattice points is defined as a discrete surface if it divides $\mathbf{Z}^3$ into two regions and the thickness of the set is as thin as possible [,];
2. the approach based on algebraic topology [ ] so that the 2-dimensional surfaces is defined as the boundary of 3-dimensional objects [, , ,];
3. the approach by using the topological structures of the 2-dimensional manifold [ ].

In fact the first approach is considered for the different problem from our problem. Our problem is construction of a polyhedron from a subset $\mathbf{V}$ in $\mathbf{Z}^3$

Fig. 1. Examples of non-manifold polyhedral surfaces

although the problem for the first approach is determining a set of lattice points in $\mathbf{Z}^3$ corresponding to a given 2-dimensional surface in $\mathbf{R}^3$. It is sometimes called digitization or discretization. The third approach only allows us to treat the surfaces which is topologically equivalent to the 2-dimensional manifold. Because we would like to treat non-manifold surfaces as shown in Figure 1, we take the second approach for our definition of discrete polyhedral surfaces.

3   Combinatorial Definition of Discrete Polyhedra

3.1   Discrete Cellular Complexes

Referring to combinatorial and algebraic topology in the n-dimensional Euclidean space $\mathbf{R}^n$ [,], we first define n-dimensional cells or n-cells and then form cellular complexes by gluing cells.

Definition 1. *An n-cell is a set whose interior is homeomorphic to the n-dimensional disc $\mathbf{U}_r(\boldsymbol{x}) = \{\boldsymbol{y} \in \mathbf{R}^n : \|\boldsymbol{x} - \boldsymbol{y}\| < r\}$ with the additional property that its boundary must be divided into a finite number of lower dimensional cells, called the faces of the n-cell. We write $\sigma < \tau$ if σ is a face of τ.*

0. A 0-cell is a point A.
1. A 1-cell is a line segment $a = AB$, and $A < a$, $B < a$.
2. A 2-cell is a polygon (often a triangle), such as $\sigma = \triangle ABC$, and then $AB, BC, AC < \sigma$. Note that $A < AB < \sigma$, so $A < \sigma$.
3. A 3-cell is a solid polyhedron (often a tetrahedron), with polygons, edges, and vertices as faces.

The point of a 0-cell, the endpoints of a 1-cell and the vertices of 2- and 3-cells are called the vertices of each cell.

Definition 2. *A complex $\mathbf{K}$ is a finite set of cells such that*

1. if $\sigma \in \mathbf{K}$, then all faces of σ are elements of $\mathbf{K}$;
2. two cells $\sigma, \tau \in \mathbf{K}$ are either disjoint or their intersection is a common face.

The dimension of $\mathbf{K}$ is the dimension of its highest-dimensional cell.

An n-dimensional complex or n-complex $\mathbf{K}$ is said to be pure if every cell with less than n dimension of $\mathbf{K}$ precedes some n-cell [ ].

Table 1. All discrete n-cells for $n = 0, 1, 2, 3$ such that the vertices are all lattice points in $\mathbf{Z}^3$ and the adjacent vertices are m-neighboring for $m = 6, 18, 26$

In this paper, we consider complexes such that the vertices of cells in complexes are all lattice points in $\mathbf{Z}^3$ and the vertices of any 1-cell in complexes are neighboring in the sense of the 6-, 18- or 26-neighborhood of (1). For constructing such complexes, we first list up all possibles cells whose vertices are all lattice points and the adjacent vertices of a cell are m-neighboring each other for $m = 6, 18, 26$. Because of the constraint of m-neighboring, we set a unit cube whose vertices are all lattice points to be the maximum size of the cells. In other words, we look for the cells which exist in a unit cube.

For each lattice point in a unit cube we assign the value of either 1 or 0 and call the point a 1- or 0-point, respectively. There are 256 arrangements of 1- and 0-points for the eight lattice points in a unit cubic region. In fact, we can reduce the number from 256 to 23 with considering the congruent arrangements by rotations. For each arrangement, we obtain a convex polyhedron such that the vertices of the polyhedron are 1-points. We then classify each polyhedron into a set of n-cell with the dimension of $n = 0, 1, 2, 3$ and with the m-neighborhood relations between the adjacent vertices for $m = 6, 18, 26$ as shown in Table 1.

From Table 1, we see that the finite number of cells are obtained under the constraints such that the vertices of cells in complexes are all lattice points in $\mathbf{Z}^3$ and any adjacent vertices are m-neighboring for $m = 6, 18, 26$. Hereafter, we call the cells in Table 1 discrete cells. We also call a complex which is a finite set of discrete cells a discrete complex.

3.2   Definition of Discrete Polyhedra: Boundary of Discrete Objects

For defining discrete polyhedra, we need some basic topological notions. Let $\mathbf{K}$ be a complex. If $\mathbf{K}_0$ is any subcomplex of $\mathbf{K}$, the complex consisting of all the elements of $\mathbf{K}_0$ and of all the elements of $\mathbf{K}$ which precedes (i.e. is less than) at least one element of $\mathbf{K}_0$ is called the combinatorial closure $Cl(\mathbf{K}_0)$ of $\mathbf{K}_0$ in $\mathbf{K}$ [ ]. Now we give the combinatorial definitions of 3-dimensional discrete objects and the boundary.

Definition 3. *Any pure discrete 3-complex is a discrete object.*

Definition 4. *Let* $\mathbf{O}$ *be a discrete object and* $\mathbf{H}$ *be the set of discrete 2-cells in* $\mathbf{O}$ *each of which precedes exactly one discrete 3-cell in* $\mathbf{O}$*. The combinatorial boundary of* $\mathbf{O}$ *is defined as*

$$\partial\mathbf{O} = Cl(\mathbf{H}).$$

Our definition is based on the definition of the boundary in algebraic topology [ ]. For the boundary of an n-complex, we need the concept of oriented $(n-1)$-cells. Each discrete 2-cell which is a face of a discrete 3-cell has a particular ordering of its vertices so that the order is counterclockwise from a viewpoint exterior the discrete 3-cell. The boundary of a discrete object is obtained by the operation which is called the "modulo 2 union" of oriented discrete 2-cells. Therefore, a discrete 2-cell which precedes two discrete 3-cells in $\mathbf{O}$ is not considered to be in $\partial\mathbf{O}$ because it means that there are two oppositely oriented discrete 2-cells and the modulo 2 union of them is empty. From Definition 4, the boundary $\partial\mathbf{O}$ is a 2-dimensional pure discrete subcomplex of a discrete object $\mathbf{O}$. Hereafter, we call the boundary of a discrete object a discrete polyhedron. Note that a discrete polyhedron can be a non-manifold as shown in Figure 1 according to the definition. Clearly, discrete complexes, objects and polyhedra are defined for each m-neighborhood system, $m = 6, 18, 26$. When we would like to insist that an m-neighborhood system is considered, they are denoted by $\mathbf{C}_m$, $\mathbf{O}_m$ and $\partial\mathbf{O}_m$ instead of $\mathbf{C}$, $\mathbf{O}$ and $\partial\mathbf{O}$.

4   Polyhedrization of a 3D Lattice Point Set

Given a finite subset $\mathbf{V}$ of $\mathbf{Z}^3$, we consider polyhedrization of $\mathbf{V}$. The procedure is divided into three stages: decomposition of $\mathbf{V}$ into discrete cells whose dimensions are from 0 to 3 such that those discrete cells constitutes a discrete complex $\mathbf{C}_m$, making $\mathbf{C}_m$ pure to obtain a discrete object $\mathbf{O}_m$, and extraction of the boundary $\partial\mathbf{O}_m$ of $\mathbf{O}_m$ for each m-neighborhood system, $m = 6, 18, 26$. We also present a direct algorithm generating $\partial\mathbf{O}_m$ from $\mathbf{V}$. In this section, we will omit m when it is clear that one of the 6-, 18- and 26-neighborhood systems is considered through the procedure for polyhedrization.

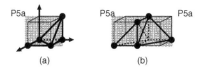

Fig. 2. (a) Another possible discrete complex of the 18-neighborhood system for the arrangement P5a and (b) an example of the problem which may be caused by the cellular decomposition (a) during construction of a cellular complex

4.1   Cellular Decomposition of a 3D Lattice Point Set

The decomposition of $\mathbf{V}$ into discrete cells is achieved in two stages. For $\boldsymbol{x} = (i, j, k)$ in $\mathbf{Z}^3$, let

$$\mathbf{D}(\boldsymbol{x}) = \{(i + \epsilon_1, j + \epsilon_2, k + \epsilon_3) \mid \epsilon_i = 0 \text{ or } 1\}.$$

We say that the points of $\mathbf{V}$ are 1-points and the points of $\mathbf{Z}^3 \setminus \mathbf{V}$ are 0-points. For each $\boldsymbol{x} \in \mathbf{Z}^3$, let $\mathbf{C}(\boldsymbol{x})$ denote the discrete complex generated by the set of 1-points in $\mathbf{D}(\boldsymbol{x})$ as shown in Table 2. Finally, let

$$\mathbf{C} = \bigcup_{\boldsymbol{x} \in \mathbf{Z}^3} \mathbf{C}(\boldsymbol{x}). \tag{3}$$

We will now prove that $\mathbf{C}$ is a discrete complex. Say that $\mathbf{C}(\boldsymbol{x})$ and $\mathbf{C}(\boldsymbol{y})$ are adjacent if $\mathbf{D}(\boldsymbol{x}) \cap \mathbf{D}(\boldsymbol{y}) \neq \emptyset$. Their adjacency types are classified into the following three cases

$$\#(\mathbf{D}(\boldsymbol{x}) \cap \mathbf{D}(\boldsymbol{y})) = 1, 2 \text{ or } 4 \text{ (and never 3)}$$

where $\#(A)$ represents the number of elements of the set A. The adjacency types and the conceivable cellular decomposition at the joint are illustrated in Table 3. For each adjacent pair of $\mathbf{C}(\boldsymbol{x})$ and $\mathbf{C}(\boldsymbol{y})$, let $\mathbf{C}(\boldsymbol{x}, \boldsymbol{y}) = \mathbf{C}(\boldsymbol{x}) \cup \mathbf{C}(\boldsymbol{y})$. We then verify, from Table 2, that $\mathbf{C}(\boldsymbol{x}, \boldsymbol{y})$ is a discrete complex satisfying the conditions of Definition 2.

For the arrangement P5a in the case of 18-neighborhood system, we do not take the discrete complex as shown in Figure 2 (a) to avoid the problem such as Figure 2 (b) which may occur in the procedure of constructing a cellular complex. Instead of Figure 2 (a), we have the discrete complex for the arrangement P5a in Table 2 (b). Consequently, we see that we can uniquely obtain $\mathbf{C}$ (or $\mathbf{C}_m$) from (3) which is a discrete complex for any m-neighborhood system, $m = 6, 18, 26$, given a finite subset $\mathbf{V} \subset \mathbf{Z}^3$.

4.2   Construction of a 3D Pure Complex

Assume that the dimension of $\mathbf{C}$ is three. Let $\mathbf{G}$ to be a set of all discrete 3-cells in $\mathbf{C}$. In order to obtain a pure discrete complex $\mathbf{O}$ from $\mathbf{C}$, we remove all discrete n-cells which are not included in any discrete 3-cells in $\mathbf{C}$ for every $n < 3$. Thus, the pure discrete complex $\mathbf{O}$ is obtained such that

$$\mathbf{O} = Cl(\mathbf{G}). \tag{4}$$

Table 2. A discrete complex $\mathbf{C}(\boldsymbol{x})$ corresponding to each arrangement of 1-points in a discrete unit cube $\mathbf{D}(\boldsymbol{x})$ for the (a) 6-, (b) 18- or (c) 26-neighborhood system

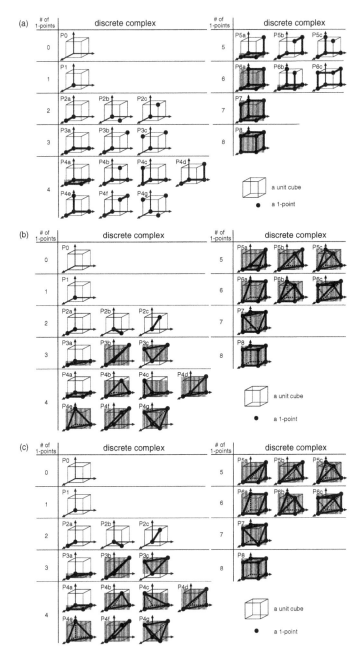

Table 3. Three adjacency types of two unit cubic regions $\mathbf{D}(x)$ and $\mathbf{D}(y)$ such that $\#(\mathbf{D}(x) \cap \mathbf{D}(y)) = 1, 2$ or 4. For each adjacency type, all possible arrangements of 1- and 0-points and the cellular decomposition are shown

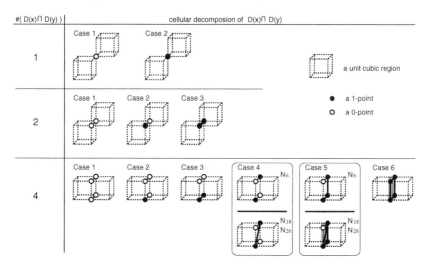

If $\mathbf{C}$ is less than three dimensions, $\mathbf{G}$ is empty and thus $\mathbf{O}$ is also empty. From (4), it is clear that a discrete object $\mathbf{O}$ is uniquely obtained from a discrete complex $\mathbf{C}$.

4.3   Boundary Extraction of a 3D Pure Complex

From Definition 4, the boundary $\partial\mathbf{O}$ of a discrete object $\mathbf{O}$ is derived from the set $\mathbf{H}$ of discrete 2-cells in $\mathbf{O}$ each of which precedes exactly one discrete 3-cell in $\mathbf{O}$. Because $\mathbf{H}$ is uniquely obtained from $\mathbf{O}$ by using the boundary operation defined in algebraic topology [ ], the boundary $\partial\mathbf{O}$ is also uniquely obtained from $\mathbf{O}$. From the above procedure for polyhedrization, we consequently obtain the following proposition.

Proposition 1. *Given a finite subset $\mathbf{V} \subset \mathbf{Z}^3$, a discrete polyhedron $\partial\mathbf{O}_m$, or the boundary of a discrete object $\mathbf{O}_m$, is uniquely obtained for any m-neighborhood system, $m = 6, 18, 26$.*

4.4   Discrete Polyhedrization Algorithm

For practical use, we present an effective algorithm such that we obtain a discrete polyhedron $\partial\mathbf{O}_m$ directly from a given finite subset $\mathbf{V} \subset \mathbf{Z}^3$ for each m-neighborhood system, $m = 6, 18, 26$. In other words, we would like to skip the conversion steps such as a discrete complex $\mathbf{C}_m$ and a discrete object $\mathbf{O}_m$ between $\mathbf{V}$ and $\partial\mathbf{O}_m$ because we do not need them. Similarly to the marching cubes method [ ], our algorithm generates $\partial\mathbf{O}_m$ directly from $\mathbf{V}$ by referring

Fig. 3. An example of the case in which two adjacent unit cubes have the common 2-cell

to Table 4 for each neighborhood system. We derived the table by looking for discrete 2-cells which may be in the boundary $\partial \mathbf{O}_m$ and by ignoring other discrete 2-cells in Table 2. The arrow of every discrete 2-cell in Table 4 shows the exterior orientation of $\partial \mathbf{O}_m$; all discrete 2-cells are oriented as we discussed in subsection 3.2. In other words, for each unit cube in Table 2, we choose only a oriented discrete 2-cell such that the exterior side is in the unit cubic region for obtaining Table 4.

Algorithm 1

input: *A subset* $\mathbf{V}$ *of* $\mathbf{W} = \{(i, j, k) \in \mathbf{Z}^3 : 1 \le i \le L, 1 \le j \le M, 1 \le k \le N\}$.
output: *A discrete polyhedron* $\partial \mathbf{O}_m$ *for an* m*-neighborhood system,* $m =$
    $6, 18, 26$.
begin
    1. **for** $1 \le k \le N - 1$ **do**
        for $1 \le j \le M - 1$ **do**
        for $1 \le i \le L - 1$ **do**
        1.1 for $\boldsymbol{x} = (i, j, k)$, *obtain a pure discrete 2-complex* $\mathbf{C}_m(\boldsymbol{x})$ *at a unit cubic region* $\mathbf{D}(\boldsymbol{x})$ *by referring to Table 4;*
        1.2 for each $\boldsymbol{y} = (i - 1, j, k), (i, j - 1, k), (i, j, k - 1)$, *check if any common discrete 2-cell* σ *of* $\mathbf{C}_m(\boldsymbol{x})$ *and* $\mathbf{C}_m(\boldsymbol{y})$ *exists as shown in Figure 3; if so, replace* $\mathbf{C}_m(\boldsymbol{x})$ *and* $\mathbf{C}_m(\boldsymbol{y})$ *with* $Cl(\mathbf{C}_m(\boldsymbol{x}) \setminus Cl(\{\sigma\}))$ *and* $Cl(\mathbf{C}_m(\boldsymbol{y}) \setminus Cl(\{\sigma\}))$, *respectively;*
    2. obtain $\partial \mathbf{O}_m = \cup_{\boldsymbol{x} \in \mathbf{W}} \mathbf{C}_m(\boldsymbol{x})$.
end

We see that discrete 2-cells which locate at the joint of two adjacent unit cubic regions remain in Table 4 such as the arrangements P3a and P4a. If such discrete 2-cells are obtained so that they are not included in any discrete 3-cell of a discrete object $\mathbf{O}_m$, then we will have pairs of the discrete 2-cells which have the same vertices but the opposite orientations as shown in Figure 3. They do not constitute $\partial \mathbf{O}_m$ due to the boundary operation. The step 1.3 in Algorithm 1 is for the removal procedure of such discrete 2-cells.

5   Relations between Border and Discrete Polyhedra

Let us define the *skeleton* $Sk(\sigma)$ of a discrete cell σ such as the set of the vertices of σ [ ]. Given $\mathbf{V} \subset \mathbf{Z}^3$, let $\partial \mathbf{O}_m$ be a discrete polyhedron constructed

Table 4. The look-up table which provides the one-to-one correspondence between the arrangements of 1- and 0-points in a discrete unit cubic region and a 2-dimensional pure discrete subcomplex constituting a discrete polyhedron for each neighborhood system

by Algorithm 1 for $m = 6, 18, 26$. We then call the union of the skeletons of all discrete cells of $\partial \mathbf{O}_m$ the skeleton of $\partial \mathbf{O}_m$ and it is denoted by $Sk(\partial \mathbf{O}_m)$. Let us compare the skeleton $Sk(\partial \mathbf{O}_m)$ with the border point set $Br_{m'}(\mathbf{V})$ at each discrete unit cube $\mathbf{D}(x)$ seeing Tables 2 and 4. For instance, the comparisons for the arrangements P5a and P7 are shown in Figure 4. After unifying the comparison at every $\mathbf{D}(x)$, we consequently obtain

$$Br_6(\mathbf{V}) = Sk(\partial \mathbf{O}_{26}) \cup Sk(\mathbf{C}_{26} \setminus \mathbf{O}_{26}) \tag{5}$$

$$= Sk(\partial \mathbf{O}_{18}) \cup Sk(\mathbf{C}_{18} \setminus \mathbf{O}_{18}) \setminus \mathbf{E}_{18}, \tag{6}$$

$$Br_{18}(\mathbf{V}) = Sk(\partial \mathbf{O}_6) \cup Sk(\mathbf{C}_6 \setminus \mathbf{O}_6) \setminus \mathbf{F}_{18}, \tag{7}$$

$$Br_{26}(\mathbf{V}) = Sk(\partial \mathbf{O}_6) \cup Sk(\mathbf{C}_6 \setminus \mathbf{O}_6). \tag{8}$$

In (6) and (7), we set $\mathbf{E}_{18}$ and $\mathbf{F}_{18}$ to be the sets of points each of which is given as shown in Figure 4. The points of $\mathbf{E}_{18}$ and $\mathbf{F}_{18}$ in fact show the difference between $Sk(\partial \mathbf{O}_{18})$ and $Sk(\partial \mathbf{O}_{26})$ of (5) and (6) and the difference between $Br_{18}(\mathbf{V})$ and $Br_{26}(\mathbf{V})$ of (7) and (8), respectively. A discrete complex $\mathbf{C}_m \setminus \mathbf{O}_m$ represent less than 3-dimensional parts in $\mathbf{C}_m$. Such parts are ignored in the procedure for polyhedrization because of the lack of the dimensions. Thus, some points of $Sk(\mathbf{C}_m \setminus \mathbf{O}_m)$ are not in $Sk(\partial \mathbf{O}_m)$ even if they are in the corresponding $Br_{m'}(\mathbf{V})$.

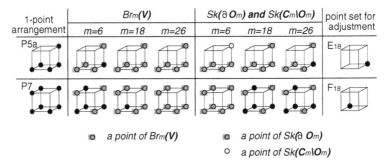

Fig. 4. For each of the arrangements P5a and P7 of 1-points (black points), the border points (gray points) $Br_m(\mathbf{V})$ for $m = 6, 18, 26$, the skeletons of the discrete polyhedra (gray points) $Sk(\partial \mathbf{O}_m)$, the points (white points) of $Sk(\mathbf{C}_m \setminus \mathbf{O}_m)$ for $m = 6, 18, 26$, and the point sets $\mathbf{E}_{18}$ and $\mathbf{F}_{18}$ for adjusting (6) and (7)

6   Conclusions

In this paper, we presented a discrete polyhedrization scheme of a given finite subset $\mathbf{V} \subset \mathbf{Z}^3$. Because we were initiated into the polyhedrization problem from the border tracking problem, we gave an overview of the border tracking problem in 3-dimensional digital images and clarified the difficulty of the problem caused by the discontinuous of the space. We then showed the proper reasons for our approach based on algebraic topology in which we mainly considered combinatorial topology. Because discrete polyhedra are defined as the boundary of 3-dimensional cellular complexes, we first defined discrete cells of the dimensions from 0 to 3 and discrete complexes which is a cellular complex consisting of discrete cells. Due to such algebraic topological definition, we have the case such that a discrete polyhedron may be non-manifold as shown in Figure 1.

Another interesting point of our approach is that we interpret a digitized three-dimensional object $\mathbf{V}$ as a cellular complex $\mathbf{C}$ which may not be pure and may not have three dimensions. Therefore, we may have the dimension reductions because of the object digitization or the limited resolution of digital images. Due to such interpretation, we can give the theoretical reason to ignore some lattice points in the border $Br(\mathbf{V})$ for a discrete polyhedron $\partial \mathbf{O}$ and succeeded to make the comparison between $Br(\mathbf{V})$ and the skeleton of $\partial \mathbf{O}$.

For the construction of a given finite subset $\mathbf{V}$ of $\mathbf{Z}^3$, we presented a similar look-up table to the marching cubes [ ]. Therefore, we do not have to go through all theoretical steps such as a 3-dimensional cellular decomposition $\mathbf{C}$ of $\mathbf{V}$, making the complex $\mathbf{C}$ pure for a discrete object $\mathbf{O}$ and obtaining the boundary $\partial \mathbf{O}$ of the discrete object $\mathbf{O}$; we directly obtain a discrete polyhedron $\partial \mathbf{O}$ from a given finite subset $\mathbf{V} \subset \mathbf{Z}^3$.

Acknowledgments

The authors thank to the reviewers for their helpful comments to improve the paper. The first author also thanks to Prof. Gilles Bertrand and Prof. Michel Couprie at ESIEE for the fruitful discussions on the topic.

References

1. Rosenfeld, A. (1970): Connectivity in digital pictures. Journal of the Association for Computing Machinery **17 1**, 146–160   150, 152
2. Artzy, E., Frieder, G., Herman, G. T. (1981): The theory, design, implementation, and evaluation of a three-dimensional boundary detection algorithm. Computer Graphics Image Processing **15** 1–24   150, 152
3. Kong, T. Y., Rosenfeld, A. (1989): Digital topology: introduction and survey. Computer Vision, Graphics, and Image Processing **48**, 357–393   150, 151
4. Hausdorff, F. (1937): Set theory. Chelsea Publishing Company   150, 151, 152
5. Alexandrov, P. S. (1956): Combinatorial topology I. Graylock Press   150, 151, 152, 153, 155, 159
6. Stillwell, J. (1993): Classical topology and combinatorial group theory. Springer   150, 151, 152, 153, 155, 158
7. Kong, T. Y., Roscoe, A. W., Rosenfeld, A. (1992): Concepts of digital topology. Topology and its Applications **46**, 219–262   150, 151, 152
8. Voss, K. (1993): Discrete images, objects, and functions in $\mathbf{Z}^3$. Springer-Verlag   150, 151, 152
9. Tourlakis, G., Mylopoulos, J. (1973): Some results in computational topology. Journal of the Association for Computing Machinery **20 3**, 439–455   151
10. Lorensen, W. E., Cline, H. E. (1987): Marching cubes: a high-resolution 3d surface construction algorithm. Computer Graphics (SIGGRAPH '87) **21 4**, 163–169   151, 158, 161
11. Kenmochi, Y., Imiya, A., Ichikawa, A. (1998): Boundary extraction of discrete objects. Computer Vision and Image Understanding **71 3**, 281–293   151, 152
12. Clifford, W. K. (1956): The postulates of the science of space. The world of mathematics, Simon and Schuster, New York   152
13. Kovalevsky, V. A. (1989): Finite topology as applied to image analyses. Computer Vision, Graphics, and Image Processing **46**, 141–161   152
14. Morgenthaler, D. G., Rosenfeld, A. (1981): Surfaces in three-dimensional images. Information and Control **51**, 227–247   152
15. Couprie, M., Bertrand, G. (1998): Simplicity surfaces: a new definition of surfaces in $\mathbf{Z}^3$. SPIE Vision Geometry VII **3454**, 40–51   152
16. Françon, J. (1995): Discrete combinatorial surfaces. CVGIP: Graphical Models and Image Processing **57 1**, 20–26   152

Digital Partitions Encoding

Joviša Žunić*

Computer Science, Cardiff University
Queen's Buildings, Newport Road, PO Box 916
Cardiff CF24 3XF, Wales, U.K.
J.Zunic@cs.cf.ac.uk

Abstract. In this paper we are interested in problems related to the encoding of partitions of discrete sets in $\mathbf{R}^n$ with the restriction to the partitions which can be realized by a number of hyperplanes.

A particular attention is given to the partitions of subsets of a planar integer grid of a given size, which is a model for a binary picture of a given size.

An application of the used encoding method to the encoding of linear threshold functions is given, as well. It turns out, that the obtained $\mathcal{O}(n^2)$ code is asymptotically optimal with respect to the required memory space.

Keywords: Linear partitions, coding, pattern recognition, threshold functions, discrete moments

1 Introduction

If A is a subset of a given set X and B equals the set difference $X \setminus A$ then the ordered pair of sets (A, B) is a partition of X. If the partitioned set X is fixed we shall write A^c instead of $X \setminus A$. If $X \subset \mathbf{R}^n$ and there exists a function of the form: $f(x_1, x_2, \ldots x_n) = a_1 \cdot x_1 + a_2 \cdot x_2 + \ldots a_n \cdot x_n + b$, such that

(i) $f(x_1, x_2, \ldots, x_n) > 0$ for $(x_1, x_2, \ldots, x_n) \in A$ and
(ii) $f(x_1, x_2, \ldots, x_n) < 0$ for $(x_1, x_2, \ldots, x_n) \in A^c = X \setminus A$,

then, we say that (A, A^c) is a *linear partition* of X. Obviously, in that case A and A^c belong to the different half-spaces determined by the hyperplane $f(x_1, x_2, \ldots, x_n) = 0$.

If $f(x_1, x_2, \ldots, x_n)$ has the form:

$$f(x_1, x_2, \ldots x_n) = \prod_{i=1}^{i=k}(a_{i,1} \cdot x_1 + a_{i,2} \cdot x_2 + \ldots + a_{i,n} \cdot x_n + b_i) \qquad (1)$$

---

* The author is also member of the Mathematical institute of Serbian Academy of Sciences, under project 04M02

and (i) and (ii) hold, then the partition (A, A^c) is called k-*linear partition* of X. Geometrically speaking, the set X can be partitioned into A and A^c by using k hyperplanes.

If a set consists of m points it will be called m-*point set*.

In this paper we are dealing only with partitions of finite sets, i.e., X is finite. Precisely, we are interested in the efficient encoding of such partitions in order to enable a fast comparison between them and to preserve their small storage complexity. For this purpose a knowledge about the number of partitions from the studied class is essential. It is not possible to give a general answer what is the number of k-linear partitions, of a planar m-point set in n-dimensional space, as a function of k, m, and n. That is not possible even in the planar case ($n = 2$) by assuming that only one line (hyperplane) is used, i.e., $k = 1$. Namely, let a pair of points from a given set is said to be a *minimal pair* if there are no other points (from the same set) belonging to the straight line segment determined by this two points. Under this definition, the number of linear partitions of a planar set, equals the number of minimal pairs of the same set ([ ]). Consequently, the number of linear partitions of a planar set with n points is somewhere between $n-1$ and $\frac{n(n-1)}{2}$. Precisely, the maximum $\frac{n(n-1)}{2}$ is reached if there are no three collinear points (i.e., the points of the set are in so called *general position*). The minimum $n - 1$ is reached if all points are collinear. So, one can conclude, that the number of linear partitions of a given set strongly depends of the structure of the partitioned set, as it is already mentioned in [ ] and [ ].

Here, we give a coding scheme which can be applied to the partitions of an arbitrary discrete planar set by using of an arbitrary number of lines, but naturally, the encoding can not be efficient in all cases. We give aá detailed performance analysis in the case when the partitioned set is an $r \times r$ integer grid (Section 2).

In Section 3, we give a coding scheme for the threshold functions defined on a fixed m-point subset of n-dimensional cube ($\{0, 1\}^n$). The coding requires an $\mathcal{O}(n \cdot \log m)$ amount of bits per coded function. If $m = 2^n$ that is an asymptotic minimum.

The encoding is based on a use of discrete moments, which are already shown to be a useful tool in the area of digital objects encoding ([ ], [ ], [ ]), as well.

2   Encoding of Planar k-Partitions

In this section we give a coding scheme for the partitions (of a finite planar set) which can be realized with k lines. A set of discrete moments is used for such coding. Precisely, for a discrete set X, from $\mathbf{R}^n$, its $(p_1, p_2, \ldots, p_n)$-*discrete moment* is defined as:

$$\mu_{p_1, p_2, \ldots, p_n}(X) = \sum_{(x_1, x_2, \ldots, x_n) \in X} x_1^{p_1} \cdot x_2^{p_2} \cdot \ldots \cdot x_n^{p_n}.$$

The order of the moment $\mu_{p_1, p_2, \ldots, p_n}(X)$ is equal to $p_1 + p_2 + \ldots + p_n$.

For the simplicity we start with $n = 2$, i.e., we study the encoding of k-linear partitions of a planar set.

Theorem 1. *Let (A_1, A_1^c) and (A_2, A_2^c) be two partitions of a planar set X, obtained by k lines. Then, the partitions are equal, i.e., $(A_1, A_1^c) = (A_2, A_2^c)$, if and only if*

$$\mu_{p,q}(A_1) = \mu_{p,q}(A_2) \quad \text{for the nonnegative integers} \quad p, \ q, \quad \text{with} \quad p + q \le k.$$

Proof. If $A_1 = A_2$ then the corresponding discrete moments are equal obviously. What we have to prove is that the equalities of the corresponded moments of the order up to k preserve the equality $A_1 = A_2$. We prove it by a contradiction.
Let

$$\sum_{(x,y) \in A_1} x^p \cdot y^q \ = \ \mu_{p,q}(A_1) \ = \ \mu_{p,q}(A_2) \ = \ \sum_{(x,y) \in A_2} x^p \cdot y^q$$

holds for the nonnegative integers p and q satisfying $p + q \le k$.
If $A_1 \neq A_2$ then we can assume $A_1 \setminus A_2$ is nonempty, for an example (else we can start with the nonempty $A_2 \setminus A_1$). Since (A_1, A_1^c) is a k-linear partition, then there exists a function $f(x,y)$ of the form (1) which satisfies the conditions (**i**) and (**ii**). In the two-dimensional case it can be expressed as:

$$f(x,y) = \sum_{p+q \le k} \alpha_{p,q} \cdot x^p \cdot y^q,$$

for some real coefficients $\alpha_{p,q}$, with $p + q \le k$.
Further we have:

$$0 < \sum_{(x,y) \in A_1 \setminus A_2} \left(\sum_{p+q \le k} \alpha_{p,q} \cdot x^p \cdot y^q \right)$$

$$= \sum_{p+q \le k} \left(\sum_{(x,y) \in A_1 \setminus A_2} \alpha_{p,q} \cdot x^p \cdot y^q \right) + \sum_{p+q \le k} \left(\sum_{(x,y) \in A_1 \cap A_2} \alpha_{p,q} \cdot x^p \cdot y^q \right)$$

$$- \sum_{p+q \le k} \left(\sum_{(x,y) \in A_1 \cap A_2} \alpha_{p,q} \cdot x^p \cdot y^q \right)$$

$$= \sum_{p+q \le k} \left(\alpha_{p,q} \cdot \sum_{(x,y) \in A_1} x^p \cdot y^q \right) - \sum_{p+q \le k} \left(\alpha_{p,q} \cdot \sum_{(x,y) \in A_1 \cap A_2} x^p \cdot y^q \right)$$

$$= \sum_{p+q \le k} \left(\alpha_{p,q} \cdot \sum_{(x,y) \in A_2} x^p \cdot y^q \right) - \sum_{p+q \le k} \left(\alpha_{p,q} \cdot \sum_{(x,y) \in A_1 \cap A_2} x^p \cdot y^q \right)$$

$$= \sum_{(x,y) \in A_2 \setminus A_1} \left(\sum_{p+q \leq k} \alpha_{p,q} \cdot x^p \cdot y^q \right) \leq 0.$$

The contradiction $\quad 0 < 0 \quad$ finishes the proof. $\quad []$

The extension of the previous theorem to the $\mathbf{R}^n$ case, with $n > 2$, can be made in a similar manner. We give the statement without a proof.

Theorem 2. *Let (A_1, A_1^c) and (A_2, A_2^c) be two k-linear partitions of an n-dimensional set X. Then, the partitions (A_1, A_1^c) and (A_2, A_2^c) are equal, if and only if*

$$\mu_{p_1,p_2,\ldots,p_n}(A_1) = \mu_{p_1,p_2,\ldots,p_n}(A_2)$$

for any choice of nonnegative integers $p_1, p_2, \ldots, p_n$ with $p_1 + p_2 + \ldots + p_n \leq k$.

The previous theorem gives an answer how many discrete moments are enough for a unique identification of a k-linear partition of a finite discrete set.

What is important for us is:

How many bits are necessary for a storage of the suggested set of moments?

Obviously, a precise and general answer can not be given as a function of cardinality of the partitioned set and, the number of the lines used. It also depends on the "topology" of the set as well as on the "nature" of the points from the set (are they integers, or irrational numbers, how large they are, etc.).

Here we give a performance analysis of the proposed coding scheme assuming that the partitioned set is a subset of a two-dimensional integer grid of a given size, say $r \times r$.

Theorem 3. *Order of magnitude of the number of bits for the coding of all k-linear partitions of an $r \times r$ integer grid is*

$$k^3 \cdot \log r \quad - \quad k^2 \cdot \log k$$

per coded partition.

Proof. Since

$$\mu_{p,q}(S) \leq \iint\limits_{0 \leq x,y \leq r} x^p y^q dx dy \quad = \quad \frac{r^{p+q+2}}{(p+1)(q+1)}$$

holds for any subset S of $r \times r$ -integer grid, it follows that the number of bits necessary for the storage of $\mu_{p,q}(S)$ is upper bounded by

$$\lceil (p+q+2) \cdot \log r \quad - \quad \log((p+1)(q+1)) \rceil.$$

So, the total bit rate for the storage of all moments having the order less or equal to k, can be upper bounded by:

$$\sum_{p=0}^{k}\left(\sum_{q=0}^{k-p}\left\lceil\log\frac{r^{p+q+2}}{(p+1)(q+1)}\right\rceil\right)$$

$$=\sum_{p=0}^{k}\left(\sum_{q=0}^{k-p}(p+q+2)\cdot\log r - \sum_{q=0}^{k-p}\log((p+1)(q+1))\right) + \mathcal{O}(k^2)$$

$$=\frac{k^3+6k^2+11k+9}{3}\cdot\log r - 2\cdot\sum_{p=0}^{k}(k-p+1)\cdot\log(p+1) + \mathcal{O}(k^2)$$

$$=\frac{k^3+6k^2+11k+9}{3}\cdot\log r - 2\cdot\frac{k+1}{\ln 2}\cdot\sum_{p=0}^{k}\ln(p+1)$$

$$+2\cdot\sum_{p=0}^{k}p\cdot\log(p+1) + \mathcal{O}(k^2)$$

$$=\frac{k^3+6k^2+11k+9}{3}\cdot\log r - 2\cdot\frac{k+1}{\ln 2}\cdot\int_0^k\ln(x+1)\,dx$$

$$+\frac{2}{\ln 2}\cdot\int_0^k x\cdot\ln(x+1)\,dx + \mathcal{O}(k^2)$$

$$=\frac{k^3+6k^2+11k+9}{3}\cdot\log r - 2\cdot\frac{k^2\cdot\ln k - k^2}{\ln 2}$$

$$+\frac{2}{\ln 2}\cdot\left(\frac{k^2\cdot\ln k}{2}-\frac{k^2}{4}\right) + \mathcal{O}(k^2)$$

$$=\frac{k^3+6k^2+11k+9}{3}\cdot\log r - \frac{k^2\cdot\ln k}{\ln 2} + \mathcal{O}(k^2) ,$$

which completes the proof. []

We want to mention that the above discussion is related to the partitions of an $r \times r$-integer grid by k lines which are in an arbitrary position.

An upper bound for the number of partitions of $r \times r$ grid which are made with parallel lines can be found in [ ]. A fast algorithm for counting of such partitions is described in the same paper.

If the partitions are made by lines parallel to the y-axis then the number of moments used can be reduced. This particular situation is studied in [ ]. An alternative for such a coding is studied in [ ].

A quick analysis of Theorem 3 shows that in the case of small number of the used straight lines the coding is efficient – by the way, if k is assumed to be a constant the coding scheme is optimal. This will be illustrated in the rest of this section.

A *discrete analytical hyperplane* $H(\alpha_0, \alpha_1, \ldots, \alpha_n, \omega)$ in dimension n is defined ([ ], [ ]) by the double Diophantine inequality:

$$(x_1, x_2, \ldots, x_n) \in H(\alpha_0, \alpha_1, \ldots, \alpha_n, \omega) \quad \Leftrightarrow \quad 0 \le \alpha_0 + \sum_{i=1}^{n} \alpha_i \cdot x_i < \omega$$

where the coefficients $\alpha_0, \alpha_1, \ldots \alpha_n$ are integers with $gcd(\alpha_0, \alpha_1, \ldots, \alpha_n) = 1$, while ω is a positive integer. Of course, $x_1, x_2, \ldots, x_n$ are integers too.

We give the following theorem showing that the encoding proposed here enables an optimal encoding of discrete analytical line segments.

Theorem 4. *The set of two-dimensional discrete analytical hyperplane segments which can be inscribed into an $r \times r$ integer grid can be coded by $\mathcal{O}(\log r)$ bits per coded segment.*

That is an asymptotically optimal bit rate.

Proof. The proof is a direct consequence of Theorem 1, and Theorem 3. Namely, any discrete analytical line (two dimensional discrete analytical hyperplane) $H(\alpha_0, \alpha_1, \alpha_2, \omega)$ can be "separated" by a function of the form

$$f(x, y) = (\alpha_0 + \alpha_1 \cdot x + \alpha_2 \cdot y - \varepsilon) \cdot (\alpha_0 + \alpha_1 \cdot x + \alpha_2 \cdot y + \omega)$$

(ε is a positive small enough real number) from the rest of $r \times r$-integer grid. So, the moments

$$\mu_{0,0}, \quad \mu_{1,0}, \quad \mu_{0,1}, \quad \mu_{2,0}, \quad \mu_{1,1}, \quad \mu_{0,2}$$

are sufficient for the coding.

The number of required bits for the storage is upper bounded by $\mathcal{O}(\log r)$ because any of the moments used is less than r^4.   □

3   Threshold Functions Encoding

A switching function $f(x_1, x_2, \ldots, x_n)$ of n binary variables is defined by assigning either 0 or 1 to each of the 2^n points $(x_1, x_2, \ldots, x_n) \in \{0, 1\}^n$. A switching function of n variables is linearly separable if there exists a hyperplane π in n-dimensional space, which strictly separates the "on" set $f^{-1}(1)$ from the "off" set $f^{-1}(0)$. In other words, $f^{-1}(1)$ lies on one side of π and $f^{-1}(0)$ on the other, and $\pi \cap \{0, 1\}^n$ is empty. Linearly separable functions are also called *threshold functions*. A threshold function simulates a neuron examining its input and making its decision as to its next state.

The following theorem shows how the set of threshold functions restricted to an arbitrary subset of $\{0, 1\}^n$ can be coded by using n integers instead of a using $n + 1$, so called, *Chow parameters* ([ ]).

Theorem 6 estimates the number of bits necessary for a coding of n-dimensional threshold functions defined on m-point subset of $\{0,1\}^n$ and proves that such a coding is optimal in an asymptotic sense if $m = 2^n$.

Theorem 5. *Any threshold function of* $f(x_1, \ldots x_n)$ *defined on a fixed set* $S \subset \{0,1\}^n$ *can be coded uniquely by:*

$$\sum_{f(x_1,\ldots,x_n)=1} 1, \quad \sum_{f(x_1,\ldots,x_n)=1} x_1, \quad \sum_{f(x_1,\ldots,x_n)=1} x_2, \quad \ldots, \quad \sum_{f(x_1,\ldots,x_n)=1} x_{n-1} .$$

Proof. Let a set $S \subset \{0,1\}^n$ be given. A threshold function $f(x_1, \ldots, x_n)$ defines a partition of S into the sets A and $A^c = S \setminus A$, where (for example) A consists of points in which $f(x_1, x_2, \ldots, x_n) = 1$, while the points from $A^c = S \setminus A$ satisfy $f(x_1, x_2, \ldots, x_n) = 0$.

Also, by the definition, A and A^c are separable by a hyperplane π_1 : $\alpha_1 \cdot x_1 + \alpha_2 \cdot x_2 + \ldots + \alpha_n \cdot x_n + \gamma_1 = 0$. For instance, let us assume

$$\alpha_1 \cdot x_1 + \alpha_2 \cdot x_2 + \ldots + \alpha_n \cdot x_n + \gamma_1 > 0 \quad \textbf{iff} \quad (x_1, x_2, \ldots, x_n) \in A$$

(let us remind that $(x_1, x_2, \ldots, x_n) \in A \Leftrightarrow f(x_1, x_2, \ldots x_n) = 1.$)

Let us assume that there is another partition (B, B^c) of S which is corresponded to another threshold function $g(x_1, x_2, \ldots, x_n)$, which has the same proposed code, i.e.,

$$\sum_{f(x_1,\ldots,x_n)=1} 1 = \sum_{g(x_1,\ldots,x_n)=1} 1, \quad \sum_{f(x_1,\ldots,x_n)=1} x_1 = \sum_{g(x_1,\ldots,x_n)=1} x_1 ,$$

$$\sum_{f(x_1,\ldots,x_n)=1} x_2 = \sum_{g(x_1,\ldots,x_n)=1} x_2, \quad \ldots, \quad \sum_{f(x_1,\ldots,x_n)=1} x_{n-1} = \sum_{g(x_1,\ldots,x_n)=1} x_{n-1} .$$

Then, we have to show that $(A, A^c) = (B, B^c)$ holds. It will be proved by a contradiction.

Since $g(x_1, x_2, \ldots, x_n)$ is another threshold function defined on S, there exist a hyperplane π_2 : $\beta_1 \cdot x_1 + \beta_2 \cdot x_2 + \ldots + \beta_n \cdot x_n + \gamma_2 = 0$ which separates B and $S \setminus B$.

Without loss of generality we can assume that α_n and β_n are both different from zero (because A and B consist of a finite number of points).

Then there exist a hyperplane of the form

$$\delta_1 \cdot x_1 + \delta_2 \cdot x_2 + \ldots + \delta_{n-1} \cdot x_{n-1} + \gamma = 0$$

which separate the set differences $A \setminus B$ and $B \setminus A$. (A possibility is $\delta_i = \frac{\alpha_i}{\alpha_n} - \frac{\beta_i}{\beta_n}$ for $i = 1, 2, \ldots, n-1$, and $\gamma = \frac{\gamma_1}{\alpha_n} - \frac{\gamma_2}{\alpha_n}$.)

Let us assume:

$$\delta_1 \cdot x_1 + \delta_2 \cdot x_2 + \ldots + \delta_{n-1} \cdot x_{n-1} + \gamma > 0 \quad \text{for} \quad (x_1, x_2, \ldots, x_n) \in A \setminus B,$$

and

$$\delta_1 \cdot x_1 + \delta_2 \cdot x_2 + \ldots + \delta_{n-1} \cdot x_{n-1} + \gamma < 0 \quad \text{for} \quad (x_1, x_2, \ldots, x_n) \in B \setminus A.$$

Since $f \neq g$ then A and B must be different. It implies that at least one of $A \setminus B$ and $B \setminus A$ is nonempty.

Let us assume that $A \setminus B$ is nonempty (else we can start with the nonempty set $B \setminus A$). The rest of the proof can be derived on a similar manner as the proof of Theorem 1. Further, we have:

$$0 < \sum_{(x_1,\dots,x_n)\in A\setminus B} (\delta_1 \cdot x_1 + \dots + \delta_{n-1} \cdot x_{n-1} + \gamma)$$

$$= \delta_1 \cdot \sum_{(x_1,\dots,x_n)\in A\setminus B} x_1 + \dots + \delta_{n-1} \cdot \sum_{(x_1,\dots,x_n)\in A\setminus B} x_{n-1} + \gamma \cdot \sum_{(x_1,\dots,x_n)\in A\setminus B} 1$$

$$+ \delta_1 \cdot \sum_{(x_1,\dots,x_n)\in A\cap B} x_1 + \dots + \delta_{n-1} \cdot \sum_{(x_1,\dots,x_n)\in A\cap B} x_{n-1} + \gamma \cdot \sum_{(x_1,\dots,x_n)\in A\cap B} 1$$

$$- \delta_1 \cdot \sum_{(x_1,\dots,x_n)\in A\cap B} x_1 - \dots - \delta_{n-1} \cdot \sum_{(x_1,\dots,x_n)\in A\cap B} x_{n-1} - \gamma \cdot \sum_{(x_1,\dots,x_n)\in A\cap B} 1$$

$$= \delta_1 \cdot \sum_{f(x_1,\dots,x_n)=1} x_1 + \dots + \delta_{n-1} \cdot \sum_{f(x_1,\dots,x_n)=1} x_{n-1} + \gamma \cdot \sum_{f(x_1,\dots,x_n)=1} 1$$

$$- \delta_1 \cdot \sum_{(x_1,\dots,x_n)\in B\cap A} x_1 - \dots - \delta_{n-1} \cdot \sum_{(x_1,\dots,x_n)\in B\cap A} x_{n-1} - \gamma \cdot \sum_{(x_1,\dots,x_n)\in B\cap A} 1$$

$$= \delta_1 \cdot \sum_{g(x_1,\dots,x_n)=1} x_1 + \dots + \delta_{n-1} \cdot \sum_{g(x_1,\dots,x_n)=1} x_{n-1} + \gamma \cdot \sum_{g(x_1,\dots,x_n)=1} 1$$

$$- \delta_1 \cdot \sum_{(x_1,\dots,x_n)\in B\cap A} x_1 - \dots - \delta_{n-1} \cdot \sum_{(x_1,\dots,x_n)\in B\cap A} x_{n-1} - \gamma \cdot \sum_{(x_1,\dots,x_n)\in B\cap A} 1$$

$$= \delta_1 \cdot \sum_{(x_1,\dots,x_n)\in B} x_1 + \dots + \delta_{n-1} \cdot \sum_{(x_1,\dots,x_n)\in B} x_{n-1} + \gamma \cdot \sum_{(x_1,\dots,x_n)\in B} 1$$

$$- \delta_1 \cdot \sum_{(x_1,\dots,x_n)\in B\cap A} x_1 - \dots - \delta_{n-1} \cdot \sum_{(x_1,\dots,x_n)\in B\cap A} x_{n-1} - \gamma \cdot \sum_{(x_1,\dots,x_n)\in B\cap A} 1$$

$$= \delta_1 \cdot \sum_{(x_1,\dots,x_n)\in B\setminus A} x_1 + \dots + \delta_{n-1} \cdot \sum_{(x_1,\dots,x_n)\in B\setminus A} x_{n-1} + \gamma \cdot \sum_{(x_1,\dots,x_n)\in B\setminus A} 1$$

$$= \sum_{(x_1,\dots,x_n)\in B\setminus A} (\delta_1 \cdot x_1 + \dots + \delta_{n-1} \cdot x_{n-1} + \gamma) \leq 0,$$

The obtained contradiction $0 < 0$ implies that (A, A^c) must be equal to (B, B^c). But, that leads to the functional equality $f(x_1, \ldots, x_n) = g(x_1, \ldots, x_n)$, for $(x_1, \ldots, x_n) \in S$. []

The coding scheme for the threshold functions defined on $\{0, 1\}^n$, proposed by the previous theorem, is asymptotically optimal with respect to the usual storage complexity criteria.

Theorem 6. *The one-to-one correspondence between the threshold functions $f(x_1, x_2, \ldots, x_n)$ defined on a fixed m-point set $S \subset \{0, 1\}^n$ and the set of corresponded moments:*

$$\sum_{f(x_1, \ldots, x_n)=1} 1, \quad \sum_{f(x_1, \ldots, x_n)=1} x_1, \quad \sum_{f(x_1, \ldots, x_n)=1} x_2, \quad \ldots, \quad \sum_{f(x_1, \ldots, x_n)=1} x_{n-1},$$

enables an encoding of such functions within an $\mathcal{O}(n \cdot \log m)$ bit rate per coded function.

If $S = \{0, 1\}^n$ (i.e., $m = 2^n$) the required $\mathcal{O}(n^2)$ amount of memory is asymptotically optimal.

Proof. The discrete moments up to the first order are used for the coding. Any of them is an integer belonging to $\{0, 1, 2, \ldots, m\}$. So, any of them can be coded with at most $\lceil \log m \rceil$ bits, and consequently, the total bit rate is upper bounded by $\mathcal{O}(n \cdot \log m)$.

In the case of $S = \{0, 1\}^n$, the storage complexity is $\mathcal{O}(n^2)$. On the other side, if $LTF(n)$ denote the number of linear threshold functions of n variables, then the following lover bound for the logarithm of $LTF(n)$ is proved in [ ]

$$\log LTF(n) > n^2 \cdot \left(1 - \frac{10}{\log n}\right).$$

It implies that the necessary bit rate has n^2 as the order of magnitude. Since this minimum is reached by our coding scheme, its optimality follows in the asymptotic sense. []

4   Comments and Conclusion

Problems related to the encoding of the partitions of a discrete sets (in arbitrary dimensions) by a given number of hyperplanes are studied. The obtained results can be of interest in different areas: pattern recognition, pattern matching, pattern classification, neural computing, image analysis, etc.

The proposed coding scheme is shown to be very efficient in some cases. Then it preserves a fast transmission and a fast comparison between partitioned sets, as well as their efficient storage, with a usage of relatively small memory space.

Let us mention that a problem of enumeration and estimation of the number of linear partitions is studied as well. For an illustration we refer to [ ], [ ], and [ ].

References

1. Andres, E., Acharya, R., Sibata C.: Discrete Analytical Hyperplanes. Graphical Models and Image Processing **59** (1997) 302-309. 168
2. Courrieu, P.: Two Methods for Encoding Clusters. Neural Networks **14** (2001) 175-183. 167
3. Cover, M. T.: Geometric and Statistical Properties of Systems of Linear Inequalities with Applications in Pattern Recognition. IEEE Trans. Electron. Comput. **EC-14** (1965) 326-334. 171
4. Klette, R., Stojmenović, I., Žunić, J.: A New Parametrization of Digital Planes by Least Squares Plane Fit and Generalizations. Graphical Models and Image Processing. **58** (1996) 295-300. 164
5. Koplowitz, J., Lindenbaum, M., Bruckstein, A.: The number of Straight Lines on an $n \times n$ grid. IEEE Trans. on Information Theory. **36** (191990) 192-197. 164
6. Krueger, R. F.: Comments on Takiyama's Analysis of the Multithreshold Threshold Element. IEEE Trans. Pattern Anal. Machine Intell. **8** (1986) 760-761. 161
7. Ngom, A.,Stojmenović, I., Žunić, J.: On the Number of Multilinear Partitions and Computing Capacity of Multiple-Valued Multiple Threshold Perceptrons. Proceedings of 29th IEEE International Symposium on Multiple-Valued Logic, Freiburg in Breisgau (Germany), (1999) 208-213. 167, 171
8. Ojha, C. P.: Enumeration of Linear Threshold Functions from the Lattice of Hyperplane Intersections. IEEE Trans. on Neural Networks **11** (2000) 839-850. 171
9. 'Olafsson, S., Abu-Mostafa, S. Y.: The Capacity of Multilevel Threshold Functions. IEEE Trans. Pattern Anal. Machine Intell. **10** (1988) 277-281. 164
10. Reveilles, J.-P.: Geometrie Discrete, calcul en nombres entiers et algorithmique. State thesis (in French), Universite Louis Pasteur, Strasbourg, Dec. 1991. 168
11. Winder, R. O.: Chow Parameters in Threshold Logic. Journal of ACM **18** (1971) 265-289. 168
12. Zuev, A. Y.: Asymptotics of the Logarithm of the Number of Threshold Functions of the Algebra of Logic. Sov. Math. Dok. **39** (1989) 512-513. 171
13. Žunić, J., Acketa, M. D.: A General Coding Scheme for Families of Digital Curve Segments. Graphical Models and Image Processing **60** (1998) 437-460. 161, 167
14. Žunić, J.: Encoding of Linear and Multi-linear Threshold Functions of n-dimesional Binary Inputs. Submitted (2001).
15. Žunić, J., Sladoje, N.: Efficiency of Characterizing Ellipses and Ellipsoids by Discrete Moments. IEEE Trans. Pattern Anal. Machine Intell. **22** (2000) 407-414. 164

Part III

Geometry

Stability and Instability in Discrete Tomography

Andreas Alpers[1], Peter Gritzmann[2*], and Lionel Thorens[3]

[1] Zentrum Mathematik, Technische Universität München
Arcisstr. 21, D-80290 München, Germany
`alpers@ma.tum.de`
[2] Zentrum Mathematik, Technische Universität München
Arcisstr. 21, D-80290 München, Germany
`gritzman@ma.tum.de`
[3] tfk GmbH, Baierbrunner Str. 39, D-81379 München, Germany
`Lionel.Thorens@tfk-gmbh.de`

Abstract. The paper gives strong instability results for a basic reconstruction problem of discrete tomography, an area that is particularly motivated by demands from material sciences for the reconstruction of crystalline structures from images produced by quantitative high resolution transmission electron microscopy. In particular, we show that even extremely small changes in the data may lead to entirely different solutions. We will also give some indication of how one can possibly handle the ill-posedness of the reconstruction problem in practice.

1 Introduction

The field of discrete tomography deals with the retrieval of information about discrete objects from typically noisy data. The given data describe (possibly weighted) incidences of the object with query sets. Typical query sets are lines, planes, or various kinds of windows. The field's numerous applications range from scheduling problems, [   ], questions of data security, [   ], tasks in image processing, [   ] and many others, [   ], [   ], to the question of the reconstruction of crystalline structures from few of their images under high resolution transmission electron microscopy, [   ], [   ], [   ]. In the present paper we focus on the question of ill-posedness of the latter *discrete inverse problem*.

The quantitative analysis of high resolution transmission electron microscopic images of a given crystal yields essentially the information how many atoms of the given object interact with sharply focused electron beams in a given viewing direction, [   ], [   ]; see also [   ]. So, in principle, we are given the information how many atoms there are on each line parallel to a given small number of directions. To be more precise, let $n \in \mathbb{N}$, $n \geq 2$, let F be a finite subset of $\mathbb{Z}^n$, let S be a line through the origin, and let $\mathcal{A}(S)$ denote the set

---

* Supported in part by the German Federal Ministry of Education and Research Grant 03-GR7TM1.

G. Bertrand et al. (Eds.): Digital and Image Geometry, LNCS 2243, pp. 175–186, 2001.
© Springer-Verlag Berlin Heidelberg 2001

of all lines of Euclidean n-space $\mathbb{E}^n$ that are parallel to S. Then the *(discrete)* *X-ray of F parallel to S* is the function

$$X_S F : \mathcal{A}(S) \to \mathbb{N}_0 = \mathbb{N} \cup \{0\}$$

defined by

$$X_S F(T) = |F \cap T| = \sum_{x \in T} \mathbb{1}_F(x),$$

for each $T \in \mathcal{A}(S)$. In the following let $\mathcal{F}^n = \{F : F \subset \mathbb{Z}^n \land F \text{ is finite}\}$ and $\mathcal{L}^n = \{\text{lin}\{u\} : u \in \mathbb{Z}^n \setminus \{0\}\}$. The elements of $\mathcal{F}^n$ and $\mathcal{L}^n$ are called *lattice sets* and *lattice lines*, respectively. Quite typically, we have some additional a priori information available. This can be modeled by considering suitable subsets $\mathcal{G}$ of $\mathcal{F}^n$.

Given m different lattice lines $S_1, \ldots, S_m$, central questions in discrete tomography are as follows. What kind of information about a finite lattice set F can be retrieved from its X-ray images $X_{S_1} F, \ldots, X_{S_m} F$? How difficult is the reconstruction algorithmically? How sensitive is the task to data errors?

The surveys [   ], [   ] and the book [   ] give an overview of the known results dealing with the first two questions and their relatives. (The surveys also include some further explanation for the restriction made here, give examples for relevant classes $\mathcal{G}$ and describe the imaging process in more detail.)

In the present paper we focus on the stability and instability of the reconstruction task. Of course, it is clear that small changes in the data can produce inconsistency. But this is not really a problem. We will, however, prove the following two theorems which show that a small change in the data can lead to a dramatic change in the image, already in the plane. Here the data is given in terms of functions

$$f_i : \mathcal{A}(S_i) \to \mathbb{N}_0, \qquad i = 1, \ldots, m$$

with finite support, hence the difference of two data functions with respect to the same line S is a function $f : \mathcal{A}(S) \to \mathbb{Z}$ whose size is measured in terms of its ℓ_1-norm

$$\|f\|_1 = \sum_{T \in \mathcal{A}(S)} |f(T)|.$$

Theorem 1. *Let $m \in \mathbb{N}$, $m \geq 3$, $S_1, \ldots, S_m \in \mathcal{L}^2$ and $\alpha \in \mathbb{N}$. Then there exist $F_1, F_2 \in \mathcal{F}^2$ with the following properties:*

> F_1 *is uniquely determined by* $X_{S_1} F_1, \ldots, X_{S_m} F_1$;
> F_2 *is uniquely determined by* $X_{S_1} F_2, \ldots, X_{S_m} F_2$;
> $\sum_{i=1}^m \|X_{S_i} F_1 - X_{S_i} F_2\|_1 = 2(m-1)$;
> $|F_1| = |F_2| \geq \alpha$;
> $F_1 \cap F_2 = \emptyset$.

For the case of three directions Theorem 1 shows that even if the X-ray data coincide on all but four single lines the corresponding solutions may still be completely disjoint. The next result shows that this kind of instability persists even if we weaken our measure of the 'distance' from F_1 to F_2 by replacing $|F_1 \triangle F_2|$ by the affinely invariant difference

$$\delta_{\mathrm{aff}}(F_1, F_2) := \min\{|F_1 \triangle h(F_2)| : h \text{ is an affine transformation}\}.$$

For practical purposes the special case of translations is particularly interesting. It refers to the situation that we regard an image the same no matter where we look at it. As a generalization of Theorem 1 we show the following instability even for the much weaker affinely invariant difference.

Theorem 2. *Let* $m \in \mathbb{N}$ *with* $m \geq 3$ *let* $S_1, \ldots, S_m \in \mathcal{L}^2$ *be* m *different lines, and let* $\alpha \in \mathbb{N}$. *Then there exist* $F_1, F_2 \in \mathcal{F}^2$ *with the following properties:*

> F_1 *is uniquely determined by* $X_{S_1} F_1, \ldots, X_{S_m} F_1;$
> F_2 *is uniquely determined by* $X_{S_1} F_2, \ldots, X_{S_m} F_2;$
> $\sum_{i=1}^{m} \|X_{S_i} F_1 - X_{S_i} F_2\|_1 = 2(m-1);$
> $|F_1| = |F_2| \geq \alpha;$
> $\delta_{\mathit{aff}}(F_1, F_2) \geq |F_1|.$

Our two main theorems will be proved in Section 2. Section 3 will contain a rather weak stability result and some remarks towards a possible regularization of the ill-posed discrete inverse problem with a view towards our prime application in semi-conductor industry.

2   Proof of Theorems 1 and 2

We will now prove our main theorems. In [   ], [   ] a construction is given showing that whenever $m \geq 3$ and $S_1, \ldots, S_m \in \mathcal{L}^2$ there exist arbitrarily large irreducible switching components. Our construction can be regarded as a modification of a generalization of this result. In the following we are not aiming at smallest possible sets F_1, F_2 but try to give the most transparent construction that covers both, Theorem 1 and Theorem 2 simultaneously.

Throughout this section, let $m \geq 3$ and $S_1, \ldots, S_m \in \mathcal{L}^2$ be m different lattice lines. Let $v_1, \ldots, v_m \in \mathbb{Z}^2$ such that $S_i = \lin\{v_i\}$ for $i = 1, \ldots, m$. We may assume without loss of generality that $v_1 = (1,0)^T$ and $v_2 = (0,1)^T$; see [   ].

2.1   Some Technical Lemmas on Polynomials

Our construction given in Subsection 2.2 is based on two functionals. Rather than giving it in all possible generality we will use two specific polynomials p and q here, defined by

$$p(t) = t^5 + t^4 + t^3 \qquad \text{and} \qquad q(t) = t^6 + t^5 + t^4.$$

In addition, we use a parameter $\omega \in \mathbb{N}$ that will be fixed later. The functionals are then of the form

$$f_\omega(x) = \omega p(x) \quad \text{and} \quad g_\omega(x) = q(x/\omega).$$

Let us begin with a simple remark.

Remark 3. *For any* $\omega \in \mathbb{N}$,

(a) $f_\omega(\mathbb{N}), g_\omega(\omega \cdot \mathbb{N}) \subset \mathbb{N}$;
(b) f_ω *and* g_ω *are strictly increasing on* $[0, \infty[$ *whence invertible;*
(c) $f_\omega(t) > g_\omega^{-1}(t)$ *for every* $t \in]1, \infty[$.

Next we prove auxilliary results on certain functional equations involving f_ω and g_ω.

Lemma 1. *Let* $\alpha, \beta, \gamma, \delta, \sigma, \tau \in \mathbb{R}$ *such that* $\alpha\delta - \beta\gamma \neq 0$.

(a) Then the equation

$$\alpha t + \beta f_\omega(t) + \sigma = g_\omega\big(\gamma t + \delta f_\omega(t) + \tau\big)$$

holds for at most 30 *different values of* t.
(b) The equation

$$f_\omega\big(\alpha t + \beta f_\omega(t) + \sigma\big) = \gamma t + \delta f_\omega(t) + \tau$$

holds for all $t \in \mathbb{R}$ *if and only if* $\alpha = \delta = 1$ *and* $\beta = \gamma = \sigma = \tau = 0$. *Otherwise it holds for at most* 25 *different values of* t.
(c) The equation

$$\alpha g_\omega(t) + \beta t + \sigma = g_\omega\big(\gamma g_\omega(t) + \delta t + \tau\big)$$

holds for all $t \in \mathbb{R}$ *if and only if* $\alpha = \delta = 1$ *and* $\beta = \gamma = \sigma = \tau = 0$. *Otherwise it holds for at most* 36 *different values of* t.

Proof. To prove assertion (a) just note that $\deg(g_\omega) = 6 > \deg(f_\omega) = 5 \geq 2$ implies that the equation cannot hold as an identity. Hence there are at most $\deg(f_\omega) \cdot \deg(g_\omega)$ many solutions.

Similarly, if (b) holds identically, $\beta = 0$, and we have the equation

$$f_\omega\big(\alpha t + \sigma\big) = \gamma t + \delta f_\omega(t) + \tau.$$

Taking the k th derivative on both sides for $k = 2, 3, 4, 5$ yields the condition

$$\alpha^k f_\omega^{(k)}\big(\alpha t + \sigma\big) = \delta f_\omega^{(k)}(t).$$

The equation for $k = 5$ yields $\alpha^5 = \delta$, which implies $\delta = 1$ if $\alpha = 1$. Suppose that $\alpha \neq 1$. By choosing $t_0 = \sigma/(1-\alpha) \neq 1/5$ we obtain $\alpha^5 = \alpha^4 = \delta$. For $t_0 = 1/5$ it follows that $f_\omega^{(4)}(t_0) = 0$, but $f_\omega^{(2)}(t_0) \neq 0$. Thus, $\alpha^5 = \alpha^2 = \delta$. In any case, this implies $\alpha = \delta = 1$. Taking the fourth derivative shows now that $\sigma = 0$. Plugging $t = 0$ into the original equation yields $\tau = 0$, and finally $\gamma = 0$.

Of course, (c) follows similarly.   □

2.2   The Basic Construction

Now let $k \in \mathbb{N}$ such that $2^{m-2} \cdot k > \alpha$, where α is the positive integer of Theorem 1 or Theorem 2. For the proof of Theorem 2 we may assume that $\alpha \geq 2^{2(m+2)}$. We begin by constructing two lattice sets U_m and V_m with $|U_m| = |V_m| = 2^{m-2} \cdot k > \alpha$ that are *tomographically equivalent* with respect to $S_1, \ldots, S_m$ i.e., they have the same X-rays parallel to $S_1, \ldots, S_m$.

Let $\omega \geq 2$, set

$$\lambda_1 = g_\omega\big(f_\omega(2)\big),$$

and define

$$\lambda_{i+1} = g_\omega\big(f_\omega(\lambda_i)\big) \qquad \text{for } i = 2, \ldots, k-1.$$

Remark 4. *(a) The sequence $(\lambda_i)_{i=1,\ldots,k}$ is independent of the specific choice of ω;*
(b) $\lambda_1, \ldots, \lambda_k \in \mathbb{N}$
(c) $\lambda_1 < \cdots < \lambda_k$.

Now let

$$B_{k-1} = \left\{ \begin{pmatrix} \lambda_i \\ f_\omega(\lambda_i) \end{pmatrix} : i = 1, \ldots, k-1 \right\}, \qquad B_k = B_{k-1} \cup \left\{ \begin{pmatrix} \lambda_k \\ f_\omega(\lambda_k) \end{pmatrix} \right\}$$

and

$$C_{k-1} = \left\{ \begin{pmatrix} \lambda_{i+1} \\ f_\omega(\lambda_i) \end{pmatrix} : i = 1, \ldots, k-1 \right\}, \qquad C_k = C_{k-1} \cup \left\{ \begin{pmatrix} \lambda_1 \\ f_\omega(\lambda_k) \end{pmatrix} \right\}.$$

Observe that the points of C_{k-1} are all of the form $(t, g_\omega^{-1}(t))^T$.

The two sets B_k and C_k are tomographically equivalent with respect to S_1 and S_2. The rest of the construction will depend on suitably chosen positive integers $\theta_3, \ldots, \theta_m \in \mathbb{N}$ which will be fixed later. For each additional line in $\mathcal{L}^2$, the sizes of the sets are doubled. For $j = 3, \ldots, m$ we define

$$\begin{aligned}
U_2 &= B_k & V_2 &= C_k \\
U_j &= U_{j-1} \cup (V_{j-1} + \theta_j v_j) & V_j &= V_{j-1} \cup (U_{j-1} + \theta_j v_j).
\end{aligned}$$

Clearly, we have the following properties.

Remark 5. *For each ω the integers $\theta_3, \ldots, \theta_m$ can be chosen suitably large, so that $U_m \cap V_m = \emptyset$, $|V_m| = |U_m| = 2^{m-2} \cdot k$, and U_m and V_m are tomographically equivalent with respect to $S_1, \ldots, S_m$.*

The required sets F_1 and F_2 are now constructed by removing a point of U_m and V_m each. In fact, we choose points $z_0 \in U_m$ and $z_1 \in V_m$ that lie on the same line parallel to S_1 and set

$$F_1 = U_m \setminus \{z_0\}, \qquad F_2 = V_m \setminus \{z_1\}.$$

Note that the construction shows that

$$F_1 \cap F_2 = \emptyset \quad \text{and} \quad |F_1| = |F_2| = 2^{m-2} \cdot k - 1.$$

Clearly,

$$\sum_{i=1}^{m} ||X_{S_i} F_1 - X_{S_i} F_2||_1 = 2(m-1).$$

Our theorems will follow if we can show that the integers ω and $\theta_3, \ldots, \theta_m$ can be chosen in such a way that no other finite lattice set is tomographically equivalent to U_m and V_m with respect to $S_1, \ldots, S_m$, and U_m and V_m are such that even affine transformations cannot place them 'too closely' on top of each other.

In fact, from the former property which will be derived in Lemma 3 it follows that F_1 is uniquely determined by its X-rays parallel to $S_1, \ldots, S_m$. Indeed, if $F \in \mathcal{F}^2$ is tomographically equivalent to F_1, then $z_0 \notin F$, and $F \cup \{z_0\}$ is tomographically equivalent to U_m. Since $F \cup \{z_0\} \neq V_m$ we must have $F \cup \{z_0\} = U_m$ hence $F_1 = F$. Analogously, F_2 is uniquely determined by $X_{S_1} F_2, \ldots, X_{S_m} F_2$.

2.3   Properties of U_m and V_m

Now we derive some crucial properties of U_m and V_m for appropriate choices of the parameters. First note that every set that is tomographically equivalent to U_j or, what is the same, to V_j is contained in the *grid* of U_j

$$G_j = \mathbb{Z}^2 \cap \bigcap_{i=1}^{j} \bigcup_{x \in U_j} (x + S_i),$$

for $j = 2, \ldots, m$. Of course,

$$G_2 = \{(\lambda_i, f_\omega(\lambda_j))^T : i, j = 1, \ldots, k\}.$$

Lemma 2. *The integer ω can be chosen in such a way that there is no line parallel to $S_3, \ldots, S_m$ which intersects G_2 in more than one point.*

Proof. Suppose there are indices $i_1, i_2, j_1, j_2 \in \{1, \ldots, k\}$ with $i_1 \neq i_2, j_1 \neq j_2$, and $v \in \{v_3, \ldots, v_m\}$, and there exists a real μ such that

$$\begin{pmatrix} \lambda_{i_1} \\ f_\omega(\lambda_{j_1}) \end{pmatrix} - \begin{pmatrix} \lambda_{i_2} \\ f_\omega(\lambda_{j_2}) \end{pmatrix} = \mu v.$$

By Remark 4 (a) this equation is equivalent to

$$\begin{pmatrix} \lambda_{i_1} - \lambda_{i_2} & 0 & -\nu_1 \\ 0 & p(\lambda_{j_1}) - p(\lambda_{j_2}) & -\nu_2 \end{pmatrix} \begin{pmatrix} 1 \\ \omega \\ \mu \end{pmatrix} = 0,$$

where $v = (\nu_1, \nu_2)^T$. The rank of the 2×3 matrix is 2, hence its nullspace is of dimension 1. On the other hand, for any two different numbers ω_1, ω_2 and arbitrary μ_1, μ_2, the vectors $(1, \omega_1, \mu_1)^T$ and $(1, \omega_2, \mu_2)^T$ are always linearly independent, whence two such vectors cannot be in the nullspace simultaneously. Taking all possible quadruples of indices $i_1, i_2, j_1, j_2 \in \{1, \ldots, k\}$ leads then to the union of $k^2(k-1)^2$ one-dimensional subspaces that we have to avoid. Hence among any set of $k^2(k-1)^2 + 1$ different values of ω there must be one for which none of the vectors $(1, \omega, \mu)^T$ lie in any of the nullspaces. $\qquad\square$

Note that the result of Lemma 2 is invariant with respect to translations of G_2. In the following we assume that ω is fixed according to Lemma 2. The next lemma implies Theorem 1 already.

Lemma 3. *The integers $\theta_3, \ldots, \theta_m$ can be chosen large enough such that no other finite lattice set is tomographically equivalent to U_m and V_m with respect to $S_1, \ldots, S_m$.*

Proof. Since G_m is the grid of U_m and V_m we clearly have $U_m \cup V_m \subset G_m$. Setting $G^{(2)} = G_2$, it is also clear that we can choose $\theta_3, \ldots, \theta_m$ large enough such that

$$G_i \subset G^{(i)} = G^{(i-1)} \cup (\theta_i v_i + G^{(i-1)})$$

for $i = 3, \ldots, m$. For a formal proof see [, Proof of Lemma 3.1].

The assertion will now follow if we make sure that $G_m = U_m \cup V_m$ since the only two subsets of $U_m \cup V_m$ that are tomographically equivalent to U_m or V_m are precisely these two sets. We will show that for each point $g \in G^{(m)} \setminus (U_m \cup V_m)$ the line $g + S_3$ does not intersect $U_m \cup V_m$. So, let $g \in G^{(m)} \setminus (U_m \cup V_m)$ and suppose $g + S_3$ intersects $U_m \cup V_m$. Of course, $U_m \cup V_m$ consists of disjoint translates of $U_2 \cup V_2$, with translation vectors of the form $\sum_{i=3}^{m} \delta_i \theta_i v_i$ with $\delta_1, \ldots, \delta_m \in \{0, 1\}$. Let t_1 be such a translation vector with $g \in t_1 + G_2$. By our assumption there is another such translation vector t_2 with $(g + S_3) \cap (t_2 + G_2) \neq \emptyset$. By Lemma 2 no line parallel to S_3 intersects $t_1 + G_2$ and $t_2 + G_2$ in more than one point each. Hence these points must be g and $(t_2 - t_1) + g$. But $g \notin t_1 + (U_2 \cup V_2)$ hence $(t_2 - t_1) + g \notin t_2 + (U_2 \cup V_2)$. This contradiction concludes the proof of the assertion. $\qquad\square$

2.4   The Affinely Invariant Difference

In order to prove Theorem 2 we need to study the power of affine transformations on F_2 to reduce the symmetric difference to F_1. The next result is a corollary to Lemma 1.

Lemma 4. *Let $A \in \mathbb{R}^{2 \times 2}$ be nonsingular, let $a \in \mathbb{R}^2$, and let h be the affine transformation defined by $h(x) = Ax + a$. Then*

(a) $|B_k \cap h(C_k)| \leq 31$;
(b) $|B_k \cap h(B_k)| \geq 26 \Rightarrow h \equiv id$;
(c) $|C_k \cap h(C_k)| \geq 39 \Rightarrow h \equiv id$.

Proof. Since the one point in $C_k \setminus C_{k-1}$ is of slightly different form than the points of $B_k \cup C_{k-1}$ we always count it as 'matched'. Then the assertion follows easily from Lemma 1. □

Lemma 4 shows already that the 'overlay' power of affine transformations is rather limited. In fact, suppose that A is not the unit matrix i.e., h is not just a translation. If we use again that $U_m \cup V_m$ consists of disjoint translates of $U_2 \cup V_2$ with translation vectors of the form $\sum_{i=3}^{m} \delta_i \theta_i v_i$ with $\delta_1, \ldots, \delta_m \in \{0, 1\}$, and of course that $U_2 = B_k$ and $V_2 = C_k$ then Lemma 4 implies that

$$
\begin{aligned}
&|h(U_m \cup V_m) \cap (U_m \cup V_m)| \\
&\leq 2^{m-2} \cdot 2^{m-2} \cdot \Big(|h(U_2) \cap U_2| + |h(U_2) \cap V_2| + |h(V_2) \cap U_2| + |h(V_2) \cap V_2| \Big) \\
&\leq 125 \cdot 2^{2m-4} \leq 125 \cdot \alpha \cdot 2^{-8} \leq \tfrac{1}{2}\alpha \leq \tfrac{1}{2}|F_1|.
\end{aligned}
$$

So it remains to study the effect of translations. To do this appropriately we require one more property of the numbers $\theta_3, \ldots, \theta_m$. It is most easily stated with the help of a geometric functional, the *breadth* $b_v(S)$ of a planar point set S in a given direction $v \in \mathbb{R}^2$, $\|v\|_2 = 1$; it is defined by

$$
b_v(S) = \max_{x \in S} v^T x - \min_{x \in S} v^T x.
$$

We choose the integers $\theta_3, \ldots, \theta_m$ so that

$$
\|\theta_i v_i\|_2 > b_{\hat{v}_i}(U_{i-1} \cup V_{i-1}) \qquad \text{for } i = 3, \ldots, m,
$$

where $\hat{v}_i = v_i / \|v_i\|_2$. Of course, this choice does not interfere with the previous requirements.

Now let t be a translation vector. Suppose first that

$$
v_m^T t \geq b_{\hat{v}_m}(U_{m-1} \cup V_{m-1}).
$$

This means on the one hand that t has the capability of moving a point of $U_{m-1} \cup V_{m-1}$ into $\theta_m v_m + (U_{m-1} \cup V_{m-1})$. But on the other hand

$$
(U_{m-1} \cup V_{m-1}) \cap \Big(t + (U_{m-1} \cup V_{m-1}) \Big) = \emptyset.
$$

Hence

$$
\Big| (U_m \cup V_m) \triangle \Big(t + (U_m \cup V_m) \Big) \Big| \geq |F_1|,
$$

which is the assertion. So we can assume that

$$
\begin{aligned}
&(U_m \cup V_m) \triangle \Big(t + (U_m \cup V_m) \Big) \\
&= \Big((U_{m-1} \cup V_{m-1}) \cap \Big(t + (U_{m-1} \cup V_{m-1}) \Big) \Big) \cup \\
&\quad \cup \Big(\Big(\theta_m v_m + (U_{m-1} \cup V_{m-1}) \Big) \cap \Big(t + \theta_m v_m + (U_{m-1} \cup V_{m-1}) \Big) \Big).
\end{aligned}
$$

Hence by an induction argument we can reduce the assertion to $B_k \cup C_k$, and it suffices to show that $|B_k \cap (t + C_k)| \leq \tfrac{1}{2}|B_k|$. This follows from Lemma 4, which concludes the proof of Theorem 2.

3   Dealing with Instability in Practice

The approximation algorithms given in [   ] show that the 'combinatorial optimization part' of the reconstruction problem of discrete tomography can be handled very efficiently in spite of its computational complexity. In particular, [   ] considered the problem BEST-INNER-FIT$(S_1, \ldots, S_m)$ [BIF] that accepts as input data functions $f_1, \ldots, f_m$, and its task is to find a set $F \in \mathcal{F}^n$ of maximal cardinality such that

$$X_{S_i} F(T) \leq f_i(T) \qquad \text{for all } T \in \mathcal{A}(S) \text{ and } i = 1, \ldots, m.$$

The results show that in spite of the underlying $\mathbb{NP}$-hardness some simple heuristics already yield good a priori approximation guarantees and behave surprisingly well in practice. In fact a dynamic greedy/postprocessing strategy actually produced solutions with small *absolute error*; for instances with 250000 variables and point density of 50% an average of +only 1.07, 23.28, 64.58 atoms were missing for 3, 4, 5 directions, respectively.

As we have seen the reconstruction task of discrete tomography is intrinsically instable. There is however the following (rather weak) stability result.

Lemma 5. *Let $F_1 \in \mathcal{F}^n$ and let $f_1, \ldots, f_m$ be given (generally noisy) data functions. Then there exists a finite lattice set F_2 with $X_{S_i} F_2(T) \leq f_i(T)$ for all $T \in \mathcal{A}(S)$ and $i = 1, \ldots, m$ such that*

$$|F_1 \triangle F_2| \leq \sum_{i=1}^{m} \|f_i - X_{S_i} F_1\|_1.$$

Proof. F_2 is constructed from F_1 by deleting at most $\sum_{i=1}^{m} \|f_i - X_{S_i} F_1\|_1$ points in order to satisfy the constraints $X_{S_i} F_2(T) \leq f_i(T)$ for all $T \in \mathcal{A}(S)$ and $i = 1, \ldots, m$.   □

Lemma 5 shows in particular that under data errors the cardinality of a solution behaves in a stable way. Also it shows that even with noisy data there is always a solution of [BIF] that is 'close' to the original object in terms of the symmetric difference. (Of course, it is not even clear how to recognize a 'good' solution if the original image is not known, let alone how to find one.) In view of Theorem 1 this means that it is generally not a good strategy in practice to try to satisfy the measured constraints as well as possible. In fact, Theorem 1 shows that the 'perfect solution' for the noisy data may be disjoint from the original image while by Lemma 5 a 'nonperfect solution' may be quite close.

In any case, it is important to utilize any additional physical knowledge and experience available in order to actually produce solutions that are close to the original physical objects. For the task of quality control in semi-conductor industry we can, e.g., utilize the fact that we know the 'perfect' object in advance. Suppose we are given a few images under high resolution transmission electron microscopy of parts of a silicon wafer that carries an etched pattern of some electrical circuit. Our instability results seem to suggest that there is not much hope for a reasonably good reconstruction of the object from these measurements since all physical measurements are noisy. On the other hand, we do not need to reconstruct the object 'from scratch'. We know what the object would look like if the production process had been exact and the measurements were correct. Hence it seems reasonable to measure the distance from the theoretical perfect chip. This means we are trying to reconstruct a finite lattice set approximately satisfying the (noisy) X-ray measurements so that its distance from the perfect template is as small as possible. If there does not exist a good enough approximation, we know that the chip is faulty. But if there exists a good enough approximation, all we know is that the data from high resolution transmission electron microscopy is not sufficient to rule out the possibility that the chip is correct. Let us look at the following simple example of two images F_1 and F_2 depicted in Figure 1.

The corresponding grid G has the size 40×40; whence the reconstruction problem involves 1600 0-1-variables. The sets F_1 and F_2 are depicted as fat black points ($\bullet$); the empty places are represented as small dots ($\cdot$). Template F_1 is regarded as proper, while template F_2 is faulty since the conductor path is blocked. Now we consider X-rays parallel to the lines $S_1 = \mathrm{lin}\{(1,0)^T\}$, $S_2 = \mathrm{lin}\{(0,1)^T\}$ and $S_3 = \mathrm{lin}\{(1,1)^T\}$. Note that F_1 and F_2 are uniquely determined by their X-rays parallel to these lines.

However, as shown in Figure 2 there are finite lattice sets F_1' and F_2' with

$$-1 \le X_{S_i} F_1(T) - X_{S_i} F_1'(T) \le 1 \quad \text{for each} \ \ T \in \mathcal{A}(S_i), \ i = 1, 2, 3,$$
$$\text{and} \quad -1 \le X_{S_i} F_2(T) - X_{S_i} F_2'(T) \le 1 \quad \text{for each} \ \ T \in \mathcal{A}(S_i), \ i = 1, 2, 3$$

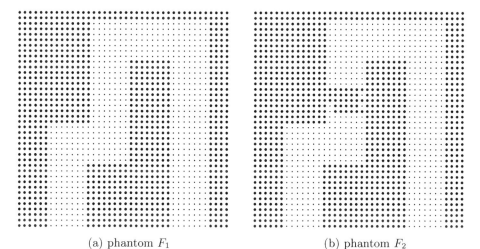

(a) phantom F_1 (b) phantom F_2

Fig. 1. Two lattice sets in a 40×40 grid

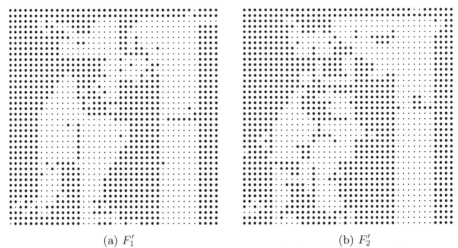

(a) F_1' (b) F_2'

Fig. 2. Reconstructions of F_1 and F_2, respectively, based on data with error at most one on each line

that do not share the same 'conductor path structure'. Also, F_1' and F_2' look similar and it does not seem justified to distinguish a 'proper chip' from a 'faulty' one based on this reconstruction. But after all, the goal is only to see whether there is a chance to detect faulty chips even in the presence of considerable data errors. And in fact, if the error on each X-ray line is reasonably small such information can be derived. Let $\epsilon \in \mathbb{N}$, and suppose that for $j = 1, 2$ the data functions f_1^j, f_2^j, f_3^j are measured with

$$-\epsilon \le X_{S_i} F_j(T) - f_i^j \le \epsilon \qquad \text{for each } T \in \mathcal{A}(S_i),\ i = 1, 2, 3,\ j = 1, 2.$$

We pursue an approach motivated by Lemma 5, choose some $\delta \in \mathbb{N}$ and determine two sets F^1 and F^2 that minimize the symmetric difference to F_1 under the constraints that

$$-\delta \le X_{S_i} F^j(T) - f_i^j \le \delta \qquad \text{for each } T \in \mathcal{A}(S_i),\ i = 1, 2, 3,\ j = 1, 2.$$

Clearly, if $\delta \ge \epsilon$, the set F^1 coincides with the template F_1. Further, if

$$\delta + \epsilon < \max_{i=1,2,3} \|X_{S_i} F_1 - X_{S_i} F_2\|_\infty,$$

then $F^2 \ne F_1$. In our specific example this is for instance the case when $\delta = \epsilon = 3$. The question is whether this trivial observation which is only based on the 'local' information of what happens on single X-ray lines can be substantially extended when all the given information is taken into account and if the error is regarded as random variable with some given underlying distribution. Also, what can be said for real-world data? These questions will require further theoretical investigations and an extended experimental study for the above and other conceivable regularization techniques.

References

[FLRS91]   P. C. Fishburn, J. C. Lagarias, J. A. Reeds, and L. A. Shepp. Sets uniquely determined by projections on axes II: Discrete case. *Discrete Math.*, 91(2):149–159, 1991. 175

[GdV01]   P. Gritzmann and S. de Vries. Reconstructing crystalline structures from few images under high resolution transmission electron microscopy. In *Mathematics: Key Technology for the Future, (ed. by W. Jäger)*. Springer, 2001. in print. 175, 176

[GdVW00]   P. Gritzmann, S. de Vries, and M. Wiegelmann. Approximating binary images from discrete X-rays. *SIAM J. Opt.*, 11:522–546, 2000. 183

[GGP99]   R. J. Gardner, P. Gritzmann, and D. Prangenberg. On the computational complexity of reconstructing lattice sets from their X-rays. *Discrete Math.*, 202:45–71, 1999. 177, 181

[GGP00]   R. J. Gardner, P. Gritzmann, and D. Prangenberg. On the computational complexity of determining polyatomic structures by X-rays. *Th. Comput. Sci.*, 233:91–106, 2000. 175

[Gri97]   P. Gritzmann. On the reconstruction of finite lattice sets from their X-rays. In *Discrete Geometry for Computer Imagery, (ed. by E. Ahronovitz and C. Fiorio)*, pages 19–32. Springer, Berlin, 1997. 176

[HK99]   G. T. Herman and A. Kuba (eds.). *Discrete Tomography: Foundations, Algorithms, and Applications*. Birkhäuser, Boston, 1999. 176

[IJ94]   R. W. Irving and M. R. Jerrum. Three-dimensional statistical data security problems. *SIAM J. Comput.*, 23:170–184, 1994. 175

[KH98]   T. Y. Kong and G. T. Herman. On which grids can tomographic equivalence of binary pictures be characterized in terms of elementary switching operations. *Int. J. Imaging Syst. Technol.*, 9:118–125, 1998. 177

[KH99]   T. Y. Kong and G. T. Herman. Tomographic equivalence and switching operations. In *Discrete tomography: foundations, algorithms, and applications (ed. by G. Herman and A. Kuba)*, pages 59–84. Birkhäuser, Boston, 1999. 177

[KSB+95]   C. Kisielowski, P. Schwander, F. H. Baumann, M. Seibt, Y. Kim, and A. Ourmazd. An approach to quantitative high-resolution transmission electron microscopy of crystalline materials. *Ultramicroscopy*, 58:131–155, 1995. 175

[Rys78]   H. J. Ryser. *Combinatorial Mathematics*, chapter 6, Matrices of zeros and ones. Mathematical Association of America and Quinn & Boden, Rahway, NJ, 1963, pp. 61–78. 175

[SG82]   C. H. Slump and J. J. Gerbrands. A network flow approach to reconstruction of the left ventricle from two projections. *Comput. Graphics Image Process.*, 18:18–36, 1982. 175

[SKB+93]   P. Schwander, C. Kisielowski, F. H. Baumann, Y. Kim, and A. Ourmazd. Mapping projected potential, interfacial roughness, and composition in general crystalline solids by quantitative transmission electron microscopy. *Phys. Rev. Lett.*, 71(25):4150–4153, 1993. 175

Point-to-Line Mappings and Hough Transforms

Prabir Bhattacharya[1], Azriel Rosenfeld[2], and Isaac Weiss[2]

[1] Dept. of Computer Science, University of Nebraska
Lincoln, NE 68588-0115, USA
[2] Center for Automation Research, University of Maryland
College Park, MD 20742-3275, USA

Abstract. Nearly 40 years ago Hough showed how a point-to-line mapping that takes collinear points into concurrent lines can be used to detect collinear sets of points, since such points give rise to peaks where the corresponding lines intersect. Over the past 30 years many variations and generalizations of Hough's idea have been proposed. Hough's mapping was linear, but most or all of the mappings studied since then have been nonlinear, and take collinear points into concurrent curves rather than concurrent lines; little or no work has appeared in the pattern recognition literature on mappings that take points into lines.

This paper deals with point-to-line mappings in the real projective plane. (We work in the projective plane to avoid the need to deal with special cases in which collinear points are mapped into parallel, rather than concurrent, lines.) We review basic properties of linear point-to-point mappings (collineations) and point-to-line mappings (correlations), and show that any one-to-one point-to-line mapping that takes collinear points into concurrent lines must in fact be linear. We describe ways in which the matrices of such mappings can be put into canonical form, and show that Hough's mapping is only one of a large class of inequivalent mappings. We show that any one-to-one point-to-line mapping that has an incidence-symmetry property must be linear and must have a symmetric matrix which has a diagonal canonical form. We establish useful geometric properties of such mappings, especially in cases where their matrices define nonempty conics.

1 Introduction

In 1962 Hough [ ] patented a method of detecting collinear sets of points in an image by applying a linear point-to-line mapping to the image. Such a mapping, which in projective geometry is called a *correlation* (see Section 3), takes collinear points (points that all lie on the same line) into concurrent lines (lines that all pass through the same point). Thus if the image contains n collinear points the mapping takes these points into n concurrent lines; hence if the mapping is additive it produces a peak of height n at the point where these lines intersect.

Over the past 30 years many variations and generalizations of Hough's idea have been proposed. These methods generally involve (not necessarily linear) mappings f from the plane into some space, such that if many points of the plane

G. Bertrand et al. (Eds.): Digital and Image Geometry, LNCS 2243, pp. 187–208, 2001.
© Springer-Verlag Berlin Heidelberg 2001

belong to a locus of a given form, their images under f give rise to a peak in the space. Such mappings are called *Hough transforms*; for reviews of the literature on these transforms see [, ,]. Hough's original mapping was linear, but as we will see in Section 2 the Hough transforms developed afterward were nonlinear and map collinear points into concurrent curves rather than concurrent lines. Thus since Hough's original patent little or no work on point-to-line mappings has appeared in the pattern recognition literature.

We will not attempt here to develop a general theory of nonlinear Hough transforms; we will restrict ourselves to (not necessarily linear) point-to-line mappings from the plane into itself. To avoid special cases in which collinear points map into parallel lines, we will work in the real projective plane. To make the paper more widely accessible, in Section 3 we will provide a brief introduction to the real projective plane and to point-to-point and point-to-line mappings of the projective plane into itself. A point-to-line mapping need not be linear even if it is one-to-one, but it can be shown that if it takes collinear points into concurrent lines it must be linear; thus any point-to-line Hough transform must in fact be linear.

In Section 4 we will review some basic properties of linear point-to-line mappings, with emphasis on point-line incidence properties of the mappings and on symmetry properties of their matrices. We will also discuss canonical forms for the matrices of such mappings and will show that Hough's mapping is only one of a large class of inequivalent mappings. We will show that a (not necessarily linear) one-to-one point-to-line mapping that has an incidence-symmetry property must be linear and must have a symmetric matrix; such mappings have diagonal canonical matrices and simple geometric properties, particularly when their matrices define nonempty conics.

2   Classical (Linear and Nonlinear) Hough Transforms

In the point-to-line mapping used by Hough in his patent [ ] the point (u, v) was mapped into the line $x = vy + u$; evidently this mapping is linear. The mapping is given in [ ] p. 123 in a slightly different form, as taking (u, v) into the line $y = -ux + v$; evidently this too is a linear mapping. (Canonical forms for these two versions of Hough's mapping will be given in Section 4.2; interestingly, they are not identical.) Note that in the first mapping the coordinates of the point determine the x-intercept of the line and the reciprocal of its slope, and in the second mapping they determine the line's y-intercept and the negative of its slope.

The first paper to introduce the term "Hough transform" [ ], published nearly 30 years ago, criticized Hough's definition because it used a parameterization of the line in which the parameters can become infinite. (The slope and y-intercept of a vertical line are infinite, and the reciprocal slope and x-intercept of a horizontal line are infinite; thus for both $x = vy + u$ and $y = -ux + v$, both parameters are infinite either for vertical lines or for horizontal lines.) It was therefore proposed in [ ] to use the so-called "normal form" for lines rather than a slope-intercept form. In the normal form a line is parameterized by the slope angle θ of its normal (which takes values in the bounded range $[0, \pi)$; the slope of the normal is $\tan\theta$) and its distance ρ from the origin (which is bounded for lines that intersect an image of bounded size). In this parameterization the equation of the line is $x\cos\theta + y\sin\theta = \rho$; the point-to-line mapping used in [ ] mapped the point (θ, ρ) into this line. Evidently this is not a linear mapping; the coefficients of x and y are sinusoidal rather than linear functions of the slope angle θ. Also, this mapping usually does not take collinear points into concurrent lines; for example, if ρ is a nonzero constant the points (θ, ρ) go into the lines that are tangent to a circle of radius ρ centered at the origin, and these lines are not concurrent. However, it can be shown that if the points x_i, y_i in the xy-plane are collinear then the sinusoidal curves $\rho = x_i\cos\theta + y_i\cos\theta$ in the $\theta\rho$-plane are concurrent; thus this mapping takes collinear points into concurrent sinusoids which give rise to a peak in the $\theta\rho$-plane where they intersect.

Several other Hough transform definitions were proposed during the 1980's; they too make use of nonlinear point-to-line mappings. In the transform defined in [ ] a line is mapped into its intercepts with the sides of an upright square (e.g., the sides of the image); evidently these intercepts are of bounded sizes. Let the sides of the square lie on the positive x and y axes, let their length be s, and let the equation of the line be $y = mx + b$. If $0 < m < \infty$ and $0 \leq b \leq s$ this line has intercepts m and $(s - b)/m$ with the left and upper sides of the square; the formulas for the intercepts are similar for other values of m and b such that the line intersects the square. Here again the mapping that takes the point $(m, (s-b)/m)$ into the line $y = mx+b$ is not linear; indeed $b = -m[(s-b)/m]+s$, so b is a linear function of the product of the intercepts. Similar remarks apply to the variant of this transform defined in [ ]. The same is true for the transform defined in [ ], which maps the line $y = mx + b$ into the coordinates of the foot of the perpendicular from the origin onto that line. It is easily seen that the coordinates of this point are $(b\sin\theta\cos\theta, b\cos^2\theta)$ where $\tan\theta = m$; here again the coefficients of the line are not linear functions of these parameters. Thus the classical Hough transforms [, , ,] all involve nonlinear point-to-line mappings.

Although these point-to-line mappings are nonlinear they are still one-to-one; a line is uniquely defined by the parameters of its normal form or by the coordinates of the foot of the perpendicular to it from the origin, and a line that crosses an image is uniquely defined by its intercepts with the sides of the image. In Section 3.3 we will show that if a (not necessarily linear) point-to-line mapping is one-to-one and takes collinear points into concurrent lines it must in fact be linear.

If a point-to-line mapping takes collinear points into concurrent lines we can use it to detect sets of collinear points even if we discard the lines and keep only the "peaks" where sets of the lines intersect. Evidently even if the point-to-line mapping is one-to-one, the mapping that takes the points into the positions and heights of the peaks is many-to-one; many different point patterns that include sets of collinear points can give rise to the same set of peaks, since the height and position of a peak depend only on the number of collinear points in a set and the equation of the line on which they lie. It can be shown [ ] that if the points all lie on the sides of a convex polygon then the polygon is uniquely determined by the set of peaks, but this is not true if the polygon is not convex. (Actually, [ ] did not deal with discrete sets of points on the sides of the polygon; it assumed that the points lying on a line segment give rise to a peak whose height is proportional to the length of the segment.) For further discussion of the non-uniqueness of the correspondence between patterns of points (or line segments) and the patterns of peaks in their Hough transforms see [, ,].

3   Linear Point-to-Point and Point-to-Line Mappings in the Real Projective Plane

In this paper we study (not necessarily linear) point-to-line mappings from the projective plane into itself. In Section 3.1 we present a brief, informal introduction to the real projective plane, and in Section 3.2 we introduce linear mappings of the projective plane into itself that take points into points ("collineations") or points into lines ("correlations"). It is known that if a one-to-one point-to-point mapping of the projective plane into itself preserves collinearity it must be linear. In Section 3.3 we use this result to prove (Theorem 1) that if a one-to-one point-to-line mapping of the projective plane into itself takes collinear points into concurrent lines it must be linear.

3.1   Points and Lines; The Projective Plane; Homogeneous Coordinates

In the Euclidean plane a point can be specified by a pair of real-number coordinates (x, y), and a line can be specified as a set of points that satisfy a linear equation $ax + by + c = 0$ where the coefficients a, b, c are real numbers and a, b are not both zero. Any two distinct points are contained in a unique line; specifically, the line containing the points (r, s) and (u, v) has the equation $(v - s)x - (u - r)y + (us - vr) = 0$. Any two distinct lines intersect in at most one point, but parallel lines (i.e. lines $ax + by + c = 0$ and $dx + ey + f = 0$ for which the coefficients a and d, b and e are proportional) do not intersect.

The relationship between points and lines can be made symmetric by adjoining "points at infinity" to the plane and specifying that any two parallel lines intersect at such a point. This can be done without introducing infinite values for the coordinates of points. Instead, we introduce a *homogeneous coordinate* system in which a point is specified by a set of proportional triples of real numbers (i.e., if (r, s, t) and (u, v, w) are proportional we say that they specify the same point) and we require that the elements of a triple not all be zero. In this coordinate system, if $t \neq 0$ the triple (r, s, t) represents the "finite" point $(x, y) = (r/t, s/t)$; if $t = 0, (r, s, t)$ represents a "point at infinity". The set of points defined by triples in this way is called the *real projective plane*.

When we use homogeneous coordinates lines are still specified by linear equations; a line is a set of triples (x, y, z), not all zero, that satisfy an equation of the form $ax + by + cz = 0$ where a, b, c are not all zero. Note that if (r, s, t) satisfies this equation so does any triple proportional to (r, s, t); thus a line is still a set of points. Moreover, if two triples of coefficients (a, b, c) and (d, e, f) are proportional they evidently define the same line; thus lines too are defined by triples of homogeneous "coordinates" (their coefficients). It is not hard to see that any two distinct (non-proportional) points are contained in a unique line and any two distinct lines (=lines whose coefficients are not proportional) contain (i.e., intersect in) a unique point.

To summarize: In homogeneous coordinates both points and lines are defined by triples of real numbers (coordinates or coefficients) not all of which are zero, where proportional triples are regarded as defining the same point or line. Because of this symmetric relationship between points and lines the projective plane can be regarded as either a set of points or a set of lines, and if a statement about points is true (e.g., that the union of any two distinct points is contained in a unique line) then the "dual" statement about lines (e.g., that the intersection of any two distinct lines contains a unique point) is also true.

3.2 Collineations and Correlations

A linear mapping of the projective plane into itself is called a *collineation*. Because of the symmetric relationship between points and lines such a mapping can be regarded as either taking points into points or lines into lines. The symmetry between points and lines also allows us to define linear mappings that take points into lines or vice versa; such a mapping is called a *correlation*. The terms "collineation" and "correlation" were introduced by Möbius in 1827. (Note that in signal processing and statistics the word "correlation" has rather different meanings.) Our main interest in this paper is in correlations, but we will occasionally make use of basic facts about collineations. Note that statements about correlations are not the "duals" of statements about collineations.

In any given homogeneous coordinate system a correlation $f : \mathbf{x} \to \mathbf{u}$ can be written in vector-matrix notation as $\mathbf{u} = \mathbf{A}\mathbf{x}$ where $\mathbf{A}$ is a 3×3 matrix and $\mathbf{x}$ and $\mathbf{u}$ are 3×1 matrices ("column vectors") representing the coordinates of $\mathbf{x}$ and the coefficients of $\mathbf{u}$ respectively. Evidently two matrices define the same correlation f (with respect to a given coordinate system) iff one is a nonzero scalar multiple of the other. If $\mathbf{A}$ defines f we should call it "a" matrix of f, but we will sometimes refer to it as "the" matrix of f if this will not cause confusion.

Evidently a collineation or correlation is one-to-one (i.e. invertible) iff its matrix (more precisely: any of its matrices) is nonsingular. The invertible collineations evidently form a group under composition of functions. The invertible correlations cannot form a group since they are not closed under composition of functions, but together with the invertible collineations they do form a group.

As an example, we saw in Section 2 that the point-to-line mapping used by Hough [ ] mapped the point (u, v) into the line $x = vy + u$. In homogeneous coordinates this mapping takes the point $[u, v, 1]^T$ into the line $[-1, v, u]^T$. Evidently this mapping is a correlation and has matrix $\begin{pmatrix} 0 & 0 & -1 \\ 0 & 1 & 0 \\ 1 & 0 & 0 \end{pmatrix}$. Note that this matrix has determinant 1, so the mapping is invertible.

Collineations and correlations are treated in detail in books on (projective) geometry such as [, , , , , , , , , , , , , , , , , , ,] [, , , , ,]; treatments of the subject intended for use by computer vision researchers can be found in [, , ,]. Correlations, in particular, are treated in [ ] pp. 186–187; [ ] pp.58, 63; [ ] pp.30-31, 74; [ ] pp.254-259, 262-263; [ ] pp.90-93; [ ] pp.55-61; [ ] pp.179-188; [ ] pp.51-53, 85; [ ] pp.165-171; [ ] pp.49, 57, 98; [ ] pp.265-270; [ ] pp.339-342; [ ] pp.168, 326-334; [ ] pp.79, 102; [ ] p.39; [ ] Ch.9, pp.362-429; [ ] p.28 (see also [ ]); [ ] pp.234-235; [ ] pp.229-235; [ ] pp.136, 268; [ ] p.483; [ ] pp.198, 280, 288; [ ] pp.224-230; [ ] p.171; [ ] pp.89-90; [ ] p.24; [ ] pp.262-267, 278-280; [ ] pp.77-86; [ ] pp.88-94.

3.3   A Point-to-Line Hough Transform Must Be Linear

Let f be a mapping (not necessarily linear) defined on the real projective plane that takes points into lines. Even if f is a one-to-one correspondence and hence is invertible, it need not be linear. For example, if f takes the point (u, v, w) into the line whose coefficients are (vw, wu, uv) (see [ ] p.40, and [ ] pp. 74-76, where this mapping is called a polarity with respect to a triangle) it is invertible but not linear; this is the simplest nontrivial example of an invertible *Cremona transformation* (see [ ] p.231). The same is true if f takes (u, v, w) into (u^3, v^3, w^3), since on the real projective plane this mapping too is invertible. However, as we shall now show, if f is a one-to-one point-to-line correspondence that takes collinear points into concurrent lines it must be linear.

It is known that a one-to-one point-to-point correspondence preserves collinearity iff it is linear; dually, the same is true for a one-to-one line-to-line correspondence that preserves concurrence. Proofs of this theorem can be found in many books on projective geometry; for examples see [ ] p.88 (where it is called the Fundamental Theorem of Projective Geometry), [ ] p.57, [ ] p.109, and [ ] p.152. Interestingly, none of these books mentions the corresponding result about point-to-line correspondences (which is NOT the dual of the point-to-point or line-to-line result); we therefore prove it here using an argument based on [ ] p.85.

Note first that if f is a one-to-one point-to-line correspondence f^{-1} is a one-to-one line-to-point correspondence, and if f and g are one-to-one point-to-line correspondences $f^{-1} \circ g$ is a one-to-one point-to-point correspondence. Let f_0 be any specific linear one-to-one point-to-line correspondence—for example, the "identity" mapping that takes the point (u, v, w) into the line whose coefficients are (u, v, w); note that this particular f_0 is defined by the 3×3 identity matrix $\mathbf{I}$. Then for any one-to-one point-to-line correspondence f we have $f = f_0 \circ (f_0^{-1} \circ f)$, where $f_0^{-1} \circ f$ is a one-to-one point-to-point correspondence. Evidently a linear point-to-line correspondence maps collinear points into concurrent lines and vice versa; hence if f maps collinear points into concurrent lines $f_0^{-1} \circ f$ preserves collinearity. Hence by the theorem stated in the preceding paragraph $f_0^{-1} \circ f$ is linear. Since f_0 is linear $f = f_0 \circ (f_0^{-1} \circ f)$ must also be linear. We have thus proved

Theorem 1. *A one-to-one point-to-line correspondence takes collinear points into concurrent lines iff it is linear.*

Since a point-to-line Hough transform must take collinear points into concurrent lines, we have thus proved that any such transform must be linear. As we saw in Section 2 the "classical" nonlinear Hough transforms take collinear points into concurrent curves, not into concurrent lines.

4   Linear and Non-linear Point-to-Line Mappings

In Section 4.1 we review some basic properties of linear point-to-line mappings (correlations), with emphasis on point-line incidence properties of the mappings and symmetry properties of their matrices; in particular, we discuss the relationship between polarities (correlations whose matrices are symmetric) and conics. In Section 4.2 we discuss how to define canonical forms for the matrices of correlations on the real projective plane. In Section 4.3 we study point-line incidence properties of point-to-line mappings that are not necessarily linear, and we prove (Theorem 3) that a one-to-one point-to-line mapping that satisfies an incidence-symmetry property must in fact be linear and must have a symmetric matrix.

4.1   Linear Correlations

Let f be an invertible linear point-to-line mapping, say $f : \mathbf{x} \to \mathbf{u} = \mathbf{A}\mathbf{x}$. The *dual* of f is the line-to-point mapping $g : \mathbf{u} \to \mathbf{x} = (\mathbf{A}^{-1})^T \mathbf{u}$, where T denotes matrix transpose. Note that g is not the same as the inverse of f unless $\mathbf{A}$ (and $\mathbf{A}^{-1}$) are symmetric. We begin this section by observing that f and its dual have a useful "incidence-preservation" property (see also Section 4.3):

Proposition 1. *A point P lies on a line L iff the point $g(L)$ lies on the line $f(P)$.*

Proof. Let P and L have homogeneous coordinates $\mathbf{x} = [x_0, x_1, x_2]^T$ and $\mathbf{u} = [u_0, u_1, u_2]^T$ respectively and let $\mathbf{A}$ be the matrix of f. Thus $f(P) = \mathbf{A}\mathbf{x}$ and $g(L) = (\mathbf{A}^{-1})^T \mathbf{u}$. Since P lies on L we have $\mathbf{u}^T \mathbf{x} = 0$; taking the transpose gives $\mathbf{x}^T \mathbf{u} = 0$. This can be rewritten as $\mathbf{x}^T [\mathbf{A}^T (\mathbf{A}^{-1})^T]\mathbf{u} = 0$, i.e. $(\mathbf{A}\mathbf{x})^T [(\mathbf{A}^{-1})^T \mathbf{u}] = 0$, which implies that $g(L)$ lies on $f(P)$. Conversely $(\mathbf{A}\mathbf{x})^T [(\mathbf{A}^{-1})^T \mathbf{u}] = \mathbf{x}^T [\mathbf{A}^T (\mathbf{A}^{-1})^T]\mathbf{u} = \mathbf{x}^T \mathbf{u}$; hence if the left side is 0 so is the right side, and taking the transpose gives $\mathbf{u}^T \mathbf{x} = 0$. $\square$

Our next two propositions are concerned with situations in which P lies on $f(P)$ where f is not necessarily invertible.

Proposition 2. *The point* $P = \mathbf{x}$ *lies on the line* $f(P)$ *iff* $\mathbf{x}$ *satisfies the homogeneous quadratic equation* $\mathbf{x}^T \mathbf{A}^T \mathbf{x} = 0$, *where* $\mathbf{A}$ *is the matrix of* f.

Proof. $P = \mathbf{x}$ and $f(P) = \mathbf{A}\mathbf{x}$; hence P lies on $f(P)$ if and only if $(\mathbf{A}\mathbf{x})^T \mathbf{x} = 0$, i.e. $\mathbf{x}^T \mathbf{A}^T \mathbf{x} = 0$. □

If f is invertible and g is the dual of f we can show similarly that the line $L = \mathbf{u}$ passes through the point $g(L)$ iff $\mathbf{u}$ satisfies the homogeneous quadratic equation $\mathbf{u}^T (\mathbf{A}^{-1})^T \mathbf{u} = 0$. Note that $\mathbf{x}^T \mathbf{A}^T \mathbf{x}$ is the transpose of $\mathbf{x}^T \mathbf{A}\mathbf{x}$; thus the quadratic loci $\mathbf{x}^T \mathbf{A}\mathbf{x} = 0$ and $\mathbf{x}^T \mathbf{A}^T \mathbf{x} = 0$ are always the same. Hence the quadratic loci $\mathbf{x}^T \mathbf{A}^T \mathbf{x} = 0$ and $\mathbf{u}^T (\mathbf{A}^{-1})^T \mathbf{u} = 0$ are the same iff $\mathbf{A} = (\mathbf{A}^{-1})^T$, i.e. $\mathbf{A}^{-1} = \mathbf{A}^T$, which means that $\mathbf{A}$ is orthogonal.

If $\mathbf{A} = (a_{ij})$ where $0 \leq i \leq 2$ and $0 \leq j \leq 2$, $\mathbf{x}^T \mathbf{A}^T \mathbf{x}$ is $a_{00}x_0^2 + a_{11}x_1^2 + a_{22}x_2^2 + (a_{01} + a_{10})x_0 x_1 + (a_{12} + a_{21})x_1 x_2 + (a_{20} + a_{02})x_2 x_0$. If $\mathbf{A}$ is skew-symmetric the coefficients of this quadratic expression are all zero; thus $\mathbf{x}^T \mathbf{A}^T \mathbf{x}$ is identically zero for every $\mathbf{x}$. Conversely, if the quadratic is identically zero, taking $\mathbf{x} = [1, 0, 0]^T$ shows that we must have $a_{00} = 0$; similarly, taking $\mathbf{x} = [0, 1, 0]^T$ and $[0, 0, 1]^T$ shows that $a_{11} = a_{22} = 0$, and taking $\mathbf{x} = [1, 1, 0]^T, [0, 1, 1]^T$ and $[1, 0, 1]^T$ shows that $(a_{01} + a_{10}) = (a_{12} + a_{21}) = (a_{20} + a_{02}) = 0$. Thus if the quadratic is identically zero $\mathbf{A}$ is skew-symmetric. Finally, note that $\mathbf{A}$ is skew-symmetric iff any nonzero scalar multiple of $\mathbf{A}$ is skew-symmetric. We have thus proved

Proposition 3. *Every point* P *lies on the line* $f(P)$ *iff the matrix of* f *is skew-symmetric.*

A correlation whose matrix is skew-symmetric is called a *null-polarity.* Such a correlation is not invertible because a 3×3 skew-symmetric matrix is always singular. A correlation whose matrix is symmetric is called a *polarity;* note that a polarity may or may not be invertible.

Any matrix $\mathbf{A}$ is the sum of a symmetric matrix and a skew-symmetric matrix. Indeed, let $\mathbf{B} = \frac{1}{2}(\mathbf{A} + \mathbf{A}^T)$ and $\mathbf{C} = \frac{1}{2}(\mathbf{A} - \mathbf{A}^T)$; then evidently $\mathbf{B}$ is symmetric, $\mathbf{C}$ is skew-symmetric, and $\mathbf{A} = \mathbf{B} + \mathbf{C}$. Moreover, $P = \mathbf{x}$ satisfies $\mathbf{x}^T \mathbf{A}^T \mathbf{x} = 0$ iff it satisfies $\mathbf{x}^T \mathbf{B}^T \mathbf{x} = 0$, since $\mathbf{x}^T \mathbf{C}^T \mathbf{x} = 0$ for all $\mathbf{x}$. Thus if $\mathbf{A}$ is not skew-symmetric, $P = \mathbf{x}$ lies on $f(P)$ iff $\mathbf{x}$ lies on the quadratic locus $\mathbf{x}^T \mathbf{B}^T \mathbf{x} = 0$. A correlation f for which both $\mathbf{B}$ and $\mathbf{C}$ are nonsingular is sometimes called a *simple* correlation ([ ] p.394).

We next show that f and its transpose (i.e., the correlation whose matrix is the transpose of that of f) have a useful "incidence-symmetry" property (see also Section 4.3):

Proposition 4. *Let* $\mathbf{A}$ *be the matrix of* f *and let* f^* *be the correlation defined by* $\mathbf{A}^T$. *If* Q *lies on* $f(P)$ *then* P *lies on* $f^*(Q)$.

Proof. Let $P = \mathbf{x}$ and $Q = \mathbf{y}$. Then Q lies on $f(P)$ implies $(\mathbf{A}\mathbf{x})^T \mathbf{y} = 0$. Taking transposes gives $\mathbf{y}^T (\mathbf{A}\mathbf{x}) = 0$, i.e. $(\mathbf{A}^T \mathbf{y})^T \mathbf{x} = 0$, which implies that P lies on $f^*(Q)$. □

If $\mathbf{A}^T$ is a scalar multiple of $\mathbf{A}$ the proof of Proposition 4 shows that f itself has the "incidence-symmetry" property: if Q lies on $f(P)$ then P lies on $f(Q)$. But evidently $\mathbf{A}^T = \lambda \mathbf{A}$ iff $\lambda^2 = 1$, i.e. $\lambda = \pm 1$; thus we have

Corollary *If the matrix $\mathbf{A}$ of f is symmetric or skew-symmetric (i.e., if f is a polarity or a null-polarity) and Q lies on $f(P)$ then P lies on $f(Q)$.*

In the remainder of this section we discuss conics and their relationship to polarities. A *conic* is the set of all points $\mathbf{x}$ in the real projective plane that satisfy a homogeneous quadratic equation $a_{00}x_0^2 + a_{11}x_1^2 + a_{22}x_2^2 + 2a_{01}x_0x_1 + 2a_{12}x_1x_2 + 2a_{20}x_2x_0 = 0$ where the a's are not all zero. Note that since we are working in the real projective plane a conic may be empty — for example, if a_{00}, a_{11}, a_{22} are all nonnegative and $a_{01} = a_{12} = a_{20} = 0$. If we let $a_{10} = a_{01}$, $a_{21} = a_{12}$, $a_{02} = a_{20}$ and let $\mathbf{A}$ be the symmetric matrix (a_{ij}), $0 \le i, j \le 2$, the equation becomes $\mathbf{x}^T \mathbf{A} \mathbf{x} = 0$. A conic is called *degenerate* if $\mathbf{A}$ is singular. It can be shown that a conic is degenerate iff $\mathbf{x}^T \mathbf{A} \mathbf{x}$ can be written as the product of two not necessarily distinct linear factors; they are distinct if the rank of $\mathbf{A}$ is 2 and are the same if its rank is 1 (see [ ] pp.81-83). It follows that a degenerate conic cannot be empty. For a discussion of the classification of conics see, e.g., [ ] pp.39-41; [ ] p.103; [ ] pp.122-125.

If C is the conic $\mathbf{x}^T \mathbf{A} \mathbf{x} = 0$ where $\mathbf{A}$ is symmetric, the correlation $f : \mathbf{x} \rightarrow \mathbf{u} = \mathbf{A} \mathbf{x}$ is called the polarity associated with C. The line $\mathbf{u}$ is called the *polar* of the point $\mathbf{x}$ with respect to C, and the point $\mathbf{x}$ is called the *pole* of the line $\mathbf{u}$ with respect to C. If C is nondegenerate, so $\mathbf{A}$ is nonsingular, the polarity is invertible and we can also define a line-to-point mapping $g : \mathbf{u} \rightarrow (\mathbf{A}^{-1})^T \mathbf{x}$. Since $\mathbf{A}$ (and therefore $\mathbf{A}^{-1}$) is symmetric this polar-pole mapping is both the inverse and the dual of the pole-polar mapping.

We will now discuss the polarities associated with nondegenerate, nonempty conics, which we call *C-polarities*. For a discussion of the polarities associated with degenerate conics see [ ] p.118, Exercise (i).

Let C be a nondegenerate, nonempty conic. A line L can intersect C in at most two points. If it does not intersect C it is called an exterior line; if it intersects C in one point P it is called a tangent (more fully: the tangent to C at P); and if it intersects C in two points it is called a secant. These concepts are illustrated in Figure 1.

For any point P at most two lines through P are tangent to C. If there are no such lines P is called an interior point of C; there is exactly one such line iff P is on C; and if there are two such lines P is called an exterior point of C.

The polar (with respect to C) of an interior point is an exterior line; the polar of a point P on C is the tangent to C at P; and the polar of an exterior point P is the line QR, where the two tangents to C through P are tangent to C at Q and R. The line QR is called the "chord of contact" of the two tangents ("secant of contact" would be better term). Thus a point P lies on its polar iff P lies on C, and its polar is then the tangent to C at P. These concepts are illustrated in Figure 2.

Every point on an exterior line, and every point on a tangent except for the "point of contact" where the tangent intersects C, is an exterior point. A secant L contains both interior and exterior points; if L intersects C at Q and R, and S, T is a pair of points on L that are separated by Q, R (see Section 5.1), then one of S, T is interior and the other is exterior. For further details see [ ] pp. 56, 86-87; [ ] pp. 83-84; [ ] pp. 196-198; [ ], pp. 74-75; [ ] pp. 138-139; [ ] p. 110.

If a Hough transform is defined by a C-polarity, the geometric relationships between points (poles) and their corresponding lines (polars) can be used to simplify the search for peaks in the transform. For example, if the P_i's are collinear points that lie on a line exterior to C, the corresponding lines L_i must intersect at a point interior to C.

If the points W, X, Y, Z are collinear, the pair of points W, X is called *harmonic* with respect to the pair Y, Z if the *cross-ratio*

$$\frac{|WY|/|YX|}{|WZ|/|ZX|}$$

$= -1$, where $|UV|$ is the signed length of the line segment UV. Two distinct points P, Q are called *conjugate* with respect to C if the line PQ intersects C in two points and P, Q are harmonic with respect to the two points of intersection. It can be shown that for any point P the locus of the points that are conjugate to P with respect to C is a line, and this line is the polar of P with respect to C (see [ ] p.107; [ ] p.73; [ ] p.108-109). Thus P is conjugate to Q iff Q lies on the polar of P.

4.2   Canonical Forms for the Matrices of Correlations

In this section we discuss how the matrix of a correlation can be put into a "canonical form" — specifically, with respect to an equivalence relation on matrices called "congruence".

Let f, f' be correlations and let $\mathbf{A}, \mathbf{A}'$ be the matrices of f, f' respectively. We call f and f' (or $\mathbf{A}$ and $\mathbf{A}'$) *congruent* if there exists a nonsingular matrix D such that $\mathbf{A}' = \mathbf{D}^T \mathbf{A} \mathbf{D}$. We can regard $\mathbf{D}$ as a matrix of a point-to-point collineation and $\mathbf{D}^T$ as a matrix of a line-to-line collineation; the effect of $\mathbf{A}'$ on any point $\mathbf{x}$ is obtained by first applying $\mathbf{D}$ to $\mathbf{x}$, then $\mathbf{A}$ to $\mathbf{D}\mathbf{x}$, and finally $\mathbf{D}^T$ to the line $\mathbf{A}(\mathbf{D}\mathbf{x})$. If $\mathbf{A}'$ is congruent to $\mathbf{A}$ the points that lie on their corresponding lines under $\mathbf{A}'$ are related to the points that lie on their corresponding lines under $\mathbf{A}$ by

Proposition 5. *Let* $\mathbf{A}' = \mathbf{D}^T \mathbf{A} \mathbf{D}$ *and let* $\mathbf{y} = \mathbf{D}^{-1}\mathbf{x}$. *If* $\mathbf{x}$ *lies on* $\mathbf{A}\mathbf{x}$ *then* $\mathbf{y}$ *lies on* $\mathbf{A}'\mathbf{y}$.

Proof: By Proposition 2 $\mathbf{x}$ lies on $\mathbf{A}\mathbf{x}$ iff $\mathbf{x}^T \mathbf{A}^T \mathbf{x} = 0$ and $\mathbf{y}$ lies on $\mathbf{A}'\mathbf{y}$ iff $\mathbf{y}^T \mathbf{A}'^T \mathbf{y} = 0$. If $\mathbf{x}$ lies on $\mathbf{A}\mathbf{x}$ we have $\mathbf{y}^T \mathbf{A}'^T \mathbf{y} = (\mathbf{D}^{-1}\mathbf{x})^T \mathbf{A}'^T (\mathbf{D}^{-1}\mathbf{x}) = \mathbf{x}^T (\mathbf{D}^{-1})^T (\mathbf{D}^T \mathbf{A}^T \mathbf{D})(\mathbf{D}^{-1}\mathbf{x}) = \mathbf{x}^T \mathbf{A}^T \mathbf{x} = 0$; hence $\mathbf{y}$ lies on $\mathbf{A}'\mathbf{y}$.   □

Congruence of matrices is evidently an equivalence relation and congruent matrices have the same rank. Congruence also preserves the property of being a polarity or null-polarity. Indeed, if $\mathbf{A}, \mathbf{A}'$ are congruent we have $\mathbf{A}' = \mathbf{D}^T \mathbf{A} \mathbf{D}$ for some nonsingular $\mathbf{D}$; hence $(\mathbf{A}')^T = \mathbf{D}^T \mathbf{A}^T \mathbf{D}$, and if $\mathbf{A}$ is symmetric or skew-symmetric so is $\mathbf{A}'$. It can be shown that a matrix is congruent to a diagonal matrix iff it is symmetric ([ ] Theorem 6.4.1). We shall now discuss how a matrix can be put into a canonical form under congruence, beginning with the cases where $\mathbf{A}$ is symmetric or skew-symmetric.

If $\mathbf{A}$ is symmetric and has rank r it can be shown (see, e.g., [ ] Theorem 5.5.4; [ ] Ch.9) that $\mathbf{A}$ is congruent to a diagonal matrix of the form

$$\begin{pmatrix} \mathbf{I}_p & \mathbf{O} & \mathbf{O} \\ \mathbf{O} & -\mathbf{I}_{r-p} & \mathbf{O} \\ \mathbf{O} & \mathbf{O} & \mathbf{O} \end{pmatrix}$$ where $\mathbf{I}_p$ and $\mathbf{I}_{r-p}$ are identity matrices of sizes $p \times p$ and

$(r-p) \times (r-p)$ respectively and the $\mathbf{O}$'s are zero matrices of appropriate sizes. (These results hold for any $n \times n$ matrix $\mathbf{A}$ over the field of real numbers. Simpler results hold for matrices over the field of complex numbers.) The uniqueness of p was discovered by Sylvester in 1852; it is called *Sylvester's law of inertia*. For example, if $\mathbf{A}$ is 3×3 and nonsingular, the only possible canonical forms are $\mathbf{I}_3$,

$\begin{pmatrix} \mathbf{I}_2 & \mathbf{O} \\ \mathbf{O} & -\mathbf{I}_1 \end{pmatrix}$, and $\begin{pmatrix} \mathbf{I}_1 & \mathbf{O} \\ \mathbf{O} & -\mathbf{I}_2 \end{pmatrix}$; $-\mathbf{I}_3$ is not on this list because it is a scalar multiple of $\mathbf{I}_3$.

Two conics $\mathbf{x}^T \mathbf{A} \mathbf{x} = 0$ and $\mathbf{x}^T \mathbf{A}' \mathbf{x} = 0$ are called congruent if $\mathbf{A}$ and $\mathbf{A}'$ are congruent. Since the matrix of a conic is symmetric and the matrix of a nondegenerate conic is nonsingular, this gives us

Proposition 6. *A nondegenerate conic* $\mathbf{x}^T \mathbf{A} \mathbf{x} = 0$ *is congruent to one of the following canonical forms: (i)* $x_0{}^2 + x_1^2 + x_2^2 = 0$, *(ii)* $x_0{}^2 + x_1^2 - x_2^2 = 0$, *or (iii)* $x_0{}^2 - x_1^2 - x_2^2 = 0$.

Note that in the real projective plane the first of these canonical forms is an empty conic and the other two are nonempty. Thus congruence preserves the property of being a C-polarity, which is equivalent to having canonical form (ii) or (iii).

Suppose next that $\mathbf{A}$ is skew-symmetric. It is well known that the rank of a skew-symmetric matrix must be even (e.g., [ ] p. 236). Since in our case $\mathbf{A}$ is 3×3 its rank must be 2. It can be shown (e.g., [ ] Theorem 5.4.1; [ ] Ch.9)

that $\mathbf{A}$ is congruent to the matrix $\begin{pmatrix} 0 & 1 & 0 \\ -1 & 0 & 0 \\ 0 & 0 & 0 \end{pmatrix}$. Thus all skew-symmetric matrices

have the same canonical form, so that all null-polarities are congruent to each other.

One way to obtain a canonical form for a general correlation is to express $\mathbf{A}$ as $\mathbf{B} + \mathbf{C}$ where $\mathbf{B}$ is symmetric and $\mathbf{C}$ is skew-symmetric (see Section 4.1) and then to *simultaneously* reduce $\mathbf{B}$ and $\mathbf{C}$ to canonical forms using a suitable congruence transformation. This approach is described in detail in [ ] p. 397, Theorem II for $n \times n$ matrices over an algebraically closed field. It applies to matrices over the field of complex numbers but not to matrices over the field of real numbers.

The canonical form of an $n \times n$ matrix over a field that is not necessarily algebraically closed is treated in detail in [ ] Chapter 7, Section 9. For a 3×3 matrix over the field of real numbers all the possible canonical forms are

$$\begin{pmatrix} d_{11} & 0 & 0 \\ d_{21} & d_{22} & 0 \\ d_{31} & d_{32} & d_{33} \end{pmatrix}, \begin{pmatrix} d_{11} & 0 & 0 \\ d_{21} & d_{22} & 0 \\ d_{31} & d_{32} & 0 \end{pmatrix}, \text{and} \begin{pmatrix} d_{11} & 0 & 0 \\ d_{21} & 0 & 0 \\ d_{31} & 0 & 0 \end{pmatrix}$$

where the d's are real numbers. It is not hard to see that if the d's are nonzero these canonical forms are not congruent to each other. They are the canonical forms of matrices that have rank 3 (nonsingular matrices), 2, and 1, respectively.

EXAMPLE: As we saw in Section 3.1 the correlation that was used in Hough's patent has matrix $\mathbf{A} = \begin{pmatrix} 0 & 0 & -1 \\ 0 & 1 & 0 \\ 1 & 0 & 0 \end{pmatrix}$. Since $\mathbf{A}$ has determinant 1 it is nonsingular,

so its canonical form must be a matrix of the form $\begin{pmatrix} d_{11} & 0 & 0 \\ d_{21} & d_{22} & 0 \\ d_{31} & d_{32} & d_{33} \end{pmatrix}$. Let $\mathbf{D}$ be

the nonsingular symmetric matrix $\begin{pmatrix} 1 & 1 & 0 \\ 1 & 1 & 1 \\ 0 & 1 & 1 \end{pmatrix}$. It can be verified that $\mathbf{D}^T \mathbf{A} \mathbf{D} =$

$\begin{pmatrix} 1 & 0 & 0 \\ 2 & 1 & 0 \\ 2 & 2 & 1 \end{pmatrix}$; this matrix is the canonical form of $\mathbf{A}$.

$\mathbf{A}$ is neither symmetric nor skew-symmetric, so Hough's correlation is neither a polarity nor a null-polarity. If we decompose $\mathbf{A}$ into $\mathbf{B}$ and $\mathbf{C}$ as described in Section 4.1 we find that $\mathbf{B} = \begin{pmatrix} 0 & 0 & 0 \\ 0 & 1 & 0 \\ 0 & 0 & 0 \end{pmatrix}$ and $\mathbf{C} = \begin{pmatrix} 0 & 0 & -1 \\ 0 & 0 & 0 \\ 1 & 0 & 0 \end{pmatrix}$; note that both $\mathbf{B}$ and $\mathbf{C}$ are singular, so Hough's correlation is not a simple correlation.

The variant form of Hough's correlation given in [ ] p.123 takes the point $[u, v, 1]^T$ into the line $ux + y - v = 0$ which has coefficients $[u, 1, -v]^T$, so it has matrix $\mathbf{A}' = \begin{pmatrix} 1 & 0 & 0 \\ 0 & 0 & 1 \\ 0 & -1 & 0 \end{pmatrix}$. It can be verified that using $\mathbf{D} = \begin{pmatrix} 1 & 1 & -1 \\ 0 & 1 & -1 \\ 1 & 0 & 1 \end{pmatrix}$

transforms $\mathbf{A}'$ into the canonical form $\mathbf{D}^T \mathbf{A}' \mathbf{D} = \begin{pmatrix} 1 & 0 & 0 \\ 2 & 1 & 0 \\ -2 & -2 & 1 \end{pmatrix}$; note that this

is not the same as the canonical form of $\mathbf{A}$.

If we give the d's (in the canonical form of a general nonsingular 3×3 matrix) values other than those in the canonical form of Hough's correlation we obtain infinitely many incongruent correlations none of which are congruent to Hough's correlation. For example, setting all the d's equal to 1 gives the canonical form

$\begin{pmatrix} 1 & 0 & 0 \\ 1 & 1 & 0 \\ 1 & 1 & 1 \end{pmatrix}$. This nonsingular correlation is not a polarity since its matrix is not diagonal (see the paragraph following Proposition 5).

Other types of transformations can also be used to put a matrix into canonical form. For example, the matrices $\mathbf{A}, \mathbf{A}'$ are called *similar* if there exists a nonsingular matrix D such that $\mathbf{A}' = \mathbf{D}^{-1}\mathbf{A}\mathbf{D}$. A canonical form for $n \times n$ matrices under similarity, where no assumption is made about the algebraic closure of the underlying field, is called the *rational canonical form*; it was first described by Frobenius (1849-1917). A simpler canonical form called the *Jordan canonical form* exists when the underlying field is algebraically closed. For details about both of these forms see, e.g., [,]. However, these canonical forms seem to be of less interest in our case; for example, if $\mathbf{A}, \mathbf{A}'$ are similar and $\mathbf{A}$ is a polarity or null-polarity $\mathbf{A}'$ is not necessarily a polarity or null-polarity. (Interestingly, it can be shown that the two variants of Hough's correlation discussed above have the same rational canonical form.) Note that if $\mathbf{D}$ is an orthogonal matrix we have $\mathbf{D}^{-1} = \mathbf{D}^T$, so similarity and congruence are the same. Canonical forms for polarities in the case where $\mathbf{D}$ is orthogonal are studied in Section 1.8.2 of [ ], but we will not consider them here.

4.3 General Point-to-Line Mappings

In this section we consider (not necessarily linear) point-to-line mappings that have the incidence-symmetry and incidence-preserving properties described in Section 4.1, and we discuss conditions under which these properties imply linearity.

Let f be a point-to-line mapping defined on the (real) projective plane. We call f *incidence-symmetric* if $Q \in f(P)$ implies $P \in f(Q)$. Similarly, if g is a mapping that takes lines into points we call g incidence-symmetric if $g(L) \in M$ implies $g(M) \in L$. As we saw in the corollary to Proposition 4, if f is a linear correlation and its matrix is symmetric or skew-symmetric (i.e., f is a polarity or null-polarity) then f is incidence-symmetric; the analogous result evidently also holds for g. It is not hard to see that if f is incidence-symmetric iff its inverse is incidence symmetric.

Proposition 7. *If f is onto and incidence-symmetric it takes collinear points into concurrent lines. If g is onto and incidence-symmetric it takes concurrent lines into collinear points.*

Proof. For any line L we have $L = f(P)$ for some P, and for any $Q \in L$, $f(Q)$ is a line through P. For any point P we have $P = g(L)$ for some L, and for any M through P, $g(M)$ is a point of L.   □

From Theorem 1 and Proposition 7 we have

Theorem 2. *A one-to-one incidence-symmetric point-to-line correspondence must be linear.*

Let f be one-to-one and incidence-symmetric. By Theorem 2 f must be linear; let $\mathbf{A} = (a_{ij})$ be the matrix of f (in a given homogeneous coordinate system). Let P and Q be the points $[x_0, x_1, x_2]^T$ and $[y_0, y_1, y_2]^T$ respectively. By incidence-symmetry $Q \in f(P)$ implies $P \in f(Q)$. Now $Q \in f(P)$ means $[x_0, x_1, x_2]\mathbf{A}^T[y_0, y_1, y_2]^T = 0$; taking transposes gives $[y_0, y_1, y_2]\mathbf{A}[x_0, x_1, x_2]^T = 0$. By incidence-symmetry this implies $P \in f(Q)$, i.e. $[y_0, y_1, y_2]\mathbf{A}^T[x_0, x_1, x_2]^T = 0$. Now $[y_0, y_1, y_2]\mathbf{A}[x_0, x_1, x_2]^T = x_0(y_0 a_{00} + y_1 a_{01} + y_2 a_{02}) + x_1(y_0 a_{10} + y_1 a_{11} + y_2 a_{12}) + x_2(y_0 a_{20} + y_1 a_{21} + y_2 a_{22})$, and similarly for $[y_0, y_1, y_2]\mathbf{A}^T[x_0, x_1, x_2]^T$. Let $[x_0, x_1, x_2] = [1, 0, 0]$; then by incidence-symmetry $y_0 a_{00} + y_1 a_{01} + y_2 a_{02} = 0$ implies $y_0 a_{00} + y_1 a_{10} + y_2 a_{20} = 0$. If $y_2 = 0$ and $y_1 \neq 0$ the first equation can be rewritten as $a_{00} y_0 / y_1 = -a_{01}$ and the second equation as $a_{00} y_0 / y_1 = -a_{10}$; hence if the first equation implies the second we must have $a_{01} = a_{10}$. By assuming instead that $y_1 = 0$ and $y_2 \neq 0$ we can similarly show that $a_{20} = a_{02}$. Similarly, by letting $[x_0, x_1, x_2] = [0, 1, 0]$ or $[0, 0, 1]$ we can prove that $a_{12} = a_{21}$; thus $\mathbf{A}$ is symmetric, so f is a polarity. Since (as pointed out at the beginning of this section) a polarity must be incidence-symmetric, we have thus proved

Theorem 3. *A one-to-one point-to-line correspondence is incidence-symmetric iff it is an invertible polarity.*

We can now show that the converse of Proposition 7 is false. If f is a one-to-one point-to-line correspondence that takes collinear points into concurrent lines, by Theorem 1 it must be linear and its matrix must be nonsingular; but since the matrix needn't be symmetric f needn't be incidence-symmetric.

Some useful properties of an invertible polarity f can be derived from the fact that f is one-to-one and incidence-symmetric:

1) Let $f(P) = L$; then f is a one-to-one correspondence between the points of L and the lines through P.
2) For all $P \neq Q$ we have $f[f(P) \cap f(Q)] = PQ$ (the line containing the points P and Q). Proof: Since f is one-to-one, $P \neq Q$ implies $f(P) \neq f(Q)$. Let $f(P) \cap f(Q) = R$; then $R \in f(P)$ and $R \in f(Q)$, so by incidence-symmetry P and Q are both on $f(R)$; hence $f(R) = PQ$. [We have not assumed that $P \in f(Q)$ or $Q \in f(P)$. Let $f(P) \cap PQ = Q'$ and $f(Q) \cap PQ = P'$; then $f(P') = QR$ and $f(Q') = PR$. Also P', P, Q, Q' are collinear and $f(P'), f(P), f(Q) f(Q')$ are concurrent (at R).]
3) If $P \in f(P)$ and $Q \in f(Q)$ then $R = f(P) \cap f(Q) \notin PQ$ ([ ] p.67; [ ] pp.60–61). Proof: Both P and R are on $f(P)$, so if R were on $PQ, f(P)$ would be the same as PQ. Similarly, both Q and R are on $f(Q)$, so if R were on $PQ, f(Q)$ would be the same as PQ. This would imply $f(P) = f(Q)$, contradicting the fact that f is one-to-one.
4) If $P \in f(P)$ then for any $Q \neq P$ such that $R = f(P) \cap f(Q) \neq P$ we have $R \notin PQ$. Proof: If $P \in f(P)$ and $R \in PQ$, P and R would both be on $f(P)$ and on PQ, so $f(P) = PQ$. Hence $Q \in f(P)$, so $P \in f(Q)$. Thus P and R would both be on $f(Q)$ and on PQ, so $f(Q) = PQ$; hence $f(Q) = f(P)$, contradicting the fact that f is one-to-one.

Let f be a mapping that takes points into lines and g a mapping that takes lines into points. We call the pair (f, g) *incidence-preserving* if $P \in L$ implies $g(L) \in f(P)$. As we saw in Proposition 1, if f is a linear correlation and g is its dual the pair (f, g) is incidence-preserving.

If f and g are inverses of each other incidence-preservation is closely related to incidence-symmetry; indeed, we have

Proposition 8. *If f and g are inverses of each other, (f, g) is incidence-preserving iff f or g is incidence-symmetric iff they are both incidence-symmetric.*

Proof: Let $f(P) = L$. If (f, g) is incidence-preserving $Q \in L = f(P)$ implies $P = g(L) \in f(Q)$, so f is incidence-symmetric. Similarly, let $g(L) = P$; then $P = g(L) \in M$ implies $g(M) \in f(P) = L$, so g is incidence-symmetric. Conversely, let f be incidence-symmetric and let $f(P) = L$. Then $Q \in L = f(P)$ implies $P = g(L) \in f(Q)$, so (f, g) is incidence-preserving; similarly, if g is incidence-symmetric, (f, g) is incidence-preserving. □

We saw at the beginning of this section that if f and g are inverses of each other, f is incidence-symmetric iff g is incidence-symmetric. Note also that by Theorem 3 if f is one-to-one and incidence-symmetric it is an invertible polarity; hence its matrix is symmetric, so its inverse g is the same as its dual, which implies that (f, g) is incidence-preserving by Proposition 1.

By Propositions 7 and 8, if f and g are inverses of each other and (f, g) is incidence-preserving, f takes collinear points into concurrent lines and g takes concurrent lines into collinear points, and by Theorem 2 f and g must both be linear. If f and g are not inverses of each other they need not be linear even if (f, g) is incidence-preserving; but they still take collinear points into concurrent lines and vice versa, as we see from

Proposition 9. *If (f, g) is incidence-preserving f takes collinear points into concurrent lines and g takes concurrent lines into collinear points.*

Proof: Let P, Q, R be collinear, say on L; then $g(L)$ must lie on $f(P), f(Q), f(R)$, so these lines are concurrent. Let K, L, M be concurrent, say at P; then $g(K), g(L), g(M)$ must lie on $f(P)$, so these points are collinear. □

5   Concluding Remarks

This paper has made many basic observations about Hough transforms and point-to-line mappings. We summarize a few of these observations here:

1) A wide variety of Hough transforms that make use of (nonlinear) point-to-curve mappings have been defined. Linear point-to-line mappings like Hough's original mapping are computationally simpler, but have been neglected.

2) A one-to-one point-to-line mapping that takes collinear points into concurrent lines, and so can be used as a Hough transform, must in fact be linear (Theorem 1).

3) A one-to-one point-to-line mapping that preserves point-line incidence must also be linear, and its matrix must be symmetric, so it must be a polarity (Theorem 3). Because of their incidence-symmetry property, invertible polarities are an especially well-behaved class of point-to-line mappings. In particular, if an invertible polarity (with respect to a nonempty conic) is used as a Hough transform, the geometrical relationships between points and their corresponding lines can be used to simplify the search for peaks in the transform.

4) The matrix of an invertible polarity is congruent to a diagonal matrix whose diagonal elements are $(1, 1, 1), (1, 1, -1)$, or $(1, -1, -1)$. In the latter two cases the polarity is a C-polarity, i.e. it defined by a nonempty conic. Hough's mapping had matrix $\begin{pmatrix} 0 & 0 & -1 \\ 0 & 1 & 0 \\ 1 & 0 & 0 \end{pmatrix}$ and was not a polarity; the C-polarity whose matrix is $\begin{pmatrix} 1 & 0 & 0 \\ 0 & 1 & 0 \\ 0 & 0 & -1 \end{pmatrix}$ might have been a better choice.

This paper could be expanded in many directions, as suggested in the following paragraphs.

a) We have considered only point-to-line mappings and their use for detecting sets of collinear points. As pointed out in Section 2 much of the work on Hough transforms [, ,] has dealt with point-to-curve mappings, and such mappings are also often used for detecting sets of points that lie on a curvilinear locus such as a circle or ellipse. It would be of interest to generalize the concepts in this paper to mappings that take points into curves and that take sets of points that lie on a curvilinear locus into sets of concurrent curves.

b) We have considered only mappings defined on the plane. The Hough transform approach has also been used to detect sets of collinear points (or sets of points that lie on other types of loci — e.g. space curves, planes, or surfaces) in higher-dimensional spaces [, ,]. It would be of interest to generalize the concepts in this paper to loci in space. Note that in n-dimensional space there is a dual relationship between points and hyperplanes, not between points and lines. Note also that if n is even a null-polarity can be invertible because a skew-symmetric matrix need not be singular. Invertible null-polarities in projective 3D space (using 4-tuples of homogeneous coordinates) have geometrical relationships to "twisted cubics" analogous to the relationships between polarities and conics in the projective plane. For references on correlations in n-dimensional spaces see [ ] pp. 422-424; [ ] Ch. 9; [ ]; [ ] pp. 373-377; [ ] pp.99-105; [ ] pp. 359-363.

c) It would be of interest to study the effects of point-to-line mappings on various types of planar loci — for example, on conics. The relationship between point-to-plane mappings and quadric surfaces might also be of interest.

d) Point-to-line mappings and twisted cubics are useful for camera calibration; see [ ], [ ], [ ]. It would be of interest to investigate other applications of the concepts studied in this paper to the interpretation of images of three-dimensional scenes.

e) For convenience, in this paper we have dealt only with the real projective plane. For some purposes it might have been advantageous to work in other types of planes. For example,

(1) We could have worked in the Euclidean plane, as Hough did in [ ]. The study of points and lines in the Euclidean plane is more complicated because of the need to deal with exceptional cases (e.g., parallel lines), but it is more faithful to the geometry of real-world images. For a discussion of Hough transforms in the Euclidean plane see [ ].

(2) As a compromise between the Euclidean and projective planes, we could have worked in the affine plane (see also [ ]). This would have allowed us to distinguish between different types of conics and between intersecting and parallel lines.

(3) We could have worked in the complex projective plane. This would have required us to introduce imaginary points and lines, but would have allowed us to give more general treatments of such topics as conics (see Section 4.1) and canonical forms (see Section 4.2). Many of the results in this paper remain valid in the complex projective plane. "Circular points" in the complex projective plane, which are discussed in many books on projective geometry, play key roles in camera calibration and in the analysis of images of 3D scenes, as discussed in [, ,].

(4) Many of the results in this paper involve only finite sets of points and lines and could possibly have been formulated as results about finite projective planes. (A finite projective plane is finite set of "points" and a finite set of subsets of the points, called "lines", such that each pair of distinct points are contained in a unique line, each pair of distinct lines intersect in a unique point, and there exist four points that are not all contained in the same line.) For treatments of finite projective planes see [, , , ,]. Note that a digital image contains only finite numbers of points and lines, but it is not a finite projective plane because digital lines can intersect in more than one point.

(5) Incidence properties of correlations were a central theme of this paper. Such properties can be studied in an abstract framework; see [ ], which presents a general theory of incidence structures and defines correlations on such structures. It would be of interest to generalize the results of this paper to such structures.

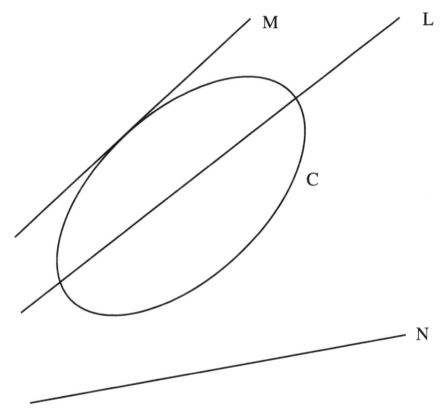

Fig. 1. Exterior lines, tangents, and secants of a conic C. L is a secant, M is a tangent, and N is an exterior line

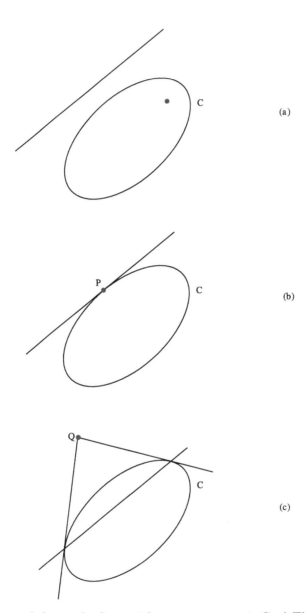

Fig. 2. Points and their polar lines with respect to a conic C: a) The polar of an interior point is an exterior line. b) The polar of a point P on C is the tangent to C at P. c) The polar of an exterior point Q is the secant joining the points at which the two tangents from Q meet C

References

1. A. S. Aguado, E. Montiel, and M. S. Nixon, On the intimate relationship between the principle of duality and the Hough transform, *Proc. Royal Soc. London* **A456**, 2000, 503–526. 190, 203

2. V. S. Alagar and L. H. Thiel, Algorithms for detecting m-dimensional objects in n-dimensional spaces, *IEEE Trans. Pattern Analysis Machine Intelligence* **3**, 1981, 245–256. 202

3. A. A. Albert and R. Sandler, *An Introduction to Finite Projective Planes*, Holt, Rinehart and Winston, New York, 1968. 203

4. E. Artin, *Geometric Algebra*, Wiley/Interscience, New York, 1988. 192

5. F. Ayers, Jr., *Theory and Problems of Projective Geometry*, Mc-Graw Hill, New York, 1967. 192

6. H. F. Baker, *Principles of Geometry*, vol. 1, Cambridge University Press, Cambridge, U. K., 1922.

7. L. M. Batten and A. Beutelspacher, *The Theory of Finite Linear Spaces: Combinatorics of Points and Lines*, Cambridge University Press, Cambridge, U. K., 1993. 203

8. L. M. Batten, *Combinatorics of Finite Geometries* (second edition), Cambridge University Press, Cambridge, U. K., 1997. 203

9. T. Beth, D. Jungnickel, and H. Lenz, *Design Theory*, Cambridge University Press, Cambridge, U. K., 1986. 203

10. P. B. Bhattacharya, S. K. Jain and S. R. Nagpaul, *Basic Abstract Algebra*, second ed., Cambridge University Press, New York, 1995. 199

11. P. Bhattacharya, H. Liu, A. Rosenfeld, and S. Thompson, Hough-transform detection of lines in 3-D space, *Pattern Recognition Letters* **21**, 2000, 843–849. 202

12. L. Bieberbach, *Projektive Geometrie*, Teubner, Leipzig and Berlin, 1931. 192

13. W. Blaschke, *Projektive Geometrie* (second edition), Wolfenbüttler Verlag, Wolfenbüttel and Hannover, Germany, 1948. 192

14. J. A. Blattner, *Projective Plane Geometry*, Holden-Day, San Francisco, 1968. 192

15. R. J. Bumcrot, *Modern Projective Geometry*, Holt, Rinehart and Winston, New York, 1969. 192

16. H. Busemann and P. J. Kelley, *Projective Geometry and Projective Metrics*, Academic Press, New York, 1953. 192

17. J. N. Cederberg, *A Course in Modern Geometry*, Springer, New York, 1989. 192

18. H. S. M. Coxeter, *Non-Euclidean Geometry*, University of Toronto Press, Toronto, 1957. 192, 196

19. H. S. M. Coxeter, *The Real Projective Plane*, Cambridge University Press, Cambridge, U. K., 1961. 192, 196, 200

20. H. S. M. Coxeter, *Projective Geometry*, Blaisdell, New York, 1964. 192, 200

21. E. R. Davies, Image space transforms for detecting straight edges in industrial images, *Pattern Recognition Letters* **4**, 1986, 185–192. 189

22. P. Dembowski, *Finite Geometries*, Springer, Berlin, 1997. 203

23. R. O. Duda and P. E. Hart, Use of the Hough transformation to detect lines and curves in pictures, *Communications of the ACM* **15**, 1972, 11–15. 188, 189

24. H. Eves, *Elementary Matrix Theory*, Dover, New York, 1980. (Originally published by Allyn and Bacon, Boston, 1966.) 197

25. O. Faugeras, *Three-Dimensional Computer Vision: A Geometric Viewpoint*, MIT Press, Cambridge, MA, 1993. 203

26. A. Forman, A modified Hough transform for detecting lines in digital images, *Proc. SPIE* **635**, *Applications of Artificial Intelligence III*, 1986, 151–160. 189

27. J. Frenkel, *Géométrie pour l'élève-professeur*, Hermann, Paris, 1973. 192

28. D. Gans, *Transformations and Geometries*, Appleton-Century-Crofts, New York, 1968. 192

29. W. C. Graustein, *Introduction to Higher Geometry*, Macmillan, New York, 1930. 192, 196, 202

30. K. W. Gruenberg and A. J. Weir, *Linear Geometry*, Van Nostrand, New York, 1967. 192

31. R. Hartley and A. Zisserman, *Multiple View Geometry in Computer Vision*, Cambridge University Press, Cambridge, U. K., 2000. 192, 195, 199, 203

32. I. N. Herstein, *Topics in Algebra* (second edition), Springer, New York, 1975. 199

33. A. Heyting, *Axiomatic Projective Geometry*, North-Holland, Amsterdam, 1963. 192

34. J. W. P. Hirschfeld, *Projective Geometries over Finite Fields* (second edition), Clarendon Press, Oxford, 1998. 203

35. W. V. D. Hodge and D. Pedoe, *Methods of Algebraic Geometry*, vol. 1, Cambridge University Press, Cambridge, U. K., 1953. 192, 194, 197, 202

36. E. J. Hopkins and J. S. Hails, *An Introduction to Plane Projective Geometry*, Clarendon Press, Oxford, 1953. 192

37. P. V. C. Hough, Method and means for recognizing complex patterns, U. S. Patent 3069654, December 18, 1962. 187, 188, 191, 203

38. J. Illingworth and J. Kittler, A survey of the Hough transform, *Computer Vision, Graphics, and Image Processing* **44**, 1988, 87–116. 188, 202

39. D. Ioannou, E. T. Dugan, and A. F. Laine, On the uniqueness of the representation of a convex polygon by its Hough transform, *Pattern Recognition Letters* **17**, 1996, 1259–1264. 190

40. K. Kanatani, *Geometric Computation for Machine Vision*, Clarendon Press, Oxford, 1993. 192

41. K. Kanatani, Computational Projective Geometry, *CVGIP: Image Understanding* **54**, 1991, 333–348. 192

42. F. Klein, *Vorlesungen über höhere Geometrie* (third edition, ed. W. Blaschke), Chelsea, New York, 1949. 192

43. P. M. Lanfar, On the Hough transform of a polygon, *Pattern Recognition Letters* **17**, 1996, 209–210. 190

44. V. F. Leavers, Which Hough transform?, *CVGIP: Image Understanding* **58**, 1993, 250–264. 188, 202

45. H. Levy, *Projective and Related Geometries*, Macmillan, New York, 1964. 192, 196

46. E. A. Maxwell, *The Methods of Plane Projective Geometry Based on the Use of General Homogeneous Coordinates*, Cambridge University Press, Cambridge, U. K., 1960. 192, 195, 196

47. S. J. Maybank, The projective geometry of ambiguous surfaces, *Phil. Trans. Royal Soc. London* **A332**, 1990, 1–47. 202

48. B. E. Meserve, *Fundamental Concepts of Geometry*, Addison-Wesley, Reading, MA, 1955. 192, 196

49. L. Mirsky, *An Introduction to Linear Algebra*, Dover, New York, 1990. (Originally published by Clarendon Press, Oxford, 1955.) 197

50. Y. Muller and R. Mohr, Planes and quadrics detection using Hough transform, *Proc. Intl. Conf. on Pattern Recognition*, 1984, 1101–1103. 202

51. J. L. Mundy, D. Kapur, S. J. Maybank, P. Gros, and L. Quan, Geometric interpretation of joint conic invariants, in *Geometric Invariance in Computer Vision* (ed. J. L. Mundy and A. Zisserman), MIT Press, Cambridge, MA, 1992, pp. 77–86.

52. J. L. Mundy and A. Zisserman, Projective geometry for machine vision, Appendix to *Geometric Invariance in Computer Vision*, MIT Press, Cambridge, MA, 1992, pp. 463–520. 192, 203

53. D. Pedoe, *Geometry: A Comprehensive Course*, Dover, New York, 1988. 202

54. P. D. Picton, Hough transform references, *International Journal of Pattern Recognition and Artificial Intelligence* **1**, 1987, 413–425. 188, 202

55. R. A. Rosenbaum, *Introduction to Projective Geometry and Modern Algebra*, Addison-Wesley, Reading, MA, 1963. 192

56. A. Rosenfeld and A. C. Kak, *Digital Picture Processing* (second edition), Academic Press, New York, 1982. 188, 198

57. A. Rosenfeld and I. Weiss, A convex polygon is uniquely determined by its Hough transform, *Pattern Recognition Letters* **16**, 1995, 305–306. 190

58. O. Schreier and E. Sperner, *Projective Geometry of n Dimensions* (vol. 2 of *Introduction to Modern Algebra and Matrix Theory*), Chelsea, New York, 1961. 202

59. J. G. Semple and G. T. Kneebone, *Algebraic Projective Geometry*, Clarendon Press, Oxford, 1952. 192, 195, 196, 202

60. C. E. Springer, *Geometry and Analysis of Projective Spaces*, Freeman, San Francisco, 1964. 192

61. D. J. Struik, *Lectures on Analytic and Projective Geometry*, Addison-Wesley, Reading, MA, 1953. 192, 195

62. J. A. Todd, *Projective and Analytic Geometry*, Pitman, London, 1947. 192

63. H. W. Turnbull and A. C. Aitken, *An Introduction to the Theory of Canonical Matrices*, Blackie, London, 1932. 198

64. O. Veblen and J. W. Young, *Projective Geometry*, Ginn and Company, Boston, 1910. 192

65. J. Verdina, *Projective Geometry and Point Transformations*, Allyn and Bacon, Boston, 1971. 192

66. R. S. Wallace, A modified Hough transform for lines, *Proc. IEEE Conf. on Computer Vision and Pattern Recognition*, 1985, 665–667. 189

67. F. S. Woods, *Higher Geometry: An Introduction to Advanced Methods in Analytic Geometry*, Ginn and Company, Boston, 1922. 192

Digital Lines and Digital Convexity

Ulrich Eckhardt

Fachbereich Mathematik, Optimierung und Approximation
Universität Hamburg
Bundesstraße 55, D-20146 Hamburg
Eckhardt@math.uni-hamburg.de

Abstract. Euclidean geometry on a computer is concerned with the translation of geometric concepts into a discrete world in order to cope with the requirements of representation of abstract geometry on a computer. The basic constructs of digital geometry are digital lines, digital line segments and digitally convex sets. The aim of this paper is to review some approaches for such digital objects. It is shown that digital objects share much of the properties of their continuous counterparts. Finally, it is demonstrated by means of a theorem due to Tietze (1929) that there are fundamental differences between continuous and discrete concepts.

Keywords: digital geometry, digital lines, digital convexity, Tietzes's theorem

1 Introduction

The concept of convexity plays an important role in mathematics and also in applications. Specifically in visual perception convex sets are of importance since almost all visible objects are either convex or else composed of a finite number of convex sets (the "convex ring", see e.g. [,]). Therefore a considerable part of books on digital geometry is devoted to convexity (see e.g. [, Chapitre 5] or [, Chapter 4.3]). In shape recognition two dimensional sets can be decomposed into perceptually meaningful parts by considering convex and concave parts of their boundaries [ ]. There exist numerous generalizations of convexity [, Chapter 6.2]. Ronse gave a bibliography on convexity which covers the years 1961 to 1988 [ ]. A very detailed history of the topic is given by Hübler [ ]. For a quite recent account of the subject including also detailed historical informations the reader is referred to Klette's survey article [ ].

Convexity is closely related to the geometry of lines. Convex sets are defined by means of line segments and on the other hand line segments are convex hulls of two-point sets. In an abstract setting a hyperplane is a nonempty convex set whose complement is also nonempty and convex [, Part II].

The aim of this article is to give a review on digital lines and digital convexity.

Most of the proofs are omitted in this paper. Only in cases where the proof is not easily available or if it involves some special constructions, a sketch of it is indicated.

G. Bertrand et al. (Eds.): Digital and Image Geometry, LNCS 2243, pp. 209–228, 2001.

2   The Digital Space

Given a linear space X over the real numbers $\mathbb{R}$, for example $\mathbb{R}^d$, the ordinary d-dimensional real vector space. In X we consider a certain subset X_Δ, the 'digital space'. In the standard example $X = \mathbb{R}^d$, we can take $X_\Delta = \mathbb{Z}^d$ which is the set of all vectors whose components have integer values. Although it is also possible to treat a good deal of the theory in irregular digital sets, we concentrate here exclusively on the case $(\mathbb{R}^d, \mathbb{Z}^d)$.

The topic of *digital geometry* is to translate continuous concepts, i.e. concepts in X into the digital world in X_Δ. There are basically two possibilities. The *pragmatic* way is to define a *discretization mapping* $X \mapsto X_\Delta$. A digital set $S_\Delta \subseteq X_\Delta$ is said to have a certain property if there is a continuous set $S \subseteq X$ having this property such that S_Δ is the image of S under the discretization mapping (for a related approach see [ ]). Of course, one has to take care that such a definition is well-defined. One disadvantage of this approach is that it is not canonic, it depends on the discretization mapping used.

The other possible approach is the *axiomatic* way. Here, suitable characteristic properties are translated into the digital setting. The main advantage of this approach is that it is mathematically more attractive than the pragmatic way and that it allows to derive properties of the digital objects in a rigorous abstract way. In some fortunate cases both approaches lead to the same concepts.

3   Digital Lines

For sake of simplicity we now concentrate on planar sets, i. e. we consider $(\mathbb{R}^2, \mathbb{Z}^2)$. In digital (plane) geometry two metrics are popular. Let $x = (\xi_1, \xi_2)$ and $y = (\eta_1, \eta_2)$ two vectors in $\mathbb{R}^2$. Then the 4-*metric* is defined as

$$d_4(x, y) = |\xi_1 - \eta_1| + |\xi_2 - \eta_2|$$

and the 8-metric is

$$d_8(x, y) = \max\left(|\xi_1 - \eta_1|, |\xi_1 - \eta_1|\right).$$

These metrics are named according to the number of $\mathbb{Z}^2$-neighbors of distance 1 of a point in $\mathbb{Z}^2$.

A digital set $S_\Delta \subseteq \mathbb{Z}^2$ is termed σ-connected, $\sigma \in \{4, 8\}$, if there exist points $x_1, x_2, \cdots, x_n$ such that $x_i \in S_\Delta$ for all i and $d_\sigma(x_i, x_{i+1}) = 1$ for $i = 1, 2, \cdots, n - 1$.

This concept of connectedness induces a 'topology' (more precisely: a graph structure, see e.g. [ ] or [ ] for details) on $\mathbb{Z}^2$. We define a *digital σ-curve* to be a σ-connected digital set S_Δ with the property that each point $x \in S_\Delta$ has exactly two σ-neighbors in S_Δ with the possible exception of at most two points, the so-called *end points* of the curve, having exactly one neighbor in S_Δ. A finite curve without end points is a *closed curve*.

For any set $S \subseteq \mathbb{R}^2$ and a point $x \in \mathbb{R}^2$ we define for $\sigma \in \{4, 8\}$

$$d_\sigma(x, S) = \inf_{y \in S} d_\sigma(x, y).$$

Then two discretization mappings can be defined

$$\Delta_\sigma(S) = \left\{ x \in \mathbb{Z}^2 \mid d_\sigma(x, S) \leq \frac{1}{2} \right\}, \qquad \sigma \in \{4, 8\}.$$

Δ_4 is sometimes referred to as *grid-intersection-discretization*.

Lemma 1. *If $S \subseteq \mathbb{R}^2$ is connnected then $\Delta_\sigma(S)$ is σ^*-connected, $\sigma \in \{4, 8\}$,*

$$\sigma^* = \begin{cases} 4 & \text{if } \sigma = 8, \\ 8 & \text{if } \sigma = 4. \end{cases}$$

By means of the discretization mappings Δ_σ we could define a digital line as the image of a ontinuous line under one of these mappings. However, the image of a line under Δ_σ is not necessarily a digital σ-curve. If one wants to get digital lines which are also digital curves, the discretization mapping must be modified such that each point in $\mathbb{R}^2$ is mapped to exactly one point in $\mathbb{Z}^2$. If we define the *influence region* of a point $x = (i, j) \in \mathbb{Z}^2$ by

$$\pi_\sigma(x) = \left\{ y \in \mathbb{R}^2 \mid d_\sigma(x, y) \leq \frac{1}{2} \right\},$$

then the union of all $\pi_8(x)$, $x \in \mathbb{Z}^2$ covers the plane $\mathbb{R}^2$ whereas for the union of all $\pi_4(x)$ this is not true. Even more important, the influence regions have boundary points in common. This accounts for the fact that there exist points in $\mathbb{R}^2$ which are mapped on more than one point in $\mathbb{Z}^2$ by means of the discretization mappings Δ_σ. In the next section we see how this latter situation can be remedied in the case $\sigma = 4$.

4   The Chord Property

The first systematic approach to define digital lines by means of a suitable discretization mapping dates back to 1974 when Rosenfeld published his paper on digital straight line segments [ ]. For the proofs of the theorems of this section the reader is referred to Rosenfeld's original paper or the paper of Ronse [ ].

For x and y in $\mathbb{R}^2$ the (continuous) line segment joining x and y is

$$[x, y] = \left\{ z \in \mathbb{R}^2 \mid z = \lambda x + (1 - \lambda)y, \quad 0 \leq \lambda \leq 1 \right\}.$$

A digital set $S_\Delta \subseteq \mathbb{Z}^2$ is said to possess the *chord property* whenever for any two points x, y in S_Δ and for any $u \in [x, y]$ there exists a $z \in S_\Delta$ such that $d_8(z, u) \leq 1$. S_Δ has the *strict chord property* if the strict inequality sign holds.

Lemma 2. *A digital set S_Δ which has the strict chord property is 8-connected.*

In accordance with the pragmatic approach to digital geometry we define: a *digital (σ-) line segment* is a digital set which is a Δ_σ discretization of a line segment in $\mathbb{R}^2$.

Theorem 1. *Each 4-line segment has the chord property.*

Remark 1. It is generally not true that a 4-line segment has the strict chord property.

We define the *modified influence region* (see figure 1) of a point $x = (i, j) \in \mathbb{Z}^2$ by

$$
\begin{aligned}
\pi'_4(x) = \Big\{ (\xi, \eta) \ \Big| \ |i - \xi| + |\eta - j| < \frac{1}{2} \Big\} \ \cup \\
\cup \Big\{ (\xi, \eta) \ \Big| \ \xi - i + \eta - j = \frac{1}{2} \quad \text{and} \quad i \leq \xi \leq i + \frac{1}{2} \Big\} \ \cup \\
\cup \Big\{ (\xi, \eta) \ \Big| \ \xi - i - \eta + j = \frac{1}{2} \quad \text{and} \quad i < \xi \leq i + \frac{1}{2} \Big\}.
\end{aligned}
$$

Note that each point in $\mathbb{R}^2$ belongs to at most one set $\pi'_4(x)$, $x \in \mathbb{Z}$, in contrast to the situation with influence regions $\pi_\sigma(x)$.

The *modified grid-intersection-discretization* then is given by the mapping

$$
\Delta'_4(S) = \{ x \in \mathbb{Z}^2 \mid \pi'_4(x) \cap S \neq \emptyset \}.
$$

Using this concept, we can formulate a stronger variant of theorem 1.

Theorem 2. *Any digital line segment generated by the modified 4-discretization mapping Δ'_4 is an 8-curve having the strict chord property.*

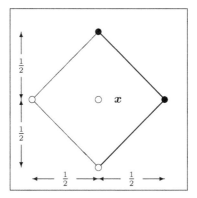

Fig. 1. The modified influence region $\pi'_4(x)$. Parts of the boundary indicated by thick lines belong to the influence region as well as the vertices marked •

Remark 2. The assertion of theorem 2 is very appealing since it connects a certain discretization method with a definition for digital lines which always leads to lines having the strict chord property and being 8-curves. The price we have to pay for this greater symmetry is lack of invariance. The mapping Δ'_4 is not invariant with respect to 90° rotations of the plane.

In this form digital lines were introduced by Rosenfeld [ ]. Usually, if we refer in this paper to digital lines, then a Δ'_4-line is meant.

Corollary 1. *Given a digital set S_Δ such that for any two points x and y in S_Δ there is a digital Δ_4- (Δ'_4-) line segment containing x and y which is contained in S_Δ.*

Then S_Δ has the (strict) chord property.

There is a converse of theorem 2:

Theorem 3. *Given a digital set $S_\Delta \subseteq \mathbb{Z}^2$ having the strict chord property. Let x and y be points in S_Δ.*

Then there exists a line segment $\gamma \in \mathbb{R}^2$ such that $\Delta'_4(\gamma) \subseteq S_\Delta$ and $x, y \in \Delta'_4(\gamma)$.

The proof of this theorem needs Helly's theorem [, Part VI] from convexity theory [,].

Corollary 2. *A digital 8-curve S_Δ having the strict chord property is a digital line segment (in the Δ'_4 sense).*

The theory sketched here has some advantages. Specifically, it translates the common concept of a line or a line segment, respectively, in a quite straightforward way to the digital case via a quite natural discretization mapping. All constructs used here can be generalized to higher dimensions [,]. However, it is not easily possible to verify the chord property for a given digital set. There exist different approaches for finding a characterization of digital lines and line segments which makes exclusively use of $\mathbb{Z}^d$-concepts. Moreover, these approaches lead to algorithms with linear time-complexity. The two main approaches in this direction are the syntactic characterization of Freeman [ ], Rosenfeld [ ] and Hübler, Klette and Voss [ ] and the arithmetic characterization of Debled-Rennesson and Reveillès [ ]. There slso exists a very interesting number-theoretic approach by Voss [ ] and Bruckstein [ ]. This latter approach, however, will not be treated here.

5   Syntactic Characterization of Digital Line Segments

The *chain code* for coding digital 8-curves was proposed by Freeman [ ]. The 8-neighbors of a point $x = (i, j) \in \mathbb{Z}^2$ are numbered according to the following scheme.

	$i-1$	i	$i+1$
$j+1$	$N_3(x)$	$N_2(x)$	$N_1(x)$
j	$N_4(x)$	x	$N_0(x)$
$j-1$	$N_5(x)$	$N_6(x)$	$N_7(x)$

A digital 8-curve can be described by means of a simple compact linearly ordered data stucture containing the coordinates of one of its end points (or of a fixed point on it if a closed curve is treated) and a sequence of code numbers in $\{0, 1, 2, 3, 4, 5, 6, 7\}$ indicating for each point on the curve which of its neighbors will be the next point on the curve.

The following theorem was proved by Rosenfeld [ ].

Theorem 4. *Given a digital 8-curve consisting of more than three points which has the strict chord property. Let $a_1, a_2, \cdots, a_n$ be the set of chain code numbers of the curve, $0 \le a_i \le 7$. Then*

1. *There are at most two different code numbers among the a_i,*
2. *If there are two different code numbers among the a_i, say α and β, then these belong to adjacent directions, i.e. $\alpha \equiv \beta \pm 1 \pmod 8$,*
3. *If there are two different code numbers, one of them is singular, i.e. it never occurs twice in succession.*

A digital 8-curve having the properties of theorem 4 is not necessarily a digital line segment.

For digital lines (with rational slope) the following theorem holds:

Theorem 5. *Let $\Gamma \subseteq \mathbb{Z}^2$ be a digital line. Then the following three assertions are equivalent:*

1. *The chain code of Γ fulfills the conditions of theorem 4 and is periodic.*
2. *There exists a line with rational slope so that Γ is the Δ'_4 discretization of this line.*
3. *All lines generating Γ by discretization have the same (rational) slope.*

For digital lines which are characterized as in theorem 5 the axiom of parallelity is not necessarily true. There do exist pairs of parallel lines which are different but not disjoint. The intersection of such parallel lines is not necessarily a digital line. Moreover, the intersection of two lines can contain more than one point.

Freeman [ ] proposed as a heuristic criterion that the singular direction should be distributed as regularly as possible in the chain code. Hübler, Klette and Voss [ ] made this heuristic statement precise by giving a syntactic characterization of digital lines and digital line segments. Assume that we are given a sequence $F = \{a_j\}_{j \in \mathbb{Z}}$ of code numbers, $0 \le a_j \le 7$. An element a_k of this sequence is *singular* whenever $a_{k-1} \ne a_k$ and $a_k \ne a_{k+1}$, otherwise a_k is *regular*. A *regular piece* of F is a finite subsequence of maximal length of successive elements of F with mutually equal code numbers.

Let $\mathcal{F} = \{\{a_k\}_{k=-\infty}^{\infty}\}$ be the set of all sequences of integers ($a_k \in \mathbb{N}$ for all k). We introduce a *reduction operator* $R : \mathcal{F} \longrightarrow \mathcal{F}$ by means of the following procedure:

- Delete all singular elements in F which are adjacent to two regular pieces.
- Replace all regular pieces of F by their lengths.
- Leave all other elements of F unchanged.

The sequence F is said to possess the *Freeman property* if

F$_1$ There exists a number $a \in \mathbb{N}$ such that $a_j \in \{a, a+1\}$ for all j.
F$_2$ Whenever F contains both numbers a and $a+1$ then at least one of them occurs only singularly.

The sequence F is said to possess the *HKV (Hübler, Klette, Voss) property* if each reduced sequence $R^k(F)$, $k = 0, 1, 2 \cdots$ has the Freeman property.

We now introduce a convexity concept. A digital set $S_\Delta \subseteq \mathbb{Z}^2$ is convex in the sense of Minsky and Papert or *MP-convex* if for any two points x and y in S_Δ and $z \in [x, y] \cap \mathbb{Z}^2$ then $z \in S_\Delta \,[\ \ ]$.

Theorem 6. *Any digital 8-curve which is unbounded in both directions and MP-convex has the HKV-property.*

Proof. The proof of this theorem is rather long and technical but straightforward. We give only a sketch of it. First it is shown that the chain code of an MP-convex curve has the Freeman property. This can be done by straightforward enumerative discussion of all possible cases.

Then it is shown that each reduced code sequence has the Freeman property. In order to do this it is first shown that the Freeman code of an MP-convex curve consists of only two different regular pieces of size k and $k+1$. It is easily seen that one of these regular pieces occurs singularly.

The converse of theorem 6 is also true [ ]:

Theorem 7. *Any digital 8-curve which is unbounded in both directions and has the HKV-property is MP-convex.*

Proof. One basic trick for proving this theorem is due to Hübler. Without loss of generality we can assume that the chain code F for the digital curve S_Δ has only code numbers 0 and 1. If the reduced sequence $R(F)$ has elements k and $k+1$ and if $k+1$ is singular then the linear mapping of the (real) plane with matrix

$$M = \begin{pmatrix} k+1 & 1 \\ 1 & 0 \end{pmatrix}$$

maps the vectors $(1,0)$ and $(1,1)$ on the elements $(k+1,1)$ and $(k+2,1)$, respectively. The mapping $S_\Delta \mapsto MS_\Delta$, if interpreted as a mapping from $\mathbb{R}^2$ to $\mathbb{R}^2$, is invertible and $M^{-1}R(S_\Delta)$ can be interpreted as the chain code of a digital curve.

The remainder of the proof is rather straightforward.

For digital line segments the HKV-property must be modified in order to cope with boundary effects. For sake of completeness we indicate this modification. For details the reader is referred to the original paper of Hübler, Klette and Voss [ ].

Denote by $\mathcal{F}_0$ the set of all finite sequences of integers. As above we define singular and regular elements in a sequence and also the Freeman property for sequences $F \in \mathcal{F}_0$.

The *modified reduction operator* $R' : \mathcal{F}_0 \longrightarrow \mathcal{F}_0$ is defined as follows: All maximal subsequences of successive regular elements of F which are between two singular elements are replaced by their lengths and all other elements are deleted. If F contains no singular elements, F is replaced by its length.

For a sequence $F \in \mathcal{F}_0$ with the Freeman property we denote by $\ell(F)$ the number of regular elements preceeding the first singular element in F and by $r(F)$ the number of regular elements following the last singular element in F.

The modified HKV-property then looks as follows:

HKV₁ All reduced sequences $R'^k(F)$, $k = 0, 1, 2, \cdots$ have the Freeman property (or are empty).

HKV₂ If $R'^k(F)$, $k = 1, 2, \cdots$, consists of identical elements a or of two different elements a and $a+1$ then $\ell(R'^{k-1}(F)) \leq a + 1$ and $r(R'^{k-1}(F)) \leq a + 1$.

HKV₃ If $R'^k(F)$, $k = 1, 2, \cdots$, contains two elements a and $a+1$ such that a ($\leq a + 1$) is nonsingular, then
$R'^k(F)$ starts with a if $\ell(R'^{k-1}(F)) = a + 1$ and
$R'^k(F)$ ends with a if $r(R'^{k-1}(F)) = a + 1$.

The following theorem characterizes digital lines by the HKV-property. The theorem was first stated by Hübler, Klette and Voss [ ]. The first proof of the theorem was given by Wu [ ] (see also [ ]).

Theorem 8. *A (finite) 8-curve S_Δ is a digital line segment if and only if it has the modified HKV-property.*

The relevance of the theorems of this section lies in the fact that the modified HKV-property can be verified by a syntactic parsing of the chain code of a given digital curve. This process obviously can be performed in time proportional to the length of the curve.

6   Arithmetic Characterization of Digital Line Segments

An alternative definition of a digital line is due to J.-P. Reveilles [ ]. Given numbers $a, b \in \mathbb{Z}$ with $b \neq 0$ such that the greatest common divisor of a and b is 1 and a number $\mu \in \mathbb{Z}$, then a digital line is the set

$$\mathcal{D}(a, b, \mu) = \left\{ (x, y) \in \mathbb{Z}^2 \mid \mu \leq ax - by < \mu + \max(|a|, |b|) \right\}. \tag{1}$$

This definition is an immediate consequence of the modified grid-intersection-discretization of a real line. To a (nontrivial) digital line segment S_Δ one can associate two lines in $\mathbb{R}^2$, namely the line

$$\left\{ (\xi, \eta) \in \mathbb{R}^2 \mid a\xi - b\eta = \mu \right\}$$

and

$$\left\{ (\xi, \eta) \in \mathbb{R}^2 \mid a\xi - b\eta = \max_{(x,y) \in S_\Delta} ax - by \right\}.$$

The strip bounded by these two lines contains the digital line segment.

If an 8-curve is to be tested for linearity, a system of Diophantine inequalities has to be solved, since for each point (x, y) on the curve an inequality of the form (1) yields a condition for the integer variables μ, a, b. The solution of such a system is straigtforward and easy and can be done in time proportional to the length of the curve.

7   Digital Lines and Translations

Hübler [, ,] proposed a theory of digital lines which is based on translations. First one can state:

Observation *The chain code of a digital line is periodic if and only if there exists a nontrivial translation of $\mathbb{Z}^2$ which leaves the digital line fixed.*

This leads immediately to the following definition: Given a translation $T : \mathbb{Z}^2 \longrightarrow \mathbb{Z}^2$. A digital line in the sense of Hübler is any set of the form $\{x_0 + T^n x \mid n \in \mathbb{Z}\}$ with fixed vectors $x_0, x \in \mathbb{Z}^2$. We assume that the translation is *irreducible* which means that there is no other translation $T' : \mathbb{Z}^2 \longrightarrow \mathbb{Z}^2$ such that $T = (T')^k$ with $k > 1$.

By means of this concept one gets digital lines which are not necessarily 8-curves but have attractive properties.

- For any two different points x and y in $\mathbb{Z}^2$ there is exactly one digital line containing these points.
- If two lines are parallel then they either coincide or else they are disjoint.
- Two lines are parallel if and only if there is a translation mapping one of them into the other.
- Two lines are parallel if and only if each translation leaving one of them fixed also leaves the other fixed.
- Two lines which are not parallel intersect in at most one point.

A *digital line segment* is a finite subset of a digital line containing consecutive points of the line.

The main advantage of Hübler's definition is its potential for generalizations to $\mathbb{Z}^d$.

Any digital Δ'_4-line as defined above always contains a line in the sense of Hübler if it is generated by a line having rational slope.

8   Digital Convexity

There exist different definitions of digital convexity in the literature. We list some of the most common definitions. Let S_Δ be a digital set in $\mathbb{Z}^2$ (or $\mathbb{Z}^d$, respectively).

The first definition of digital convexity was stated in 1969 by Minsky and Papert [ ] (see above; MP-convexity). The main reason for introducing this concept was to give an example of a predicate which has parallel order 3 (i.e. it can be verified by looking at triples of points in a digital set) and which is not 'local'.

MP-convexity For x and y in S_Δ and $z \in [x, y] \cap \mathbb{Z}^2 \implies z \in S_\Delta$.
H-convexity The convex hull of S_Δ is the set

$$\mathrm{conv} S_\Delta = \left\{ \sum_{j=1}^n \lambda_j x_j \mid \sum_{j=1}^n \lambda_j = 1, \lambda_j \geq 0 \text{ and } x_j \in S_\Delta \right\}.$$

S_Δ is H-convex if $S_\Delta = \mathrm{conv} S_\Delta \cap \mathbb{Z}^d$.
D-convexity S_Δ is *digital convex* or D-convex if for $x, y \in S_\Delta$ the Δ_4'-line segment joining x and y belongs to S_Δ.
DH-convexity S_Δ is *digital convex in the sense of Hübler* or DH-convex if for $x, y \in S_\Delta$ the line segment (according to Hübler) joining x and y belongs to S_Δ.

Remark 3. If one considers irregular digital spaces $X_\Delta \subseteq \mathbb{R}^d$ then the definitions of MP- and H-convexity carry over without any difficulty. However, if no three different points of X_Δ are collinear, each set $S_\Delta \subseteq X_\Delta$ is MP-convex. On the other hand, D- and DH-convexity depend on a geometry of lines in X_Δ which is compatible with lines in $\mathbb{R}^d$.
In figure 2 examples are given for different convexity definitions.

All these definitions have in common that the different convex sets have some of the properties which are known from ordinary convex sets. For example, the intersection of two MP- (H-, DH-) convex sets is always an MP- (H-, DH-) convex set. However, as can be seen by the conterexample in figure 3, the intersection of two D-convex sets is not always a D-convex set.

Let S_Δ be a set which is convex according to one of the definitions above. An *interior point* of S_Δ is a point $x \in S_\Delta$ such that the direct neighbors $N_0(x)$, $N_2(x)$, $N_4(x)$ and $N_6(x)$ all belong to S_Δ. The *interior* of S_Δ is the set of all interior points. It can easily be shown that the interior of a convex digital set is also convex.

We state some assertions on the mutual relations of these convexity concepts.

Theorem 9. *A digital set is DH-convex if and only if it is MP-convex.*

The proof of this theorem is obvious.

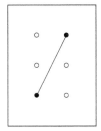 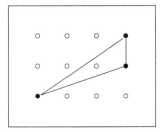

Fig. 2. Examples of convex sets. The digital set (•) in the left picture is MP-convex and H-convex but not 8-connected, hence not D-convex. The digital set in the right picture is MP-convex but not H-convex

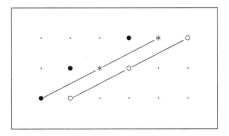

Fig. 3. Two D-convex sets whose intersection is not D-convex. The origin of the coordinate system is at the leftmost point •. The real line segment joining points $(0,0)$ and $(4,2)$ generates the digital Δ'_4-line segment consisting of all points • and *. Similarly, the line segment joining points $(1,0)$ and $(5,2)$ generates the digital line segment consisting of points ○ and *. The intersection of both digital line segments (points *) is not 8–connected, hence not D-convex

Theorem 10. *A digital set has the chord property if and only if it is 8-connected and H-convex.*

Theorem 11. *Any MP-convex and 8-connected digital set is D-convex.*

The proof of this theorem is very similar to the proof of theorem 6.

The relations of all these convexity definitions are illustrated in Figure 4. It is an interesting fact that under the assumption of 8-connectedness all these different concepts coincide. In the sequel we will always assume that the sets under consideration are 8-connected so that there is no need to distinguish different convexity concepts.

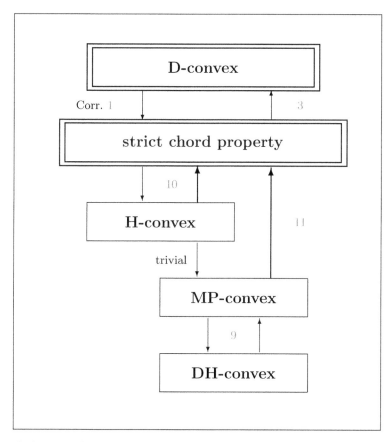

Fig. 4. Relations of different convexity definitions. Double frames indicate convexity definitions which imply 8-connectivity. Bold arrows mean that the corresponding relation holds under the assumption of 8-connectivity. Numbers at the arrows denote the theorems where the corresponding assertions are formulated

9   Convexity Verification

Given a finite set $S \subseteq \mathbb{R}^2$ with n elements, the convex hull of S can be determined in time $O(n \log n)$ [ ].

The situation is more favourable if convexity of a polygonal set P in $\mathbb{R}^2$ is investigated. Such a set can be understood as an ordered set of boundary vertices such that each pair of two consecutive vertices is joined by a line segment. Convexity of a polygonal set in $\mathbb{R}^2$ can be verified in a time which is proportional to the number of vertices of the polygon. By moving along the oriented boundary of the polygonal set in such a way that the interior of the set is always on the left-hand side, one determines for each vertex the direction change of the adjacent line segments. The set S_Δ is convex if and only if at each vertex these line segments form a left turn.

Given a (bounded) digital set $S_\Delta \in \mathbb{Z}^2$, it is very easy to determine its boundary points (which are all points in S_Δ having at least one 4-neighbor not in S_Δ). In $\mathbb{Z}^2$ the set of boundary points of a simply connected digital set (i.e. a digital set which is 8-connected and its complement is 4-connected) is a closed 8-curve. If consecutive points on the digital boundary curve are joined by line segments, a closed polygonal curve is obtained. Thus, to each (simply connected) digital set in $S_\Delta \subseteq \mathbb{Z}^2$ a polygonal set $P(S_\Delta)$ can be associated in a canonical way such that $S_\Delta = \mathbb{Z}^2 \cap P(S_\Delta)$.

Unfortunately, if S_Δ is convex in $\mathbb{Z}^2$, $P(S_\Delta)$ is not necessarily convex in $\mathbb{R}^2$. It was noted by Hübler, Klette and Voss [ ] that for a digital set in $\mathbb{Z}^2$ convexity can be verified by traversing the boundary along digital line segments and monitoring the direction changes at each intersection point of two consecutive digital lines. This yields a linear time algorithm for convexity verification in $\mathbb{Z}^2$ (see also [ ]).

The approach of Debled-Rennesson and Reveillès [ ] also leads to a linear time algorithm for detecting convexity.

The situation is more complex in higher dimensions. First of all, the complexity assertions for constructing convex hulls in the continuous space are no longer true, in dimensions ≥ 3 the picture is more complicated. Ronse [,] was able to generalize the chord property to arbitrary dimensions. However, for algorithmic purposes the chord property is not very helpful. Hübler's approach is independent of the dimension [ ], however, it is also not immediately suitable for convexity detection. Debled-Rennesson [ ] was able to carry over the main ideas of the approach proposed by her and by Reveillès [ ] to three dimensions.

10   Tietze's Theorem

In continuous convexity theory (see e.g. [ ]) the boundary points of a convex set are classified by means of separation arguments. It is a remarkable fact that such arguments play virtually no role in digital convexity theory. Only in a more recent paper of Latecki and Rosenfeld digital supportedness was investigated [ ].

Assume that $\mathbb{R}^d$ is equipped with Euclidean geometry, i. e. to each pair of vectors $x = (\xi_1, \xi_2, \cdots, \xi_d)$ and $y = (\eta_1, \eta_2, \cdots \eta_d)$ the *inner product*

$$\langle x, y \rangle = \sum_{j=1}^{d} \xi_j \eta_j$$

is assigned. Given a nonzero vector $x^* \in \mathbb{R}^d$ and a real number α, a *hyperplane* is a set

$$[x^* : \alpha] = \left\{ x \in \mathbb{R}^d \mid \langle x, x^* \rangle = \alpha \right\}.$$

The following Characterization Theorem holds

Theorem 12. *A digital set $S_\Delta \subseteq \mathbb{Z}^d$ is digital convex if and only if for any point $x \in \mathbb{Z}^d \setminus S_\Delta$ there exists a hyperplane $[x^* : \alpha]$ such that $\langle x, x^* \rangle = \alpha$ and $\langle y, x^* \rangle > \alpha$ for all $y \in S_\Delta$.*

This theorem is an obvious consequence of the Separation Theorem for Convex Sets [ ]. A digital variant of the theorem may be found in [ ]. A point of a set $S \subseteq \mathbb{R}^d$ is termed an *exposed point* of S if there exists a hyperplane $[x^* : \alpha]$ such that $\langle x, x^* \rangle = \alpha$ and $\langle y, x^* \rangle \geq \alpha$ for all $y \in S$. Using this concept, the assertion of the following theorem is also known from convexity theory [ ]:

Theorem 13. *If each boundary point of a digital set is an exposed point then the set is (digital) convex.*

In 1929 H. Tietze [ ] proved a remarkable theorem. A set $S \subseteq \mathbb{R}^d$ is termed *weakly supported locally* if for each boundary point x of S there exists an (open) neighborhood $U(x)$ and a hyperplane $[x^* : \alpha]$ such that $\langle x, x^* \rangle = \alpha$ and

$$\langle y, x^* \rangle < \alpha \text{ for } y \in U(x), \ y \neq x \quad \Longrightarrow \quad y \notin S.$$

Tietze's theorem now is [, Theorem 4.10].

Theorem 14. *Let S be an open connected set in $\mathbb{R}^d$. If each point of the boundary of S is one at which S is weakly supported locally, then S is convex.*

The assertion of Tietze's theorem means that in $\mathbb{R}^d$ each set can be tested for convexity by looking only locally at boundary points. It is surprising that in Tietze's theorem one needs topological assumptions (S is required to be open). It would be very attractive to translate this theorem into the digital world since it would allow to test any digital set for convexity in a time proportional to the number of its boundary points. Unfortunately, however, Tietze's theorem does not hold in general in digital spaces.

For a point $x \in \mathbb{Z}^d$ denote by $\mathcal{N}_8(x)$ the set consisting of x together with its 8-neighbors. First we define: Given a set $S_\Delta \subseteq \mathbb{Z}^d$. A point $x \in S_\Delta$ is a *locally exposed point* of S_Δ if there exists a hyperplane $[x^* : \alpha]$ such that $\langle x, x^* \rangle = \alpha$ and $\langle y, x^* \rangle \geq \alpha$ for all $y \in (S_\Delta \cap \mathcal{N}_8(x))$.

A set $S_\Delta \subseteq \mathbb{Z}^d$ is said to meet the *interior point condition* at point $x \in S_\Delta$ if the digital set $(\mathcal{N}_8(x) \cap S) \setminus \{x\}$ is 4-connected and contains at least two points.

Now the following theorem is true:

Theorem 15. *Let $S_\Delta \subseteq \mathbb{Z}^d$ be a digital set and let x be a boundary point in S_Δ which is locally exposed. Assume further that the interior point condition does not hold at x.*

Then there exists a digital set S'_Δ with $\mathcal{N}_8(x) \cap S_\Delta = \mathcal{N}_8(x) \cap S'_\Delta$ such that each boundary point of S'_Δ is a locally exposed point and S'_Δ is not convex.

The meaning of this theorem is that without the interior point condition nothing can be said about convexity of a set only by looking at neighborhoods. The theorem can be proved by a rather lengthy discussion of all possible cases. We treat only one of these cases.

Since x was assumed to be a boundary point, one of the direct neighbors of x, say $N_6(x)$, must not belong to S_Δ. We consider the case that also one of the indirect neighbors, say $N_1(x)$, is not in S_Δ. Furthermore we assume that $N_0(x)$ and $N_7(x)$ are in S_Δ. By convexity of S_Δ is $N_5(x) \notin S_\Delta$. Therefore we have the following situation ($\bullet$ denotes a point in S_Δ, $\circ$ a point not in S_Δ and $\cdot$ denotes a point whose status with respect to S_Δ is left open):

If the interior point condition does not hold, one of the remaining points in $\mathcal{N}_8(x)$ belongs to S_Δ. If e.g. $N_2(x)$ were in S_Δ, then we choose S'_Δ as follows (all points which are not marked $\bullet$ are assumed not to belong to S'_Δ):

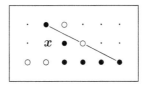

Obviously all points of S'_Δ are locally exposed points and S'_Δ is not convex.

The situation becomes less pessimistic if one considers digital sets fulfilling the interior point condition. Each of the eight neighbors of a point either belongs to S_Δ or not, hence there are $2^8 = 256$ different possible neighborhood configurations. These configurations can be obtained from 51 generating configurations by rotations and reflections. If one is interested in configurations fulfilling the interior point condition, there remain 8 generating configurations (these are exactly the "simple strict boundary point-configurations" of [, p. 160]). 7 of these configurations are digitally convex and 6 of them are configurations of locally exposed points. These latter are given in Figure 5.

There remains one configuration which fulfills the interior point condition and is convex but not locally exposed:

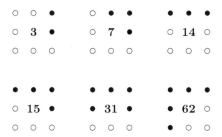

Fig. 5. Configurations of exposed points. ● denotes points in S_Δ and ○ denotes points not in S_Δ. The number in the center of each configuration is the configuration number $\#(\mathcal{N}_8) = \sum_{j=0}^{7} c_j \cdot 2^j$ where $c_j = 1$ for points in S_Δ and $c_j = 0$ for points not in S_Δ. It is assumed that the center point belongs to S_Δ

We formulate an encouraging result sharpening the assertion of theorem 13:

Theorem 16. *Given a simply connected digital set S_Δ such that all boundary points of S_Δ fulfill the interior point condition and are locally exposed.*
Then S_Δ is convex.

Proof. Assume that S_Δ is simply connected and that all of its boundary points fulfill the interior point condition. Then the polygonal set $P(S_\Delta)$ in $\mathbb{R}^2$ associated canonically to S_Δ as above is "regular", i.e. the (topological) closure of its interior is the set itself (This can be seen easily by inspection of the neighborhood configurations in Figure 5). So, $P(S_\Delta)$ fulfills all conditions of Tietze's theorem, which completes the proof.

The relevance of the latter theorem is rather limited since digitally convex sets rarely have only locally exposed boundary points, which means that configuration **63** above (as well as its rotated or reflected versions) occurs quite often in realistic digital sets. If this happens, neighborhood information alone is not sufficient for testing convexity. However, it is possible to monitor the occurence of critical configurations and to remedy the situation. One possibility, of course, for treating this situation is to segment the boundary into digital line segments in the manner of Hübler, Klette and Voss [ ] or of Debled-Rennesson and Reveillès [ ]. These approaches, however, are not local in the sense of Tietze's theorem.

The problems encountered in connection with application of Tietze's principle are sketched here by means of a special application. For retrieving shapes from a pictoral database it is necessary to simplify the contours of the objects under consideration. This can be done by removing "irrelevant" points from the contour [ ]. Such a procedure has the advantage that the original data structure (a linearly ordered list of points in $\mathbb{Z}^2$) is not changed. Moreover, this approach also has a favourable stability behaviour [ ]. A first step in the method could be to replace the polygonal set $P(S_\Delta)$ obtained canonically from the chain code by a set $Q(S_\Delta)$ having the properties:

1. $Q(S_\Delta)$ is a polygonal set obtained from $P(S_\Delta)$ by deleting vertices,
2. $S_\Delta = Q(S_\Delta) \cap \mathbb{Z}^2$.
3. $Q(S_\Delta)$ has the same "convexity behaviour" of the boundary as S_Δ.

It can be shown that such a set $Q(S_\Delta)$ cannot be found in a canonic way from S_Δ. Of course, there is some freedom of choice in interpreting the third requirement which is not mathematically precise.

In Figure 6 a specific digital set is shown. The set was reduced by deleting points from the boundary as long as this is possible in a unique way under the first two conditions indicated above.

Further reduction under the conditions stated above is no longer possible in a well-defined way. The problem is that a part of the boundary may have the following form (notations as in Figure 6, left image):

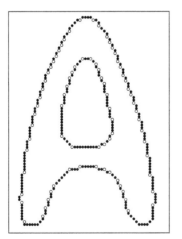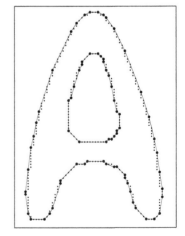

Fig. 6. Construction of an enclosing polyhedral set. In the left picture the boundary of a digital set is shown. Points having neighborhood configuration **63** (up to rotation and reflection) are indicated by ∘. In the right picture the enclosing polyhedral set after detection of digital lines is given. The original boundary contains 297 points, the reduced polygonal set has 75 vertices

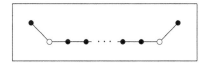

It is assumed here that the interior of the set is below the curve. It is possible to delete one of the critical points ∘, but this is not an invariant action. There are other pragmatic possibilities to cope with this situation, e.g. one can introduce additional points not in $\mathbb{Z}^2$ thus relaxing the first two conditions above. Another pragmatic approach which is based on a "relevance measure" associated to the vertices of the polygon is given in [ ].

11 Conclusions

The concepts of digital lines and digital convexity were introduced thirty years ago in a pragmatic way by means of the grid-intersection discretization. It turned out that these concepts share a number of properties with lines and convex sets in continuous spaces. It could be shown [ ] that these pragmatic concepts can be embedded into a formal axiomatic theory. Moreover, different definitions of digital convexity coincide if in addition 8–connectedness is required. Thus, at present digital lines and digital convexity are well understood theoretically and the different approaches to the subject can be unified within the framework of a nice theory.

Due to the practical relevance of the subject, a number of algorithmic methods were developed to verify convexity of digital sets. They can be classified as syntactic [ ], arithmetic [ ] and number-theoretic characterizations [ ]. Typically, these methods allow to verify convexity in a time proportional to the length of the boundary of the set under investigation.

There do exist differences between digital and continuous convexity. In contrast to continuous sets, it is not possible to verify digital convexity by local inspections only.

References

1. Bruckstein, A. M.: Self-similarity properties of digitized straight lines. 1989, Contemp. Mathem. **119** (1991) 1–20 213
2. Chassery, J.-M., Montanvert, A.: Géométrie discrète en analyse d'images. Traite des Nouvelles Technologies. Serie Images. Hermes, Paris (1991) 209
3. Debled-Rennesson, I.: Etude et reconnaisance des droites et plans discrets. Theses, Université Strasbourg, 1995 221
4. Debled-Rennesson, I., Reveillès, J.-P.: A linear algorithm for segmentation of digital curves. Int. J. Pattern Recogn. Artif. Intell. **9** (1995) 635–662 213, 216, 221, 224, 226
5. Eckhardt, U., Maderlechner, G.: Invariant thinning. Int. J. Pattern Recognition Artif. Intell. **7** (1993) 1115–1144 223
6. Freeman, H.: On the encoding of arbitrary geometric configurations. IRE Trans. **EC-10** (1961) 260–268 213

7. Freeman, H.: Boundary encoding and processing. In: Rosenfeld, A., Lipkin, B. S. (eds.): Picture Processing and Psychopictorics. Academic Press, New York (1970) 241–266 213, 214

8. Hübler, A.: A theoretical basis for digital geometry — an axiomatic approach. In: Eckhardt, U. (ed.): Mathematical Methods in Image Processing. Proceedings of a Minisymposium at ECMI '90 in Lahti. Hamburger Beiträge zur Angewandten Mathematik, Reihe B, Bericht 15 (1991) 5–20 217

9. Hübler, A.: Geometrische Transformationen in der diskreten Ebene. In: Burkhardt, H., Höhne, K. H., Neumann, B. (eds.): Mustererkennung 1989. 11. DAGM-Symposium, Hamburg, 2.–4. Oktober 1989. Proceedings. Informatik-Fachberichte, Vol. 219. Springer-Verlag, Berlin Heidelberg New York (1989) 36–43 217

10. Hübler, A.: Diskrete Geometrie für die digitale Bildverarbeitung. Dissertation B, Universität Jena (1989) 209, 215, 216, 217, 221, 226

11. Hübler, A., Klette, R., Voss, K.: Determination of the convex hull of a finite set of planar points within linear time. Elektronische Informationsverarbeitung und Kybernetik EIK **17** (1981) 121–139 213, 214, 216, 221, 224, 226

12. Klette, R.: Digital Geometry — The Birth of a New Discipline. (Festschrift for Azriel Rosenfeld, to appear 2001). CITR-TR-79 209

13. Klette, R., Stojmenovic, I., Žunić, J.: A parametrization of digital planes by least-squares fits and generalizations. Graphical Models and Image Processing **58** (1996) 295–300 210

14. Klette, R.: On the approximation of convex hulls of finite grid point sets. Pattern Recognition Letters **2** (1983) 19–22 221

15. Latecki, L. J., Ghadially, R.-R., Lakämper, R., Eckhardt, U.: Continuity of discrete curve evolution. J. Electron. Imaging **9** (2000) 317–326 225

16. Latecki, L. J., Lakämper, R.: Convexity rule for shape decomposition based on discrete contour evolution. Computer Vision and Image Understanding **73** (1999) 441–454 209, 225, 226

17. Latecki, L. J., Rosenfeld, A.: Supportedness and tameness. Differentialless geometry of plane curves. Pattern Recognition, **31** (1998) 607–622 221, 222

18. Latecki, L. J.: Discrete Representation of Spatial Objects in Computer Vision. Computational Imaging and Vision. Kluwer Academic Publishers, Dordrecht Boston London (1998) 209, 210

19. Matheron, G.: Random Sets and Integral Geometry. John Wiley & Sons, New York London Sydney Toronto (1975) 209

20. Minsky, M., Papert, S.: Perceptrons. An Introduction to Computational Geometry. The MIT Press, Cambridge London (1969) 215, 218

21. Preparata, F. P., Shamos. M. I.: Computational Geometry. An Introduction. Texts and Monographs in Computer Science. Corr. 3rd printing, Springer-Verlag, New York Berlin Heidelberg (1985) 221

22. Ronse, C.: A note on the approximation of linear and affine functions: The case of bounded slope. Arch. Math. **54** (1990) 601–609 213, 221

23. Ronse, C.: Criteria for approximation of linear and affine functions. Arch. Math., **46** (1986) 371–384 213, 221

24. Ronse, C.: A bibliography on digital and computational convexity (1961-1988). IEEE Trans. **PAMI-11** (1989) 181–190 209

25. Ronse, C.: A simple proof of Rosenfeld's characterization of digital straight line segments. Pattern Recognition Letters **3** (1985) 323–326 211, 213

26. Rosenfeld, A.: Digital straight line segments. IEEE Trans. **C-23** (1974) 1254–1269 211, 213, 214

27. Stoyan, D., Kendall, W. S., Mecke, J.: Stochastic Geometry and Its Applications. John Wiley & Sons, Chichester New York (1987) 209
28. Tietze, H.: Bemerkungen über konvexe und nicht-konvexe Figuren. J. Reine Angew. Math **160** (1929) 67–69 222
29. Valentine, F. A.: Convex Sets. McGraw-Hill Series in Higher Mathematics. McGraw-Hill Book Company, New York San Francisco Toronto London (1964) 209, 213, 221, 222
30. Voss, K.: Coding of digital straight lines by continued fractions. Comput. Artif. Intell., **10** (1991) 75–80 213, 226
31. Voss, K.: Discrete Images, Objects, and Functions in $\mathbb{Z}^n$. Algorithms and Combinatorics, Vol. 11. Springer-Verlag, Berlin Heidelberg New York (1993) 209, 210
32. Wu, L.-D.: On the chain code of a line. IEEE Trans., **PAMI-4** (1982) 347–353 216

Curvature Flow in Discrete Space

Atsushi Imiya

Media Technology Division, Institute of Media and Information Technology
Chiba University
1-33 Yayoi-cho, Inage-ku, 263-8522, Chiba, Japan
imiya@media.imit.chiba-u.ac.jp

Abstract. In this paper, we first define the curvature indices of vertices of discrete objects. Second, using these indices, we define the principal normal vectors of discrete curves and surfaces. Third, we define digital curvature flow as a digital version of curvature flow in discrete space. Finally, these definitions of curvatures in a discrete space derives discrete snakes as a discrete variational problem since the minimization criterion of the snakes is defined using the curvatures of points on the discrete boundary.

1   Introduction

The decomposition of the three-dimensional neighborhood to a collection of two-dimensional neighborhoods reduces the combinatorial properties of a discrete object to the collections of combinatorial properties of planar patterns [,]. Using this geometric property of neighborhood, we define the vertex angle and curvature of a discrete surface as an extension of the chain code of digital curves. These angles on the boundary provide a relation between the point configurations on the boundary and the Euler characteristic of a discrete object. In reference [ ], Toriwaki and coworkers applied this idea for the definition of the Euler characteristic of a discrete object. Furthermore, they derived a Boolean function for the computation of the Euler characteristic [ ]. Their method requires all points in the region of interest since they defined the topological properties of discrete object using all points in the region of interest. Our method only requires the point configurations on the boundary. We second introduce a new transform for the binary digital set, which we call "digital curvature flow." Digital curvature flow is a digital version of curvature flow [ ]. We have already proposed a discrete version of curvature flow which describes the geometry of curvature flow for polyhedrons and polytopes [ ]. Digital curvature flow describes the geometric flow which is controlled by the curvature of the boundary of binary digital images in a discrete space.

Distance transform and skeletization based on distance transform are the classical transformations for the derivation of global features of binary digital images by constructing the medial axes of images. On the other hand, the AND-pyramid extracts features by shrinking the original images. The transformation provides the hierarchical expression of global and local features of images. There are continuous versions for both medial axes extraction and shrinking for feature extraction. The former and latter are the medial axes equation and the scale-space analysis. These transforms are expressed as evolution equations with respect to scale parameters. They permit the multiscale expression of features. There is one more diffusion-like transformation for shapes, namely, curvature flow, which also permits multiscale description of geometrical and topological features of shapes. In this paper, we deal with a complete digital version of this third type of transform. Using the Euler characteristic of discrete point set, we can detect collapses of topology and control the topology of shapes and objects during the deformation by flow-based processing.

The method of snakes extracts the boundaries of the regions by deforming the boundaries dynamically. For practical computation of the boundary using snakes [ ], we must solve a variational problem using an appropriate numerical method [ ]. In digital image analysis, image data are expressed in digital space [ ]. Therefore, in this paper, we propose a digital version of this variational problem for boundary detection using the point configurations on the boundary. Furthermore, we prove that the digital boundary detected based on mathematical morphology [ ] is derived as the solution of this digital variational problem.

2   Connectivity and Neighborhood

Setting $\mathbf{R}^2$ and $\mathbf{R}^3$ to be two- and three-dimensional Euclidean spaces, we express vectors in $\mathbf{R}^2$ and $\mathbf{R}^3$ as $\boldsymbol{x} = (x, y)^\top$ and $\boldsymbol{x} = (x, y, z)^\top$, respectively, where $\top$ is the transpose of the vector. Setting $\mathbf{Z}$ to be the set of all integers, the two- and three-dimensional discrete spaces $\mathbf{Z}^2$ and $\mathbf{Z}^3$ are sets of points such that both x and y are integers and all x, y, and z are integers, respectively. For $n = 2, 3$ we define the dual set for the set lattice points $\mathbf{Z}^n$ as

$$\overline{\mathbf{Z}^n} = \{\boldsymbol{x} + \frac{1}{2}\boldsymbol{e} | \boldsymbol{x} \in \mathbf{Z}^n\}, \tag{1}$$

where $\boldsymbol{e} = (1, 1)^\top$ and $\boldsymbol{e} = (1, 1, 1)^\top$ for $n = 2$ and $n = 3$, respectively. We call $\mathbf{Z}^n$ and $\overline{\mathbf{Z}^n}$ the lattice and the dual lattice, respectively.

On $\mathbf{Z}^2$ and in $\mathbf{Z}^3$,

$$\mathbf{N}^4((m, n)^\top) = \{(m \pm 1, n)^\top, (m, n \pm 1)^\top\} \tag{2}$$

and

$$\mathbf{N}^6((k, m, n)^\top) = \{(k \pm 1, m, n)^\top, (k, m \pm 1, n)^\top, (k, m, n \pm 1)^\top\} \tag{3}$$

are the planar 4-neighborhood of point $(m, n)^\top$ and the spatial 6-neighborhood of point $(k, m, n)^\top$, respectively. We assume the 4-connectivity on $\mathbf{Z}^2$ and the 6-connectivity in $\mathbf{Z}^3$.

Setting one of x, y, and z to be a fixed integer, we obtain two dimensional sets of lattice points $\mathbf{Z}_1^2((k, m, n)^\top)$, $\mathbf{Z}_2^2((k, m, n)^\top)$, and $\mathbf{Z}_3^2((k, m, n)^\top)$ which pass through point $(k, m, n)^\top$ and are perpendicular to vectors, $e_1 = (1, 0, 0)^\top$, $e_2 = (0, 1, 0)^\top$, and $e_3 = (0, 0, 1)^\top$, respectively. These two-dimensional discrete spaces are mutually orthogonal. Denoting $\mathbf{N}_i^4((k, m, n)^\top)$ to be the 4-neighborhood of point $(k, m, n)^\top$ on plane $\mathbf{Z}_i^2((k, m, n)^\top)$ for $i = 1, 2, 3$, we have the relationship

$$\mathbf{N}^6((k, m, n)^\top) = \mathbf{N}_1^4((k, m, n)^\top) \cup \mathbf{N}_2^4((k, m, n)^\top) \cup \mathbf{N}_3^4((k, m, n)^\top). \quad (4)$$

Equation (4) implies that the 6-neighborhood is decomposed into three mutually orthogonal 4-neighborhoods.

A pair of points $(m, n)^\top$ and $x \in \mathbf{N}^4((m, n)^\top)$ is a unit line segment in $\mathbf{Z}^2$. A pair of points $(k, m, n)^\top$ and $x \in \mathbf{N}^6((k, m, n)^\top)$ is a unit line segment in $\mathbf{Z}^3$. Four 4-connected points which form a circle define a 4-connected discrete 2-simplex [ ]. Four 6-connected points which form a circle define a unit plane segment in $\mathbf{Z}^3$ with respect to the 6-connectivity. Six 6-connected points in a $2 \times 2 \times 2$ cube form a 6-connected discrete 3-simplex in a discrete space [ ]. Since an object is a complex, n-simplexes of an object share $(n - 1)$-simplexes. However, some $(n - 1)$-simplexes are not shared by a pair of simplexes.

We assume that our object on a plane is a complex of 2×2 sqare which share at least one edges with each other and that our object in a space is a complex of $2 \times 2 \times 2$ cubes which share at least one face with each other [ ]. The boundary of an object is a collection of $(n - 1)$-simplexes which are not shared by a pair of n-simplexes of an object. Thus, the boundary of our object on a plane is a a collection of unit line segments which are parallel to lines $x = 0$ and $y = 0$. The surface of an object is a collection of unit squares which are parallel to planes $x = 0$, $y = 0$, and $z = 0$. This definition agrees with that of Kovalevsky [ ] for planar objects.

In Figure 1, we show simplexes, objects and the boundaries of objects in $\mathbf{Z}^n$ for $n = 2, 3$. On the top columun of Figure 1, we show a 1-simplex (a), a 2-simplex (b), an object (c), and the boundary of this object on a plane. Furthermore, in the bottom columun of the same figure, we show a 2-simplex (a), a 3-simplex (b), an object (c), and the boundary of this object (d) in a space.

Since in $\mathbf{Z}^n$ there exist k-simplexes for $k = 1, 2, \cdots, n$, we define the collection of these simplexes as a discrete quasi-object; that is, a collection of connected k-simplexes for $1 \leq k \leq n$ in an n-dimensional discrete space [ ]. If both point sets $\mathbf{A}$ and $\overline{\mathbf{A}}$ are not quasi-objects, we call $\mathbf{A}$ a normal object.

3   Curvature and Normal Vector

3.1   Normal Vector on Discrete Plane

Since we are concerned with a binary discrete object, we assign values of 0 and 1 to points in the background and in objects, respectively. On $\mathbf{Z}^2$, three types of point configurations which are illustrated in Figure 2, exist in the neighborhood

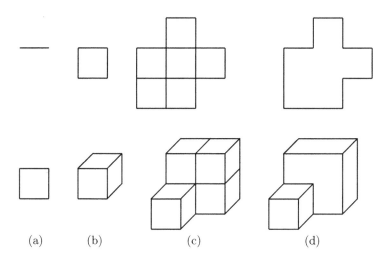

Fig. 1. In the top columun, the 1-simplex (a), the 2-simplex (b), an object (c), and the boundary of this object (d) on a plane, and in the bottom columun, the 2-simplex (a), the 3-simplex (b), an object (c), and the boundary of this object in a space

of the point with symbol × on the boundary. In Figure 2, symbols • and ○ indicate points on the boundary and in the background, respectively. Setting $f_i \in \{0, 1\}$ to be the value of point x_i such that $x_0 = (m, n)^\top$, $x_1 = (m + 1, n)^\top$, $x_3 = (m, n + 1)^\top$, $x_5 = (m - 1, n)^\top$, and $x_7 = (m, n - 1)^\top$, the curvature of point x_0 is defined by

$$r(x_0) = 1 - \frac{1}{2} \sum_{k \in N} f_k + \frac{1}{4} \sum_{k \in N} f_k f_{k+1} f_{k+2}, \qquad (5)$$

where $N = \{1, 3, 5, 7\}$ and $k + 8 = k \ [\ ]$.

Setting $C = \{x_j\}_{j=1}^N$ to be the set of points on the boundary, we assume that for point x_j, only two points $x_{j=1}$ and x_{j+1} are connected, and a triplet x_{j-1}, x_j and x_{j+1} lies in the counterclockwise direction on the boundary. Using a triplet of points, x_{j-1}, x_j and x_{j+1}, we compute the normal vector of each point. For planar point $x_j = (x_j, y_j)^\top$, we express this vector as a complex number $z_j = x_j + i y_j$. We define the outward normal vector n_j as

$$n_j = \lambda a_j, \ a_j = (\alpha_j, \beta_j)^\top, \ \alpha_j + i\beta_j = \left(\frac{z_{j+1} - z_j}{z_{j-1} - z_j} \right)^{\frac{1}{2}}, \ \lambda_j = \begin{cases} \frac{1}{2} & \text{if } \alpha_j \beta_j = 0, \\ \frac{1}{\sqrt{2}} & \text{otherwise,} \end{cases}$$
$$(6)$$

for point x_j, since from the local configurations of triplets on the boundary, there exist three possibilities for the combination of α_j and β_j such that both $|\alpha_j|$ and $|\beta_j|$ are $1/\sqrt{2}$, and one of $|\alpha_j|$ and $|\beta_j|$ is zero and the other is one.

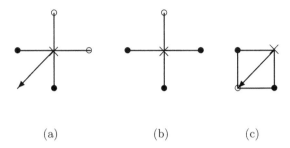

(a)   (b)   (c)

Fig. 2. Directions of the normal vectors for the positive and negative points on the boundary

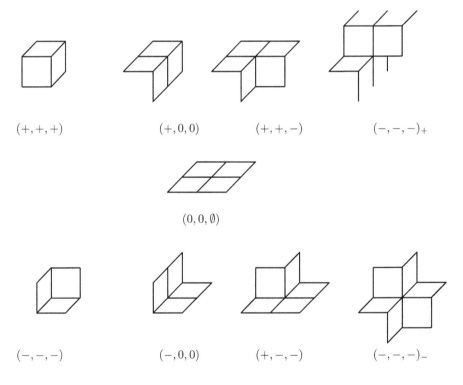

$(+,+,+)$ $(+,0,0)$ $(+,+,-)$ $(-,-,-)_+$

$(0,0,\emptyset)$

$(-,-,-)$ $(-,0,0)$ $(+,-,-)$ $(-,-,-)_-$

Fig. 3. Angles and configurations on the boundary of three-dimensional 6-connected objects

Figure 2 shows the normal vectors of three configurations on the 4-connected boundary shown in Figure 2.

The curvature indices of configurations (a), (b) and (c) , which is defined by $r(x_0)$ are positive, zero, and negative, respectively, when the directions of the normal vectors are inward, neutral, and outward, respectively. Therefore, we call these configurations convex, flat, and concave, respectively, and affix the indices $+$, 0, and $-$, respectively.

3.2   Normal Vector in Discrete Plane

Using combinations of planar curvature indices on three mutually orthogonal planes which pass through a point x_0, we define the curvature index of a point x_0 in $\mathbf{Z}^3$ since the 6-neighborhood is decomposed into three 4-neighborhoods. Setting α_i to be the planar curvature index on plane $\mathbf{Z}_i^2(x_0)$ for $i = 1, 2, 3$, the curvature index of a point in $\mathbf{Z}^3$ is a triplet of two-dimensional curvature indices $(\alpha_1, \alpha_2, \alpha_3)$ such that $\alpha_i \in \{+, -, 0, \emptyset\}$. Here, if $\alpha_i = \emptyset$, the curvature index of a point on plane $\mathbf{Z}_i^2(x_0)$ is not defined. Therefore, for the boundary points, seven configurations, $(+, +, +), (+, +, -), (+, 0, 0), (0, 0, \emptyset), (-, -, -), (+, -, -),$ $(-, 0, 0)$, and their permutations are possible.

Since a triplet of mutually orthogonal planes separates a space into eight parts, we call one eighth of the space an octspace. The number of octspaces determines the configurations of points in a $3 \times 3 \times 3$ cube. There exist nine configurations in the $3 \times 3 \times 3$ neighborhood of a point on the boundary, since these configurations separate $\mathbf{Z}^3$ into two parts which do not share any common points. The same configurations have also been introduced by Françon [ ] for the analysis of discrete planes. The curvature analysis of discrete surfaces also yields these configurations.

For a spatial curvature index α, setting $n(\alpha)$ to be the number of octspaces in the $3 \times 3 \times 3$ neighborhood of a point, the relationships $n((+, +, +)) = 1$, $n((+, 0, 0)) = 2$, $n((+, +, -)) = 3$, $n((-, -, -)_+) = 4$, $n((0, 0, \emptyset)) = 4$, $n((-, -, -)_-) = 4$, $n((+, -, -)) = 5$, $n((-, 0, 0)) = 6$, and $n((-, -, -)) = 7$ are held.

The curvature indices correspond to nine configurations of points which are illustrated in Figure 3. Since $\frac{i}{8} = 1 - \frac{8-i}{8}$, we define the vertex angles of configurations as $\gamma((+, +, +)) = 1/8$, $\gamma((+, 0, 0)) = 2/8$, $\gamma((+, +, -)) = 3/8$, $\gamma((-, -, -)_+) = 4/8$, $\gamma((0, 0, \emptyset)) = 0$, $\gamma((-, -, -)_-) = -4/8$, $\gamma((+, -, -)) = -3/8$, $\gamma(-, 0, 0) = -2/8$, and $\gamma((-, -, -)) = -1/8$. Considering congruent transformations in $\mathbf{Z}^3$, since a code uniquely corresponds to a configuration, the vertex angle for a point on a surface is an extension of the chain code for a point on a curve.

Point configurations on the boundary are defined by the discrete lines which lie on three or two mutually orthogonal vectors. Therefore, we define three normal vectors n_1, n_2, and n_3 which lie on planes perpendicular to vectors e_1, e_2, and e_3, respectively. According to these definitions, for $n = (\alpha, \beta)^\top$ on each plane, we have $n_1 = (0, \alpha, \beta)^\top$, $n_2 = (\alpha, 0, \beta)^\top$, and $n_3 = (\alpha, \beta, 0)^\top$. Using vector n_α, $\alpha = 1, 2, 3$, we define normal vector $n(x)$ for point x on the boundary $\mathbf{S}$ in digital space $\mathbf{Z}^3$. There are nine combinations of vertex configurations in the $3 \times 3 \times 3$ neighborhood on the boundary.

The directions of the normal vectors for the boundary configurations shown in Figure 3 are determined according to the combinations of two and three normal vectors on mutually orthogonal three planar discrete boundaries. Therefore, we express the normal vector $\boldsymbol{n}(\boldsymbol{x})$ of point $\boldsymbol{x}$ as

$$\boldsymbol{n}(\boldsymbol{x}) = \frac{1}{2}\boldsymbol{N}\boldsymbol{a}(\boldsymbol{x}) = \frac{1}{2}(a_1\boldsymbol{n}_1 + a_2\boldsymbol{n}_2 + a_3\boldsymbol{n}_3) \qquad (7)$$

for matrix $\boldsymbol{N} = (\boldsymbol{n}_1, \boldsymbol{n}_2, \boldsymbol{n}_3)$. For codes $+$, $-$, 0, and $\emptyset$, setting $s(\pm) = \pm 1$, $s(0) = 0$, and $s(\emptyset) = 0$, we define a function as

$$f(\alpha_1, \alpha_2, \alpha_3) = (s(\alpha_1) + s(\alpha_2) + s(\alpha_3)) \times (|s(\alpha_1)| + |s(\alpha_2)| + |s(\alpha_3)|). \qquad (8)$$

This function takes values of $\{0, \pm 1, \pm 3, \pm 9\}$ according to the configurations on the boundary. We define the relation between $\alpha(\boldsymbol{x}) = (\alpha_1, \alpha_2, \alpha_3)$ and trinary coefficients $\boldsymbol{a}(\boldsymbol{x}) = (a_1, a_2, a_3)^\top$ for point $\boldsymbol{x}$ as

$$\begin{array}{lllll}
\alpha(\boldsymbol{x}) & (\pm, 0, 0) & (0, \pm, 0) & (0, 0, \pm) & (\pm, \pm, \pm) \\
\boldsymbol{a}(\boldsymbol{x}) & (1, 1, 1) & (1, 1, 1) & (1, 1, 1) & (1, 1, 1) \\
\alpha(\boldsymbol{x}) & (\pm, \pm, \mp) & (\pm, \mp, \pm) & (\mp, \pm, \pm) \\
\boldsymbol{a}(\boldsymbol{x}) & (1, 1, -1) & (1, -1, 1) & (-1, 1, 1) \\
\alpha(\boldsymbol{x}) & (0, 0, \emptyset) & (0, \emptyset, 0) & (\emptyset, 0, 0) \\
\boldsymbol{a}(\boldsymbol{x}) & (1, 1, 0) & (1, 0, 1) & (0, 1, 1).
\end{array} \qquad (9)$$

Then, a positive or negative of the sign of function $f(\alpha_1, \alpha_2, \alpha_3)$ indicates the direction of normal vector $\boldsymbol{n}(\boldsymbol{x})$ at point $\boldsymbol{x}$ on the boundary to be outward or inward, respectively.

4   Curvature and Topology

4.1   Euler Characteristics of 4- and 6-Connected Objects

When a closed simple curve C on a plane is expressed as vector $\boldsymbol{x}(s)$ for $0 \leq s \leq S$, where S is the length of the curve, $\dot{\boldsymbol{x}} = (\cos\theta(s), \sin\theta(s))^\top$ denotes the tangent vectors of this curve at point $\boldsymbol{x}(s)$. We assume that vector $\boldsymbol{x}(s)$ moves on the boundary so that vector $\dot{\boldsymbol{x}}^\perp = (-\sin\theta(s), \cos\theta(s))^\top$, which is perpendicular to $\dot{\boldsymbol{x}}$, directs inward. The function $\theta(s)$ such that $-\pi \leq \theta(s) \leq \pi$ satisfies the Gauss-Bonnet formula

$$\int_0^S \theta(s)ds = 2\pi(1 - g), \qquad (10)$$

where g is the number of holes of this object. Setting n_+ and n_- to be the numbers of positive and negative points for point configurations on the 4-connected boundary on a plane, respectively, we obtain the following theorem as discrete version of the Gauss-Bonnet formula.

Theorem 1. *Let g be the number of holes of a planar discrete object. If a discrete object is 4-connected, the relation $n_+ - n_- = 4(1 - g)$ is held.*

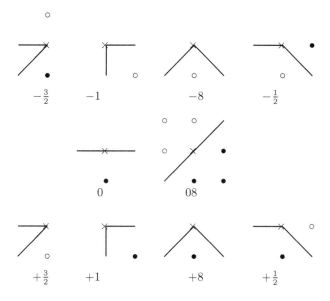

Fig. 4. Configurations in a 3×3 region on the 8-connected boundary graphs. We show the code of the point with the symbol $\times$, and points with symbols $\bullet$ and $\circ$ belong to the object and the background, respectively

Setting n_α to be the number of configurations on the boundary whose angle is $\alpha/8$, for vectors $\boldsymbol{n} = (n_{-4}, n_{-3}, n_{-2}, n_{-1}, n_0, n_1, n_2, n_3, n_4)^\top$, and $\boldsymbol{a} = (-2, -1, 0, 1, 0, 1, 0, -1, -2)^\top$, we obtain the following theorem as the three-dimensional analogous of this theorem for planar sets.

Theorem 2. *For an object with g tunnels, the object holds the relationship $\boldsymbol{a}^\top \boldsymbol{n} = 8(1-g)$.*

4.2   Euler Characteristic of 8-Connected Objects

Next, we deal with 8-connected objects on $\mathbf{Z}^2$. A point set

$$\mathbf{N}^8((m, n)^\top) = \{(m \pm 1, n \pm 1)^\top, (m \pm 1, n \pm 1)^\top\} \tag{11}$$

is the planar 8-neighborhood of point $(m, n)^\top$. Triangle with vertices $(m, n)^\top$, $(m+1, n)^\top$, and $(m, n+1)^\top$, and its congruent triangles in $\mathbf{Z}^2$ are two-dimensional simplexes on 8-connected discrete plane [ ]. 8-connected object is a complex of these simplexes. Therefore, in the 3×3 regions along the 8-connected boundary, there exist 10 configurations, as illustrated in Figure 4. We assign the curvature codes $-3/2, -1/2, -1, -8, 0, 08, +8, +1, +1/2, +3/2$ to these configurations and we express the numbers of these configurations on the boundary of an object as p_-, m_-, n_-, s_-, n_0, s_0, s_+, n_+, m_+, and p_+. For point configurations on the 8-connected boundary, we have a discrete version of the Gauss-Bonnet formula.

Theorem 3. *Let g be the number of holes for an 8-connected object. Then, points on the boundary satisfy the relation*

$$(m_+ + 2n_+ + 2s_+ + 3p_+) - (m_- + 2n_- + 2s_- + 3p_-) = 8(1 - g). \qquad (12)$$

5   Digital Curvature Flow

5.1   Deformation on Discrete Plane

Using the outward normal vector for each point on the boundary, we define a transform from point set $\mathbf{C}$ on $\mathbf{Z}^2$ to point set $\overline{\mathbf{C}}$ on $\overline{\mathbf{Z}}^2$ as

$$\overline{\boldsymbol{x}}_j = \begin{cases} \boldsymbol{x}_j - \boldsymbol{n}_j, & \text{if } \gamma(\boldsymbol{x}_j) \neq 0, \\ \boldsymbol{x}_j - \boldsymbol{n}_j \pm \frac{1}{2}\boldsymbol{n}_j^\top, & \text{if } \gamma(\boldsymbol{x}_j) = 0. \end{cases} \qquad (13)$$

This transform controls the deformation of points based on the local geometrical properties of the discrete boundary. Therefore, we derive a transformation using the global information of concavity of the discrete boundary. First, using the curvature code $\gamma(\boldsymbol{x}_j)$ of each point $\boldsymbol{x}_j$, we classify points on boundary $\mathbf{C}$ into types $\mathbf{N}_+$, and $\mathbf{N}_-$. First we define

$$\mathbf{N}_-(j) = \{\boldsymbol{x}_\beta | \gamma(\boldsymbol{x}_\beta) = 0,\ j < \beta < j + m,\ s(\gamma(\boldsymbol{x}_j)) \times s(\gamma(\boldsymbol{x}_{j+m})) = -1\}, \quad (14)$$

where $s(\gamma(\boldsymbol{x})) = 1$ and $s(\gamma(\boldsymbol{x})) = -1$ for $(\gamma(\boldsymbol{x}_j), \gamma(\boldsymbol{x}_{j+m})) = (+, -)$ and $(\gamma(\boldsymbol{x}_j), \gamma(\boldsymbol{x}_{j+m})) = (-, +)$, respectively. Second, we set

$$\mathbf{N}_- = \bigcup_j \mathbf{N}_-(j),\ \mathbf{N}_+ = \mathbf{C} \setminus \mathbf{N}_-. \qquad (15)$$

Each $\mathbf{N}_-^j$ is a sequence of flat points whose one endpoint is the concave point. Furthermore, $\mathbf{N}_-$ is the union of these sequences on the boundary. Then, we have the relation

$$\mathbf{C} = \mathbf{N}_+ \bigcup \mathbf{N}_-,\ \mathbf{N}_+ \bigcap \mathbf{N}_- = \emptyset. \qquad (16)$$

Points in $\mathbf{N}_+$ and $\mathbf{N}_-$ lie in convex and concave parts on the original boundary. The configurations of positive and negative end points and the lines determined by the signs of the endpoints are shown in Figure 5.

Using these geometrical properties of point sets $\mathbf{N}_+$ and $\mathbf{N}_-$ on the boundary, we define a global transform from point set $\mathbf{C}$ on $\mathbf{Z}^2$ to point set $\overline{\mathbf{C}}$ on $\overline{\mathbf{Z}}^2$ as

$$\overline{\boldsymbol{x}}_j = \begin{cases} \boldsymbol{x}_j - \boldsymbol{n}_j, & \text{if } \gamma(\boldsymbol{x}_j) \neq 0, \\ \boldsymbol{x}_j - \boldsymbol{n}_j \pm \frac{1}{2}\boldsymbol{n}_j^\top, & \text{if } \gamma(\boldsymbol{x}_j) = 0,\ \boldsymbol{x}_j \in \mathbf{N}_+, \\ \boldsymbol{x}_j + \boldsymbol{n}_j \pm \frac{1}{2}\boldsymbol{n}_j^\top, & \text{if } \gamma(\boldsymbol{x}_j) = 0,\ \boldsymbol{x}_j \in \mathbf{N}_-. \end{cases} \qquad (17)$$

In Figure 5, we show the direction of motion for flat points during the deformation based on the curvature vectors of a discrete curve. Although this transformation transforms a point on a corner to a point, a flat point is transformed to a pair of points. Therefore, for a corner, this transformation acts as curvature flow, though for flat points this transformation acts as diffusion on the boundary curve. We call the motion of boundary points caused by the successive application of this transformation digital curvature flow. Therefore, odd and even steps of digital curvature flow transform points on the lattice to points on the dual lattice and points on the dual lattice to points on the lattice, respectively. Pyramid transform is also defined using the lattice and the dual lattice. The first and second steps of pyramid transformation transform functions defined on the lattice to functions defined on the dual lattice, and functions defined on the dual lattice to functions defined on the lattice, respectively. More generally, the odd and even steps of pyramid transformation transform functions defined on the lattice to functions defined on the dual lattice, and functions defined on the dual lattice to functions defined on the lattice, respectively.

From vectors $e_1 = (1,0)^\top$ and $e_2 = (0,1)^\top$, we define vectors $d_0 = \frac{1}{2}e_0$, $d_1 = \frac{1}{2}e_2$, $d_{\bar{0}} = -d_0$, and $d_{\bar{1}} = -d_1$. Furthermore, setting $\bar{\alpha} = 1 - \alpha$ and $\bar{\beta} = 1 - \beta$, we define vector $d_{\alpha\beta} = d_\alpha + d_\beta$. Vector $d_{\alpha\beta}$ satisfies the relation $d_{\bar{\alpha}\bar{\beta}} = -d_{\alpha\beta}$. According to the definition of normal vector n_j at each point, n_j is an element of vector set $\mathbf{D} = \mathbf{D}_0 \cup \mathbf{D}_\pm$, where $\mathbf{D}_0 = \{d_0, d_1, d_{\bar{0}}, d_{\bar{1}}\}$, and $\mathbf{D}_\pm = \{d_{00}, d_{10}, d_{11}, d_{01}\}$. Later in this section, we express the sequence of curvature codes as $\Gamma = \langle \gamma_1 \gamma_2 \cdots \gamma_N \rangle$ for $\gamma_i = \gamma_{i+N}$. Setting the transformation from curvatures to normal vectors as

$$D : \Gamma = \langle d_1 d_2 \cdots d_N \rangle, \tag{18}$$

we have the relations in the 3×3 neighborhood as

$$\begin{aligned}
D : \langle \cdots 0 + 0 \cdots \rangle &= \langle \cdots d \cdots \rangle, \ d \in \mathbf{D}_\pm, \\
D : \langle \cdots 0 - 0 \cdots \rangle &= \langle \cdots - d \cdots \rangle, \ d \in \mathbf{D}_\pm, \\
D : \langle \cdots 000 \cdots \rangle &= \langle \cdots d \cdots \rangle, \ d \in \mathbf{D}_0.
\end{aligned} \tag{19}$$

As seen in Figures 5 and 6, the transformation defined by eq. (17) does not locally change the numbers n_+ and n_-, where n_+ and n_- are the numbers

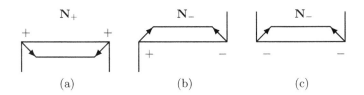

(a) (b) (c)

Fig. 5. Motions of the positive and negative lines on the boundary of convex and concave points. For three configurations (a), (b) and (c), shown in Figure 5, the transformation from a lattice to a dual lattice transforms the sequences of curvature codes as

$$F : \langle \cdots 0 + \underbrace{0 \cdots 0}_{N} + 0 \cdots \rangle = \langle \cdots 0 + \underbrace{0 \cdots 0}_{N-1} + 0 \cdots \rangle,$$

$$F : \langle \cdots 0 + \underbrace{0 \cdots 0}_{N} - 0 \cdots \rangle = \langle \cdots 0 + \underbrace{0 \cdots 0}_{N-1+\delta} - 0 \cdots \rangle, \qquad (20)$$

$$F : \langle \cdots 0 - \underbrace{0 \cdots 0}_{N} - 0 \cdots \rangle = \langle \cdots 0 - \underbrace{0 \cdots 0}_{N-1} - 0 \cdots \rangle,$$

for $\delta \in \{0, 1\}$. The first and third configurations shorten the curve one units. However, the second configuration preserves the local length of the boundary curve, or shortens one units. Therefore, considering the relation $n_+ - n_- = 4$, in each step, four positive points move inward, this motion shortens the curve by $4 \times 1 = 4$ units.

Third, we show configurations in regions of intermediate size. For the configuration $\langle \cdots 00 - 0 + 0 + 0 - 00 \cdots \rangle$, we have the relations

$$F^4 : \langle \cdots 00 - 0 + 0 + 0 - 00 \cdots \rangle = \langle \cdots 00 \cdots \rangle, \qquad (21)$$

where F^k for $k \geq 1$ expresses that operation F is applied k times.

In Figure 7 (a), we show configurations of points derived by the deformation based on eq (17). In the figure, the configuration on top is the original configuration, and the second one is the configuration for the next step. Furthermore, successive application of the transformation yields the configuration shown at the bottom. This example suggests that the boundary curve on a digital plane is smoothed by digital curvature flow.

Finally, we examine asymptotic properties of boundary-curve evolution under digital curvature flow. In Figure 7 (b), we show an example of the deformation process for a digital curve. If we consider discrete quasi-objects in digital space [ ], a line segment shrinks to a point under digital curvature flow. From the

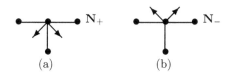

(a)   (b)

Fig. 6. Local motion of the positive and negative lines on the boundary

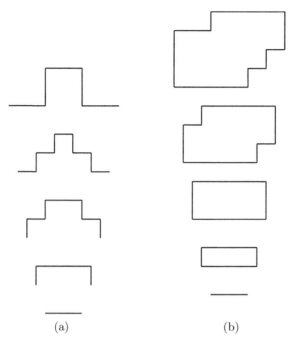

Fig. 7. Examples of a local deformation process (a) and a global deformation process (b)

definition of the transformation, points in $\mathbf{N}_+$ and $\mathbf{N}_-$ move inward and outward, respectively. Furthermore, convex and concave points also move inward and outward, respectively. Moreover, the relation $n_+ - n_- = 4$ for corner points is preserved. Therefore, n_- finally becomes 0, since all line segments which connect convex points and concave points are eliminated from the boundary. If we apply curvature flow to a rectangle, we obtain a line segment. These considerations lead to the following theorems.

Theorem 4. *For planar curves, in each step, digital curvature flow shortens the boundary curve by 4 unit lengths if there is no hole.*

Theorem 5. *Digital curvature flow preserves topology if there is no hole.*

Theorem 6. *Digital curvature flow transforms a boundary curve into a rectangle.*

Theorem 7. *The final form produced by digital curvature flow is a line segment.*

We also examine geometric properties of digital curvature flow for planar objects with holes. For the computation of outward normal vectors for the inner boundary curves, we should follow the curves in the clockwise direction. The directions of outward normal vectors are the same as the directions of the inward normal vectors of curves if we assume that these curves are the outer boundary curves of appropriate regions. The motion of sequences of flat points on the inner boundaries are the same as the motion of flat points on the outer boundary curve. Therefore, each inner boundary converges to a line segment. This property leads to the following theorem.

Theorem 8. *Inner boundaries of a planar digital object converge to line segments.*

The boundary curve encircles a collection of inner boundary curves. For a pair of curves, the following property is held.

Property 1 *Assuming that one curve encircles the other curve, the inner curve converges faster to a line segment.*

This property implies the next theorem since the convergence speed of small inner boundaries is faster than the convergence speed of outer boundaries.

Theorem 9. *If sizes of holes are small, the number of line segments in the rectangle is equivalent to the number of holes of the original object.*

5.2   Deformation in Discrete Space

As mentioned in the previous section, we can define the signs of vertices and the points on edges. Using the codes for vertices and edges, we define codes for flat points on the boundary surface.

If an end of a line segment is negative, we say that this line segment is a negative line segment. Furthermore, if both ends are positive, we say that this line segment is positive line segment. For boundary points whose two-dimensional codes are zero, we affix curvature codes with the same codes with the codes of line segments on which points lie.

This rule derives the code $(\alpha, \beta_1, \beta_2)$, $(\beta_1, \alpha, \beta_2)$, and $(\beta_1, \beta_2, \alpha)$, where $\beta_i \in \{+, -\}$, for a point whose codes are $(\alpha, 0, 0)$, $(0, \alpha, 0)$, and $(0, 0, \alpha)$, where $\alpha \in \{+, -, \emptyset\}$. These notations lead to codes $(+, +, +)$, $(+, +, -)$, $(+, -, -)$, $(-, -, -)$, $(\emptyset, +, +)$, $(\emptyset, +, -)$, $(\emptyset, -, -)$, and their permutations. Using these codes, we define the code of the flat points and points on edges as

$$g(\alpha, \beta_1, \beta_2) = \begin{cases} +, \text{ if } f(\alpha, \beta_1, \beta_2) = 9, \\ -, \text{ otherwise,} \end{cases} \tag{22}$$

for $(\alpha, \beta_1, \beta_2)$ and its permutations.
Setting

$$h(\beta_i, \beta_2) = (s(\beta_1) + s(\beta_2)) \times (|s(\beta_1)| + |s(\beta_2)|), \tag{23}$$

we set the codes of flat points and points on edges as positive and negative, if $h(\beta_1, \beta_2) > 0$, and otherwise, respectively. For these curvature codes, we define that the code of a flat point is negative if the number of the negative codes in the curvature code is positive. Furthermore, we move flat points outward and inward, if the codes are positive and negative, respectively.

Two line segments, which are mutually orthogonal, pass through on the flat points with codes $(\emptyset, 0, 0)$, $(0, \emptyset, 0)$, and $(0, 0, \emptyset)$. Therefore, using the codes of end points of these two line segments on a plane, we can affix the codes of flat points. If at least one end point is negative, we assgin negative to a point on this line. There are six possibilities for the configurations of end points of two mutually orthogonal line segments, and the codes of common point of these two line segments are shown in Figure 8. These definitions for codes also conclude that the codes of points with curvature codes $(+, 0, 0)$, $(0, +, 0)$, and $(0, 0, +)$, on an edge are negative if one end or both ends are negative. In Figure 9, we show examples of the curvature codes $(+, 0, 0)$ on edges and a flat point which are derived using the six configurations in Figure 8.

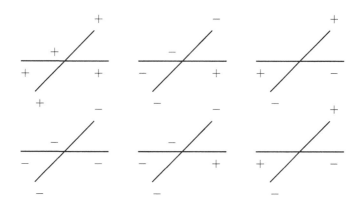

Fig. 8. The configurations of end points of mutually orthogonal line segments and the signs of flat points

The normal vectors on the discrete surface defined in section 3.2 are in the form

$$n(x) = \frac{1}{2}(a_1 e_1 + a_2 e_2 + a_3 e_3), \tag{24}$$

where $a_i \in \{-1, 0, 1\}$. Therefore, if $a_i = 1$ and $a_i = -1$, the vector $(x + n(x))$ determines the transformation from $x \in \mathbf{Z}^3$ to $y \in \overline{\mathbf{Z}}^3$. Moreover, if and only if the codes are $(+, +, +)$, $(-, -, -)$, and $(-, -, -)$ and (α, β, γ) for $\alpha, \beta, \gamma \in$

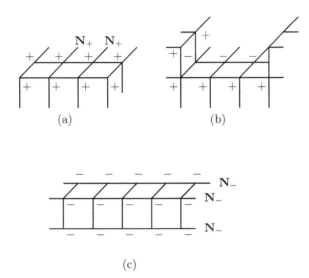

Fig. 9. Examples of the signs of flat points and edge points

$\{+,-\}$, the vector $(\boldsymbol{x} + \boldsymbol{n}(\boldsymbol{x}))$ determines the transformation from $\boldsymbol{x} \in \mathbf{Z}^3$ to $\boldsymbol{y} \in \overline{\mathbf{Z}^3}$.

For the code $(-,-,-)_+$, $(\alpha,0,0)$, $(\alpha,0,0)$, and $(0,0,\alpha)$ for $\alpha \in \{+,-,\emptyset\}$, the normal vectors hold the relations

$$
\begin{array}{cccc}
\alpha(\boldsymbol{x}) \ (\emptyset,0,0) & (0,\emptyset,0) & (0,0,\emptyset) & (-,-,-)_+ \\
\boldsymbol{n}(\boldsymbol{x}) \ \mathbf{0} & \mathbf{0} & \mathbf{0} & a_i e_i + a_j e_j \\
\alpha(\boldsymbol{x}) \ (\pm,0,0) & (0,\pm,0) & (0,0,\pm) & \\
\boldsymbol{n}(\boldsymbol{x}) \ \pm\frac{1}{2}(a_2 e_2 + a_3 e_3) & \pm\frac{1}{2}(a_1 e_1 + a_3 e_3) & \pm\frac{1}{2}(a_1 e_1 + a_2 e_2).
\end{array}
\tag{25}
$$

However, these normal vectors do not define the transformation from points in $\mathbf{Z}^3$ to $\overline{\mathbf{Z}^3}$, Therefore, setting

$$
\begin{array}{cccc}
\alpha(\boldsymbol{x}) \ (\emptyset,0,0) & (0,\emptyset,0) & (0,0,\emptyset) & (-,-,-)_+ \\
\boldsymbol{m}(\boldsymbol{x}) \ \pm\frac{1}{2}e_1 & \boldsymbol{n}(\boldsymbol{x}) \pm \frac{1}{2}e_2 & \boldsymbol{n}(\boldsymbol{x}) \pm \frac{1}{2}e_3 & \frac{1}{2}\boldsymbol{n}(\boldsymbol{x}) \pm \frac{1}{2}e_k \\
\alpha(\boldsymbol{x}) \ (\pm,0,0) & (0,\pm,0) & (0,0,\pm) & \\
\boldsymbol{m}(\boldsymbol{x}) \ \pm\frac{1}{2}(a_2 e_2 + a_3 e_3) & \pm\frac{1}{2}(a_1 e_1 + a_3 e_3) & \pm\frac{1}{2}(a_1 e_1 + a_2 e_2),
\end{array}
\tag{26}
$$

we define the transformation from $\boldsymbol{x} \in \mathbf{Z}^3$ to $\boldsymbol{y} \in \overline{\mathbf{Z}^3}$ as

$$
\boldsymbol{y} + \boldsymbol{x} + \epsilon \boldsymbol{m}(\boldsymbol{x}),
\tag{27}
$$

where ϵ is 1 for codes $(+,+,+)$, $(-,-,-)$, $(-,-,-)_-$, $(+,+,-)$, $(+,-,+)$, and $(-,+,+)$ and ϵ is -1 for points where $g(\alpha, \beta_1, \beta_2)$ is negative. Therefore, the negative sign of ϵ is depends on the configuration of two mutually orthogonal line segments which pass through the points. Since the saddle points on a discrete dambbell move inward, our method preserves the topology of the dambbell. This is the significant difference from the curvature-flow-based deformation in the three-dimensional Euclidean space $\mathbf{R}^3$. However, according to the definitions of the directions of the motion of points in curvature-based flow, positive points on edges with both end points negative move outward. These configuration defines the outword motion for points on the bar of a dumbbel. Therefore, these definitions of the codes of flat points and edges preserve the topology of dumbbells.

Curvature flow transforms each planar curve segment which passes through lattice points on a polyhedral boundary into a straight line segment which passes through lattice points on a polyhedral boundary. Furthermore, each closed curve on a plane is transformed to a rectangle. If $\alpha(\boldsymbol{x}) = (+, +, +)$, on each step, curvature flow eliminates a 3/4 unit area from the corners. Moreover, positive and negative parts on the boundary moves outward and inward, respectively. These considerations lead to the following theorems assuming that an object does not contain any tunnels.

Theorem 10. *For a closed surface, in each step, digital curvature flow shrinks the boundary curves on each plane six unit areas from corners such that $\gamma(\boldsymbol{x}) = (+, +, +)$.*

Theorem 11. *Digital curvature flow preserves topology if there is no tunnel.*

Theorem 12. *Digital curvature flow transforms a boundary into a rectangle, if there is no tunnel.*

Theorem 13. *The final form of digital curvature flow is a spatial rectangle if there is no tunnel.*

In Figures 10 and 11, we show examples of discrete deformation. These examples show that this deformation preserves the the topology of a discrete dambell.

6   Discrete Variational Method

Setting $f(\boldsymbol{x})$ to be a gray-valued image in a three-dimensional space, a three-dimensional version of snakes detects a boundary surface $\mathbf{S}$ by minimizing the criterion,

$$J = \int_S (\alpha + \beta|\boldsymbol{K}|)d\boldsymbol{s} - \gamma \int_S F(|\nabla f(\boldsymbol{x})|_{\boldsymbol{x}=\boldsymbol{s}})d\boldsymbol{s} \tag{28}$$

for positive constants α, β, and γ, where $\boldsymbol{K}$ and $d\boldsymbol{s}$ are Gauss curvature on the surface S, and the infinitesimal area on the surface S, respectively. It is possible to select positive function $F(\cdot)$ such that if $\mathbf{S}$ passes through locally

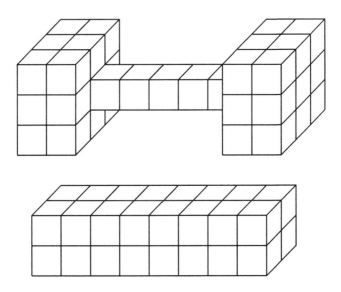

Fig. 10. An example of deformation. The discrete dambbell at the top is deformed into a parallelepiped at the bottom

steepest points, the second term becomes large. We set $F(x) = x$. For gray-valued images, all points on the boundary should have the same gray value. Therefore, we adopt the minimization criterion

$$E = J + \delta \int_S |\nabla_S f(\boldsymbol{x})|\, d\boldsymbol{s}, \qquad (29)$$

where δ is a positive constant and ∇_S is the gradient on surface S. For quantized digital images, the second term of eq. (29) becomes the minimum value zero if $\mathbf{S}$ lies in one of two regions $\mathbf{B}$ and $\mathbf{W} = \overline{\mathbf{B}}$ for binary images, where $\mathbf{B}$ is the region with value 1, namely, the black region. We call $\mathbf{W}$ the white region.

If we assume 6-connectivity, the first and second terms are αn and βk, respectively, where n and k are the total numbers of points on the boundary and of nonflat points, respectively. For $f_{ijk} = f(i,j,k)$ for integers i, j and k, we define the discrete gradient as $(\nabla f)_{ijk} = (u_{ijk}, v_{ijk}, w_{ijk})^\top$, for $u_{ijk} = f_{i+1\,j\,k} - f_{i-1\,j\,k}$, $v_{ijk} = f_{i\,j+1\,k} - f_{i\,j-1\,k}$, and $w_{ijk} = f_{i\,j\,k+1} - f_{i\,j-1\,k-1}$. These equations lead to the relationship, for binary images,

$$|u_{ijk}|^2 + |v_{ijk}|^2 + |w_{ijk}|^2 = c_1 + c_2 + c_3, \qquad (30)$$

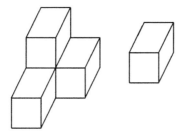

Fig. 11. An example of deformation. The object in the left is tansformed into the cube on the right

where, for $\alpha = 1, 2, 3$,

$$c_\alpha = \begin{cases} 1, \text{ if } \gamma_\alpha(\boldsymbol{p}) \geq 0 \\ 0, \text{ otherwise,} \end{cases} \tag{31}$$

where $\gamma_\alpha(\boldsymbol{p})$ is the two-dimensional curvature code of point $\boldsymbol{p}$ on the plane perpendicular to vector $\boldsymbol{e}_\alpha$. Setting m_1 and m_0 to be the numbers of black points and white points on the boundary surface of binary images, we obtain a criterion

$$g(S) = \begin{cases} 0, & \text{if } \mathbf{S} \in \mathbf{B}, \\ 0, & \text{if } \mathbf{S} \in \mathbf{W}, \\ |m_1 - m_0|, \text{ otherwise.} \end{cases} \tag{32}$$

These considerations lead to the relation

$$E_D(\boldsymbol{v}) = \alpha n + \beta k - \gamma f + \delta g, \tag{33}$$

where f is the total sum of $|u_{ijk}|^2 + |v_{ijk}|^2 + |w_{ijk}|^2$ on the boundary surface.

Theorem 14. E_D *takes the minimum value if* $\mathbf{S}$ *is a 6-connected digital surface.*

(Proof) Let the numbers of points and vertexes in $\mathbf{A}$ be $n' + 1$ and k', respectively. Assuming that the boundary surface is 6-connected surface, the local deformation of this surface is described as $\mathbf{S}' = \mathbf{S} \setminus \{p\} \bigcup \mathbf{A}$, where a point $\boldsymbol{p}$ and a point set $\mathbf{A}$ are a point on $\mathbf{S}$ and a subset of $\mathbf{B}$ or $\mathbf{W}$, respectively. There are two possibilities for the selection of a point $\boldsymbol{p}$, since there are flat points and nonflat points on the boundary. Furthermore, for nonflat points, there are 8 configurations listed in Figure 3. There are two possibilities for the selection of a set $\mathbf{A}$ which is replaced to a point $\boldsymbol{p}$. Therefore, setting

$$\Delta E_D = E_D(\mathbf{S}') - E_D(\mathbf{S}), \tag{34}$$

the combinations of $\{p\}$ and $\mathbf{A}$, based on nine configurations of a point on the boundary shown in Figure 3 and two possibilities for the selection of a set $\mathbf{A}$, lead to the relation

$$\Delta E_D = \alpha n' + \beta(k' + a - \epsilon) - \gamma\sigma + \delta(\theta|n - n' - 2|) \tag{35}$$

for

$$\epsilon = \begin{cases} 0, \text{ if } \gamma(\boldsymbol{p}) = (\emptyset, 0, 0), (0, \emptyset, 0), (0, 0, \emptyset), \\ 1, \text{ otherwise,} \end{cases} \tag{36}$$

$$a = \begin{cases} 6, \text{ if } i \text{ is even for vertex angle } i/8, \\ 8, \text{ otherwise,} \end{cases} \tag{37}$$

$$\sigma = \begin{cases} 0, \text{ if } \gamma(\boldsymbol{p}) = (\emptyset, 0, 0), (0, \emptyset, 0), (0, 0, \emptyset), \\ m(\boldsymbol{p}), \text{ otherwise,} \end{cases} \tag{38}$$

where $m(\boldsymbol{p})$ is the difference between the total of the gradients of the boundaries $\mathbf{S}'$ and $\mathbf{S}$, and

$$\theta = \begin{cases} 0, \text{ if } \mathbf{A} \subset \mathbf{B}, \\ 1, \text{ otherwise.} \end{cases} \tag{39}$$

Converting a point set $\{\boldsymbol{p}\}$ to a set of points $\mathbf{A}$, the total sum of gradient on the surface decreases, since the gradient is zero on each point in $\mathbf{A}$. Therefore, $m(\boldsymbol{p})$ is always negative. These relations lead to the conclusion that ΔE_D is always positive. Therefore, a locally optimal boundary surface is 6-connected. (Q.E.D.)

The 6-connected surface is computed by $\mathbf{S} = \mathbf{B} \setminus (\mathbf{B} \ominus \mathbf{N}^{26})$, where $\mathbf{A} \ominus \mathbf{B}$ is the Minkowski subtraction [ ] of point set $\mathbf{B}$ from point set $\mathbf{A}$ and $\mathbf{N}^{26}$ is the 26-neighborhood of the origin which is defined as $\mathbf{N}^{26}((k, m, n)^{\top}) = \{(k \pm \epsilon_1, m \pm \epsilon_2, n \pm \epsilon_3)^{\top}\}$ for $\epsilon_1, \epsilon_2, \epsilon_3 \in \{0, 1\}$.

Setting $\bar{n}$, $\bar{k}$, $\bar{f}$, and $\bar{g}$ to be the parameters for $\bar{\mathbf{S}}$, we have the equalities and inequalities $\bar{n} > n$, $\bar{k} = k$, $\bar{f} < f$, and $\bar{g} = g = 0$, if both $\mathbf{S}$ and $\bar{\mathbf{S}}$ are not quasi-objects. Here, both $\mathbf{S}$ and $\bar{\mathbf{S}}$ are not quasi-objects, if both $\mathbf{B}$ and $\mathbf{W} = \bar{\mathbf{B}}$ are not quasi-objects. Therefore, we have the following theorem.

Theorem 15. *If* $\mathbf{B}$ *is a normal object, we have the inequality* $E_D(\mathbf{S}) < E_D(\bar{\mathbf{S}})$.

Theorem 15 implies that a 6-connected boundary surface is the optimal solution for the minimization criterion E_D if we select an initial surface in the neighborhood of $\mathbf{B}$ in $\mathbf{W}$.

For a surface in the interior of $\mathbf{B}$, the third and fourth terms are always zero. It is possible to select a surface such that $n_{+1} = 8$ and $n_{-\alpha} = 0$ for $\alpha = 1, 2, 3, 4$, which is a parallelepiped. This surface $\mathbf{S}_i$ satisfies the relation $E_D(\mathbf{S}) > E_D(\mathbf{S}_i)$. Furthermore, in the exterior of $\mathbf{W}$, we can also select a surface $\boldsymbol{v}_e$ such that $n_{+1} = 8$ and $n_{-\alpha} = 0$ for $\alpha = 1, 2, 3, 4$, which is a parallelepiped, satisfying the relation $E_D(\mathbf{S}) > E_D(\mathbf{S}_e)$. Although the existence of these surfaces depends on function F and the selection of parameters α, β, γ, and δ, these properties lead to the following theorem.

Theorem 16. *The minimization criterion* E_D *for binary images is not globally convex.*

7   Conclusions

In this paper, we first defined the curvature indices of vertices of discrete objects [ ]. Second, using these indices, we defined the principal normal vectors of discrete curves and surfaces. Third, we defined digital curvature flow as a digital version of curvature flow in a discrete space. Our treatment of curvature flow is also considered as a numerical method for partial differential equation of curvature flow in a grid space. Finally, these definitions of curvatures in a discrete space derived discrete snakes as a discrete variational problem since the minimization criterion of the snakes is defined using the curvatures of points on the discrete boundary. Furthermore, we proved that the digital boundary curve detected using mathematical morphology is derived as the solution of this digital variational problem.

Images and objects in computers are expressed in digital form. Therefore, in computer vision, we first describe problems for the variational forms using continuous functions, and second discretize the problems for the numerical computation. In this paper, we directly converted a variational problem of a continuous functional into an optimization problem of a discrete functional which is defined in the lattice space.

References

1. Imiya, A.: Geometry of three-dimensional neighborhood and its applications, Transactions of Information Processing Society of Japan, **34**, 2153-2164, 1993 (in Japanese). 229
2. Imiya, A. and Eckhardt, U.: The Euler characteristics of discrete objects and discrete quasi-objects, CVIU, **75**, 307-318, 1999. 229, 231, 239
3. Imiya, A. and Eckhardt, U.: Discrete curvature flow, Lecture Notes in Computer Science, **1682**, 477-483, 1999. 229, 248
4. Sapiro, G.:*Geometric Partial Differencial Equations and Image Analysis*, Cambridge University Press: Cambridge, 2001. 229
5. Yonekura, T., Toriwaki, J.-I., and Fukumura, T., and Yokoi, S.: On connectivity and the Euler number of three-dimensional digitized binary picture, Transactions on IECE Japan, **E63**, 815-816, 1980. 229
6. Toriwaki, J.-I., Yokoi, S., Yonekura, T., and Fukumura, T.: Topological properties and topological preserving transformation of a three-dimensional binary picture, Proc. of the 6th ICPR, 414-419, 1982. 229, 232
7. Toriwaki, J.-I.: *Digital Image processing for Image Understanding,* Vols.1 and 2, Syokodo: Tokyo, 1988 (in Japanese).
8. Kenmochi, Y., Imiya, A., and Ezuquera, R.: Polyhedra generation from lattice points, Lecture Notes in Computer Science, **1176**, 127-138, 1996. 231, 236
9. Françon, J.: Sur la topologie d'un plan arithmétique, Theoretical Computer Sciences, **156**, 159-176, 1996. 234
10. Kovalevsky, V. A.: Finite topology as applied to image analysis, Computer Vision, Graphics and Image Processing, **46**, 141-161, 1989. 231
11. Kass, M., Witkin, A., and Terzopoulos, D.: Snakes:Active contour model, International Journal of Computer Vision, **1**, 321-331, 1988. 230

12. Schnörr, Ch.: Variational methods for adaptive image smoothing and segmentation, 451-484, Jähne, B., Haußecker, H., and Geißler, P. eds. *Handbook of Computer Vision*, **2**, Academic Press: London, 1999. 230

13. Mumford, D. and Shah, J.: Optimal approximations by piecewise smooth functions as associated variational problems, Communications on Pure and Applied Mathematics, **42**, 577-685, 1989. 230

14. Serra, J.: *Image Analysis and Mathematical Morphology*, Academic Press: San Diego, 1982. 230, 247

Hausdorff Sampling of Closed Sets into a Boundedly Compact Space

Christian Ronse and Mohamed Tajine

LSIIT UPRES-A 7005, Université Louis Pasteur, Dépt. d'Informatique
Boulevard Sébastien Brant, 67400 Illkirch (FRANCE)
{tajine,ronse}@dpt-info.u-strasbg.fr
http://lsiit.u-strasbg.fr/

Abstract. Our theory of Hausdorff discretization has been given in the following framework [ ]. Assume an arbitrary metric space (E, d) (E can be a Euclidean space) and a nonvoid proper subspace D of E (the discrete space) such that: (1) D is *boundedly finite*, that is every bounded subset of D is finite, and (2) the distance from points of E to D is bounded; we call this bound the *covering radius*, it is a measure of the resolution of D. For every nonvoid compact subset K of E, any nonvoid finite subset S of D such that the Hausdorff distance between S and K is minimal is called a *Hausdorff discretizing set* (or *Hausdorff discretization*) of K; among such sets there is always a greatest one (w.r.t. inclusion), which we call the *maximal Hausdorff discretization of K*. The distance between a compact set and its Hausdorff discretizing sets is bounded by the covering radius, so that these discretizations converge to the original compact set (for the Hausdorff metric) when the resolution of D tends to zero. Here we generalize this theory in two ways. First, we relax condition (1) on D: we assume simply that D is *boundedly compact*, that is every closed bounded subset of D is compact. Second, the set K to be discretized needs not be compact, but boundedly compact, or more generally closed (cfr. [ ] in the particular case where $E = \mathbf{R}^n$ and $D = \mathbf{Z}^n$).

Keywords: Distance, metric space, compact set, proximinal set, boundedly compact set, discretization, dilation, Hausdorff metric.

1 Introduction

In [ ] we indroduced a new approach to discretization, based on the Hausdorff metric: given a nonvoid compact subset K of the "Euclidean" metric space E, a possible discretization of K is a finite subset S of the "discrete" nonvoid proper subspace D of E, whose Hausdorff distance to K is minimal. Since there can be several such sets S, we called any such S a *Hausdorff discretizing set* (or *Hausdorff discretization*) of K, and used the term *maximal Hausdorff discretization of K* for the greatest among those Hausdorff discretizing sets. (Note that the latter always exists, because the family of Hausdorff discretizing sets is nonvoid, finite, and closed under union).

G. Bertrand et al. (Eds.): Digital and Image Geometry, LNCS 2243, pp. 250–271, 2001.
© Springer-Verlag Berlin Heidelberg 2001

For example, take $E = \mathbf{R}^n$, $D = \mathbf{Z}^n$. We associate to each $p \in D$ the closed cell $C(p)$ consisting of all Euclidean points $x \in E$ which are at least as close to p as to any other point in D:

$$\forall p \in D, \qquad C(p) = \{x \in E \mid \forall q \in D, \ d_2(x,p) \le d_2(x,q)\} \ , \qquad (1)$$

where d_2 denotes the Euclidean distance. The *supercover discretization* Δ_{SC} associates to every $X \subseteq E$ the set of all $p \in D$ such that $C(p)$ intersects X (see Fig. 1):

$$\forall X \subseteq E, \qquad \Delta_{SC}(X) = \{p \in D \mid C(p) \cap X \ne \emptyset\} \ . \qquad (2)$$

Then given a metric on E induced by a norm, for every nonvoid compact subset K of E, $\Delta_{SC}(K)$ is a Hausdorff discretizing set of K [ ].

Such a discretization follows a global approach, in the sense that whether a point $p \in D$ belongs to a Hausdorff discretizing set of K or not, may depend on points of K which can be at an arbitrary distance from p (see Fig. 2). We showed in fact that the maximal Hausdorff discretization of K consists of all points $p \in D$ such that $B_{r_H(K)}(p)$, the closed ball of radius $r_H(K)$ centered about p, intersects K; here that radius $r_H(K)$, called the *Hausdorff radius of K*, depends globally on K (and on D): it is the maximal distance from a point of K to the closest point of D. It is precisely the dependence of $r_H(K)$ on K that makes Hausdorff discretization global.

However, as explained in [,], this does not mean that our theory is incompatible with the usual practice of discretizing objects by means of local operations. We have already given above the example of the supercover discretization, which is purely local, but nevertheless gives always a Hausdorff discretizing set when the distance is based on a norm. In [ ], we have also shown that in any case, all Hausdorff discretizing sets lie in a precisely defined neighbourhood of the supercover discretization.

The main interest of our approach is that the Hausdorff distance between a nonvoid compact set K and its Hausdorff discretizing sets is equal to the Hausdorff radius $r_H(K)$. Now a measure of the resolution (or grid spacing) of the "discrete" space D is what we call the *covering radius*, namely the supremum r_c of the distances of points of E to D. We assume that $r_c < \infty$, and we have always $r_H(K) \le r_c$. A consequence of this is that when the resolution D tends

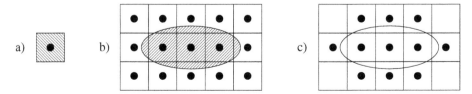

Fig. 1. a) The structuring element A is $C(o)$. b) A Euclidean set X overlayed with the discrete points p and their cells $C(p)$. c) The supercover discretization $\Delta_{SC}(X)$

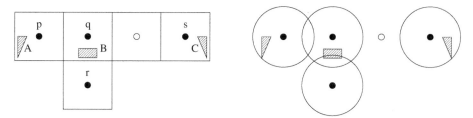

Fig. 2. Left: A compact set $K = A \cup B \cup C$ overlayed with discrete points p, q, r, s and their cells $C(p), C(q), C(r), C(s)$. Right: For $d = d_2$ (the Euclidean distance), the maximal Hausdorff discretizations of A, B and C are $\{p\}$, $\{q\}$, and $\{s\}$ respectively, while the maximal Hausdorff discretization of K is $\{p, q, r, s\}$. We show the circles of radius $r_H(K)$ centered about these points. The additional point r arises because its Hausdorff distance to B is less than the Hausdorff distance of p to A. The unique other Hausdorff discretizing set of K is $\{p, q, s\}$

to zero, so does $r_H(K)$, and the Hausdorff discretization tends to the compact set in the Hausdorff metric sense.

Many theories consider the discretization of closed subsets of the Euclidean space $\mathbf{R}^n$ into subsets of the discrete space $\mathbf{Z}^n$. This is the case of the morphological theory of discretization by dilation [ ]. In [ ] we also presented the theory of Hausdorff discretization of closed sets in the case of $E = \mathbf{R}^n$ and $D = \mathbf{Z}^n$.

As we will see, the main results of our abstract framework [ ] can be extended to closed sets, in particular the fact that the Hausdorff distance between K and its Hausdorff discretizing sets is equal to its Hausdorff radius $r_H(K)$, and so is bounded by the resolution of D. Our characterization of Hausdorff discretization takes exactly the same form as for compact sets in [ ], for a class of closed sets called *proximinal* sets, that is those where for every point outside the set, some point in the set minimizes the distance to it. Note that if the space E is *boundedly compact*, that is every bounded closed set is compact, then every closed set is proximinal. For example $\mathbf{R}^n$ ($n \in \mathbf{N}$) is boundedly compact. In fact, a normed vector space is boundedly compact iff it is isomorphic to $\mathbf{R}^n$ for some $n \in \mathbf{N}$.

Also we extend our theoretical framework by relaxing one of the two axioms for D postulated in [ ]. There we assumed that every bounded subset of D is finite. Now we make the weaker assumption that every bounded closed subset of D is compact, in other words D is boundedly compact. In some way, the space D is not necessarily discrete, so instead of a discretization we may speak of a *sampling*.

What is the interest of this generalization, namely of taking closed sets instead of compact ones, and of sampling in a space D that is not necessary discrete? First, taking into account unbounded objects allows one to give models for the discretization of standard unbounded shapes like straight lines, parabolas, etc. Second, by assuming the space D to be boundedly compact instead of discrete (i.e., boundedly finite), we can give a complete theory of sampling on any kind of grid with variable resolution, especially if that grid has accumulation points. For example, in the log-polar model for circular images, where pixels are positioned by sampling the angle and the logarithm of the radius, the origin is an accumulation point of such a grid. Although the accumulation point is excluded from the digital model, it has to be taken into account in any theory that deals with all possibles resolutions (in the same way as the theory of real numbers is necessary to understand digital numbers with arbitrary precision). Another use of our theory is when one discretizes an object only in some of its coordinates, so one associates to a closed subset F of $E = \mathbf{R}^{a+b}$ a closed subset S of $D = \mathbf{R}^a \times \mathbf{Z}^b$ such that $H_d(S, F)$ is minimal; again D will be boundedly compact, but not boundedly finite. A related problem is the sampling without quantization of numerical functions, which transforms a function $\mathbf{R}^n \to \mathbf{R}$ into a function $\mathbf{Z}^n \to \mathbf{R}$; a geometrical approach to such a sampling will consider the graph (or hypograph) of the original function as a subset of $E = \mathbf{R}^{n+1}$, and will derive from it the graph (or hypograph) of the sampled function as a subset of the boundedly compact subspace $D = \mathbf{Z}^n \times \mathbf{R}$.

The approach of minimizing the Hausdorff distance has also been used by Sendov [ ] to approximate a given function f by a function in some class $\mathcal{S}$: one takes the function $g \in \mathcal{S}$ such that the Hausdorff distance between the graphs of f and g is the smallest possible.

Paper Organization

In Sect. 2 we recall certain known facts about closed and compact sets in a metric space, as well as related types of sets: boundedly compact, boundedly finite, and proximinal sets. We recall also the Hausdorff metric on compact sets and its extension to closed sets, and prove two results on compact families (in Hausdorff spaces) of compact sets, which are probably known, but do not appear in standard textbooks. Then in Sect. 3 we recall the main results of [ ] and extend them to the case where the subspace D of E is boundedly compact (instead of boundedly finite). Section 4 extends further the theory by considering the sampling of closed sets (instead of compact ones in [ ]), with particular results for boundedly compact sets. The Conclusion describes the place taken by these results in our research programme.

2   Some Families of Closed Sets in a Metric Space

We recall first some elementary facts from topology, especially in the case of metric spaces. Our survey goes deeper than in [ ], and the reader is referred to [, , ,] for more details. Then in Subsect. 2.1 we give the definition and main properties of boundedly compact, boundedly finite, and proximinal subsets of a metric space. The definition and basic properties of the Hausdorff metric for these sets are given in Subsect. 2.2. Finally in Subsect. 2.3 we give two important results on compact families of compact sets in a metric space.

We adopt a systematic terminological convention. We will define several families of subsets of a topological or metric space E; if we denote $\mathcal{S}$ the family of subsets of E having some property σ, we write $\mathcal{S}'$ for the family of nonvoid elements of $\mathcal{S}$; when the space E in which it is defined must be specified, we will write $\mathcal{S}(E)$ and $\mathcal{S}'(E)$. Now for $X \subseteq E$, we will write $\mathcal{S}(X)$ (resp., $\mathcal{S}'(X)$) for the analogous family in the relative topology or metric of X, namely the one of subsets (resp., nonvoid subsets) of X having property σ in that relative topology or metric.

Let $(E, \mathcal{G})$ be a topological space; here $\mathcal{G}$ designates the family of open subsets of E. Write $\mathcal{F}$ for the family of closed subsets of E. For $X \subseteq E$, write $\overline{X}$ for the closure of X.

We say that E is T_1 if for two distinct $x, y \in E$, there is some open set G such that $x \in G$ but $y \notin G$; equivalently, every singleton in E is closed. When E is T_1, for every $X \subseteq E$ and $y \in E$, the following two statements are equivalent: a) for every neighbourhood V of y, $V \setminus \{y\}$ contains a point of X; b) for every neighbourhood V of y, $V \setminus \{y\}$ contains infinitely many points of X. We say then that y is a *limit point of* X.

Given a sequence $(x_n)_{n \in \mathbf{N}}$ in E and $y \in E$, we say that y is *adherent* to the sequence, if every neighbourhood V of y verifies $x_n \in V$ for infinitely many values of n. We say that the sequence *converges* to y, and write $x_n \to y$, if for every neighbourhood V of y, there is some $n_V \in \mathbf{N}$ such that for every $n \geq n_V$, $x_n \in V$.

E is *compact* if from every covering of E by a family of open sets, one can extract a covering by a finite sub-family of those open sets: if $E = \bigcup_{i \in I} G_i$ for a family G_i, $i \in I$, of open sets, then there is a *finite* $J \subseteq I$ such that $E = \bigcup_{j \in J} G_j$. E is *sequentially compact* if from every covering of E by a *countable* family of open sets, one can extract a covering by a finite sub-family of those open sets. The following is well-known:

- E is sequentially compact iff for every sequence in E there is a point in E adherent to it.
- If E is T_1, E is sequentially compact iff for every infinite $X \subseteq E$, some $y \in E$ is a limit point of X.

A subset X of E is compact if the relative topology on X is compact. Every finite set is compact, and the intersection of a compact set with a closed set is compact.

Write $\mathcal{K}$ for the family of compact subsets of E. In [ ], one writes $\mathcal{H}$ instead of $\mathcal{K}'$, and we followed this convention in [ ]. Given $Y \subseteq X$, Y is compact in the relative topology on X iff it is compact in the topology on E, that is:

$$\forall X \subseteq E, \qquad \mathcal{K}(X) = \mathcal{K}(E) \cap \mathcal{P}(X) . \tag{3}$$

On the other hand, if Y is closed in E, it is closed in the relative topology on X, that is

$$\forall X \subseteq E, \qquad \mathcal{F}(X) \supseteq \mathcal{F}(E) \cap \mathcal{P}(X) ,$$

but for X closed the reciprocal holds, so

$$\forall X \in \mathcal{F}, \qquad \mathcal{F}(X) = \mathcal{F}(E) \cap \mathcal{P}(X) \ . \tag{4}$$

Assume now that (E, d) is a metric space. For $r \geq 0$ and $x \in E$ we define $B_r(x)$ and $\overset{\circ}{B}_r(x)$, the *closed ball* and the *open ball*, respectively, *of radius r centered about x*, by

$$B_r(x) = \{y \in E \mid d(x, y) \leq r\} \quad \text{and} \quad \overset{\circ}{B}_r(x) = \{y \in E \mid d(x, y) < r\} \ . \tag{5}$$

Here the open sets are all unions of open balls.

Given a subset X of E, for every point $p \in E$ we define the distance between p and X as

$$d(p, X) = \inf\{d(p, x) \mid x \in X\} \ . \tag{6}$$

Note that the distance to X is continuous, because we have

$$|d(p, X) - d(q, X)| \leq d(p, q) \ .$$

For $r \geq 0$ we define the *dilation of radius r* as the map $\delta_r : \mathcal{P}(E) \to \mathcal{P}(E)$ given by

$$\delta_r(X) = \bigcup_{x \in X} B_r(x) \ , \tag{7}$$

and for $r > 0$ we define δ_r°, the *open dilation of radius r*, by

$$\delta_r^{\circ}(X) = \bigcup_{x \in X} \overset{\circ}{B}_r(x) \ . \tag{8}$$

We have always

$$\delta_r^{\circ}(X) = \{p \in E \mid d(p, X) < r\} \ . \tag{9}$$

Note that $\delta_r^{\circ}(X)$, being a union of open balls, is open. On the other hand we have in general the inclusion

$$\delta_r(X) \subseteq \{p \in E \mid d(p, X) \leq r\} \ ; \tag{10}$$

the condition for having equality will be discussed in Subsect. 2.1. However we have always:

$$\{p \in E \mid d(p, X) \leq r\} = \bigcap_{s > r} \delta_s(X) = \bigcap_{s > r} \delta_s^{\circ}(X) \ . \tag{11}$$

Note that for X bounded, $\delta_r(X)$ is also bounded: indeed, if $X \subseteq B_s(p)$ for some $p \in E$ and $s \geq 0$, then $\delta_r(X) \subseteq \delta_r(B_s(p)) \subseteq B_{r+s}(p)$.

A *Cauchy sequence* is a sequence $(x_n)_{n \in \mathbf{N}}$ in E, such that for any $\varepsilon > 0$ there is some $M \in \mathbf{N}$ such that for all $m, n \geq M$ we have $d(x_m, x_n) < \varepsilon$. We say that the metric space (E, d) is *complete* if every Cauchy sequence in E converges to some point in E.

There are several equivalent formulations for compactness of a metric space:

Property 1. The following properties are equivalent in a metric space E:

1. E is compact.
2. E is sequentially compact.
3. For every infinite $X \subseteq E$, some point of E is a limit point of X.
4. E is complete, and for every $r > 0$, there is a finite subset $X(r)$ of E such that $E = \bigcup_{p \in X(r)} B_r(p)$.
5. Every sequence in E contains a subsequence converging to a point in E.

Given $X \subseteq E$, (X, d) is a metric space, and the metric topology of (X, d) coincides with the relative topology on X induced by the metric topology of (E, d); in other words, the open sets of (X, d) are all sets $X \cap G$, where G ranges over the open sets of (E, d).

2.1   Boundedly Compact, Boundedly Finite, and Proximinal Sets

Note that every compact subset of a metric space is always bounded and closed. We consider here the case where the reverse holds.

Lemma 1. *The following properties are equivalent in a metric space E:*

1. *For every $p \in E$ and $r > 0$, $B_r(p)$ is compact.*
2. *Every bounded and closed (infinite) subset of E is compact.*
3. *For every infinite and bounded $X \subseteq E$, some $y \in E$ is a limit point of X.*

Proof. 1 implies 3.   Since X is bounded, we have $X \subseteq B_r(p)$ for some $p \in E$ and $r > 0$. As $B_r(p)$ is compact, by Property 1, some $y \in B_r(p)$ is a limit point of X.

3 implies 2.   Let Y be bounded and closed. If Y is finite, then Y is trivially compact, so we can assume that Y is infinite. Given an infinite $X \subseteq Y$, X is bounded, and so some $y \in E$ is a limit point of X; but then $y \in \overline{X}$, and as Y is closed, $\overline{X} \subseteq Y$, and hence $y \in Y$. By Property 1, Y is compact.

2 implies 1.   For every $p \in E$ and $r > 0$, $B_r(p)$ is bounded and closed.   □

We say then that E is *boundedly compact*. In [ ], one says *finitely compact*, the terminology is borrowed from Busemann [ ]. A subset X of E is said to be boundedly compact if the metric subspace (X, d) is boundedly compact.

A subset X of E is called *proximinal* [,] if either $X = \emptyset$, or for every $y \notin X$, there is some $x \in X$ minimizing the distance to y, in other words

$$\forall y \notin X, \ \exists x \in X \quad d(y, x) = d(y, X) \ .$$

Write $\mathcal{F}_{bc}$ and $\mathcal{F}_p$ for the families of respectively boundedly compact and proximinal subsets of E.

Proposition 1. *In a metric space E:*

1. $\mathcal{K} \subseteq \mathcal{F}_{bc} \subseteq \mathcal{F}_p \subseteq \mathcal{F}$.

2. *If E is boundedly compact, then every closed subset of E is boundedly compact, that is $\mathcal{F}_{bc} = \mathcal{F}_p = \mathcal{F}$.*

Proof. 1) If K is compact, every closed subset of K is compact, so K is boundedly compact. Let X be boundedly compact; if $X = \emptyset$, it is by definition proximinal; otherwise for $p \notin X$ we take $x \in X$ and set $r = d(p, x)$, then $B_r(p) \cap X$ is nonvoid, bounded and closed (relatively to X), so it is compact and there is some $y \in B_r(p) \cap X$ such that

$$d(p, y) = \min\{d(p, z) \mid z \in B_r(p) \cap X\} = \min\{d(p, z) \mid z \in X\} ,$$

that is X is proximinal. If Y is proximinal, either $Y = \emptyset$, which is closed, or for $p \notin Y$ we take $y \in Y$ such that $d(p, y) = d(p, Y)$, so $d(p, Y) > 0$, and for $0 < r < d(p, Y)$, $\overset{\circ}{B}_r(p) \subseteq E \setminus Y$, and Y is closed.

2) Let F be a closed subset of E. Given a subset K of F which is bounded and closed relative to F, by (4) K is closed in E, Assuming E to be boundedly compact, K will be compact in E, so by (3) K is compact relatively to F. Hence F is boundedly compact. By item 1, we deduce that $\mathcal{F}_{bc} = \mathcal{F}_p = \mathcal{F}$.   □

Proposition 2. *In a metric space E, a set X is proximinal iff for every $r \geq 0$ we have*

$$\delta_r(X) = \{p \in E \mid d(p, X) \leq r\} . \tag{12}$$

In particular, $\delta_r(X)$ is closed.

Proof. If $X = \emptyset$, X is proximinal, and both sides of (12) are the empty set. Assume now that $X \neq \emptyset$. For $r \geq 0$ we have

$$\delta_r(X) = \{p \in E \mid \exists x \in X, \ d(p, x) \leq r\} .$$

If X is proximinal, for every $p \in E$, there is some $x \in X$ such that $d(p, x) = d(p, X)$ (indeed, either $p \in X$ and we take $x = p$, or $p \notin X$ and this is true by definition). So the above equality becomes (12). If X is not proximinal, there exists $y \notin X$ such that for every $x \in X$ we have $d(y, x) > d(y, X)$; setting $r = d(y, X)$, we have $y \notin \delta_r(X)$ but $y \in \{p \in E \mid d(p, X) \leq r\}$.

Being the inverse image of the closed interval $[0, r]$ by the continuous map $E \to \mathbf{R} : p \mapsto d(p, X)$, $\{p \in E \mid d(p, X) \leq r\}$ is closed. So for X proximinal, $\delta_r(X)$ is closed by (12).   □

Corollary 1. *The following properties are equivalent in a metric space E:*

1. *E is boundedly compact.*
2. *For every nonvoid compact K and $r > 0$, $\delta_r(K)$ is compact.*
3. *For every nonvoid closed F and $r > 0$, $\delta_r(F)$ is boundedly compact.*

Proof. 1 implies 3.   Let E be boundedly compact, take a nonvoid closed F and $r > 0$; by Proposition 1, F is proximinal, and by Proposition 2, $\delta_r(F)$ is closed; thus $\delta_r(F)$ is boundedly compact by Proposition 1.

3 implies 2.   Let K be a nonvoid compact; as K is closed, $\delta_r(K)$ is bound-edly compact. But K is also bounded, hence $\delta_r(K)$ is bounded. Being a bounded closed subset of a boundedly compact set (itself), $\delta_r(K)$ is compact.

2 implies 1.   If $\delta_r(K)$ is compact for every nonvoid compact K and $r > 0$, take $p \in E$, and as $\{p\}$ is compact, $B_r(p) = \delta_r(\{p\})$ is compact; thus E is boundedly compact.   □

Now we consider a generalization of boundedly compact sets:

Lemma 2. *The following properties are equivalent in a metric space E:*

1. *For every $p \in E$ and $r > 0$, $B_r(p)$ is finite.*
2. *Every bounded subset of E is finite.*
3. *Every bounded subset of E is compact.*
4. *E is boundedly compact, and every subset of E is closed.*

Proof. That *1 implies 2* is trivial.

2 implies 4.   Every bounded and closed subset of E is finite, and hence compact; thus E is boundedly compact. For any $p \in E$, as the balls $\overset{\circ}{B}_r(p)$ are finite for all $r > 0$, there is some $r(p) > 0$ such that $\overset{\circ}{B}_{r(p)}(p) = \{p\}$, so that every singleton in E is open; therefore every subset of E is closed.

4 implies 3.   Les X be a bounded subset of E; by hypothesis, X is closed, and as E is boundedly compact, X is thus compact.

3 implies 1.   Suppose that for some $p \in E$ and $r > 0$, $B_r(p)$ is infinite. Then it contains a sequence $(x_n)_{n \in \mathbf{N}}$ whose terms are mutually distinct ($x_n \neq x_m$ for $n \neq m$). As $B_r(p)$ is bounded, it is compact, so by Property 1, this sequence contains a subsequence $(x_{i_m})_{m \in \mathbf{N}}$ converging to some $y \in B_r(p)$. As the x_{i_m} are pairwise distinct, there is some $k \in \mathbf{N}$ such that for $m \geq k$, $x_{i_m} \neq y$. Let $K = \{x_{i_m} \mid m \geq k\}$; as K is bounded, it is compact, so by Property 1 the sequence $(x_{i_m})_{m \geq k}$ contains a subsequence converging to some $x_{i_t}, t \geq k$; however such a subsequence must converge to y, and $y \neq x_{i_t}$, so we have a contradiction. Therefore $B_r(p)$ must be finite.   □

We say then that E is *boundedly finite*. A subset X of E is said to be bound-edly finite if the relative topology on X is boundedly finite. Write $\mathcal{F}_{bf}$ for the family of boundedly finite subsets of E. Clearly every boundedly finite set is boundedly compact (and hence proximinal and closed). A set is boundedly finite and compact iff it is finite. Note that if E is boundedly finite, then every subset of E is boundedly finite (and hence proximinal and closed).

2.2   Hausdorff Metric on Compact, Closed, and Proximinal Sets

Let X and Y be two nonvoid compact subsets of E. The set of $d(x, Y)$ for $x \in X$ attains a maximum, so we define:

$$h_d(X, Y) = \max\{d(x, Y) \mid x \in X\} , \tag{13}$$

which we call the *oriented Hausdorff distance from X to Y*. Note that for $x \in X$ the set of $d(x, y)$ for $y \in Y$ admits a minimum, and so

$$\exists x^* \in X, \ \exists y^* \in Y, \qquad d(x^*, y^*) = d(x^*, Y) = h_d(X, Y) \ . \qquad (14)$$

We define the *Hausdorff distance between X and Y* as:

$$H_d(X, Y) = \max\big(h_d(X, Y), h_d(Y, X)\big) \ . \qquad (15)$$

It is well-known [ ] that H_d is indeed a distance function on the space $\mathcal{K}'$ of nonvoid compact subsets of E.

We have the following characterization [ ] of h_d and H_d:

Property 2. For every $X, Y \in \mathcal{K}'$ and for every $r \geq 0$:

$$
\begin{aligned}
h_d(X, Y) \leq r &\iff X \subseteq \delta_r(Y), \\
H_d(X, Y) \leq r &\iff X \subseteq \delta_r(Y) \quad \text{and} \quad Y \subseteq \delta_r(X) \ .
\end{aligned}
\qquad (16)
$$

In particular

- $h_d(X, Y)$ is the least $r \geq 0$ such that $X \subseteq \delta_r(Y)$.
- $H_d(X, Y)$ is the least $r \geq 0$ such that both $X \subseteq \delta_r(Y)$ and $Y \subseteq \delta_r(X)$.

In the case where $E = \mathbf{R}^n$ and d is the Euclidean distance, in the above formulas $\delta_r(Y)$ and $\delta_r(X)$ are replaced by the Minkowski additions $Y \oplus \mathcal{B}_r$ and $X \oplus \mathcal{B}_r$ respectively, where $\mathcal{B}_r$ is the closed ball of radius r centered about the origin; such an expression of the Euclidean Hausdorff distance in terms of Minkowski additions was considered in [ ].

A well-known theorem [ ] states that:

Property 3. If (E, d) is complete, then $(\mathcal{K}'(E), H_d)$ is also complete.

We can extend the Hausdorff metric from $\mathcal{K}'$ to $\mathcal{F}'$ [ ]. Given two nonvoid closed sets X and Y, we set

$$h_d(X, Y) = \sup\{d(x, Y) \mid x \in X\} \ , \qquad (17)$$

but here (14) does not hold; now we define the Hausdorff distance $H_d(X, Y)$ as in (15). Then H_d is a *generalized metric* on $\mathcal{F}$; by this we mean that H_d satisfies the axioms of a metric, with the only difference that it can take infinite values. Nevertheless $(\mathcal{F}', H_d)$ will define a topology in the same way as a metric space.

Using (11), we see easily that Property 2 becomes here:

Property 4. For every $X, Y \in \mathcal{F}'$ and for every $r \geq 0$:

$$
\begin{aligned}
h_d(X, Y) \leq r &\iff X \subseteq \bigcap_{s > r} \delta_s(Y), \\
H_d(X, Y) \leq r &\iff X \subseteq \bigcap_{s > r} \delta_s(Y) \quad \text{and} \quad Y \subseteq \bigcap_{s > r} \delta_s(X) \ .
\end{aligned}
\qquad (18)
$$

In particular

- $h_d(X, Y) = \inf\{r > 0 \mid X \subseteq \delta_r(Y)\}$.
- If $h_d(X, Y) < \infty$, it is the least $r \geq 0$ such that $X \subseteq \bigcap_{s>r} \delta_s(Y)$.
- $H_d(X, Y) = \inf\{r > 0 \mid X \subseteq \delta_r(Y) \text{ and } Y \subseteq \delta_r(X)\}$.
- If $H_d(X, Y) < \infty$, it is the least $r \geq 0$ such that both $X \subseteq \bigcap_{s>r} \delta_s(Y)$ and $Y \subseteq \bigcap_{s>r} \delta_s(X)$.

However, from (12) it follows that Property 2 remains true for two nonvoid proximinal sets X, Y.

We have an alternative definition for the Hausdorff distance on $\mathcal{F}'$ [ ]:

$$\forall X, Y \in \mathcal{F}', \qquad H_d(X, Y) = \sup_{p \in E} |d(p, X) - d(p, Y)| \ . \tag{19}$$

Note that it is also possible to define the Hausdorff distance when one of the sets is empty, using formulas (6,17), and setting an empty supremum to 0 and an empty infimum to ∞:

$$\forall F \in \mathcal{F}', \qquad \begin{array}{ll} h_d(\emptyset, \emptyset) = h_d(\emptyset, F) = 0 & \text{and} \quad h_d(F, \emptyset) = \infty \ ; \\ H_d(\emptyset, \emptyset) = 0 & \text{and} \quad H_d(F, \emptyset) = \infty \ . \end{array} \tag{20}$$

Then Property 4 remains true in this particular case.

2.3   Compact Families of Compact Sets

Consider a metric space (E, d). Recall the characterization of compact sets in a metric space given in Property 1. The set $\mathcal{K}'(E)$ of nonvoid compact subsets of the metric space (E, d), provided with the Hausdorff distance H_d is a metric space $(\mathcal{K}'(E), H_d)$. We can consider the topology on $\mathcal{K}'(E)$ induced by H_d, and define compact sets in it; we write thus $\mathcal{K}'(\mathcal{K}'(E))$ for the set of nonvoid compact subsets of $\mathcal{K}'(E)$, in other words the set of all nonvoid compact families (for H_d) of compact subsets (for d) of E.

From (3) we deduce that for every $X \subseteq E$,

$$\mathcal{K}'(\mathcal{K}'(X)) = \mathcal{K}'(\mathcal{K}'(E)) \cap \mathcal{P}(\mathcal{K}'(X)) = \mathcal{K}'(\mathcal{K}'(E)) \cap \mathcal{P}(\mathcal{P}(X)) \ . \tag{21}$$

We will generalize to compacts the well-known facts that: a) the union of a finite family of finite subsets of E is finite, and b) the set of parts of a finite set is finite. The following result was proven in [ ]:

Proposition 3. *For $\mathcal{K} \in \mathcal{K}'(\mathcal{K}'(E))$, $\bigcup \mathcal{K} \in \mathcal{K}'(E)$.*

Proof. Let $(x_n)_{n \in \mathbf{N}}$ be any sequence in $\bigcup \mathcal{K}$. Then for each $n \in \mathbf{N}$ there is some $K_n \in \mathcal{K}$ such that $x_n \in K_n$. Since $\mathcal{K}$ is compact in the metric space $(\mathcal{K}'(E), H_d)$, the sequence $(K_n)_{n \in \mathbf{N}}$ contains a subsequence $(K_{i_n})_{n \in \mathbf{N}}$ converging to $K \in \mathcal{K}$. Since K is compact in (E, d), for each $n \in \mathbf{N}$ there is some $y_n \in K$ such that $d(x_{i_n}, K) = d(x_{i_n}, y_n)$, and the sequence $(y_n)_{n \in \mathbf{N}}$ contains a subsequence $(y_{j_n})_{n \in \mathbf{N}}$ converging to $y \in K$. For each $n \in \mathbf{N}$ we have

$$d(x_{i_n}, y_n) - d(x_{i_n}, K) \leq h_d(K_n, K) \leq H_d(K_n, K) \ ,$$

and as $H_d(K_n, K) \to 0$ for $n \to \infty$, we get $d(x_{i_n}, y_n) \to 0$ for $n \to \infty$. Thus $d(x_{i_{j_n}}, y_{j_n}) \to 0$ and $d(y_{j_n}, y) \to 0$ for $n \to \infty$, so we deduce that the subsequence $(x_{i_{j_n}})_{n \in \mathbf{N}}$ converges to y; now $y \in K \in \mathcal{K}$, so $y \in \bigcup \mathcal{K}$, and we have thus shown that any sequence in $\bigcup \mathcal{K}$ contains a subsequence converging to an element of $\bigcup \mathcal{K}$, so $\bigcup \mathcal{K}$ is compact. □

Proposition 4. *For $K \in \mathcal{K}'(E)$, $\mathcal{K}'(K) \in \mathcal{K}'(\mathcal{K}'(E))$.*

Proof. Let $(K_n)_{n \in \mathbf{N}}$ be a sequence in $\mathcal{K}'(K)$. Take any $\varepsilon > 0$; we will first show that there is an infinite sequence T_ε in $\mathbf{N}$ such that for all $n, n' \in T_\varepsilon$ we have $H_d(K_n, K_{n'}) \leq 3\varepsilon$. Since K is compact, there exists a finite number of points $x_1, \ldots, x_m \in K$ such that $K \subseteq B_\varepsilon(x_1) \cup \cdots \cup B_\varepsilon(x_m)$. For each $n \in \mathbf{N}$ and $j = 1, \ldots, m$ we choose $x_{n,j} \in K_n$ as follows: $x_{n,j} \in K_n \cap B_\varepsilon(x_j)$ if $K_n \cap B_\varepsilon(x_j) \neq \emptyset$, while $x_{n,j}$ is any element of K_n if $K_n \cap B_\varepsilon(x_j) = \emptyset$. For $K_n \cap B_\varepsilon(x_j) \neq \emptyset$, since $x_{n,j} \in B_\varepsilon(x_j)$, the triangular inequality implies that $B_\varepsilon(x_j) \subseteq B_{2\varepsilon}(x_{n,j})$, so that $K_n \cap B_\varepsilon(x_j) \subseteq B_{2\varepsilon}(x_{n,j})$; on the other hand for $K_n \cap B_\varepsilon(x_j) = \emptyset$, we have $K_n \cap B_\varepsilon(x_j) \subseteq B_{2\varepsilon}(x_{n,j})$ anyway. Now since $K_n \subseteq K \subseteq B_\varepsilon(x_1) \cup \cdots \cup B_\varepsilon(x_m)$, we get

$$K_n = \left(K_n \cap B_\varepsilon(x_1)\right) \cup \cdots \cup \left(K_n \cap B_\varepsilon(x_1)\right)$$

with $K_n \cap B_\varepsilon(x_j) \subseteq B_{2\varepsilon}(x_{n,j})$ for each $j = 1, \ldots, m$. Hence

$$\forall n \in \mathbf{N}, \qquad K_n \subseteq B_{2\varepsilon}(x_{n,1}) \cup \cdots \cup B_{2\varepsilon}(x_{n,m}) \ . \tag{22}$$

For $t = 0, \ldots, m$, we obtain by induction an infinite sequence S_t in $\mathbf{N}$ such that for $1 \leq j \leq t$ the sequence $(x_{n,j})_{n \in S_t}$ converges. Indeed, for $t = 0$ we have $S_0 = \mathbf{N}$; given S_t for $0 \leq t < m$, since K is compact the sequence $(x_{n,t+1})_{n \in S_t}$ contains a converging subsequence $(x_{n,t+1})_{n \in S_{t+1}}$ (where S_{t+1} is a subsequence of S_t), and clearly for $1 \leq j \leq t$, as $(x_{n,j})_{n \in S_t}$ converges, $(x_{n,j})_{n \in S_{t+1}}$ still converges to the same limit. Therefore for $j = 1, \ldots, m$, $(x_{n,j})_{n \in S_m}$ converges, and in particular it is a Cauchy sequence. Thus there is an infinite sequence T_ε in $\mathbf{N}$, which is a "tail" of S_m, such that for $j = 1, \ldots, m$ and $n, n' \in T_\varepsilon$, $d(x_{n,j}, x_{n',j}) < \varepsilon$. By the triangular inequality, we get then $B_{2\varepsilon}(x_{n,j}) \subseteq B_{3\varepsilon}(x_{n',j})$. Then (22) gives:

$$\forall n, n' \in T_\varepsilon, \qquad K_n \subseteq B_{3\varepsilon}(x_{n',1}) \cup \cdots \cup B_{3\varepsilon}(x_{n',m}) \ .$$

As $x_{n',1}, \ldots, x_{n',m} \in K_{n'}$, this means that $K_n \subseteq \delta_{3\varepsilon}(K_{n'})$; reciprocally, we have $K_{n'} \subseteq \delta_{3\varepsilon}(K_n)$, so that by (16) we have $H_d(K_n, K_{n'}) \leq 3\varepsilon$ for all $n, n' \in T_\varepsilon$.

For $0 < \eta < \varepsilon$, writing $T_\varepsilon = \{i_n \mid n \in \mathbf{N}\}$, the above argument applied to the sequence $(K_{i_n})_{n \in \mathbf{N}}$ gives an infinite subsequence T_η which is a subsequence of T_ε, such that $H_d(K_n, K_{n'}) \leq 3\eta$ for all $n, n' \in T_\eta$. Now we build the sequences $T_{1/3n}$ $(n \geq 1)$ and the infinite sequence $(s_n)_{n \geq 1}$ inductively as follows:

- s_1 is the least elements of $T_{1/3}$.
- For $n > 1$, build $T_{1/3n}$ from $T_{1/3(n-1)}$, and s_n is the least $m \in T_{1/3n}$ such that $m > s_{n-1}$.

As $T_{1/3u}$ is a subsequence of $T_{1/3n}$ for $u \geq n$, for any $n \geq 1$ and $u, v \geq n$ we have $s_u, s_v \in T_{1/3n}$, so that $H_d(K_{s_u}, K_{s_v}) \leq 1/n$, and hence $(K_{s_n})_{n \geq 1}$ is a Cauchy sequence in $\mathcal{K}'(K)$. By Property 1, (K, d) is complete, so that by Property 3, $(\mathcal{K}'(K), H_d)$ is complete, and hence the Cauchy sequence $(K_{s_n})_{n \geq 1}$ converges in $\mathcal{K}'(K)$.

We have thus shown that from every sequence $(K_n)_{n \in \mathbf{N}}$ in $\mathcal{K}'(K)$ one can extract a subsequence $(K_{s_n})_{n \geq 1}$ converging for H_d, which means thus that $\mathcal{K}'(K)$ is a compact subset of $\mathcal{K}'(E)$. Obviously $\mathcal{K}'(K) \neq \emptyset$, since $K \in \mathcal{K}'(K)$. Therefore $\mathcal{K}'(K) \in \mathcal{K}'(\mathcal{K}'(E))$. □

A well-known result (see [ ], ex. 6.16, p. 33) states that for $X_1, X_2, Y_1, Y_2 \in \mathcal{K}'(E)$,
$$H_d(X_1 \cup X_2, Y_1 \cup Y_2) \leq \max\big(H_d(X_1, Y_1), H_d(X_2, Y_2)\big) .$$
We generalize it here:

Proposition 5. *Let $r \geq 0$ and $\mathcal{X}, \mathcal{Y} \in \mathcal{K}'(\mathcal{K}'(E))$ such that for every $X \in \mathcal{X}$ there is $Y_X \in \mathcal{Y}$ with $H_d(X, Y_X) \leq r$, and for every $Y \in \mathcal{Y}$ there is $X_Y \in \mathcal{X}$ with $H_d(Y, X_Y) \leq r$. Then $H_d(\bigcup \mathcal{X}, \bigcup \mathcal{Y}) \leq r$.*

Proof. Note that $\bigcup \mathcal{X}, \bigcup \mathcal{Y} \in \mathcal{K}'(E)$ by Proposition 3. For each $X \in \mathcal{X}$ we have $Y_X \in \mathcal{Y}$ with $H_d(X, Y_X) \leq r$, so that $X \subseteq \delta_r(Y_X)$ by (16), and as $Y_X \subseteq \bigcup \mathcal{Y}$, we have $\delta_r(Y_X) \subseteq \delta_r(\bigcup \mathcal{Y})$; thus every $X \in \mathcal{X}$ verifies $X \subseteq \delta_r(\bigcup \mathcal{Y})$, and we deduce that $\bigcup \mathcal{X} \subseteq \delta_r(\bigcup \mathcal{Y})$. Reciprocally, as for every $Y \in \mathcal{Y}$ there is $X_Y \in \mathcal{X}$ with $H_d(Y, X_Y) \leq r$, we deduce that $\bigcup \mathcal{Y} \subseteq \delta_r(\bigcup \mathcal{X})$. Then $H_d(\bigcup \mathcal{X}, \bigcup \mathcal{Y}) \leq r$ by (16). □

3   Hausdorff Sampling into a Boundedly Compact Space

In [ ] we gave a new theory of discretization of compact sets; we choose as possible discretization of a nonvoid compact K any nonvoid discrete set S such that the Hausdorff distance between S and K is minimal. Formally, we have a "Euclidean" metric space (E, d) and a "discrete" subspace D, where $\emptyset \neq D \subset E$, on which we impose only the following two axioms:

1. D is boundedly finite.
2. The number $r_c = \sup_{x \in E} d(x, D)$ is finite.

The number r_c is called the *covering radius* (of D for the distance d). The first axiom implies that $\mathcal{K}'(D)$ consists of all nonvoid finite subsets of D; this axiom allowed us to show that for $K \in \mathcal{K}'(E)$, the set
$$\mathcal{M}_H(K) = \{S \in \mathcal{K}'(D) \mid \forall T \in \mathcal{K}'(D), \ H_d(K, S) \leq H_d(K, T)\} \tag{23}$$
is nonvoid, finite, and closed under union, so that $\mathcal{M}_H(K)$ has a greatest element, namely
$$\Delta_H(K) = \bigcup \mathcal{M}_H(K) . \tag{24}$$

We called any element of $\mathcal{M}_H(K)$ a *Hausdorff discretizing set* (or *Hausdorff discretization*) *of* K, while $\Delta_H(K)$ was called the *maximal Hausdorff discretization of* K. Defining the *Hausdorff radius of* K (w.r.t. D for the distance d) as the nonnegative number $r_H(K) = \max_{k\in K} d(k, D)$, we showed that for $S \in \mathcal{M}_H(K)$, $H_d(S, K) = r_H(K)$. We gave a characterization of $\Delta_H(K)$ and of elements of $\mathcal{M}_H(K)$ (we will consider it again in the next section). We have $r_H(K) \leq r_c$, so that the second axiom allows us to have a bound on the Hausdorff radiuses $r_H(K)$ in terms of the resolution of D, of which r_c is a measure.

We will see here that one can loosen the first axiom by requiring that

1. D is boundedly compact,

instead of boundedly finite. Then most results of [ ] will remain valid, provided that we replace "finite" by "compact" in each statement. We concentrate particularly on the results of Subsect. 3.3 of [ ]. We will discuss those of Subsect. 3.4 of [ ] in more detail in the next section, devoted to the extension of the theory to closed sets.

First Lemma 13 of [ ] becomes:

Lemma 3. *For every* $x \in E$ *and* $r \geq 0$, $B_r(x) \cap D$ *is compact. More generally, for every compact* $K \subseteq E$, $\delta_r(K) \cap D$ *is compact.*

Proof. Let K be compact. As K is proximinal, by Proposition 2 $\delta_r(K)$ is closed, and as K is bounded, $\delta_r(K)$ is bounded. Hence $\delta_r(K) \cap D$ is bounded and closed in the relative topology of D; as D is boundedly compact, $\delta_r(K) \cap D$ is compact in D, and therefore it is compact in E (cfr. (3)). Taking for K the singleton $\{x\}$, we have $\delta_r(K) = B_r(x)$, so $B_r(x) \cap D$ is compact. □

In particular, taking $r = 0$, for any $K \in \mathcal{K}(E)$ we have $K \cap D$ compact. Under the assumption that D is boundedly finite, we had Corollary 14 of [ ], which stated that the compact subsets of D coincide with its finite subsets. Here such a result does not apply, we simply have by (3): $\mathcal{K}(D) = \mathcal{K}(E) \cap \mathcal{P}(D)$.

As D is proximinal, we have as in equation (17) of [ ] that

$$\forall x \in E, \ \exists p \in D, \qquad d(x, p) = d(x, D) \ .$$

In particular for $x \notin D$ we have $d(x, D) > 0$, and since $D \neq E$, we have $r_c > 0$.

Lemma 16 of [ ] states that r_c is the least $r > 0$ such that $E = \delta_r(D)$. This remains true here, the proof given in [ ] remains valid, but this result is in fact a consequence of Proposition 2.

The next result is a generalization of Proposition 18 of [ ]:

Proposition 6. *For* $K \in \mathcal{K}'(E)$,

1. $\mathcal{M}_H(K) \in \mathcal{K}'(\mathcal{K}'(D))$, *in particular* $\mathcal{M}_H(K) \neq \emptyset$;
2. *given* $\mathcal{X} \in \mathcal{K}'(\mathcal{K}'(D))$ *such that* $\mathcal{X} \subseteq \mathcal{M}_H(K)$, *we have* $\bigcup \mathcal{X} \in \mathcal{M}_H(K)$; *in particular:*

  – $\bigcup \mathcal{M}_H(K) \in \mathcal{M}_H(K)$, and
  – for a closed $\mathcal{Y} \subseteq \mathcal{K}'(E)$ such that $\mathcal{Y} \cap \mathcal{M}_H(K) \neq \emptyset$, $\bigcup(\mathcal{Y} \cap \mathcal{M}_H(K)) \in$ $\mathcal{M}_H(K)$;
3. *given a decreasing sequence $(S_n)_{n \in \mathbf{N}}$ of elements of $\mathcal{M}_H(K)$, $\bigcap_{n \in \mathbf{N}} S_n \in$ $\mathcal{M}_H(K)$.*

Proof. 1) Let $p \in D$ and set $r = \max_{k \in K} d(p, k)$. Write $Z = \delta_r(K) \cap D$; clearly $p \in Z$, so that $Z \neq \emptyset$; also Z is compact by Lemma 3; thus $Z \in \mathcal{K}'(D)$. By Proposition 4, $\mathcal{K}'(Z) \in \mathcal{K}'(\mathcal{K}'(D))$. As the function $H_d(\cdot, K) : \mathcal{K}'(D) \to \mathbf{R}^+ :$ $S \mapsto H_d(S, K)$ is continuous, it reaches a minimum s on the nonvoid compact $\mathcal{K}'(Z)$, that is: a) every $S \in \mathcal{K}'(Z)$ gives $H_d(S, K) \geq s$, and b) there exists $S \in \mathcal{K}'(Z)$ such that $H_d(S, K) = s$. Take the nonvoid set

$$\mathcal{M} = \{S \in \mathcal{K}'(Z) \mid H_d(S, K) = s\};$$

Clearly $\{p\} \in \mathcal{K}'(Z)$, and so $H_d(K, \{p\}) = \max_{k \in K} d(p, k) = r$ (see Property 8 of [ ]); thus $s \leq r$. For $S \in \mathcal{K}'(D)$, if $H_d(S, K) \leq r$, then by (16) we have $S \subseteq$ $\delta_r(K)$, so that $S \subseteq \delta_r(K) \cap D = Z$ and hence $S \in \mathcal{K}'(Z)$; thus $H_d(S, K) > r \geq s$ for $S \notin \mathcal{K}'(Z)$. Therefore s is the minimum of all $H_d(S, K)$ for $S \in \mathcal{K}'(D)$, and $\mathcal{M} = \mathcal{M}_H(K)$. Thus $\mathcal{M}_H(K) \neq \emptyset$. Now $\mathcal{M}_H(K)$ is the inverse image of the singleton $\{s\}$ by the continuous function $H_d(\cdot, K) : \mathcal{K}'(D) \to \mathbf{R}^+$, and so it is closed. Thus, being a closed subset of the compact set $\mathcal{K}'(Z)$ in the space $\mathcal{K}'(D)$, $\mathcal{M}_H(K)$ is compact, and as it is nonvoid, we have $\mathcal{M}_H(K) \in \mathcal{K}'(\mathcal{K}'(D))$.

2) Let $\mathcal{X} \in \mathcal{K}'(\mathcal{K}'(D))$ such that $\mathcal{X} \subseteq \mathcal{M}_H(K)$. By Proposition 3, $\bigcup \mathcal{X} \in$ $\mathcal{K}'(D)$. For each $S \in \mathcal{X}$ we have $H_d(S, K) = s$, so applying Proposition 5 with $\mathcal{Y} = \{K\}$, we get $H_d(\bigcup \mathcal{X}, K) \leq s$. But s is the least $H_d(S, K)$ for $S \in \mathcal{K}'(D)$, and we deduce that $H_d(\bigcup \mathcal{X}, K) = s$, so that $\bigcup \mathcal{X} \in \mathcal{M}_H(K)$. In particular, taking $\mathcal{X} = \mathcal{M}_H(K)$, we get $\bigcup \mathcal{M}_H(K) \in \mathcal{M}_H(K)$. Also, for a closed $\mathcal{Y} \subseteq \mathcal{K}'(E)$, as $\mathcal{M}_H(K)$ is compact in $\mathcal{K}'(D)$ (and so in $\mathcal{K}'(E)$, see (21)), $\mathcal{Y} \cap \mathcal{M}_H(K)$ is a compact subset of $\mathcal{K}'(E)$; if we have also $\mathcal{Y} \cap \mathcal{M}_H(K) \neq \emptyset$, then $\mathcal{Y} \cap \mathcal{M}_H(K) \in$ $\mathcal{K}'(\mathcal{K}'(E)) \cap \mathcal{P}(\mathcal{K}'(D)) = \mathcal{K}'(\mathcal{K}'(D))$. Taking $\mathcal{X} = \mathcal{Y} \cap \mathcal{M}_H(K)$, $\bigcup(\mathcal{Y} \cap \mathcal{M}_H(K)) \in$ $\mathcal{M}_H(K)$.

3) Consider a decreasing sequence $(S_n)_{n \in \mathbf{N}}$ of elements of $\mathcal{M}_H(K)$, and let $S = \bigcap_{n \in \mathbf{N}} S_n$. For each $n \in \mathbf{N}$, $H_d(S_n, K) = s$, and as S_n is proximinal, for every $x \in K$ $B_s(x) \cap S_n \neq \emptyset$; as the $B_s(x) \cap S_n$ ($n \in \mathbf{N}$) constitute a decreasing sequence of nonvoid closed subsets of the compact $B_s(x) \cap S_0$, by the finite intersection property, $B_s(x) \cap S = \bigcap_{n \in \mathbf{N}}(B_s(x) \cap S_n) \neq \emptyset$. Thus $S \neq \emptyset$, and so $S \in \mathcal{K}'(D)$, and as $B_s(x) \cap S \neq \emptyset$ for every $x \in K$, we have $h_d(K, S) \leq s$. Now as $S \subseteq S_0$, $h_d(S, K) \leq h_d(S_0, K) \leq s$. Therefore $H_d(S, K) \leq s$, from which we conclude that $S \in \mathcal{M}_H(K)$. □

Note in particular that any finite subset of $\mathcal{M}_H(K)$ is compact; thus for $S_1, \ldots, S_n \in \mathcal{M}_H(K)$, we have $S_1 \cup \cdots \cup S_n \in \mathcal{M}_H(K)$.

Now the rest of our theory, in particular the characterization given in Subsect. 3.4 of [ ], remains valid. Note that here the discretization of radius r of K, namely $\Delta_r(K) = \delta_r(K) \cap D$ (see (26) below), is compact (instead of finite, as in [ ]) for every compact K (this follows from Lemma 3).

4   Hausdorff Sampling of Closed Sets

We still assume that D is boundedly compact and that $r_c < \infty$. To stress the fact that D is not necessarily discrete, we will systematically say "sampling" instead of "discretization" or "discretizing set". We will now define Hausdorff sampling for a closed subset of E instead of a compact one.

For $F \in \mathcal{F}'(E)$, a Hausdorff sampling of F is any $S \in \mathcal{F}'(D)$ which minimizes the Hausdorff distance $H_d(F, S)$. Here (23) becomes

$$\mathcal{M}_H(F) = \{S \in \mathcal{F}'(D) \mid \forall T \in \mathcal{F}'(D), \ H_d(F, S) \leq H_d(F, T)\} \ . \tag{25}$$

Note that when F is compact, it is bounded, and so $H_d(F, S) < \infty$ implies that S is bounded, hence S is compact (because D is boundedly compact). Thus for compact sets, (25) is equivalent to (23), in other words the two definitions of a Hausdorff sampling coincide.

The argument of items 1 and 2 of Proposition 6 does not extend naturally to the non-compact case. We can only show the following:

Proposition 7. *For $F \in \mathcal{F}'(E)$,*

1. *$\mathcal{M}_H(F)$ is closed in the topology of $(\mathcal{F}'(D), H_d)$;*
2. *given $\mathcal{X} \subseteq \mathcal{M}_H(F)$ such that $\bigcup \mathcal{X} \in \mathcal{F}'(E)$, we have $\bigcup \mathcal{X} \in \mathcal{M}_H(F)$;*
3. *given a decreasing sequence $(S_n)_{n \in \mathbf{N}}$ of elements of $\mathcal{M}_H(F)$, $\bigcap_{n \in \mathbf{N}} S_n \in \mathcal{M}_H(F)$.*

Indeed, item 1 holds because $\mathcal{M}_H(F)$ is the inverse image by the continuous function $H_d(\cdot, F) : \mathcal{F}'(D) \to \mathbf{R}^+$ of the singleton $\{s\}$, where s is the infimum of all $H_d(F, T)$ for $T \in \mathcal{F}'(D)$). Now item 2 is proved in a similar way as in Propositions 5 and 6. Finally item 3 has exactly the same proof as in Proposition 6.

However, we cannot prove by a similar argument that $\mathcal{M}_H(F)$ is nonvoid and has a greatest element $\Delta_H(F)$. We will thus apply here the method used in Subsect. 3.4 of [ ], where an explicit formula was given for $\Delta_H(F)$. Having such a formula in our case will prove the existence of the maximal Hausdorff sampling, and hence $\mathcal{M}_H(F)$ will be nonvoid.

For $r \geq 0$, the *sampling of radius r* [ ] is the map $\Delta_r : \mathcal{P}(E) \to \mathcal{P}(D)$ defined by

$$\forall X \subseteq E, \qquad \Delta_r(X) = \delta_r(X) \cap D = \{p \in D \mid B_r(p) \cap X \neq \emptyset\} \ , \tag{26}$$

while the *open sampling of radius r* is the map $\Delta_r^\circ : \mathcal{P}(E) \to \mathcal{P}(D)$ defined by

$$\forall X \subseteq E, \qquad \Delta_r^\circ(X) = \delta_r^\circ(X) \cap D = \{p \in D \mid \overset{\circ}{B}_r(p) \cap X \neq \emptyset\} \ . \tag{27}$$

Finally, we define the *upper sampling of radius r* as the map $\Delta_r^+ : \mathcal{P}(E) \to \mathcal{P}(D)$ given by

$$\Delta_r^+(X) = \bigcap_{s > r} \Delta_s(X) \ . \tag{28}$$

Lemma 4. *For $X \in \mathcal{P}(E)$ and $r \geq 0$,*

$$\Delta_r^+(X) = \{p \in D \mid d(p, X) \leq r\} ,\tag{29}$$

and this set is closed. For $F \in \mathcal{F}_p(E)$, $\Delta_r^+(F) = \Delta_r(F)$.

Proof. By (11),

$$\{p \in D \mid d(p, X) \leq r\} = \left(\bigcap_{s>r} \delta_s(X)\right) \cap D = \bigcap_{s>r}(\delta_s(X) \cap D) = \bigcap_{s>r} \Delta_s(X) .$$

So (29) follows. Now this set is closed because it is the inverse image of the closed interval $[0, r]$ by the continuous map $D \to \mathbf{R} : p \mapsto d(p, X)$.

When F is proximinal, (12) gives:

$$\{p \in D \mid d(p, F) \leq r\} = \delta_r(F) \cap D = \Delta_r(F) .$$

Comparing this to (29), we get $\Delta_r^+(F) = \Delta_r(F)$. □

We extend the definition of the *Hausdorff radius* from a compact set to a closed one, by replacing the maximum by a supremum in the formula:

$$\forall F \in \mathcal{F}'(E), \qquad r_H(F) = \sup_{x \in F} d(x, D) = h_d(F, D) .\tag{30}$$

Obviously $r_c = r_H(E)$ and $r_H(F) \leq r_c$ for $F \in \mathcal{F}'(E)$. Lemma 22 of [ ] remains true: $r_H(F)$ is the least $r > 0$ such that $F \subseteq \delta_r(D)$. The proof given in [ ] remains valid, but this is also a consequence of Proposition 2.

Now Proposition 23 of [ ] becomes:

Proposition 8. *Let $F \in \mathcal{F}'(E)$ and $S \in \mathcal{F}'(D)$. Then $H_d(F, S) \geq r_H(F)$, and for any $r \geq 0$ we have:*

- *$H_d(F, S) \leq r$ iff both $S \subseteq \Delta_r^+(F)$ and $F \subseteq \delta_r(S)$.*
- *Assuming F to be proximinal: $H_d(F, S) \leq r$ iff both $S \subseteq \Delta_r(F)$ and $F \subseteq \delta_r(S)$.*

Proof. As $S \subseteq D$, we have $H_d(F, S) \geq h_d(F, S) \geq h_d(F, D) = r_H(F)$. Let $r \geq 0$. As D is boundedly compact, by Proposition 1 S is proximinal, so we can apply (16): $h_d(F, S) \leq r$ iff $F \subseteq \delta_r(S)$. By (29), $h_d(S, F) \leq r$ iff

$$S \subseteq \{p \in D \mid d(p, F) \leq r\} = \Delta_r^+(F) .$$

In the case where F is proximinal, by Lemma 4 this becomes $S \subseteq \Delta_r(F)$. Since $H_d(F, S) \leq r$ iff both $h_d(F, S) \leq r$ and $h_d(S, F) \leq r$, the two items are verified. □

We get then the following generalization of Proposition 33 of [ ]:

Corollary 2. *For $F \in \mathcal{F}'(E)$ and $r \geq r_H(F)$, we have $H_d\big(F, \Delta_r^+(F)\big) \leq r$ and $H_d\big(F, \overline{\Delta_r(F)}\big) \leq r$.*

Proof. Here we have $F \subseteq \delta_r(D)$, and from Proposition 10 of [ ] (cfr. the argument at the beginning of the second paragraph of the proof of Theorem 24 of [ ]), this means that $F \subseteq \delta_r(\Delta_r(F))$. As $\Delta_r(F) \subseteq \overline{\Delta_r(F)} \subseteq \Delta_r^+(F)$, $F \subseteq \delta_r(\overline{\Delta_r(F)}) \subseteq \delta_r(\Delta_r^+(F))$. Now $\Delta_r^+(F)$ is closed (by Lemma 4). Hence the result follows by taking in Proposition 8 successively $S = \Delta_r^+(F)$, and $S = \overline{\Delta_r(F)}$. □

We can now give the generalization to closed sets of the characterization of the maximal Hausdorff discretization and Hausdorff discretizing sets given in Theorem 24 of [ ]:

Theorem 1. *For $F \in \mathcal{F}'(E)$,*

1. $\Delta_{r_H(F)}^+(F)$ *is the greatest Hausdorff sampling of F;*
2. *the Hausdorff samplings of F are the closed subsets of $\Delta_{r_H(F)}^+(F)$, whose dilation of Hausdorff radius covers F:*
   $\mathcal{M}_H(F) = \{S \in \mathcal{F}'(D) \mid S \subseteq \Delta_{r_H(F)}^+(F), \text{ and } F \subseteq \delta_{r_H(F)}(S)\};$
3. *the Hausdorff radius of F minimizes the Hausdorff distance beween F and nonvoid closed subsets of D:*
   $r_H(F) = \min\{H(F,S) \mid S \in \mathcal{F}'(D)\}$, *that is $H_d\big(F, \Delta_{r_H(F)}^+(F)\big) = r_H(F)$.*

Note: If F is proximinal, then $\Delta_{r_H(F)}^+(F) = \Delta_{r_H(F)}(F)$ (by Lemma 4).

Proof. By Proposition 8, $H_d(F,S) \geq r_H(F)$ for all $S \in \mathcal{F}'(D)$.

By Lemma 4, $\Delta_{r_H(F)}^+(F)$ is closed. By Corollary 2, $H_d\big(F, \Delta_{r_H(F)}^+(F)\big) \leq r_H(F)$, so by the preceding paragraph we have the equality $H_d\big(F, \Delta_{r_H(F)}^+(F)\big) = r_H(F)$, and $r_H(F)$ is indeed the minimum of all $H_d(F,S)$ for $S \in \mathcal{F}'(D)$. Thus item 3 holds.

Now for any $S \in \mathcal{F}'(D)$, we have $S \in \mathcal{M}_H(F)$ iff $H_d(F,S) = r_H(F)$, and as $H_d(F,S) \geq r_H(F)$ anyway, this holds iff $H_d(F,S) \leq r_H(F)$. By Proposition 8 with $r = r_H(F)$, this is equivalent to having both $S \subseteq \Delta_{r_H(F)}^+(F)$ and $F \subseteq \delta_{r_H(F)}(S)$. Hence item 2 holds.

We saw in the second paragraph (proof of item 3) that $H_d\big(F, \Delta_{r_H(F)}^+(F)\big) = r_H(F)$. Thus $\Delta_{r_H(F)}^+(F) \in \mathcal{M}_H(F)$, and as $S \subseteq \Delta_{r_H(F)}^+(F)$ for every $S \in \mathcal{M}_H(F)$, so $\Delta_{r_H(F)}^+(F)$ is the greatest element of $\mathcal{M}_H(F)$, and item 1 holds. □

In the case where F is proximinal, the above result takes exactly the same form as Theorem 24 of [ ] in the case of compact sets; the only difference is that we consider here $S \in \mathcal{F}'(D)$ instead of $S \in \mathcal{K}'(D)$.

For any $F \in \mathcal{F}'(E)$, we have two possibilities:

1. there is some $p \in F$ such that $d(p, D) = r_H(F)$;
2. for every $p \in F$, $d(p, D) < r_H(F)$.

In the first case, we say that F is r_H-*reached*, in the second that F is r_H-*unreached*. Note that every compact is r_H-reached, because the continuous function $d(\cdot, D)$ reaches a maximum on it.

The following two results illustrate the meaning of this distinction.

Proposition 9. *Define* $\Delta : \mathcal{P}'(E) \to \mathcal{P}(D)$ *by*

$$\Delta(X) = \bigcup_{K \in K'(X)} \Delta_H(K) \ .$$

Then for $F \in \mathcal{F}'(E)$ we have:

1. *if F is r_H-reached, $\Delta(F) = \Delta_{r_H(F)}(F)$;*
2. *if F is r_H-unreached, $\Delta(F) = \Delta^{\circ}_{r_H(F)}(F)$.*

Proof. 1) Take F r_H-reached. For $K \in K'(F)$, we have $r_H(K) \leq r_H(F)$; as $\Delta_r(X)$ is increasing both in r and in X, we get:

$$\Delta_H(K) = \Delta_{r_H(K)}(K) \subseteq \Delta_{r_H(F)}(K) \subseteq \Delta_{r_H(F)}(F) \ .$$

Thus $\Delta(F) \subseteq \Delta_{r_H(F)}(F)$. Now let $p \in F$ such that $d(p, D) = r_H(F)$. For any $x \in F$ we have $r_H(\{p, x\}) = r_H(F)$, and so

$$\Delta_{r_H(F)}(\{x\}) \subseteq \Delta_{r_H(F)}(\{p, x\}) = \Delta_H(\{p, x\}) \subseteq \Delta(F) \ .$$

But Δ_{r_H} is a dilation, so that

$$\Delta_{r_H(F)}(F) = \bigcup_{x \in F} \Delta_{r_H(F)}(\{x\}) \ ,$$

and the previous equation gives thus $\Delta_{r_H(F)}(F) \subseteq \Delta(F)$. The equality follows.

2) Take F r_H-unreached. For $K \in K'(F)$, K is r_H-reached, so $r_H(K) < r_H(F)$; as $\Delta_r(X) \subseteq \Delta^{\circ}_s(X)$ for $r < s$, and Δ°_s is increasing, we get:

$$\Delta_H(K) = \Delta_{r_H(K)}(K) \subseteq \Delta^{\circ}_{r_H(F)}(K) \subseteq \Delta^{\circ}_{r_H(F)}(F) \ .$$

Thus $\Delta(F) \subseteq \Delta^{\circ}_{r_H(F)}(F)$. Now for every $r < r_H(F)$ there is some $p \in F$ such that $d(p, D) \geq r$. For any $x \in F$ we have $r_H(\{p, x\}) \geq r$, and so

$$\Delta_r(\{x\}) \subseteq \Delta_r(\{p, x\}) \subseteq \Delta_H(\{p, x\}) \subseteq \Delta(F) \ .$$

But

$$\Delta_r(F) = \bigcup_{x \in F} \Delta_r(\{x\}) \ ,$$

so the previous equation gives $\Delta_r(F) \subseteq \Delta(F)$ for every $r < r_H(F)$. From the equality $\overset{\circ}{B}_{r_H(F)}(x) = \bigcup_{r < r_H(F)} B_r(x)$, we deduce easily from the definitions (7, 8, 26, 27) that

$$\delta^{\circ}_{r_H(F)}(X) = \bigcup_{r < r_H(F)} \delta_r(x) \quad \text{and} \quad \Delta^{\circ}_{r_H(F)}(X) = \bigcup_{r < r_H(F)} \Delta_r(X) \ ,$$

so that $\Delta^{\circ}_{r_H(F)}(F) \subseteq \Delta(F)$. The equality follows.   □

Proposition 10. For $F \in \mathcal{F}'(E)$,

1. $\overline{\Delta_{r_H(F)}(F)} \in \mathcal{M}_H(F)$.
2. If F is r_H-unreached, then $\overline{\Delta^\circ_{r_H(F)}(F)} \in \mathcal{M}_H(F)$.

Proof. Item 1 follows from Corollary 2. Suppose now that F is r_H-unreached. Thus $d(x, D) < r_H(F)$ for every $x \in F$. Take $x \in F$ and let $r = d(x, D)$; then $r < r_H(F)$ and there is some $p \in D$ such that $d(x, p) = r$; thus $x \in \overset{\circ}{B}_{r_H(F)}(p)$ and $p \in \Delta^\circ_{r_H(F)}(F)$. As this is true for any $x \in F$, we have $F \subseteq \delta_{r_H(F)}\big(\Delta^\circ_{r_H(F)}(F)\big)$. Hence $F \subseteq \delta_{r_H(F)}\big(\overline{\Delta^\circ_{r_H(F)}(F)}\big)$. As $\overline{\Delta^\circ_{r_H(F)}(F)} \subseteq \Delta^+_{r_H(F)}(F)$, applying Proposition 8, item 2 holds. □

We illustrate in Fig. 3 the distinction between $\Delta_{r_H(F)}(F)$ and $\Delta^\circ_{r_H(F)}(F)$ for a r_H-unreached closed set F.

When D is boundedly finite (as was assumed in [ ]), every subset of D is closed, so that $\Delta_r(F)$ is always closed in Corollary 2, while $\Delta_{r_H(F)}(F)$ and $\Delta^\circ_{r_H(F)}$ are always closed in Proposition 10, so we don't need to take the closure of these sets.

Finally, it is possible to extend our theory from $\mathcal{F}'(E)$ to $\mathcal{F}(E)$, in other words to consider Hausdorff sampling for the empty set (up to now, we always assumed that $F \in \mathcal{F}'(E)$). From (20) we can derive that:

$$\mathcal{M}_H(\emptyset) = \{\emptyset\}, \qquad \Delta_H(\emptyset) = \emptyset, \qquad \text{and} \quad r_H(\emptyset) = 0 . \tag{31}$$

Then all the above results extend trivially (and $\emptyset$ is r_H-unreached).

5   Conclusion

In [ ] we presented our theory for the discretization of nonvoid compact sets of a metric space (E, d) into a discrete subspace D. The possible discretizations of a nonvoid compact $K \subseteq E$ are all nonvoid compact $S \subseteq D$ minimizing the Hausdorff distance to K. Such discretizations were characterized, and compared with the morphological approach using discretization by dilation [ ].

Fig. 3. Let $E = \mathbf{R}^2$, $D = \mathbf{Z}^2$, with the Euclidean distance. The unbounded curve F is a closed set verifying $r_H(F) = \sqrt{2}/2$, and it is r_H-unreached. Filled circles designate discrete points in $\Delta^\circ_{r_H(F)}(F)$, and the hollow circle is a point in $\Delta_{r_H(F)}(F) \setminus \Delta^\circ_{r_H(F)}(F)$

This paper generalizes [ ] in two ways. First, the space D is supposed to be boundedly compact instead of boundeldy finite, so D is not necessarily "discrete". Second, the discretization applies to closed sets, not only non-void compact sets.

Apart from the fact that D is boundedly compact, we simply assumed that the covering radius is finite (i.e., the distance from a point of E to the nearest point of D is bounded). In [ ] this assumption was not necessary to prove the existence of Hausdorff discretization and to characterize it, but only to give a bound on the Hausdorff distance between a compact and its discretization, and so to derive the convergence of the discretization to the original compact set when the resolution tends to zero. Here in our extension to closed sets, this assumption of a finite covering radius is necessary for the existence of a Hausdorff discretization, because the Hausdorff distance between two closed sets can be infinite. However in practice one usually takes $E = \mathbf{R}^n$ and $D = \prod_{i=1}^{n}(\rho_i \mathbf{Z})$ (where ρ_i is the resolution or grid spacing along the i-th axis), so here the covering radius is finite.

This paper is part of an ongoing project [,] to analyse the discretization of objects (sets or images), and of operators transforming objects, in terms of the Hausdorff distance. In a next paper [ ], we will extend the analysis in [ ] of the Hausdorff metric properties of the morphological discretization by dilation [ ], and consider in detail the topologies associated to it.

References

1. A. J. Baddeley. Hausdorff metric for capacities. *CWI Report* BS-R9127, Dec. 1991. 259, 260
2. M. F. Barnsley. *Fractals Everywhere*. Academic Press, second edition, 1993. 253, 254, 259, 262
3. J. M. Borwein and S. Fitzpatrick. Existence of nearest points in Banach spaces. *Canadian Journal of Mathematics*, 51(4):702–720 (1989). 256
4. H. Busemann. *The Geometry of Geodesics*. Academic Press, New York, 1955. 253, 256
5. G. Choquet. *Topology*, Academic Press (1966). 253
6. F. Deutsch, and J. Lambert. On continuity of metric projections. *Journal of Approximation Theory*, 29:116–131 (1980). 256
7. S. Duval and M. Tajine. Digital geometry and fractal geometry. In D. Richard, editor, *Proc. 5th Digital Geometry and Computer Imagery (DGCI) Conference*, pp. 93–104, Clermont-Ferrand (France), 25–27 Sept. 1995. 260
8. H. J. A. M. Heijmans. *Morphological Image Operators*. Academic Press, Boston, 1994. 252, 256, 269, 270
9. J. G. Hocking and G. S. Young. *Topology*. Dover Publications Inc., New York, 1988. 253
10. C. Ronse and M. Tajine. Discretization in Hausdorff space. *Journal of Mathematical Imaging and Vision*, 12(3):219–242 (2000). 250, 251, 252, 253, 254, 262, 263, 264, 265, 266, 267, 269, 270
11. C. Ronse and M. Tajine. Hausdorff discretization for cellular distances, and its relation to cover and supercover discretizations. Accepted for publication in *Journal of Visual Communication and Image Representation*. 251, 270

12. C. Ronse and M. Tajine. Morphological sampling of closed sets, and related topologies. *In preparation.* 270

13. Bl. Sendov. *Hausdorff Approximations.* Kluwer Academic Publishers, 1990. 253

14. J. Serra. *Image Analysis and Mathematical Morphology.* Academic Press, London, 1982. 259

15. M. Tajine and C. Ronse. Preservation of topology by Hausdorff discretization, and comparison to other discretization schemes. Accepted for publication in *Theoretical Computer Science.* 250, 252

Cell Complexes and Digital Convexity

Julian Webster *

Dept. of Computing, Imperial College
London SW7 2BZ, UK
jw4@doc.ic.ac.uk

Abstract. Abstract cell complexes (ACC's) were introduced by Kovalevsky as a means of solving certain connectivity paradoxes in graph-
theoretic digital topology, and to this extent provide an improved theoretical basis for image analysis. We argue that ACC's are a very natural setting for digital convexity, to the extent that their use permits
simple, almost trivial formulations of major convexity results such as
Caratheodory's, Helly's and Radon's theorems. ACC's also permit the
use in digital geometry of axiomatic combinatorial geometries such as
oriented matroids. We give a brief indication of how standard convexity algorithms from computational geometry applied to the points of an
ACC can form a substantial part of digital convexity algorithms.

1 Introduction

Abstract cell complexes (ACC's) were introduced by Kovalevsky [ ] as a means
of solving certain connectivity paradoxes in graph-theoretic digital topology,
and to this extent provide an improved theoretical basis for image analysis.
A very similar approach is that of Khalimsky [ ]. As a datatype for digital
geometry, however, it seems that ACC's have hardly been considered, apart from
by Kovalevsky [ ]. (The basic definitions in this work are different to those in [ ].)
In this work we argue that ACC's are a very natural theoretical setting for digital
geometry, to the extent that they permit simple, almost trivial formulations of
major convexity results such as Caratheodory's, Helly's and Radon's theorems.

ACC's are non-homogeneous digital spaces in that they contain elements
of different dimension, including both points (0-cells) and pixels (2-cells). We
first develop a geometry of points of an ACC, which is nothing other than an
oriented matroid structure.[1] The geometry of cells, or "digital geometry", is then
constructed out of the point geometry. Computation of the point convex hull
operation can be done using standard algorithms from computational geometry,
and these algorithms can then play a substantial role in the computation of
digital convex hulls.

This work is part of an overall project to develop axiomatically defined digital geometric spaces. The need for work on the mathematical foundations of

---

* This work has been supported by the EPSRC project "Digital Topology and Geometry: an Axiomatic Approach with Applications to GIS and Spatial Reasoning."

[1] No knowledge of oriented matroids is assumed here, but the standard reference is [ ]

G. Bertrand et al. (Eds.): Digital and Image Geometry, LNCS 2243, pp. 272–282, 2001.
© Springer-Verlag Berlin Heidelberg 2001

digital geometry has been argued for by, among others, Françon [ ], Klette [ ], Smyth [ ], Voss [ ]. One of the main intended applications (and indeed motivations) of axiomatic digital geometry is the development of robust geometric algorithms. It is a fundamental issue in computational geometry that $\mathbb{R}^n$ is often the wrong theoretical medium with which to describe the behaviour of geometric algorithms, in that the mismatch between the infinite precision of $\mathbb{R}^n$ and the finite precision of computation can cause rounding errors. Our overall aim is to replace $\mathbb{R}^n$ by equally valid - meaning, roughly, axiomatically defined spaces that have a geometry rich enough to describe and reason about geometric computation - but discrete geometric spaces. The axiomatic foundations of computational geometry is the subject of Knuth's monograph [ ]. Knuth's "CC systems" are a special class of oriented matroids.

For simplicity we will only present a theory of 2-dimensional digital spaces here. The theory does generalize quite smoothly to arbitrary dimensions, however, and this work is presented in [ ].

2   Digital Planes

For any $n \geq 1$, let I_n denote the set $\{0, \ldots, n\}$. The following definition is a model of Kovalevsky's axioms [ ].

Definition 1. *A 2-dimensional Cartesian ACC consists of a point set $I_n \times I_m$ together with the following subsets, called* digital cells:

- *0-cells are the singleton subsets;*
- *1-cells are the subsets $\{(x, y), (x, y + 1)\}$ and $\{(x, y), (x + 1, y)\}$;*
- *2-cells are the subsets $\{(x, y), (x, y + 1), (x + 1, y), (x + 1, y + 1)\}$.*

We will here call such structures *digital planes*. When the context is clearly that of a digital plane, digital cells will often be referred to simply as *cells*. The definition is in fact a description of a particular model of Kovalevsky's axioms [ ], one which we find convenient to use here. Our choice of description is based on the intuition that an ACC is an abstract version of the classical notion of a cell complex in combinatorial topology. We want to think of a cell as a polytope in an abstract point geometry, and an ACC as a tiling of cells.

To see how the definition formally satisfies the ACC axioms, let $\mathcal{D}$ denote the set of digital cells in a digital plane, and for any n-cell D let $dim(D) = n$. Then, where $\subset$ denotes the relation of proper subset inclusion on $\mathcal{D}$, clearly $(\mathcal{D}, \subset, dim)$ is an ACC exactly according to the definition in [ ]. It is also very straightforward to check that it is a 2-dimensional Cartesian ACC according to [ ], although our coordinate system is different to the one considered in that work. We choose to think of only the 0-cells as having coordinates, as these are the only digital cells that represent points rather than subsets of the Euclidean plane. Moreover, Kovalevsky's definitions of digital half-spaces and convex digital sets are expressed in terms of his choice of coordinates, but our definitions are different.

A cell C is a *face* of a cell D if $C \subseteq D$. A *vertex* of a cell is any of its faces that is a 0-cell. A *digital set* is any set of cells. By abuse of terminology we will often identify 0-cells with points.

The points in a digital plane will be considered as points in $\mathbb{R}^2$ in the obvious way. Digital cells are then considered as indexing subsets of $\mathbb{R}^2$ as follows. We say that the *continuous part* of a digital cell D is the relative interior of the polytope in $\mathbb{R}^2$ whose vertices are the vertices of D. In other words:

- The continuous part of the 0-cell $\{p\}$ is $\{p\}$
- The continuous part of the 1-cell $\{p, q\}$ is the open interval in $\mathbb{R}^2$ between p, q;
- The continuous part of the 2-cell $\{p, q, r, s\}$ is the interior in $\mathbb{R}^2$ of the filled-in square with vertices p, q, r, s.

Definition 2. *In any digital plane, the* digital image *of $P \subseteq \mathbb{R}^2$ is the set of all digital cells whose continuous part intersects P.*

Example 3. The diagram on the right shows the digital image of the shaded Euclidean set on the left. The diagram is supposed to represent a collection of digital cells, but it is far clearer to draw a digital cell as its continuous part, and we use this convention throughout.

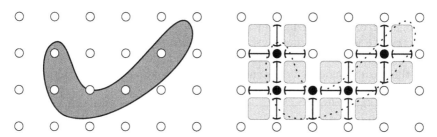

That each point in a digital plane has integer coordinates allows for robust computation, but might seem unnecessarily restrictive in that it appears to fix the resolution. This is not so, however, as resolution can be adjusted using a simple scaling operation. Implicit in the definitions above is the embedding $x \mapsto x$ of the point set E of a digital plane into $\mathbb{R}^2$, but for a different resolution we consider an embedding $e : E \to \mathbb{R}^2$, $x \mapsto \lambda x$. We then think of the continuous part of a digital cell D as the relative interior of the polytope with vertices $e(D)$, and digital images are defined accordingly. All the results concerning digital images given here are true for whatever resolution is considered; a more formal account is given in [ ].

2.1   Point Geometry

In this section we forget about the cells in a digital plane and consider only the points. The point geometry of a digital plane is simply the geometry its point set E inherits as a subset of $\mathbb{R}^2$. We say that:

- A *straight line* in E is the intersection with E of any straight line L in $\mathbb{R}^2$ that contains ≥ 2 points of E;
- The *open half-planes* determined by a straight line $L \cap E$ are the intersections with E of the open half-planes determined by L in $\mathbb{R}^2$.

This construction gives the structure of an *oriented matroid* on E (straight lines are hyperplanes and pairs of open half-planes are cocircuit orientations). The point geometry can therefore be regarded as a highly developed axiomatic structure in combinatorics. The point geometry is also very close to being a "CC-System" introduced by Knuth [ ], except that Knuth does not allow triples of collinear points.

The open half-planes determined by a straight line $L \subseteq E$ are denoted L^+, L^-. A *closed half-plane* is the union of a straight line with either of its open half-planes. The closed half-plane $L \cup L^+$ is denoted L^{0+}. The *convex hull*, denoted $[T]$, of a point set $T \subseteq E$ is the intersection of all the closed half-planes in E that contain T. Let $[T]_{\mathbb{R}}$ denote the convex hull of T in $\mathbb{R}^2$. The following result is a corollary of Proposition 10.

Proposition 4. $[T] = [T]_{\mathbb{R}} \cap E.$

2.2   Digital Geometry

The digital geometry of a digital plane is constructed entirely in terms of its point geometry. We will show that this geometry then reflects Euclidean geometry to the extent that digital convex sets are digital images of Euclidean convex sets. Once this correspondence has been established it is almost trivial to prove digital versions of major convexity theorems.

To distinguish clearly between the point and the digital geometries of a digital plane, we will sometimes speak of *point half-planes, convex point sets* for the point geometry, and *digital half-planes, convex digital sets* for the digital geometry. To distinguish between these definitions and those of classical geometry, we often refer to half-planes, convex sets in $\mathbb{R}^2$ as *Euclidean half-planes, convex Euclidean sets*.

Definition 5. *The* closed digital half-plane *determined by a closed point half-plane* L^{0+} *is the set of all digital cells D such that $D \subseteq L$ or $D \cap L^+ \neq \emptyset$.*

This definition and our subsequent definition of "digital convex hull" are different to those in [ ].

Example 6. The diagram on the left shows a straight line (the set of crosses) and one of its open half-planes. The diagram on the right shows the determined closed digital half-plane.

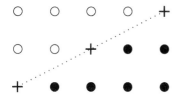 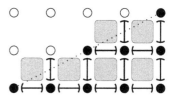

From the diagram on the right we can see, informally, that the closed digital half-plane is the digital image of the corresponding closed Euclidean half-plane. Closed digital half-planes could therefore have been *defined* as digital images of (certain) closed Euclidean half-planes, but we want to emphasize that once the point geometry of a digital plane has been set up, the digital geometry can be constructed entirely in terms of this point geometry, with no further reliance on Euclidean space.

Closed point half-spaces are convex point sets that are sufficiently well-behaved to allow their digital structure to be derived directly from their point structure. But not all convex sets have this property:

Example 7. The set of dark points in the diagram is a convex point set. We want the corresponding digital convex set to be the digital image of the Euclidean convex hull. But that the 2-cell drawn lies in this image cannot be ascertained merely by consideration of its vertices, as none of these lie in the convex point set.

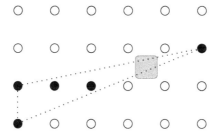

The example is meant to motivate the following definition, in which we use the abuse (which comes from identifying 0-cells with points) that a point set T is a *subset* of a digital set $\mathcal{D}$ if, for each $x \in T$, the 0-cell $\{x\} \in \mathcal{D}$.

Definition 8. *The* digital convex hull *of a point set T is the intersection of all the closed digital half-planes that contain T.*

Example 9. The diagram shows the intersection of three closed digital half-planes, which is the digital convex hull of the point set $\{a, b, c\}$ *together* with the 2-cell C. This cell must be separated from the digital convex hull using a further closed half-plane.

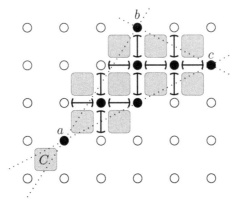

The point of the example is to illustrate the sort of problem that may arise in developing a theory of digital convexity. Using Proposition 4 is it simple to prove that the point convex hull operation has the basic, fundamental properties of a convex hull operation in axiomatic convexity theory - see [ ], for example. In particular, convex hulls and convex sets are interchangeable: a point set T is convex if $T = [T]$, and the point convex hull of a set is then the smallest convex set that contains it. Digital convexity, on the other hand, is not so well-behaved, as the collection of convex digital sets is not even going to be closed under intersections. The intersection of the three closed digital half-planes in the example is not the digital image of any Euclidean convex set, and so is probably not what we would want to count as a convex digital set.

Proposition 10. *The digital convex hull of a point set T is the digital image of* $[T]_\mathbb{R}$.

Proof. Let L^{0+} be any closed point half-plane, and let M^{0+} be the closed Euclidean half-plane such that $L = M \cap E$ and $L^+ = M^+ \cap E$, where E is the point set of the digital plane in question. Let D be any digital cell and let C denote its continuous part. From the theory of Euclidean support hyperplanes – see [ ], for example – we have that $C \cap M^{0+} = \emptyset$ if and only if $D \subseteq M^{0-}$ and $D \not\subseteq M$. The closed digital half-plane determined by L^{0+} is therefore the digital image of M^{0+}. It follows easily that the digital image of $[T]_\mathbb{R}$ is contained in the digital convex hull of T. For the converse, suppose D is not in the digital image of $[T]_\mathbb{R}$. Say that a closed Euclidean half-space M^{0+} is an E-half-space if M contains ≥ 2 points of E. To show that D is not in the digital convex hull of T we need to find an E-half-space M^{0+} that contains T such that $C \subseteq M^-$. It is simple to prove (by considering its facets) that $[T]_\mathbb{R}$ is the intersection of a collection $M_1^{0+}, \ldots, M_k^{0+}$ of E-half-spaces. Now C and $[T]_\mathbb{R}$ are disjoint, so by Helly's theorem (which says that if the intersection of a collection of convex sets in the Euclidean plane is empty, then the intersection of some three of them is empty), there exist i, j such that $C \cap M_i^{0+} \cap M_j^{0+} = \emptyset$. If either $C \cap M_i^{0+} = \emptyset$ or $C \cap M_j^{0+} = \emptyset$ we are done, which is so when D is a 0-cell. Assuming otherwise, if D is a 1-cell then there is some E-half-space parallel to the straight line through D that suffices. If D is a 2-cell then there exist E-half-spaces $N_1^{0-}, \ldots, N_4^{0-}$ such that C is the intersection of $N_1^-, \ldots, N_4^-$. Then by Helly's theorem there is some l such that $N_l^- \cap M_i^{0+} \cap M_j^{0+} = \emptyset$, so then $T \subseteq N_l^{0+}$.

The result but not the proof given here generalizes to higher dimensions. A more coherent proof (which there is not room for here) of the n-dimensional result is given in [ ].

Proposition 11 (Caratheodory's theorem). *For any point set T, $[T]$ is the union of the convex hulls of the subsets of T having cardinality ≤ 3, and the digital convex hull of T is the union of digital convex hulls of such subsets.*

Proof. Given Caratheodory's theorem on $\mathbb{R}^2$, the first part of the result follows easily from Proposition 4 and the second from Proposition 10.

Proposition 12 (Radon's theorem). *Any point set T that contains ≥ 4 points admits a partition into two sets whose respective digital convex hulls intersect.*

Proof. By Radon's theorem for $\mathbb{R}^2$, T can be partitioned into two sets T_1, T_2 whose respective Euclidean convex hulls intersect. For any point x that lies in this intersection, let D be the digital cell whose continuous part contains x. Then D lies in the digital convex hulls of both T_1, T_2.

Proposition 13 (Helly's theorem). *Let $\mathcal{T}$ be any collection of convex point sets. If the intersection of any three members of $\mathcal{T}$ is non-empty then the intersection of all the digital convex hulls of members of $\mathcal{T}$ is non-empty.*

Proof. By Helly's theorem for $\mathbb{R}^2$, the intersection of the collection of Euclidean convex hulls of members of $\mathcal{T}$ is non-empty. For any point x that lies in this intersection, let D be the digital cell whose continuous part contains x. Then D lies in the intersection of all the digital convex hulls of members of $\mathcal{T}$.

The following example indicates the advantage of ACC's over graphs with regard to Helly's theorem.

Example 14. In graph-theoretic digital topology the digital plane is composed entirely of pixels. A pixel is in effect 2-dimensional, and for digitization purposes is regarded as a square in the Euclidean plane. The diagram on the left shows the digital image[2] of the filled-in Euclidean triangle with vertices $(0,0), (0,4), (4,0)$. This image is a digital convex set according to any of the standard definitions discussed, for example, in [ ]. Rotate the image through 90 degrees successively to get four digital convex sets: the intersection of any three of these is non-empty,

---

[2] The digital image being the set of all squares (pixels) whose *interior* intersects the triangle. There are of course several different "digitization" definitions in the literature, but the point is that the digital triangle is uncontroversially a convex set in graph-theoretic digital topology.

but no pixel lies in the intersection of all four. The diagram on the right shows the digital image in an ACC of the same filled-in triangle, and the intersection of the four rotations contains the middle point.

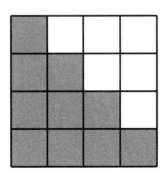 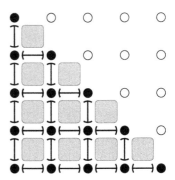

3   Computation of Digital Convex Hulls

In this section we give a brief indication of how a digital convex hull algorithm can be derived from the convex hull algorithm in computational geometry. We only go so far as to outline a computation of whether a digital cell is in the digital convex hull of a point set; our aim here is merely to indicate a potential tie-up between computational and digital geometry. There is perhaps a potential parallel to be made between this tie-up and the work of Winter [ ] of using ACC's in geographical information systems to resolve the so-called "raster/vector" (pixel/point) debate.

We begin with digital straight lines. Let p, q be distinct points of a digital plane and let L be the unique point straight line that contains them. We say that the *digital line* through p, q is the intersection of the digital closed half-spaces determined by L^{0+}, L^{0-}, which is the set of all digital cells D such that $D \subseteq L$ or D intersects both L^+, L^-.

Proposition 15. *The digital line determined by p, q is the digital image of the straight line through p, q in $\mathbb{R}^2$.*

Proof. ¿From Proposition 10 we have that the digital image of the Euclidean line through p, q is a subset of the digital line through p, q. Conversely, if D is a digital cell in the digital line through p, q then its continuous part C is either a subset of the Euclidean line through p, q or intersects both the half-spaces of the Euclidean line. In the latter case C then intersects the Euclidean line because C is convex.

Computation of whether or not a digital cell D lies on the digital line through p, q therefore amounts to computation of which side of L each of its vertices lies. A standard method (used, for example, in [ ]) is calculation of the determinant of the relevant matrix: where $p = (x_1, y_1)$ and $q = (x_2, y_2)$, on which side of L the point $r = (x_3, y_3)$ lies is given by the sign of the determinant of

$$\begin{pmatrix} x_1 & y_1 & 1 \\ x_2 & y_2 & 1 \\ x_3 & y_3 & 1 \end{pmatrix}$$

i.e. L, L^+, L^- are the sets of points for which the determinant is, respectively, zero, positive, negative.

We say that the *digital straight line segment* between p, q is the digital convex hull of the set $\{p, q\}$. Clearly, this is the set of all digital cells that lie on the digital straight line through p, q each of whose vertices r is such that $min\{x_1, x_2\} \leq x_3 \leq max\{x_1, x_2\}$ and $min\{y_1, y_2\} \leq y_3 \leq max\{y_1, y_2\}$.

Digital line segments can be used in a simple computation of whether a digital cell D lies in the digital convex hull of a point set T. It is a standard result that $[T]_\mathbb{R}$ is the disjoint union of its topological interior and its topological boundary.

Proposition 16. *A digital cell D is in the digital image of $[T]_\mathbb{R}$ if and only if it satisfies at least one of the conditions:*

1. *Some vertex of D lies in the interior of $[T]_\mathbb{R}$;*
2. *Every vertex of D lies in the boundary of $[T]_\mathbb{R}$;*
3. *Some point in the continuous part of D lies in the boundary of $[T]_\mathbb{R}$.*

Proof. ($\Leftarrow$): If D has a vertex in the interior of $[T]_\mathbb{R}$ then there is some neighbourhood of the vertex that is a subset of $[T]_\mathbb{R}$, and this neighbourhood intersects the continuous part of D. If every vertex of D is in $[T]_\mathbb{R}$ then its continuous part is a subset of $[T]_\mathbb{R}$. ($\Rightarrow$): If D is in the digital image of $[T]_\mathbb{R}$ then the continuous part of D intersects the interior or the boundary. If it does not intersect the boundary then it is a subset of the interior, and then every vertex of D is in $[T]_\mathbb{R}$.

The boundary of $[T]_\mathbb{R}$ can be computed using a convex hull algorithm from computational geometry. We refer to Knuth's convex hull algorithm [ ], which in effect will, given input T, return the list $v_1, \ldots, v_k, v_{k+1} = v_1$ of vertices of $[T]_\mathbb{R}$ such that, for each i, the straight line segment between v_i, v_{i+1} is a facet of $[T]_\mathbb{R}$. Each $v_i \in T$, and the boundary of $[T]_\mathbb{R}$ is the union of its facets, so whether a digital cell satisfies the third condition of the result amounts to whether it lies on the digital straight line segment between some v_i, v_{i+1}.

The second condition in the result can also be determined by consideration of digital straight line segments between the vertices v_i, v_{i+1}.

Whether a vertex x of a digital cell lies in the interior of $[T]_\mathbb{R}$ amounts, for example, to whether, for all i, x lies on the same side of the point line through v_i, v_{i+1} as each of the other vertices v_j, which can be computed using determinants as above.

4   Conclusions and Further Work

We have argued that ACC's are a natural setting for digital convexity in that they allow simple digital formulations of classical convexity theorems and allow a link between digital geometry and existing axiomatic finite point geometries such as oriented matroids. We have given a brief indication of how computation in digital geometry can be derived from standard methods in computational geometry.

This work is really only the beginning of a theory of digital convexity for ACC's. Our version of Helly's theorem could perhaps be strengthened (although it is analogous to the version of Helly's theorem for oriented matroids - see Theorem 9.2.1 in [ ]); we don't know whether, for example, the following stronger result holds.

Proposition 17. *Let S be any collection of point sets in a digital plane such that the intersection of the digital convex hulls of any three is non-empty. Then the intersection of the digital convex hulls of all the sets in S is non-empty.*

At any rate, for general theory of computation over $\mathbb{R}^n$ we need to consider digital convex hulls of sets of cells rather than just sets of points as considered here, which we illustrate with the following example. Let p, q be distinct points of a digital plane and suppose we want to determine whether a point $x \in \mathbb{R}^2$ lies on the Euclidean straight line segment L between them. It is well-known that this cannot necessarily be decided using any finite number of decimal places of x, so computation of the answer might not be robust. According to our definition of digital images, the point x is represented by the digital cell D whose continuous part contains x. The question therefore amounts, in this context, to whether D lies in the digital straight line segment between p, q. Uncertainty of computation is then expressed by saying that if D is not on this line segment then definitely $x \notin L$, and if D is on this line segment then possibly $x \in L$. But we have not in this work considered the more general case of when p, q are arbitrary points of $\mathbb{R}^2$, to be represented by digital cells D_p, D_q that are not necessarily 0-cells. In this case the question amounts to whether D lies on the "digital convex hull" of D_p, D_q, which is an operation not defined here. One possible definition is to consider the union of the sets of vertices of D_p, D_q and consider the digital convex hull of this point set.

References

1. A. Bjorner, M. Las Vergnas, B. Sturmfels, N. White, and G. Ziegler, *Oriented Matroids,* Cambridge University Press, 1993. 272, 281
2. J. Françon, *On recent trends in discrete geometry in computer science,* Discrete geometry for Computer imagery, 6th international workshop (A. Montanvert S. Miguet and S. Ubeda, eds.), LNCS 1176, Springer, 1996. 273
3. W. A. Coppel, *Foundations of convex geometry,* Cambridge University Press, 1998. 277
4. R. Kopperman, E. Khalimsky and P. Meyer, *Computer graphics and connected topologies an finite ordered sets,* Topology and its Applications 36 (1990), 1-17. 272
5. C. E. Kim, *On the cellular convexity of complexes,* IEEE PAMI 3 (1981), no. 6, 617-625. 278

6. R. Klette, *Digital geometry - the birth of a new discipline,* Tech. report, University of Auckland, 2001. 273
7. D. E. Knuth, *Axioms and hulls,* Lecture Notes in Computer Science 606, SpringerVerlag, 1991. 273, 275, 279, 280
8. V. A. Kovalevsky, *Finite topology as applied to image analysis,* Computer vision, graphics, and image processing 46 (1989), 141-161. 272, 273
9. V. A. Kovalevsky, *A new concept for digital geometry,* Shape in Picture: Mathematical Description of Shape in Grey-level Images (Ying lie O et al, ed.), Springer-Verlag, 1992, pp. 37-51. 272, 273, 275
10. M. B. Smyth, *Region-based discrete geometry,* Journal of Universal Computer Science 6 (2000), no. 4, 447-459. 273
11. K. Voss, *Discrete images, objects, and functions in Z^n,* Springer-Verlag, 1993. 273
12. J. Webster, *Cell complexes, oriented matroids and digital geometry,* Submitted to Theoretical Computer Science, 2001. 273, 274, 278
13. R. Webster, *Convexity,* Oxford University Press, 1994. 277
14. S. Winter, *Bridging Vector and Raster Representations in GIS,* Proceedings of the 6th International Symposium an Advances in GIS (R. Laurini et al, ed.), ACM Press, 1998, pp. 57-62. 279

Part IV

Multigrid Convergence

Approximation of 3D Shortest Polygons in Simple Cube Curves

Thomas Bülow[1] and Reinhard Klette[2]

[1] Department of Computer and Information Science
University of Pennsylvania, Philadelphia, USA
[2] CITR, University of Auckland
Tamaki Campus, Building 731, Auckland, New Zealand

Abstract. One possible definition of the length of a digitized curve in 3D is the length of the shortest polygonal curve lying entirely in a cube curve. In earlier work the authors proposed an iterative algorithm for the calculation of this minimal length polygonal curve (MLP). This paper reviews the algorithm and suggests methods to speed it up by reducing the set of possible locations of vertices of the MLP, or by directly calculating MLP-vertices in specific situations. Altogether, the paper suggests an in-depth analysis of cube curves.

1 Introduction

The analysis of digital curves in 3D space is of increasing practical relevance in volumetric image data analysis, see cited examples of applications in [,]. A digital curve is the result of a process (path on a 3D surface, 3D thinning etc.) which maps captured 'curve-like' objects into well-defined digital curves (see definition below). The length of such a simple digital curve may be defined based on local chain-code configurations in digital space [ ], or on (global) curve approximations in Euclidean space. The latter approach has the potential of multi-grid convergence of estimated length toward the true length assuming that a digitization model such as cube intersection [ ] is used for obtaining test data by digitizing curves of known length. For global length estimation algorithms there are (at least) two options, the use of

1. the length of a 3D NSS (naive straight segment) approximation [ ] of a 26-connected curve, or
2. the length of the 3D MLP (minimum-length polygonal) curve fully contained and complete in the tube of a simple 6-connected closed curve [,].

In earlier work the authors started with studying an algorithmic solution for the 3D MLP approach and proved that the only possible positions of MLP-vertices are the so-called *critical edges* which are incident with three cubes of the simple cube-curve [ ], and presented an iterative algorithm for approximating the MLP of a simple cube-curve in 3D [ ]. It is expected [ ] that the length estimation behavior of 3D NSS compares to 3D MLP similarly as that of DSS to MLP

G. Bertrand et al. (Eds.): Digital and Image Geometry, LNCS 2243, pp. 285–298, 2001.

in the two-dimensional case [ ]. However, to support such experiments we have to continue to improve the time complexity of the 3D MLP approximation as specified so far in [ ] for making statistically relevant performance evaluations feasible.

The difficulty of the subject may be illustrated by the fact that the *Euclidean shortest path problem* (given a finite collection of polyhedral obstacles in 3D space, and a source and a target point, find a shortest obstacle-avoiding path from source to target) is known to be NP-hard [ ]. However, there are polynomial-time algorithms solving the *approximate Euclidean shortest path problem* in 3D, see [ ]. Our iterative algorithm in [ ] is not yet known to be always convergent to the exact 3D MLP, or whether it may only approximate in well-defined cases, up to some error etc. the correct MLP. All our experiments so far suggest that the algorithm presented in [ ] is always convergent to the correct 3D MLP, and time measurements also support the hypothesis that its run-time behavior is asymptotically linear in the number of input cubes even if a very small threshold is used for termination of the algorithm.

In this paper we review this algorithm and discuss three ways to speed it up: 1) Not all critical edges contain MLP-vertices. We identify a subset of these irrelevant critical edges which can then be excluded from further calculation. 2) Certain critical edges can be identified which can only contain MLP-vertices – if at all – at their end-points. 3) The positions of MLP-vertices belonging to flat arcs of the MLP can under certain conditions be calculated in closed form. Therefore the present paper may also be seen as a step towards an in-depth analysis of the geometry of 3D MLP's in simple cube curves.

Although we are not able yet to provide a closed form solution for a 3D MLP algorithm of a given simple cube-curve, these three steps lead in this direction by replacing parts of the iterative procedure by direct computation.

In the following section the basic notions are introduced. We define the length of a simple cube-curve. In Section 3 we summarize our previous algorithm for the calculation of this length. In Section 4 the three items mentioned above are elaborated.

2   The Length of Simple Cube Curves

We start with the definition of simple cube curves, see Fig. 1 for two examples. Any grid point $(i, j, k) \in \mathbb{R}^3$ is assumed to be the center point of a *grid cube* with *faces* parallel to the coordinate planes, with *edges* of length 1, and *vertices* at its corners. *Cells* are either cubes, faces, edges or vertices. The intersection of two cells is either empty or a joint *side* of both cells. A *cube-curve* is a sequence $g = (f_0, c_0, f_1, c_1, ..., f_n, c_n)$ of faces f_i and cubes c_i, for $0 \leq i \leq n$, such that faces f_i and f_{i+1} are sides of cube c_i, for $0 \leq i \leq n$ and $f_{n+1} = f_0$. It is *simple* iff $n \geq 4$, and for any two cubes c_i, c_k in g with $|i - k| \geq 2 \ (mod \ n)$ it holds that if $c_i \cap c_k \neq \emptyset$ then either $|i - k| = 2 \ (mod \ n)$ and $c_i \cap c_k$ is an edge, or $|i - k| = 3 \ (mod \ n)$ and $c_i \cap c_k$ is a vertex. A *tube* $\mathbf{g}$ is the union of all cubes contained in a cube-curve g. It is a polyhedrally-bounded compact set in $\mathbb{R}^3$,

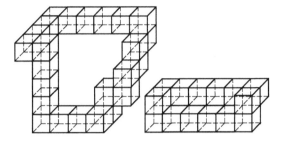

Fig. 1. Two cube-curves in 3D space

and it is homeomorphic with a torus in case of a simple cube-curve. The cube-curve on the left of Fig. 1 is simple, and the cube-curve on the right is not. Analogously, *edge-curves* or *face-curves* may be defined in 3D space. This paper deals exclusively with simple cube-curves. A curve $\mathcal{P}$ in 3D Euclidean space is *complete in* **g** iff it has a non-empty intersection with any cube contained in g. Following [,], the *length* of a simple cube-curve g is defined to be the length $l(\mathcal{P})$ of a shortest polygonal simple curve $\mathcal{P}$ which is contained and complete in tube **g**. A simple cube-curve g is *flat* iff the center (grid) points of all cubes contained in g are in one plane parallel to one of the coordinate planes.

A non-flat simple cube-curve in $\mathbb{R}^3$ specifies exactly one minimum-length polygonal simple curve (MLP) which is contained and complete in its tube [ ]. The MLP is not uniquely specified in flat simple cube-curves. Flat simple cube-curves may be treated as square-curves in the plane, and square-curves in the plane are extensively studied, see, e.g. [ ]. It seems there is no straightforward approach to extend known 2D MLP algorithms to the 3D case.

3   The Iterative Algorithm

This section contains fundamentals used in our algorithm presented in the conference paper [ ] for calculating the length of a simple cube-curve. Let g be a simple cube-curve, and $\mathcal{P} = (p_0, p_1, ..., p_m)$ be a polygonal curve complete and contained in **g**, with $p_0 = p_m$.

Lemma 1. *It holds $m \geq 3$ for any polygon $\mathcal{P} = (p_0, p_1, ..., p_m)$ complete and contained in a simple cube-curve.*

The case $m = 3$ is possible. During a traversal along the curve $\mathcal{P}$ we leave cubes, and we enter cubes. The traversal is defined by the starting vertex p_0 of the curve and the given orientation. Let $\mathcal{C}_{\mathcal{P}} = (c_0, c_1, ..., c_n)$ be the sequence of cubes in the order how they are entered during this curve traversal. Because $\mathcal{P}$ is complete and contained in **g** it follows that $\mathcal{C}_{\mathcal{P}}$ contains all cubes of g, and no further cubes are in g.

Lemma 2. *For an MLP $\mathcal{P}$ of a simple cube-curve g it holds that $\mathcal{C}_\mathcal{P}$ contains each cube of g just once.*

Now we consider a special transformation of polygonal curves. Let $\mathcal{P} = (p_0, p_1, ..., p_m)$ be a polygonal curve contained in a tube **g**. A polygonal curve $\mathcal{Q}$ is a **g**-*transform* of $\mathcal{P}$ iff $\mathcal{Q}$ may be obtained from $\mathcal{P}$ by a finite number of steps, where each step is a replacement of a triple a, b, c of vertices by a polygonal sequence $a, b_1, ..., b_k, c$ such that the polygonal sequence $a, b_1, ..., b_k, c$ is contained in the same set of cubes of g as the polygonal sequence a, b, c. The case $k = 0$ characterizes the deletion of vertex b, the case $k = 1$ characterizes a move of vertex b within **g**, and cases $k \geq 2$ specify a replacement of two straight line segments by a sequence of $k + 1$ straight line segments, all contained in **g**.

Lemma 3. *Let $\mathcal{P}$ be a polygonal curve complete and contained in the tube **g** of a simple cube-curve g such that $\mathcal{C}_\mathcal{P}$ is without repetitions of cells. Then it holds that any **g**-transform of $\mathcal{P}$ is also complete and contained in **g**.*

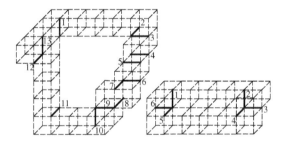

Fig. 2. Critical edges of two cube-curves

An edge contained in a tube **g** is *critical* iff this edge is the intersection of three cubes contained in the cube-curve g. Figure 2 illustrates all critical edges of the cube-curves shown in Fig. 1. Note that simple cube-curves may only have edges contained in three cubes at most. For example, the cube-curve consisting of four cubes only (note: there is one edge contained in four cubes in this case) was excluded by the constraint $n \geq 4$. Based on these lemmata it was possible to prove the following theorem [ ]:

Theorem 1. *Let g be a simple cube-curve. Critical edges are the only possible locations of vertices of a shortest polygonal simple curve contained and complete in tube **g**.*

Note that this theorem also covers flat simple cube-curves with a straightforward corollary about the only possible locations of MLP vertices within a simple square-curve in the plane: such vertices may be convex vertices of the inner

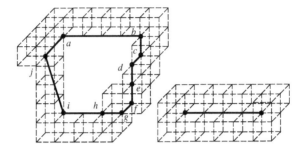

Fig. 3. Curve initializations ('clockwise')

frontier or concave vertices of the outer frontier only because these are the only vertices incident with three squares of a simple square-curve.

Our algorithm is based on the following model: Assume a rubber band is laid through the tube **g**. Letting it move freely it will contract to the MLP which is contained and complete in **g**. The algorithm consists of two subprocesses: at first **(A)** an initialization process defining a simple polygonal curve $\mathcal{P}_0$ contained and complete in the given tube **g** and such that $\mathcal{C}_{\mathcal{P}_0}$ contains each cube of g just once (see Lemma 2), and second **(B)** an iterative process (a **g**-transform, see Lemma 3) where each completed run transforms $\mathcal{P}_t$ into $\mathcal{P}_{t+1}$ with $l(\mathcal{P}_t) \geq l(\mathcal{P}_{t+1})$, for $t \geq 0$. Thus the obtained polygonal curve is also complete and contained in g.

(A) The initial polygonal curve will only connect vertices which are end points of consecutive critical edges. For curve initialization, we scan the given curve until the first pair (e_0, e_1) of consecutive critical edges is found which are not parallel or, if parallel, not in the same grid layer (see Fig. 2 (right) for a non-simple cube-curve showing that searching for a pair of non-coplanar edges would be insufficient in this case). For such a pair (e_0, e_1) we start with vertices (p_0, p_1), p_0 bounds e_0 and p_1 bounds e_1, specifying a line segment $p_0 p_1$ of minimum length (note that such a pair (p_0, p_1) is not always uniquely defined). This is the first line segment of the desired initial polygonal curve $\mathcal{P}_0$.

Now assume that $p_{i-1} p_i$ is the last line segment on this curve $\mathcal{P}_0$ specified so far, and p_i is a vertex which bounds e_i. Then there is a uniquely specified vertex p_{i+1} on the following critical edge e_{i+1} such that $p_i p_{i+1}$ is of minimum length. Length zero is possible with $p_{i+1} = p_i$. In this case we skip p_{i+1}, i.e. we do not increase the value of i. Note that this line segment $p_i p_{i+1}$ will always be included in the given tube because the centers of all cubes between two consecutive critical edges are collinear.

The process stops by connecting p_n on edge e_n with p_0 (note that it is possible that a minimum-distance criterion for this final step may actually prefer a line between p_n and the second vertex bounding e_0, i.e. not p_0). This initialization process calculates a polygonal curve $\mathcal{P}_0$ which is always contained and complete in the given tube.

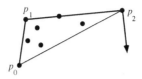

Fig. 4. Intersection points with edges

(**B**) In this iterative procedure we move pointers addressing three consecutive vertices of the (so far) calculated polygonal curve around the curve, until a completed run $t + 1$ does only lead to an improvement which is below an a-priori threshold τ, i.e. $l(\mathcal{P}_t) - \tau < l(\mathcal{P}_{t+1})$. In all our experiments the algorithm converges fast for a practically reasonable value of τ.

Assume a polygonal curve $\mathcal{P}_t = (p_0, p_1, ..., p_m)$, and three pointers addressing vertices at positions $i - 1$, i, and $i + 1$ in this curve. There are three different options that may occur which define a specific **g**-transform.

(**O$_1$**) Point p_i can be deleted iff $p_{i-1}p_{i+1}$ is a line segment within the tube. Then subsequence (p_{i-1}, p_i, p_{i+1}) is replaced in our curve by (p_{i-1}, p_{i+1}). In this case we continue with vertices $p_{i-1}, p_{i+1}, p_{i+2}$. (**O$_2$**) The closed triangular region $\triangle(p_{i-1}p_ip_{i+1})$ intersects more than just the three critical edges of p_{i-1}, p_i and p_{i+1} (see Fig. 4), i.e. simple deletion of p_i would not be sufficient anymore. This situation is solved by calculating a convex arc (note: a convex polygon is the shortest curve encircling a given finite set of planar points [ ]) and by replacing point p_i by the sequence of vertices $q_1, ..., q_k$ on this convex arc between p_{i-1} and p_{i+1} iff the sequence of line segments $p_{i-1}q_1, ..., q_kp_{i+1}$ lies within the tube. Because the vertices are ordered we may use a fast linear-time convex hull routine in case of (**O$_2$**). Barycentric coordinates with basis $\{p_{i-1}, p_i, p_{i+1}\}$ may be used to decide which of the intersection points is inside the triangle or not.[1] In this case we continue with a triple of vertices starting with the calculated new vertex q_k. If (**O$_1$**) and (**O$_2$**) do not lead to any change, the third option may lead to an improvement.

(**O$_3$**) Point p_i may be moved on its critical edge to obtain an optimum position p_{new} minimizing the total length of both line segments $p_{i-1}p_{new}$ and $p_{new}p_{i+1}$. That's a *move on a critical edge* and an $\mathcal{O}(1)$ solution is given below. Then subsequence (p_{i-1}, p_i, p_{i+1}) is replaced in our curve by $(p_{i-1}, p_{new}, p_{i+1})$. In this case we continue with vertices $p_{new}, p_{i+1}, p_{i+2}$.

We consider situation (**O$_3$**), i.e. $\boldsymbol{p}_i$ lies on a critical edge, say e, and is not collinear with $\boldsymbol{p}_{i-1}\boldsymbol{p}_{i+1}$. Let l_e be the line containing the edge e. First, we find the point $\boldsymbol{p}_{opt} \in l_e$ such that

$$|\boldsymbol{p}_{opt} - \boldsymbol{p}_{i-1}| + |\boldsymbol{p}_{i+1} - \boldsymbol{p}_{opt}| = \min_{\boldsymbol{p} \in l_e} L(\boldsymbol{p})$$

---

[1] In the majority of such cases we found $k = 1$, i.e. p_i is replaced by q_1.

with
$$L(\boldsymbol{p}) = (|\boldsymbol{p} - \boldsymbol{p}_{i-1}| + |\boldsymbol{p}_{i+1} - \boldsymbol{p}|).$$

If $\boldsymbol{p}_{opt}$ lies on the closed critical edge e we simply replace $\boldsymbol{p}_i$ by $\boldsymbol{p}_{opt}$. If it does not, we replace $\boldsymbol{p}_i$ by that vertex bounding e and lying closest to $\boldsymbol{p}_{opt}$.

We give a slightly simpler solution for finding the point $\boldsymbol{p}_{opt}$ then provided in [ ]. W.l.o.g. let us assume that l_e is parallel to the x-axis:

$$l_e = \{(t, y_e, z_e)^T | t \in \mathbb{R}\}$$

for some fixed y_e and z_e. If $x_{i-1} = x_{i+1}$ we find $\boldsymbol{p}_{opt} = (x_{i-1}, y_e, z_e)^T$. Otherwise

$$\frac{\partial L}{\partial x}(\boldsymbol{p}_{opt}) = 0$$

leads to a quadric equation in x_{opt},

$$(\alpha_{i+1} - \alpha_{i-1})x_{opt}^2 + 2(\alpha_{i-1}x_{i+1} - \alpha_{i+1}x_{i-1})x_{opt} + \alpha_{i+1}x_{i-1}^2 - \alpha_{i-1}x_{i+1}^2 = 0,$$

with $\alpha_i = (y_e - y_i)^2 + (z_e - z_i)^2$.

Figure 5 illustrates a given simple cube curve by its critical edges only, the initial polygon and the correct MLP as calculated by our algorithm. Note that some of the vertices subdivide critical edges in a rational ratio specified by the heights of consecutive linear cube segments.

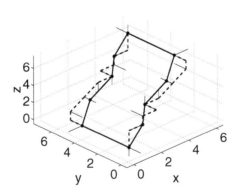

Fig. 5. Initial polygon (dashed) and correctly calculated MLP. Critical edges are shown as short line segments. The rest of the tube is not shown

We illustrate the performance of the algorithm with respect to length measurement on circles in $\mathbb{R}^3$. We generate circles by

$$\boldsymbol{c}(t) = R_x(\phi_1)R_z(\phi_2)R_x(\phi_3)\begin{pmatrix} r\cos(t) \\ r\sin(t) \\ 0 \end{pmatrix}, \quad t \in [0, 2\pi[.$$

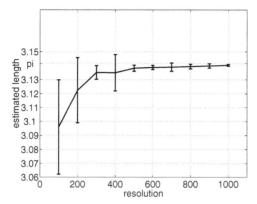

Fig. 6. Results of the length estimation of digitized circles with resolution 100^3 to 1000^3 using $\tau = l(\mathcal{P}_t) \cdot 10^{-7}$

Here, R_x and R_z are the 3×3 rotation matrices about the x- and the z-axis, respectively. The angles ϕ_1, ϕ_2 and ϕ_3 are randomly chosen from a uniform distribution in the interval $[0, \pi]$. The radius r is randomly chosen from the interval $[0.5, 1]$. In each experiment a circle is generated and digitized on a 3D grid. The grid size varies between 100^3 and 1000^3. We performed experiments using different thresholds τ, and Fig. 6 illustrates the case of $\tau = l(\mathcal{P}_t) \cdot 10^{-7}$ and resolutions $(100 \cdot n)^3, n \in \{1, 2, \ldots, 10\}$. For each combination we estimated the length of 25 digitized circles. We normalized the results such that the true circumference of the underlying circles is π.

4 Analysis of Cube Curves

In it's current form the above algorithm is slower than necessary. It can be sped up by taking into account more of the available information about the possible positions of vertices of the MLP. Here we propose three ways of doing so: 1) Not all critical edges are relevant, i.e., not every critical edge can contain a vertex of the MLP. Removing irrelevant critical edges reduces the number of necessary computations. 2) In certain situations a critical edge can contain a vertex of the MLP only – if at all – at one of its endpoints. This information can be used for the initialization of the algorithm. 3) Consider a subsequence of the MLP which is contained in one fixed grid-layer. Assume the correct endpoints are known as well as the critical edges which contain the intermediate vertices of the MLP. Under these conditions the positions of the intermediate vertices can be calculated in closed form.

Considerations as started in this section may lead to a better understanding whether it is possible to derive a non-iterative, i.e. closed form of an algorithm calculating 3D MLPs within simple cube curves.

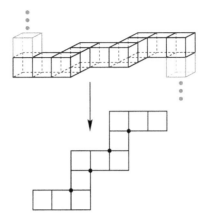

Fig. 7. A subsequence of a simple cube curve lying entirely in one cube-layer (top). Its projection to a 2D face-curve. The vertices marked by dots indicate the corresponding critical edges

4.1   Removing Irrelevant Critical Edges

Consider a flat arc of a simple cube curve, i.e. an arc all of which cubes lie within one cube-layer. As shown in Fig. 7 this arc can be projected to a 2D plane, where the cube-curve becomes a face-curve. The 2D MLP algorithm reported in [ ] may be used for such a planar segment of a simple cube curve. The following theorem also shows how in this case irrelevant critical edges can be identified.

Theorem 2. *Let $E = \{e_i, e_{i+1}, e_{i+2}, e_{i+3}, e_{i+4}\}$ be a sequence of consecutive, parallel critical edges. Let e_i, e_{i+2} and e_{i+4} be coplanar. Consider the projection of the cube-curve segment containing E as in Fig. 7. Then e_{i+2} contains no vertex of the MLP if line segment $e_i e_{i+4}$ intersects the boundary of the face-curve at e_{i+2} only.*

Figure 8 illustrates the situation and no further proof is needed.

4.2   Vertices on Endpoints of Critical Edges

We again investigate flat arcs of the cube-curve. The following theorem shows that in a situation like the one shown in Fig. 9 the only possible positions of MLP-vertices are the endpoints of critical edges.

Lemma 4. *Let $E = (e_i, e_{i+1}, \ldots, e_{i+n-1})$ be a sequence of consecutive critical edges. If all these e_j's are pairwise parallel then they lie within the same grid layer (cube-layer).*

Proof. Moving into another grid-layer results in a critical edge in the plane between both grid layers, i.e. which is not parallel to the e_j's in the sequence E before.

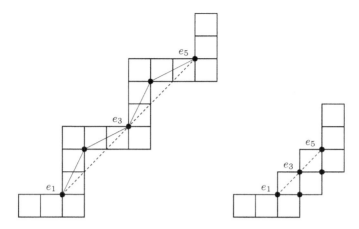

Fig. 8. Illustration of Theorem 2. Left: e_3 contains a vertex of the MLP. Right: e_3 does not contain a vertex of the MLP and can be removed from the list of critical edges

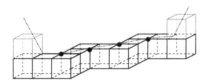

Fig. 9. Entering and leaving in same direction

It holds that $E = (e_i, e_{i+1}, \ldots, e_{i+n-1})$ is a *maximum-length sequence of parallel critical edges* iff all the e_j's contained in E are pairwise parallel and e_{i-1} and e_{i+n} are not parallel to e_i (and thus not parallel to any edge contained in E).

Theorem 3. *Let $E = (e_i, e_{i+1}, \ldots, e_{i+n-1})$ be a maximum-length sequence of parallel critical edges. If e_{i-1} and e_{i+n} are both in the upper (lower) face-layer, the only possible vertex positions on the edges $e_i, \ldots, e_{i-n-1}$ are the upper (lower) endpoints of these edges.*

Note, that by speaking about "upper" and "lower" face-layer, we assumed a horizontal layer. The same theorem holds if we replace "upper" and "lower" by "left" and "right" or by "front" and "back".

Proof. The shortest path connecting an edge-sequence $E = (e_i, e_{i+1}, \ldots, e_{i+n-1})$ lies within a horizontal plane, i.e., the positions of all vertices have the same z-coordinate (x-coordinate for "left" and "right", y-coordinate for "front" and "back"). Since the left and right neighbors of E are both above (below) the cube-layer, it is optimal to move the vertices to the upper (lower) endpoints of the edges $e_i, \ldots, e_{i-n-1}$.

4.3 Closed Form Evaluation of Vertex Positions

Let $(e_i, e_{i+1} \ldots, e_{i+n})$ be a sequence of consecutive parallel critical edges and $E = (e_i, e_j \ldots, e_k)$ the sequence which results from the former sequence by removing all the critical edges containing no vertex of the MLP. W.l.o.g. we assume that E lies within a horizontal cube-layer.

Consider the sequence of the vertices $(p_a, p_{e_i}, p_{e_j}, \ldots, p_{e_k}, p_b)$ being a subsequence of the MLP. Here p_a and p_b are assumed to be known, and p_e denotes the vertex lying on e. Define

$$l = |p_a p_{e_i}| + |p_{e_i} p_{e_j}| + \ldots + |p_{e_k} p_b|,$$

and let $(p)_z$ be the z-coordinate of the vertex p.

Theorem 4. *For the specified situation it holds that the z-coordinates of $p_{e_i}, p_{e_j}, \ldots, p_{e_k}$ are given by linear interpolations*

$$(p_{e_i})_z = (p_a)_z + \frac{(p_b)_z - (p_a)_z}{l} |p_a e_i|,$$

$$(p_{e_j})_z = (p_{e_i})_z + \frac{(p_b)_z - (p_a)_z}{l} |e_i e_j|, \ \ldots$$

Proof. We represent e_i by the set of points in $\mathbb{R}^3$ which lie on e_i:

$$e_i = \{(x_i, y_i, z + \lambda_i) | \lambda_i \in [0,1]\}.$$

We introduce a function σ acting on the indices such that

$$E = (e_{\sigma(1)}, e_{\sigma(2)} \ldots, e_{\sigma(m)}) .$$

Further we denote the positions of the vertices p_a and p_b by

$$p_a = (x_{\sigma(0)}, y_{\sigma(0)}, \lambda_{\sigma(0)}) \quad \text{and} \quad p_b = (x_{\sigma(m+1)}, y_{\sigma(m+1)}, \lambda_{\sigma(m+1)}) .$$

The length of the sequence $(p_a, p_{e_i}, p_{e_j}, \ldots, p_{e_k}, p_b)$ is thus given by

$$L(\lambda_{\sigma(1)}, \ldots, \lambda_{\sigma(m)}) = \sum_{i=1}^{m} \sqrt{(|e_{\sigma(i)} e_{\sigma(i+1)}|)^2 + (\lambda_{\sigma(i)} - \lambda_{\sigma(i+1)})^2} \, ,$$

where $(|e_{\sigma(i)} e_{\sigma(i+1)}|)^2 = (x_{\sigma(i)} - x_{\sigma(i+1)})^2 + (y_{\sigma(i)} - y_{\sigma(i+1)})^2$. Since L does not depend on the x_i and y_i directly but merely on the horizontal distances between consecutive critical edges $|e_{\sigma(i)} e_{\sigma(i+1)}|$ we can replace E by E' with edges

$$e_i' = \{(x_i', 0, \lambda_i) | \lambda_i \in [0,1]\}$$

such that

$$x_{\sigma(i)}' - x_{\sigma(i+1)}' = |e_{\sigma(i)} e_{\sigma(i+1)}|$$

which leads to

$$L = \sum_{i=1}^{m} \sqrt{(x'_{\sigma(i)} - x'_{\sigma(i+1)})^2 + (\lambda_{\sigma(i)} - \lambda_{\sigma(i+1)})^2}.$$

Minimizing L with respect to $\lambda_{\sigma(1)}, \ldots, \lambda_{\sigma(m)}$ leads to

$$\lambda_{\sigma(i)} = \lambda_{\sigma(0)} + \frac{\lambda_{\sigma(m+1)} - \lambda_{\sigma(0)}}{x'_{\sigma(m+1)} - x'_{\sigma(0)}}(x'_{\sigma(i)} - x'_{\sigma(0)}).$$

Converting to the previous notation yields the formulae in the theorem.

Figure 10 shows a simple situation in which Theo. 4 can be applied. We finally

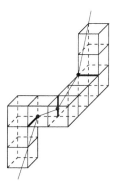

Fig. 10. An example for the application of Theo. 4: The position of the middle MLP-vertex can be calculated using the theorem. In this case $\lambda = 1/3$. An arc of the MLP is represented by the thin line. The bold edges of the cube curve are the critical edges

present an example of a cube-curve with an MLP which can be calculated in closed form (see Fig. 11). The vertices in the left and right arcs, which lie in the Y-Z plane, can by Theo. 3 found to lie on the endpoints of critical edges. Which of the critical edges in these arcs are effective can be found by a 2D method for face-curves. The latter is also true for the two other arcs which lie in the X-Y plane. The locations of the vertices on the effective critical edges of these arcs can be found by Theo. 4.

5 Conclusion

In this paper we reviewed an algorithm for iterative approximation of 3D MLP of a simple cube curve and initiated a discussion whether it is possible or not to derive a closed-form, i.e. non-iterative algorithm for this computational problem. We have shown that certain irrelevant critical edges can be identified, and

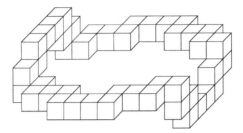

Fig. 11. The MLP of the shown cube-curve can be calculated in closed form using the presented methods. See text for details

how the positions of some MLP-vertices can directly be calculated without an iterative procedure. These first steps towards a better understanding of simple cube curves are based on a segmentation of 3D cube curves into 2D face curves.

Triplets of run-length sequences may play a crucial role in analyzing simple cube curves. Starting with one cube of a simple cube curve we may assign one sequence of numbers to any of the three coordinate axes defined by consecutive run length' parallel to this coordinate axis (i.e. 'how many consecutive cubes are in the same layer with respect to the coordinate axis?'). These sequences may be calculated during a first run through the given cube curve. For example, starting at the lower left cube of the simple cube curve shown in Fig. 1 on the left, and going to the right, we have sequences

$$1, 1, 1, 3, 6, 1, 1, 1, 1, 4, 4+$$

(4+ indicates that this run is in the same layer as the first cube) for the X-coordinate axis (left to right),

$$4, 4, 2, 7, 1, 1, 5+$$

(again continuation of the same layer as the first 4 cubes) for the foreground-to-background Y-axis, and

$$5, 2, 2, 2, 10, 1, 1, 1$$

for the bottom-to-top Z-axis. This shows that the longest run of 10 cubes appears in one Z-layer, matching the situation of Theorem 3, followed by a run of 9 cubes in one Y-layer. Such a longest run might be a good start for a 3D MLP algorithm. An analysis of these *run-length sequences* is also expected to allow a more direct calculation of (possible) positions of MLP vertices on critical edges. Figure 12 shows that these may be at irrational positions.

The evaluation of the practical relevance of algorithmic proposals (as discussed for the speed-up of the iterative algorithm) is particularly difficult, since the effect of these depends highly on the class of specific cube curves under consideration. For example, it is easily possible to construct curves with no irrelevant critical edges at one hand, and curves containing almost only irrelevant critical

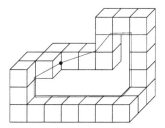

Fig. 12. A cube curve with an MLP-vertex at an irrational position: As a consequence of Theo. 4 the vertex marked by the dot is at $\lambda = \sqrt{2}/(2 + \sqrt{2})$

edges, on the other hand. Similarly, the applicability of Theorem 4 depends on the length of flat arcs in the cube-curve.

The effect of methods such as presented here on the complexity of the discussed iterative algorithm will be subject to future research, in close relation to methods modeling or generating specific classes of cube curves. For example, it will be worthwhile to investigate whether Theorem 3 allows the identification of at least two MLP-vertices for any given cube curve. This would allow a segmentation of a cube curve into two cube arcs, and the independent treatment of both arcs.

References

1. Th. Bülow and R. Klette. Rubber Band Algorithm for Estimating the Length of Digitized Space-Curves. In: Proc. 15th Intern. Conf. *Pattern Recognition*, 2000, vol. 3:551–555. 285, 286, 287, 291
2. R. Busemann and W. Feller. Krümmungseigenschaften konvexer Flächen. *Acta Mathematica*, 66:27–45, 1935. 290
3. J. Canny and J. H. Reif. New lower bound techniques for robot motion planning problems. *IEEE Foundations of Computer Science*, 28:49–60, 1987. 286
4. J. Choi, J. Sellen, C.-K. Yap. Approximate Euclidean shortest path in 3-space. In: Proc. 10th ACM Conf. *Computat. Geometry*, 41–48, 1994. 286
5. D. Coeurjolly, I. Debled-Rennesson, O. Teytaud. Segmentation and length estimation of 3D curves. Chapter in this book. 285
6. A. Jonas, N. Kiryati. Length estimation on 3-D using cube quantization. *JMIV*, 8:215–238, 1998. 285
7. R. Klette, V. Kovalevsky, and B. Yip. On the length estimation of digital curves. In: Proc. SPIE Conf. *Vision Geometry VIII*, 3811:52–63, 1999. 286, 287, 293
8. R. Klette and Th. Bülow. Critical edges in simple cube-curves. In: Proc. *DGCI'2000*, Uppsala December 2000, LNCS **1953**, 467-478. 285, 288
9. F. Sloboda, B. Zaťko, and P. Ferianc. Minimum perimeter polygon and its application. in: *Theoretical Foundations of Computer Vision* (R. Klette, W. G. Kropatsch, eds.), Mathematical Research **69**, Akademie Verlag, Berlin, 59-70, 1992. 285, 287
10. F. Sloboda and Ľ. Bačík. *On one–dimensional grid continua in R³*. Report of the Institute of Control Theory and Robotics, Bratislava, 1996. 285, 287

Segmentation and Length Estimation of 3D Discrete Curves

David Coeurjolly[1], Isabelle Debled-Rennesson[2], and Olivier Teytaud[3]

[1] Laboratoire ERIC
5, avenue Pierre-Mendes-France, F69676 Bron CEDEX
dcoeurjo@univ-lyon2.fr
[2] LORIA (Laboratoire LOrrain de Recherche en Informatique et ses Applications)
Campus Scientifique, B.P. 239, F54506 Vandœuvre-les-Nancy
debled@loria.fr
[3] ISC
67 Bd Pinel, F69675 Bron CEDEX
oteytaud@univ-lyon2.fr

Abstract. We propose in this paper an arithmetical definition of 3-D discrete lines as well as an efficient construction algorithm. From this notion, an algorithm of 3-D discrete lines segmentation has been developed. It is then used to calculate the length of a discrete curve. A proof of the multigrid convergence of length estimators is presented.

Keywords: 3D discrete line, Segmentation, Length estimation, Discrete curve.

1 Introduction

A formal definition of a 3-D discrete straight line is mandatory in numerous applications using objects based on voxels [, , , ,]. For example, drawing 3-D lines between two points of the discrete space is fundamental in discrete ray tracing. The existing methods may be sorted in two categories, a summary is given in [,]:

- "by projections" methods, where two projections of the 3-D segment are independently computed on two basic planes [ ]
- "by direct calculus" methods obtained by extension of the 2-D and 3-D algorithms [,]

Several authors have proposed definitions of 3-D 26-connected lines [,] and, as in 2D, it seems to be interesting to propose a simple arithmetical definition of 3-D discrete lines including the different structures obtained by the known drawing algorithms and suggesting a controlled thickness [ ].
A more general arithmetical definition is given in [ ] but the proposed definition in this paper is sufficient for the presented work and allows to get a simple algorithm for segmentation and 3D discrete curves length estimation. In the second section of this paper, we give an arithmetical definition of 3-D lines as well

G. Bertrand et al. (Eds.): Digital and Image Geometry, LNCS 2243, pp. 299–317, 2001.
© Springer-Verlag Berlin Heidelberg 2001

as a structure theorem allowing the control of the connectivity of 3-D discrete lines so defined. Moreover, a very simple scanning algorithm is proposed which is equivalent to those already known using "by projections" methods. It only relies on calculations on integers. In the third section, a recognition algorithm of naive 3-D lines segments is given ; it relies on the functionality property of a naive 3-D line and the recognition algorithm of 2-D naive lines segments [ ]. Then, a linear segmentation algorithm of 3-D discrete curves is directly deduced from the recognition algorithm of 3-D lines segments. In the last section, an application of this algorithm to the calculation of 3-D discrete curves length is proposed as well as the comparison with other methods. The convergence of this length estimation technique is demonstrated and several examples are given.

2   Definitions and Main Properties

2.1   3-D Discrete Line

To simplify the writing, we give a definition of a 3-D discrete line whose main vector is $(a, b, c) \in \mathbb{Z}^3$ with $a \geq b \geq c$. The definitions for coefficients ordered in a different way are obtained by permuting x, y, z as well as the coefficients. We suppose in the following that $a \geq b \geq c$, a condition to which it is easy to come back by symmetries.

Definition 1. *The **3-D discrete line**, called $\mathcal{D}$, whose main vector is $V(a, b, c)$, with $(a, b, c) \in \mathbb{Z}^3$, such that $a \geq b \geq c$ is defined as the set of points (x, y, z) of $\mathbb{Z}^3$ verifying the following diophantine inequalities:*

$$\mathcal{D} \begin{cases} \mu \leq cx - az < \mu + e & (1) \\ \mu' \leq bx - ay < \mu' + e' & (2) \end{cases}$$

*with $\mu, \mu', e, e' \in \mathbb{Z}$. μ and μ' are called the **lower bounds** of $\mathcal{D}$, moreover, e and e' are named the **arithmetical thickness** of $\mathcal{D}$. Such a line is noted $\mathcal{D}(a, b, c, \mu, \mu', e, e')$.*

Remarks:

1. For a given main vector, there is a large number of different representations according to the lower bounds and thicknesses chosen (see Fig. 1. and Fig. 3.).
2. In the plane Oxz (resp. Oxy), the double inequality (1) (resp. (2)) represents a 2-D arithmetical line [ ] with lower bound μ (resp. μ') and arithmetical thickness e (resp. e'), noted $\mathcal{D}_{2D}(c, a, \mu, e)$ (resp. $\mathcal{D}_{2D}(b, a, \mu', e')$).
3. The double diophantine inequalities (1) and (2) define two discrete planes of $\mathbb{Z}^3$ ([, ,]).

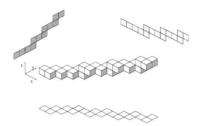

Fig. 1. Representation in voxels of $\mathcal{D}(10, 7, 3, 0, -9, 13, 17)$, a 6-connected line where a point (x, y, z) of $\mathcal{D}$ is represented by a unit cube centered at the integer point (x, y, z)

Definition 2. *We call* **3-D naive lines**, *whose main vector is* $V(a, b, c)$, *with* $(a, b, c) \in \mathbb{Z}^3$, *such that* $a \geq b \geq c$, *the 3-D discrete lines whose thickness verifies* $e = e' = a$, *they are named* $\mathcal{D}(a, b, c, \mu, \mu')$.

Remark: A 3-D naive line is projected into the plane Oxz and Oxy in two 2-D naive lines $[\quad, \quad]$, noted $\mathcal{D}_{2D}(c, a, \mu)$ and $\mathcal{D}_{2D}(b, a, \mu')$.

Theorem 1. *A 3-D naive line* $\mathcal{D}(a, b, c, \mu, \mu')$ *with* $a \geq b \geq c$ *is functional in* x.

Proof. Indeed, let us consider the naive line $\mathcal{D}(a, b, c, \mu, \mu')$, $\mathcal{D}$ is defined by

$$\begin{cases} \mu \leq cx - az < \mu + a \\ \mu' \leq bx - ay < \mu' + a \end{cases} \tag{3}$$

which is equivalent to

$$\begin{cases} z = \left\lceil \frac{cx - \mu}{a} \right\rceil \\ y = \left\lceil \frac{bx - \mu'}{a} \right\rceil \end{cases} \tag{4}$$

where $[\ldots]$ denotes integer part of x.

2.2   Relation between Arithmetical Thicknesses and Connectivity of 3-D Discrete Lines

In order to get a better understanding of the 3-D discrete line notion, here is the structure theorem which establishes a link between the arithmetical thicknesses and the connectivity of a 3-D discrete line.

Theorem 2. *Let us consider* $\mathcal{D}(a, b, c, \mu, \mu', e, e')$ *with* $a > b > c$,
If $e \geq a + c$ *and* $e' \geq a + b$, $\mathcal{D}$ *is 6-connected (see Fig 1.).*
If $e \geq a + c$ *and* $a \leq e' < a + b$ *or* $e' \geq a + b$ *and* $a \leq e < a + c$, $\mathcal{D}$ *is 18-connected.*
If $a \leq e < a + c$ *and* $a \leq e' < a + b$, $\mathcal{D}$ *is 26-connected (see Fig 3.).*
If $e < a$ *or* $e' < a$, $\mathcal{D}$ *is disconnected.*

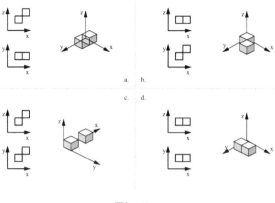

Fig. 2.

Proof. In order to demonstrate these results, we use the fact that projections of $\mathcal{D}$ in the planes Oxy and Oxz are 2-D discrete lines whose connectivity is known [ ].

For example, let us consider the third case ; $a \le e < a + c$ and $a \le e' < a + b$, in the plane Oxz, we move from a pixel to the next one in the discrete line $\mu \le cx - az < \mu + e$ through a 4-connected or 8-connected move, the same in the plane Oxy for the line $\mu' \le bx - ay < \mu' + e'$. All possible combinations between two pixels of planes Oxy and Oxz to get two voxels of $\mathbb{Z}^3$ have been studied on the figure 2.

a. an 8-connected move in the plane Oxz and a 4-connected move in the plane Oxy lead to two voxels linked by an 18-connected move,
b. a 4-connected move in the plane Oxz and an 8-connected move in the plane Oxy lead to two voxels linked by an 18-connected move,
c. an 8-connected move in the plane Oxz and an 8-connected move in the plane Oxy lead to two voxels linked by a 26-connected move,
d. a 4-connected move in the plane Oxz and a 4-connected move in the plane Oxy lead to two voxels linked by a 6-connected move.

Therefore we show that the line $\mathcal{D}$ is 26-connected.

The other results can be demonstrated in a similar way by watching all combinations of two voxels obtained from the possible ones of two pixels in the planes Oxz and Oxy.

Remark: The naive lines are 26-connected and we go from a floor to the next one only through an edge or a vertex, we can see it on the figure 3.

The next theorem states that there is a link between the 3-D GIQ (Grid Intersect Quantization) [ ] and the 3-D straight line defined above. We prove that this digitization process digitizes an Euclidean straight line into a 26-connected set of voxels such that 2 of its 3 canonical projections are 2-D discrete straight

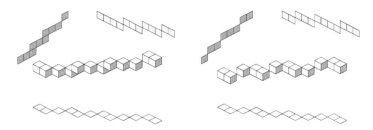

Fig. 3. On the left $\mathcal{D}(10,7,3,0,0)$, on the right $\mathcal{D}(10,7,3,-5,0)$

lines. Note that in a general case, Jonas et al. [ ] show that none of the classical digitizations into 26-connected curve has the *three*-projection property (*i.e* projections of a 3-D chain onto each axis planes are identical to the digitization of the continuous curve projections).

Theorem 3. *The 3-D GIQ (Grid Intersect Quantization) of an Euclidean straight line is a 3-D discrete straight line defined above.*

Proof. Without loss of generality, we consider $\mathcal{D}_{euc}$ a 3-D Euclidean straight line whose main vector is $(a, b, c) \in \mathbb{Z}^3$ with $a \geq b \geq c$, defined by:

$$\mathcal{D}_{euc} = \begin{cases} z = \frac{cx-r}{a} \\ y = \frac{bx-r'}{a} \end{cases} \tag{5}$$

We consider the Euclidean planes $\mathcal{P}_1 : z = \frac{cx-r}{a}$ and $\mathcal{P}_2 : y = \frac{bx-r'}{a}$. The GIQ-digitization of these planes gives the following arithmetical naive planes:

$$\begin{aligned} GIQ(\mathcal{P}_1) &= P(c, 0, -a, r + \left[\frac{a}{2}\right]) \\ GIQ(\mathcal{P}_2) &= P(b, -a, 0, r' + \left[\frac{a}{2}\right]) \end{aligned} \tag{6}$$

where $P(u, v, w, \mu) : \mu \leq ux + vy + wz < \mu + max(|u|, |v|, |w|)$ [,]. In the following, we consider a free ambiguity GIQ process [ ] (we assume a global strategy to remove pathological cases).

The intersection between this two discrete naive planes defines a 26-connected discrete line according to definition 1 and the structure theorem 2:

$$\mathcal{D} = \begin{cases} r + \left[\frac{a}{2}\right] \leq cx - az < r + \left[\frac{a}{2}\right] + a \\ r' + \left[\frac{a}{2}\right] \leq bx - ay < r' + \left[\frac{a}{2}\right] + a \end{cases} \tag{7}$$

Let v be a voxel of $\mathcal{D}$, we consider $GIQ(\mathcal{D}_{euc})$ the GIQ-digitization of $\mathcal{D}_{euc}$. If v belongs to $\mathcal{D}$, according to the GIQ process of the planes, we have (d denotes the classical Euclidean distance):

$$d(v, \mathcal{P}_1) \leq \sqrt{2} \text{ and } d(v, \mathcal{P}_1) \leq \sqrt{2} \Leftrightarrow d(v, \mathcal{D}_{euc}) \leq \sqrt{2} \tag{8}$$

This is equivalent to:

$$v \in \mathcal{D} \Rightarrow v \in GIQ(\mathcal{D}_{euc}) \tag{9}$$

and thus

$$\mathcal{D} \subset GIQ(\mathcal{D}_{euc}) \tag{10}$$

To prove the symmetric inclusion we consider $v' \in GIQ(\mathcal{D}_{euc})$, if $v' \notin \mathcal{D}$ we have two possibilities:

- $d(v', \mathcal{P}_1) > \sqrt{2}$ or $d(v', \mathcal{P}_2) > \sqrt{2}$ that is equivalent with $d(v', \mathcal{D}_{euc}) > \sqrt{2}$ which leads to a contradiction with the GIQ process of $\mathcal{D}_{euc}$;
- $d(v', \mathcal{D}_{euc}) \leq \sqrt{2}$ but since the GIQ is free ambiguity, v' disconnects $\mathcal{D}$ which leads to a contradiction with the structure theorem 2.

Finally:

$$\mathcal{D} = GIQ(\mathcal{D}_{euc}) \tag{11}$$

$\square$

2.3   A Scanning Algorithm of 3-D Naive Discrete Lines

Let us consider $A(x_A, y_A, z_A)$ and $B(x_B, y_B, z_B)$ two points of $\mathbb{Z}^3$. The algorithm given below draws a naive line segment between the two points A and B whose main vector is $\boldsymbol{AB} = (v_x, v_y, v_z)$. According to the choice of the lower bound, different structures are obtained. In particular, for $\mu = \mu' = -\lceil \frac{max(|v_x|,|v_y|,|v_z|)}{2} \rceil + 1$, the generated voxels are the same as those defined by the algorithm of Bresenham extended to 3D [,].
The principle of the algorithm consists in computing in parallel in both planes Oxy and Oxz the pixels of the 2-D naive lines which are the projections of the 3-D line in these two planes. So we recalculate the three coordinates (x, y, z) of the corresponding voxel.

Algorithm for the construction of a naive discrete line segment with lower bounds μ and μ' between two points $A(x_A, y_A, z_A)$ and $B(x_B, y_B, z_B)$ of $\mathbb{Z}^3$ such that $x_B - x_A > y_B - y_A$ and $y_B - y_A > z_B - z_A$.

```
vx = xB − xA; vy = yB − yA; vz = zB − zA;
r1 = vz * xA − vx * zA − μ;
r2 = vy * xA − vx * yA − μ';
x = xA; y = yA; z = zA;
While x < xB repeat
    x = x + 1;
    r1 = r1 + vz;
    r2 = r2 + vy;
    If r1 < 0 or r1 ≥ vx then
        z = z + 1;
        r1 = r1 − vx;
    Endif
    If r2 < 0 or r2 ≥ vx then
```

$y = y + 1$;
    $r_2 = r_2 - v_x$;
Endif
Print the point (x, y, z);
EndRepeat

This algorithm is linear in the number of points to be computed and very simple.

3   Algorithm for 3-D Naive Line Segment Recognition

The recognition of 3-D naive line segments is a direct outcome of the recognition of 2-D naive line segments [,]. Indeed, a 3-D naive line segment is bijectively projected into at least two orthogonal planes as two 2-D naive discrete straight lines. Consequently, recognizing a 3-D naive line segment consists in recognizing both 2-D naive line segments of its functional projections. Hence the linear algorithm proposed in [,] is particulary adapted to this problem of 3D curve segmentation into naive lines, even more than other linear algorithms of 2D curve segmentation like [,].

Algorithm for 3-D naive line segment recognition

Input: 26-connected sequence of voxels to be analysed, named $\mathcal{S}$
Output: True if $\mathcal{S}$ is 3-D naive line segment, False if not

If the voxels of $\mathcal{S}$ may not be bijectively projected on at least two orthogonal planes in order to create two curves of pixels C_1 and C_2 **then** $\mathcal{S}$ is not a 3-D line segment, return False; **Endif**
Apply the algorithm of 2-D naive line segments recognition on C_1 and C_2;
If C_1 and C_2 are two 2-D naive line segments **then**
    $\mathcal{S}$ is a 3-D naive line segment, return True;
Else $\mathcal{S}$ is not a 3-D naive line segment, return False;
Endif

This algorithm is linear in number of points of the sequence to be analysed and all necessary calculations are exclusively done using integers.

The characteristics of the recognized line segment are determined as a function of the characteristics of the 2-D line segments obtained. For example, let us suppose that $\mathcal{S}$ is bijectively projected on the plane Oxy curves C_1 and C_2. The recognition algorithm of 2-D naive line segments recognizes C_1 as a segment of $\mathcal{D}_{2D}(c_1, a_1, \mu_1)$ and C_2 as a segment of $\mathcal{D}_{2D}(b_2, a_2, \mu_2)$. Let us consider m the smallest common multiple of a_1 and a_2 such that $m = k_1 a_1 = k_2 a_2$ then $\mathcal{S}$ is a 3-D naive line segment whose main vector is $V(m, k_2 b_2, k_1 c_1)$ and with lower bounds μ_1 and μ_2.

4   Segmentation of 3D Discrete Curves

The segmentation of a 26-connected discrete curve, named $\mathcal{C}$, consists in "splitting" $\mathcal{C}$ into maximal length segments of 3-D naive lines. To do so, we use the same principle as the algorithm presented in the previous section, but, to achieve a linear complexity, the search of functional projection planes is not executed in the same way. We proceed by increment ; three 2-D discrete curves are built and segmented along the recognition, they correspond to the projections of the curve part already scanned in the three coordinates planes.

The algorithm given below uses the same bases as the segmentation algorithm of 2-D discrete curves presented in [,]. A sequence $\mathcal{C}$ of 26-connected voxels is given as input.

Algorithm of segmentation of 26-connected discrete curves

M = first voxel of $\mathcal{C}$
While $\mathcal{C}$ has not entirely been scanned **repeat**
    In each coordinate planes Oxy, Oxz, Oyz, init the characterictics of new segments S_1, S_2 and S_3;
    SEGMENT=true;
    While the added voxel M possesses at least two projections in which the projected points of M can be added to the current 2-D segments [,] and SEGMENT
    repeat
        If the projections of the following voxel of $\mathcal{C}$ in the two coordinates planes used belong to the same octants as the current 2-D segments **then**
            M = following point of $\mathcal{C}$;
            Add by symmetry the projected pixels of M, to the 2-D segments concerned;
        Else
            SEGMENT = false;
        Endif
    End repeat
According to symmetries, compute the characteristics of the naive lines of the two 2-D segments obtained in the initial octants, then deduce the main vector of the recognized naive line segment;
End repeat

Remarks:
1. The first point of a new segment is the last of the previous segment.
2. An algorithm for 4-connected 2D line segment recognition is presented in [ ] and it's easy to use it (as presented before) for the segmentation of 18-connected discrete curves.

Example: On the figure 4 we can see an example of the segmentation of a discrete curve, the starting and ending segment voxels are in black. The projection planes used are Oxy and Oyz, the segments resulting from the 2-D recognition are drawn in light gray on the projections in these two planes of the considered voxels set. Two naive line segments are obtained:

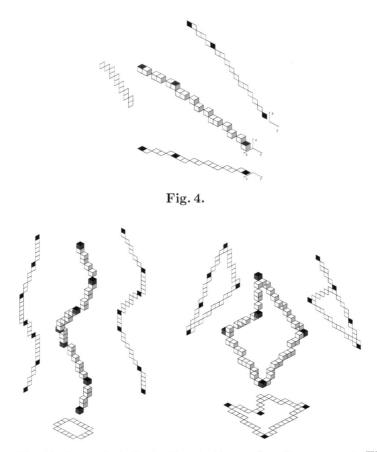

Fig. 4.

Fig. 5. The black voxels indicate the starting and ending segment. The 2-D segmented curves are in light gray

Segment 1 which contains 13 points. The inequalities of the discrete naive lines of the 2-D segments obtained are $-4 \leq -4y - 5z < 1$ and $-2 \leq -2y - 5x < 3$. Consequently, the 3-D naive line of the segment 1 has $V(2, -5, 4)$ as main vector. Segment 2 length is 6. The inequalities of the discrete naive straigh lines of the 2-D segments obtained are $0 \leq y - 2z < 2$ and $0 \leq y - 2x < 2$. Consequently, the 3-D naive line of the segment 2 has $V(1, -2, 1)$ as main vector.

Other examples of segmentation of 26-connected curves are given in the figure 5.

5   Length Estimation Algorithm

In 2D, a lot of approaches have been proposed for the length estimation of a discrete curve. These approaches can be *sorted* as follows:

- Local approaches [,]: estimators are based on local or fixed size window characteristics of the discrete curve. An example of such estimators is based

on counting the number of odd and even codes in the Freeman chain code
of the curve and returning the length estimation $l = \alpha n_{odd} + \beta n_{even}$ where
α and β are computed using statistical analysis (e.g $(\alpha, \beta) = (1.183, 1.059)$
in [ ]). Such estimators exist for 3D curves [,] in a same approach : find
the optimal weights associated to different chain codes. In the following, in
spite of advantages of such approaches, we only deal with global algorithm.

- Discrete Straight Segments (DSS) approaches [ ]: these methods are defined
  by a curve partitioning into digital straight line segments of maximum length
  each. Hence, the length estimator is based on the obtained polyline length.
- Minimum Length Polygon approaches (MLP) [, ,]: these methods compute a minimum length polygon that belongs to an open boundary of a digital
  object.
- Euclidean Path approaches: in these approaches, a semi-continuous representation of the discrete curve is computed using Euclidean points located
  in a open square of size one centered on each discrete points. Since this Euclidean path provides a good approximation of the underlying real curve,
  a length estimator on this path can be defined either as the path polyline
  length [ ] or embedded in an MLP process [ ].

In the 3D case, Klette and Bulow [ ] propose a 3D version of the MLP
approach. The input of the algorithm is a closed *cube*-curve that is a strictly 6-
connected closed curve, the minimum length polygons is computed in two steps
(see figure 5):

1. The vertices of the candidate polygon are first initialized to the *inner* vertex
   of edges of the cube-curve that are adjacent to three distinct cubes; such
   edges are called critical edges;
2. local operations (move or delete) are computed on the polygon vertices to
   converge to the minimum length polygon.

This first 3D global algorithm for the length estimation leads to several problems: the first one is purely algorithmic since this algorithm is not linear in the
number of points of the discrete curve. This non-linear complexity is due to
the local optimization processes that introduce backtrack on the polygon vertex
chain. Since a vertex of a candidate ML-polygon can be moved along a critical
edge, the calculated vertices of the polygon might not belong to the discrete grid

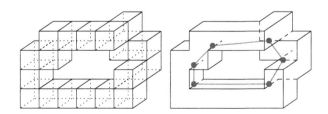

Fig. 6. An example of a cube-curve with its ML-polygon

and we have to introduce an τ-MLP that is the minimum length polygon up to a threshold τ (*i.e.* $|l_{\tau-MLP}(C) - l_{MLP}(C)| \leq \tau$ where C denotes the cube-curve).

Furthermore, there is no multigrid convergence proof of this MLP 3D version.

In the following, we propose a length estimator based on the 3D naive discrete straight line segmentation. The interesting points of such an algorithm are:

- a purely discrete and optimal in time algorithm;
- multigrid convergence of the length estimator whatever the curve is closed or not and whatever the dimension;
- since the segmentation process holds whatever *thickness* has the curve, we have a general length estimator whatever the topology of the curve.

We first describe the algorithm based on the Naive Straigth line Segmentation (NSS):

NSS length estimation algorithm

Input: 26-connected sequence of voxels to be analysed, named $\mathcal{S}$
Output: the estimated length of $\mathcal{S}$

Compute the segmentation using the above algorithm of $\mathcal{S}$
Let $\mathcal{P} = \{S_i\}_{0\dots n}$ denotes the polyline returned by the segmentation. The vertices of this polyline are the starting and the ending points of the naive line segments.
Return $\sum_{i=0}^{n} l(S_i)$ (where $l(S_i)$ denotes the Euclidean length of the segment S_i)

For the sake of clarity, this algorithm is designed as a post-processing of the segmentation but it can easily be embedded in the segmentation process since it is purely greedy.

In the next section, we present a proof of the multigrid convergence of a slight variant of NSS.

5.1 Proof of the Multigrid Convergence

In [,], Kovalevsky and Klette prove the multigrid convergence of the 2D DSS approach. The main result of this proof is that the error between the boundary length, denoted by $\mathcal{B}_C$, of an Euclidean convex object and the estimated length using a DSS segmentation of the digitized boundary, denoted by B_{C_δ} on a grid of size δ, is linear with δ:

$$|l(\mathcal{B}_C) - l_{DSS}(B_{C_\delta})| \leq O(\delta) \qquad (12)$$

In the following, we propose a proof of this asymptotic convergence under the curvature hypothesis without the convexity one, and that remains true in any dimension. In the sequel we assume the following:

1. $\Gamma : [0,1] \rightarrow \mathbb{R}^d$ denotes the underlying continuous curve whose curvature is bounded by $C = \frac{1}{R}$.

2. Γ is included in the δ-enlargements of tubes $T_1,\ldots,T_N$ of diameter δ and lengths $L(T_1),\ldots,L(T_N)$. The δ-enlargement of a tube is the union of this tube and two parts as described in figure 7 (for negative values of δ, the notion of enlargement is similarly defined).
3. T_1 starts at $\Gamma(0)$, within $K_1\delta$ error.
4. T_N ends at $\Gamma(1)$, within $K_2\delta$ error.
5. T_i's end is T_{i+1}'s start.
6. $L(T_i) \geq K_3\sqrt{\frac{\delta}{C}}$, for any $i < N$.

We approximate the length of Γ by the length sum of the T_i's. This tube approach is a general framework for discrete polygonalization schemes.

The begin and end of a tube are the intersection of the axis with the lateral faces (extremities). The pertinence of the last hypothesis is discussed later in the case of particular discretizations. Moreover, we assume that δ is small enough to avoid pathological cases such as half-turns in a tube ($\frac{\delta}{2} < R$). The goal is a bound on the error, linear in δ. Notice that the hypothesis (2) implies that Γ gets out of the $(-\delta)$-enlargements of the T_i's through lateral faces.

We first give an overview of the proof: Lemma 1 shows that the angle of Γ within a tube is small enough to achieve a linear error in δ *in* a tube (Corollary 1). In Lemma 2, we show an upper bound on the angle between two adjacent tubes that leads to an upper bound on the error while *joining* two tubes. Under the sixth hypothesis, we have a sufficiently small number of tubes to achieve a linear bound on the summed error of joinings (Corollary 2). Corollaries 1, 2 lead to a global linear bound on the error.

Lemma 1. *The angle between Γ^1 and the axis of T_i in the $(-\delta)$-enlargement of T_i is lower than $B = \arcsin(\frac{\delta}{L-\delta}) + L(T_i)/R$. If α such that $\tan\alpha = \sqrt{\frac{\delta}{2R-\delta}}$ verifies $R\sin(\alpha) \leq L(T_i)$, then the angle is bounded by $B = \alpha$.*

---

[1] This angle is the maximal one between a tangent of Γ and the tube axis.

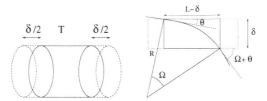

Fig. 7. *Left*: δ-enlargement of tube T. *Right*: schema for proof of lemma 1: L is the tube length, the dashed line is the tube contour, the solid line is the $(-\delta)$-enlargement of the tube

Proof. See figure 7: this is the case of largest negative derivative for a projection on a plane including the direction of the tube. The following holds:

$$\Omega \le L/R \tag{13}$$
$$\sin \theta \le \tfrac{\delta}{L-\delta} \tag{14}$$

This yields the first result. The particular case in which $R \sin(\alpha) \le L(T_i)$ for $\tan \alpha \le \sqrt{\tfrac{\delta}{2R-\delta}}$ is illustrated in figure 8; in this case, L is large, and the derivative must be small enough to keep Γ in the enlargement of the tube. This proves that the angle is bounded by $\arctan(\sqrt{\tfrac{\delta}{2R-\delta}})$. □

This bound is $O(\sqrt{\delta C})$ under the above assumptions. Moreover, this proves that the derivative of Γ in a parametrization colinear to the axis of the tube is bounded by $\sqrt{1 + (d-1)p^2}$ with $p = sin(B)$. This bounds the difference between the length of Γ in T_i and the length of T_i by $L(T_i)(\sqrt{1 + (d-1)p^2} - 1)$:

Corollary 1. *The difference between the length of T_i and the length of Γ in T_i is bounded by $L(T_i)(\sqrt{1 + (d-1)p^2} - 1)$. This is linear in $L(T_i)$ and δC.*

This is linear in δ, but does not complete the proof: we have a linear error in δC for segmentations, but "joining" the segments leads to small errors (see figure 8): some parts could be taken into account zero or two times. These errors have to be bounded too. For this we will need bounds on the derivative of parametrizations of Γ "later" than in the $(-\delta)$-enlargements; in the following lemma we work on δ enlargements, with similar results.

Lemma 2. *If the angle between Γ and the axis is lower than θ in the $-\delta$-enlargement, then it is lower than $\theta' = \arcsin(\sin(\theta) + \delta C)$ in the δ-enlargement. Moreover, the distance to the axis is lower than $R \cos(\theta) - R \cos(\theta')$ if Γ gets out of the $-\delta$ enlargement at the lateral faces (extremities).*

Proof. See Fig. 9. This is the 2D projection of the worst case. One can easily derive the followings:

$$R \cos(\theta') + A = R \cos(\theta) \tag{15}$$
$$R \sin(\theta') = R \sin(\theta) + \delta \tag{16}$$

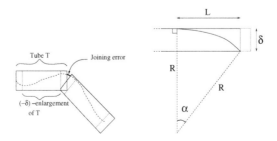

Fig. 8. *Left*: Errors out of segments. *Right*: Case L large in Lemma 1

This yields the result.   □

   This implies that Γ has variations of angle within θ'. The angle Ω between T_i and T_{i+1} is bounded by $2\theta'$. Thus the length lost or counted two times between T_i and T_{i+1} is bounded by the Γ length between getting out of T_i and getting in T_{i+1}, which is bounded by $\sqrt{1 + (d-1)p^2}\frac{\Omega\delta}{2}$, with Ω angle between T_i and T_{i+1}, with $p = sin(\theta')$.

   This leads to the following important corollary:

Corollary 2. *The error committed in joining two tubes is $O(\delta\Omega) = O(\delta^{3/2}\sqrt{C})$. As the number of such errors is bounded by $O(\frac{C}{\delta})^{\frac{1}{2}}$, the total error resulting of joins is $O(\delta C)$.*

   Hence, using corollary 1, we obtain the expected result. The error is linear under the above hypothesis. Constants have not been explicited as they are far from optimal by the proof below. More refined results would be useful. All hypothesizes are verified except the last one for usual segmentations, with δ linear in the resolution. We now have to study the hypothesis according to which $L(T_i) \geq K_3\sqrt{\delta/C}$, for any $i < N$.

   This hypothesis is particularly true whenever the discretization and polygonalization verify that any curve which lies in a tube of radius ϵr with the resolution r is included in a segment (this can easily be proved by considering the fact that for bounding curvature C, the minimal length of Γ before ϵr-deviation from the tangent is $\Theta(\sqrt{\frac{\epsilon r}{C}})$). This will be refered in the sequel as an *ϵ-tolerant algorithm.* Experiments below have been made with a straightforward algorithm which doesn't have the algorithmic properties of the initial algorithm.

   Figure 9 illustrates the fact that with 8-connectivity and with segments only 8-connected, large errors can be made. One can have similar bad behaviour with 4-connectivity; as explained above, a solution to avoid such bad results is adding a tolerance in the algorithm (this is simply an increase of the width parameters of bands assimilated to segments). As such pathological cases can be considered as negligible, perhaps the initial NSS algorithm would have a similar behavior in practical cases. Notice that such counter-examples do not mislead MLP. We present experiments illustrating proved results; Fig. 10 shows linear convergence for a circle (2D, convex, convergence proved in [,]), and Fig. 12 shows linear convergence of the ϵ-tolerant variant for an helix (3D, with bounded curvature, proof above). Then Fig. 13 shows that in practical 3D-cases, linear convergence can be achieved with an ellipse even without ϵ-tolerance; however, convergence is much less regular than with ϵ-tolerance.

   Note that, in $\mathbb{R}^2$, the proof using convexity (see [,]) can be extended to a finite number of inflexions (or finite number of areas with null curvature), with bounded curvature. This is possible without the hypothesis above about the length of segments. Convexity avoids particular cases as in Fig. 9 (right).

5.2   Experiments

In our experimentations, we compute the NSS length of 3D mathematical curves (circles, ellipses and helixes) on which the theoretical length is known, using the

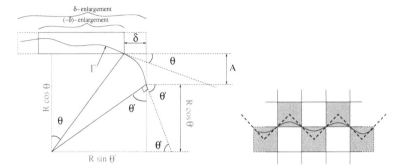

Fig. 9. *Left*: Figure for lemma 2. *Right*: $\Gamma(t) = (t, r^2 cos(> pt/r))$. The curvature is bounded by 1 for any value of the resolution r. The length computed (the length of the dashed curve) by a discretization algorithm in this case does not converge to the length of the curve for small values of r

above algorithm. Hence we consider a digitization of those Euclidean objects, using the GIQ process, on grids of increasing resolution in order to check the asymptotic convergence. In Fig. 10, we present the multigrid error graph and the execution time graph in 2D for a circle, and in Fig. 12, we present the multigrid error graph with an ϵ-tolerant algorithm for an helix in 3D. In Fig. 13, we show that in the particular case of a 3D ellipse, the NSS length algorithm converges. The interest of such a test is to check if it is possible, in concrete cases, to achieve a linear decrease of the error without ϵ-tolerance and thus use the algorithmic advantages of the NSS approach. The error is defined as the percentage of the absolute value of the difference between the true length and estimated length, to the true length. First, these results show an experimental asymptotic convergence since the error decreases with the resolution. If we compute the error inverse, we verify the linear nature of the convergence. Since the starting point of a closed curve in the segmentation algorithm may lead to different polygonalizations, we have plot for the circles curve the 95% confidence intervals (vertical bars) when we compute all possible segmentations. Hence, we can see that this starting point problem does not affect the results. The execution graph shows the linear complexity of the segmentation process in the number of points of the discrete curve.

In [ ], Klette et al. have already compaired the DSS and the MLP approaches in the 2D case and the comparisons both in execution time, speed of convergence and accuracy show that the DSS approach gives better results. In the 3D case, since the 3D MLP algorithm suffers from several problems, and since the 3D DSS does not, theoretically and practically, differ to the 2D one, we are convinced that the comparison will lead to the same results.

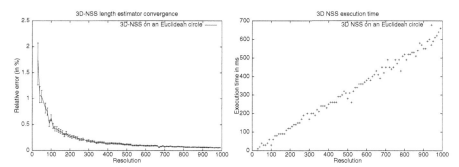

Fig. 10. These graphs present the error of the estimator on both 3D circles and the execution time in ms on a Sun Ultra SparcWorkstation

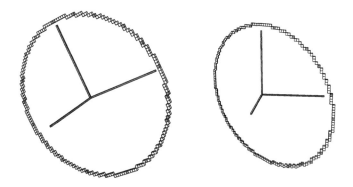

Fig. 11. An example of a NS-segmentation of a circle and an ellipse

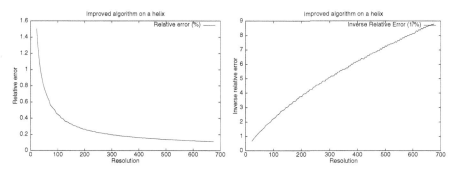

Fig. 12. *Left*: multigrid convergence of the ϵ-tolerant algorithm in the case of an helix in $\mathbb{R}^3$. *Right*: illustration of the error decrease

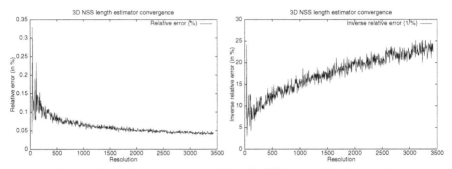

Fig. 13. *Left*: multigrid convergence of the 3D-NSS on a 3D ellipse of parameter $(1, \sqrt{2})$. *Right*: illustration of the linear decrease of the error

6   Conclusion

In this article, we have first presented an arithmetical definition of 3D discrete lines and an extension to the third dimension of the discrete naive straight line recognition algorithm. Moreover we have shown a segmentation algorithm which is purely discrete and optimal in time ($O(n)$ in time and $O(1)$ in space). Based on this algorithm, we then proposed a 3D discrete curve length estimator with the same complexity. In the last section, we have presented a proof of the multigrid convergence of length estimators based on a general *tube* framework in any dimension. We have shown that the discrete naive straight line segmentation verifies the proof hypothesis under a slight refinement and therefore the length estimator based on this algorithm also converges. The error bound we found is the same as in the literature concerning the DSS approaches but with less hypothesis on the underlying Euclidean curve. We suggest that this general framework could be used in the case of the 3D MLP approach: bounding the number of critical edges as in hypothesis (6) would be sufficient.

References

1. J. Amanatides and A. Woo. A fast voxel traversal algorithm for ray tracing. In *Eurographic's 87*, pages 3–12, 1987. 299
2. E. Andres. Le plan discret. In *Colloque en géométrie discrète en imagerie: fondements et applications*, Septembre 1993. 300, 303
3. T. Asano, Y. Kawamura, R. Klette, and K. Obokata. A new approximation scheme for digital objects and curve length estimations. Technical Report CITR-TR-65, Computer Science Departement of The University of Auckland, September 2000. 308
4. J. E. Bresenham. Algorithm for computer control of a digital plotter. In *IBM System Journal*, volume 4, pages 25–30, 1965. 304
5. D. Cohen and A. Kaufman. Scan-conversion algorithms for linear and quadratic objects. In *IEEE Computer Society Press, Los Alamitos, Calif*, pages 280–301, 1991. 299

6. D. Cohen-Or and A. Kaufman. 3d line voxelization and connectivity control. In *IEEE Computer Graphics and Applications*, pages 80–87, 1997. 299
7. I. Debled-Rennesson. *Etude et reconnaissance des droites et plans discrets*. PhD thesis, Thèse. Université Louis Pasteur, Strasbourg, 1995. 299, 300, 305, 306
8. I. Debled-Rennesson and J. P. Reveillès. A new approach to digital planes. In *In Vision Geometry III, SPIE, Boston*, volume 2356, pages 12–21, 1994. 300, 303
9. I. Debled-Rennesson and J. P. Reveillès. A linear algorithm for segmentation of digital curves. In *International Journal of Pattern Recognition and Artificial Intelligence*, volume 9, pages 635–662, 1995. 300, 301, 305, 306
10. L. Dorst and A. W. M. Smeulders. Length estimators for digitized contours. *Computer Vision, Graphics, and Image Processing*, 40(3):311–333, December 1987. 307, 308
11. L. Dorst and A. W. M. Smeulders. Decomposition of discrete curves into piecewise straight segments in linear time. In *Contemporary Mathematics*, volume 119, 1991. 305
12. O. Figueiredo and J. P. Reveillès. A contribution to 3d digital lines. In *Proc. DCGI'5*, pages 187–198, 1995. 299
13. A. Jonas and N. Kiryati. Digital representation schemes for 3d curves. *Pattern Recognition*, 30(11):1803–1816, 1997. 299, 302, 303
14. A. Kaufman. An algorithm for 3-d scan conversion of polygons. In *Proc. Eurographic's 87*, pages 197–208, 1987. 299
15. A. Kaufman. Efficient algorithms for 3-d scan conversion of parametric curves, surfaces, volumes. In *Computer Graphic 21, 4*, pages 171–179, 1987. 299
16. A. Kaufman and E. Shimony. 3-d scan conversion algorithms for voxel-based graphics. In *ACM Workshop on Interactive 3D Graphics, ACM Press, NY*, pages 45–75, 1986. 299
17. C. E. Kim. Three-dimensional digital line segments. In *IEEE Transactions on Pattern Analysis and Machine Intelligence*, volume 5, pages 231–234, 1983. 299
18. N. Kiryati and A. Jonas. Length estimation in 3d using cube quantization. *Journal of Mathematical Imaging and Vision*, 8:215–2138, 98. 308
19. N. Kiryati and O. Kubler. On chain code probabilities and length estimators for digitized three dimensional curves. 1995. 308
20. R. Klette and T. Bülow. Minimum-length polygons in simple cube-curves. *Discrete Geometry for Computer Imagery*, pages 467–478, 2000. 308
21. R. Klette, V. Kovalevsky, and B. Yip. Lenth estimation of digital curves. In *Vision Geometry VIII - SPIE*, pages 117–129, Denver, July 1999. 308, 313
22. R. Klette and J. Zunic. Convergence of calculated features in image analysis. Technical Report CITR-TR-52, Computer Science Departement of The University of Auckland, 1999. 309, 312
23. J. Koplowitz and A. M. Bruckstein. Design of perimeter estimators for digitized planar shapes. *IEEE Transactions on Pattern Analysis and Machine Intelligence*, PAMI-11(6):611–622, jun 1989. 307
24. V. Kovalevsky and S. Fuchs. Theoritical and experimental analysis of the accuracy of perimeter estimates. In *Robust Computer Vision*, pages 218–242, 1992. 308, 309, 312
25. M. Lindenbaum and A. Bruckstein. On recursive, o(n) partitioning of a digitized curve into digital straigth segments. *IEEE Transactions on PatternAnalysis and Machine Intelligence*, PAMI-15(9):949–953, september 1993. 305
26. J. P. Reveillès. *Géométrie discrète, calculs en nombre entiers et algorithmique*. PhD thesis, Thèse d'état, Université Louis Pasteur, Strasbourg, 1991. 300, 301, 302

27. F. Sloboda, B. Zatko, and P. Ferianc. *Advances in Digital and Computational Geometry*, chapter On approximation of planar one-dimensional continua. Springer, 1998. 308

28. F. Sloboda, B. Zatko, and R. Klette. On the topology of grid continua. In *Vision Geometry VII*, volume SPIE Volume 3454, pages 52–63, San Diego, July 1998. 308

29. I. Stojmenović and R. Tosić. Digitization schemes and the recognition of digital straight lines, hyperplanes and flats in arbitrary dimensions. In *Vision Geometry,contemporary Mathematics Series*, volume 119, pages 197–212, American Mathematical Society, Providence, RI, 1991. 299

30. A. Vialard. Geometrical parameters extraction from discrete paths. *Discrete Geometry for Computer Imagery*, 1996. 308

31. B. Vidal. *Vers un lancer de rayon discret*. PhD thesis, Thèse de Doctorat, Lille, 1992. 299, 304

Multigrid Convergence of Geometric Features

Reinhard Klette

CITR Tamaki, The University of Auckland, Tamaki campus
Morrin Road, Glen Innes, Auckland 1005, New Zealand
r.klette@auckland.ac.nz

Abstract. Jordan, Peano and others introduced digitizations of sets in the plane and in the 3D space for the purpose of feature measurements. Features measured for digitized sets, such as perimeter, contents etc., should converge (for increasing grid resolution) towards the corresponding features of the given sets before digitization. This type of multigrid convergence is one option for performance evaluation of feature measurement in image analysis with respect to correctness.

The paper reviews work in multigrid convergence in the context of digital image analysis. In 2D, problems of area estimations and lower-order moment estimations do have "classical" solutions (Gauss, Dirichlet, Landau et al.). Estimates of moments of arbitrary order are converging with speed $\kappa(r) = r^{-15/11}$. The linearity of convergence is known for three techniques for curve length estimation based on regular grids and polygonal approximations. Piecewise Lagrange interpolants of sampled curves allow faster convergence speed. A first algorithmic solution for convergent length estimation for digital curves in 3D has been suggested quite recently. In 3D, for problems of volume estimations and lower-order moment estimations solutions have been known for about one-hundred years (Minkowski, Scherrer et al.). But the problem of multigrid surface contents measurement is still a challenge, and there is recent progress in this field.

1 Introduction

Geometric image analysis approaches are normally motivated by concepts in Euclidean geometry. A common strategy is: approximate picture subsets in 2D by *polygons* or in 3D by *polyhedrons* and use Euclidean geometry from that moment on for any further object analysis or manipulation step. A theoretical motivation is given by the fact that rectifiable curves and measurable surfaces can be approximated by polygonal curves or polyhedral surfaces up to any desired accuracy. This means that if we consider grid resolution as a potentially improvable parameter, then polygonal or polyhedral approximations appear to converge (for a set-theoretic metric) to the original preimage of the given object. The important question arises: does a convergence toward the true value also hold for calculated properties? For example, if we measure the length of a digital curve then the calculated value should converge to the correct length of a preimage in Euclidean space digitized with increasing grid resolution.

G. Bertrand et al. (Eds.): Digital and Image Geometry, LNCS 2243, pp. 318–338, 2001.

In ancient mathematics, Archimedes and Liu Hui [39] estimated the length $\mathcal{L}(\gamma)$ of a circular curve γ. Liu Hui used regular n-gon approximations, with $n = 3, 6, 12, 24, 48, 96, ...$, see left of Fig. 1. In case of $n = 6$ it follows $3 \cdot d < \mathcal{L}(\gamma) < 3.46 \cdot d$ for diameter d, and for $n = 96$ it follows that

$$223/71 < \pi < 220/70 , \quad \text{i.e.} \quad \pi \approx 3.14 .$$

The used method is mathematically correct because the perimeters of inner and outer regular n-gons converge towards the circle's perimeter for $n \to \infty$. For example, for the inner 3×2^n-gons, having perimeters

$$p_{2n} = 2n \cdot r\sqrt{2r^2 - r\sqrt{4r^2 - p_n^2}} ,$$

it follows
$$\kappa(n) = |p_n - 2\pi r| \approx 2\pi r/n , \quad \text{for} \quad n \geq 6 .$$

The function $\kappa(n)$ defines the *speed of convergence*, which is linear in this case.

The perimeters of digitized circles have been calculated in image analysis using sometimes *graph-theoretical concepts* such as the length of a 4-curve, or of an 8-curve where diagonal steps are weighted by $\sqrt{2}$, see, for example, [13,27]. Such graph-theoretical concepts of path-length measurements are not related to digitized Euclidean geometry. Grid-intersection digitizations of line segments having a slope of 45° (for 4-paths) or of 22.5° (for 8-paths) provide simple examples for illustrating this. Convergence of digital curves toward a preimage with respect to the Hausdorff-Chebyshev-distance does not imply convergence of length calculated for these digital curves, toward the true length, but a proper preprocessing step (e.g. polygonal approximation of digital curves) may ensure such a desirable property as will be shown below.

We recall another historic example [35] cited in [18]. Assume that the lateral face **L** of a straight circular cylinder of radius ρ and of height h is cut by $(k-1)$ planes, $k \geq 2$, which are parallel to the base circles and which segment the cylinder into k congruent parts. Furthermore assume a regular n-gon, $n \geq 3$, in

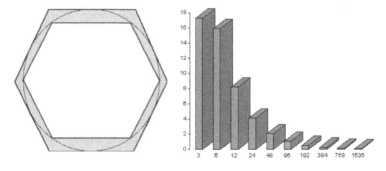

Fig. 1. Inner and outer hexagon approximating a circle (left), and percentage errors for perimeter estimation using inner n-gons

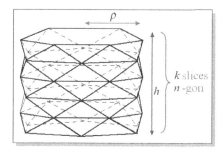

Fig. 2. Triangulation of the lateral face of a straight circular cylinder [35]

every cross section including both base circles, see Fig. 2 for $k = 4$ and $n = 6$. The axis of the cylinder and any vertex of such an n-gon defines a halfplane, which bisects an edge of the n-gon in the neighboring cross section or base circle. Now we connect for two neighboring n-gons each edge in one n-gon with those vertex of the other n-gon closest to this edge. This results into a triangulation $\mathbf{T}_{k,n}$ (i.e. a specific polyhedrization) of the lateral face $\mathbf{L}$ of the cylinder into $2kn$ congruent triangles having a surface area equal to

$$A(\mathbf{T}_{k,n}) = 2\pi\rho \cdot \frac{\sin(\pi/n)}{\pi/n}\sqrt{\frac{1}{4}\pi^4\rho^2\left(\frac{\sin(\pi/2n)}{\pi/2n}\right)^4\left(\frac{k}{n^2}\right)^2 + h^2} \; .$$

If k and n go to infinity then the length of the edges of the triangular faces of $\mathbf{T}_{k,n}$ converges to zero. However, the surface area of $\mathbf{T}_{k,n}$ does not necessarily converge towards the surface area $A(\mathbf{L}) = 2\pi\rho h$ of the lateral face! This is only true if k and n go to infinity such that k/n^2 converges to zero. If k/n^2 converges to $g > 0$ then $A(\mathbf{T}_{k,n})$ converges to

$$2\pi\rho \cdot \frac{\sin(\pi/n)}{\pi/n}\sqrt{\frac{1}{4}\pi^4\rho^2 g^2 + h^2} \; .$$

It may even happen that k/n^2 goes to infinity, e.g. $k = n^3$, and then it follows that $A(\mathbf{T}_{k,n})$ goes to infinity as well! Note that this example is based on sampling of surface points which cannot be assumed in image analysis. Digitization of sets provides an even less accurate input for subsequent steps of feature measurement.

The paper specifies the concept of multigrid convergence and reviews related results for measurements of moments, the length of curves in 2D or 3D, and surface area.

2   Multigrid Convergence

First we recall three digitization models frequently used in image analysis: Gauss digitization, grid-intersection digitization (for 2D only), and inner or outer Jordan digitization.

Let $r > 0$ be a real number called *grid resolution*. The *dilation* of a set $S \subset \Re^n$ by factor r is defined to be

$$r \cdot S = \{(r \cdot x_1, \ldots, r \cdot x_n) : (x_1, \ldots, x_n) \in S\},$$

for $n \geq 1$. Following [ ], this is a dilation with respect to the origin $(0, \ldots, 0)$, and other points in the Euclidean space $\Re^n$ could be chosen to be the fixpoint as well.

In studies on multigrid convergence sometimes it may be more appropriate to consider sets of the form $r \cdot S$ (the approach preferred, e.g., by Jordan and Minkowski) digitized in the orthogonal grid with unit grid length, instead of sets S digitized in r-grids with grid length $1/r$. The study of $r \to \infty$ corresponds to the increase in grid resolution, and this may be either a study of repeatedly dilated sets $r \cdot S$ in the grid with unit grid length, or of a given set S in repeatedly refined grids. This is a general *duality principle for multigrid studies* [ ]. We choose the repeatedly refined grid approach for this paper which is of common use in numerics. An *r-grid point* $g^r_{i_1,\ldots,i_n} = (i_1/r, \ldots, i_n/r)$ is defined by integers $i_1, \ldots, i_n$.

Definition 1. *For a set $S \subset \Re^n$, $n \geq 1$, its* Gauss digitization $G_r(S)$ *is defined to be the set of all r-grid points contained in S. When $r = 1$ the Gauss digitization is denoted by $G(S)$.*

For example, consider $\Re^2$ and all r-grid points as centers of isothetic squares with edge length $1/r$. Then the set $\mathbf{G}_r(S)$ is defined to be the union of all those squares having their center points in $G_r(S)$.

If the given set is a curve γ in the plane then the grid-intersection model [,] is of common use in digital geometry. Of course, this scheme can be adapted to r-grid points for any value of $r > 0$, and the resulting sequence of r-grid points is the *grid intersection digitization* $I_r(\gamma)$, which can be characterized by a start point and a chain code (i.e. a sequence of directional codes). A *digital straight line* $I_r(\gamma)$ is an 8-curve of r-grid points resulting from the grid-intersection digitization of the straight line γ in the Euclidean plane, excluding the straight lines $y = x + i/2$, where i is an integer. A *digital straight line segment* (DSS) is a finite 8-connected subsequence of a digital straight line.

The important problem of *volume estimation* was studied in [ ] based on gridding techniques. Any grid point (i, j, k) in the Euclidean space $\Re^3$ is assumed to be the center point of a cube with faces parallel to the coordinate planes and with edges of length 1. The boundary is part of this cube (i.e. it is a closed set). Let S be a set contained in the union of finitely many such cubes. Dilate the set S with respect to an arbitrary point $p \in \Re^3$ in the ratio $r : 1$. This transforms S into S^p_r. Let $l^p_r(S)$ be the number of cubes completely contained in the interior of S^p_r, and let $u^p_r(S)$ be the number of cubes having a non-empty intersection with S^p_r. In [ ] it is shown that $r^{-3} \cdot l^p_r(S)$ and $r^{-3} \cdot u^p_r(S)$ always converge to limit values $L(S)$ and $U(S)$, respectively, for $r \to \infty$, independently of the chosen point p. Jordan called $L(S)$ the *inner volume* and $U(S)$ the *outer volume* of set S, or the *volume* $\mathcal{V}(S)$ of S if $L(S) = U(S)$. The volume definition based on gridding techniques was further studied, e.g., in [,].

The following definition is about an n-dimensional situation. For $n = 3$, a *regular Euclidean cell complex* consists of (topologically closed) r-cubes, r-squares, r-edges and r-vertices (see [ ] for a review on cell complexes), and generalizations to higher dimensions, as well as a restriction to the two-dimensional case are straightforward.

Definition 2. *For a set $S \subset \Re^n$, $n \geq 1$, its* Jordan digitizations $J_r^-(S)$ *and* $J_r^+(S)$ *are defined as follows: the set $J_r^-(S)$ (also called the* inner *digitization) contains all n-dimensional cells completely contained in the interior of set S, and the set $J_r^+(S)$ (also called the* outer *digitization) contains all n-dimensional cells having a non-empty intersection with set S.*

The unions of all cells contained in $J_r^-(S)$ or $J_r^+(S)$ are isothetic polyhedra $\mathbf{J}_r^-(S)$ or $\mathbf{J}_r^+(S)$, respectively. The Hausdorff-Chebyshev distance, generated by the d_∞ metric, between the polyhedral boundaries $\partial \mathbf{J}_r^-(S)$ and $\partial \mathbf{J}_r^+(S)$ is greater than or equal to $1/r$ for any non-empty closed set S, and it holds that

$$\mathbf{J}_r^-(S) \subset S \subseteq \mathbf{J}_r^+(S)$$

in this case.

Gauss and Jordan digitizations have been used in gridding studies in mathematics (geometry of numbers, number theory, analysis). The model of grid-intersection digitization has been introduced for computer images, and it may be applied to planar curves.

A general scheme for comparing results obtained for picture subsets with the true quantities defined by the corresponding operation on the preimage in Euclidean space has been formalized in [ ]. The following definition [ ] specifies a measure for the speed of convergence toward the true quantity.

Definition 3. *Let F be a family of sets S in $\Re^n$, and $dig_r(S)$ a digital image of set S, defined by a digitization mapping dig_r. Assume that a quantitative property $\mathcal{P}$, such as area, perimeter, or a moment, is defined for all sets in family F. An estimator $E_\mathcal{P}$ is* multigrid convergent *for this family F and this digitization model dig_r iff there is a grid resolution $r_S > 0$ for any set $S \in F$ such that the estimator value $E_\mathcal{P}(dig_r(S))$ is defined for any grid resolution $r \geq r_S$, and*

$$|E_\mathcal{P}(dig_r(S)) - \mathcal{P}(S)| \leq \kappa(r)$$

for a function κ defined for real numbers, having positive real values only, and converging toward 0 if $r \to \infty$. The function κ specifies the convergence speed.

Gauss and Dirichlet knew already that the number of grid points inside a planar convex curve γ estimates the area of the set bounded by the curve within an order of $O(\mathcal{L}(\gamma))$, where $\mathcal{L}(\gamma)$ is the length of γ.

Theorem 1. (Gauss/Dirichlet ca. 1820) *For the family of planar convex sets, the number of r-grid points contained in a set approximates the true area with at least linear convergence speed, i.e. $\kappa(r) = r^{-1}$.*

Today we know that the convergence speed of this estimator is actually at least $r^{-1.3636}$ [ ] for planar, bounded, *3-smooth* (i.e. continuous 3rd derivatives with positive curvature at all boundary points except a finite number of arc endpoints) convex sets, and it cannot be better than $r^{-1.5}$, which is a trivial lower bound.

Theorem 2. (Huxley 1990) *For the family of planar, bounded, 3-smooth convex sets, the number of r-grid points contained in a set approximates the true area with a convergence speed of $\kappa(r) = r^\alpha$, for $-1.5 \le \alpha < -1.3636$.*

Closing the gap is an open problem which is a famous subject in number theory [ ], and is closely related to digital geometry [ ].

3   Moments and Moment-Based Features in 2D

In this section we cite worst-case error bounds from [ ] in estimating real moments (and related features) of sets $S \subset \Re^2$ from corresponding discrete moments. Note that in case of 3-smooth sets the claimed positive curvature excludes straight boundary segments. Throughout this section we assume that S is a planar convex set whose boundary consists of a finite number of C^3 arcs, also allowing straight line segments if not otherwise stated. The (p,q)-*moments* of set S are defined by

$$\mathcal{M}_{p,q}(S) = \iint\limits_S x^p y^q \, dx \, dy \, ,$$

for integers $p, q \ge 0$. The moment $\mathcal{M}_{p,q}(S)$ has the *order* $p+q$. In image analysis, the exact values of moments $\mathcal{M}_{p,q}(S)$ remain unknown. They are estimated by *discrete moments* $\mu_{p,q}(S)$ where

$$\mu_{p,q}(S) = \sum_{(i,j) \in G(S)} i^p \cdot j^q$$

which can be calculated from the corresponding digitized set $G(S)$ of set S. The grid resolution r has to be used as scaling factor if the approach involves repeatedly refined grids. The moment-concept has been introduced into image analysis in [ ].

The *contents* or *area* $\mathcal{A}(S)$ of a planar set S, i.e. the moment $\mathcal{M}_{0,0}(S)$ of order zero, is estimated by the number of grid points in $G(S)$, i.e. by the discrete moment $\mu_{0,0}(S)$. For the *center of gravity* of a set S,

$$\left(\frac{\mathcal{M}_{1,0}(S)}{\mathcal{M}_{0,0}(S)}, \frac{\mathcal{M}_{0,1}(S)}{\mathcal{M}_{0,0}(S)} \right)$$

the estimate

$$\left(\frac{\mu_{1,0}(S)}{\mu_{0,0}(S)}, \frac{\mu_{0,1}(S)}{\mu_{0,0}(S)} \right)$$

is calculated from its digital set $G(S)$. The orientation of a set S can be described by its axis of the least second moment. That is the line for which the integral of the squares of the distances to points of S is a minimum. That integral is

$$I(S, \varphi, \rho) = \iint\limits_{S} r^2(x, y, \varphi, \rho)dxdy \ ,$$

where $r(x, y, \varphi, \rho)$ is the perpendicular distance from the point (x, y) to the line given in the form

$$x \cdot \cos \varphi - y \cdot \sin \varphi = \rho \ .$$

We are looking for the value of φ for which $I(S, \varphi, \rho)$ takes its minimum, and by this angle we define the *orientation* of the set S. This φ-value will be denoted by $\mathcal{D}(S)$, i.e.

$$\min_{\varphi, \rho} I(S, \varphi, \rho) \ = \ I(S, \mathcal{D}(S), \overline{\rho}), \text{ for some value of } \overline{\rho} \ .$$

Again, this feature is estimated by replacing integration and set S by a discrete addition and a digital set $G(S)$, respectively. With respect to applications note that this feature requires sets with "a main orientation", i.e. $\mathcal{M}_{2,0}(S) \neq \mathcal{M}_{0,2}(S)$. Finally, we also mention the *elongation* of S (see [,]) in direction φ which is the ratio of maximum and minimum values of $I(S, \varphi, \rho)$, i.e.

$$\mathcal{E}(S) \ = \ \frac{\max\limits_{\varphi, \ \rho} I(S, \varphi, \rho)}{\min\limits_{\varphi, \ \rho} I(S, \varphi, \rho)} \ .$$

It may be estimated by digital approximations of the I-function values as in case of the orientation of set S.

The curvature of the boundary of the considered set plays an important role. It makes an essential difference whether at least one straight section on the boundary is allowed or not.

Theorem 3. (Klette/Žunić 2000) *Let S a convex set whose boundary consists of a finite number of C^3 arcs, then $\mathcal{M}_{p,q}(S)$ can be estimated by $r^{-(p+q+2)} \cdot \mu_{p,q}(r \cdot S)$ within an error of $O(r^{-1})$, and this error term is the best possible.*

However, if S is 3-smooth and convex, i.e. the boundary does not possess any straight segment, then the application of *Huxley's* theorem leads to a reduced upper error bound.

Theorem 4. (Klette/Žunić 2000) *Let a planar bounded 3-smooth convex set S be given. Then $\mathcal{M}_{p,q}(S)$ can be estimated by $r^{-(p+q+2)} \cdot \mu_{p,q}(r \cdot S)$ within an error of $O\left((\log r)^{\frac{47}{22}}\right) \cdot r^{-\frac{15}{11}}\right) \approx O\left(r^{-1.3636...}\right)$.*

The following theorem specifies how progress in the estimation of the "basic difference" $|\mathcal{M}_{0,0}(r \cdot S) - r^2 \cdot \mathcal{A}(S)|$ by $O(\kappa(r))$ can be used to improve error bounds for higher-order estimates $|\mathcal{M}_{p,q}(S) - r^{-(p+q+2)} \cdot \mu_{p,q}(r \cdot S)|$, for a set S being n-smooth, for some integer $n = 3, 4, \ldots$, including $n = \infty$. For a function $\kappa(r) \geq 0$, for $r \geq 0$, a family $F_{\kappa(r)}$ of classes C of planar sets is defined such that the the following conditions are satisfied:

(i) C is nonempty;

(ii) if $S \in C$ then it satisfies

$$|\mu_{0,0}(r \cdot S) - r^2 \cdot \mathcal{A}(S)| = O(\kappa(r)) \;;$$

(iii) if a set S belongs to C then any isometric transformation of S belongs to C as well;

(iv) any set which can be represented by a finite number of unions, intersections and set-differences of sets from C also belongs to C.

This definition allows a formulation of the following theorem which 'translates' possible future progress in number theory into a formulation of related error bounds for moments of arbitrary order. Let F_0 be the smallest family of sets which contains all n-smooth planar convex bounded sets, and which is closed with respect to finite numbers of intersections, unions and set-theoretical differences.

Theorem 5. (Klette/Žunić 2000) *Let S be a planar n-smooth convex set and let $\kappa(r)$ be such that F_0 is contained in the family $F_{\kappa(r)}$. It follows that $M_{p,q}(S)$ can be estimated by $r^{-(p+q+2)} \cdot \mu_{p,q}(r \cdot S)$ within an error of $O\left(\kappa(r) \cdot r^{-2}\right)$.*

Let S be a set in a class in F_0. The given theorems allow a derivation of the following upper error bounds for feature estimations [ ]. An upper error bound for area estimates $\frac{1}{r^2} \mu_{0,0}(r \cdot S)$ is directly given by *Huxley*'s theorem, i.e.

$$\left|\mathcal{A}(S) - \frac{1}{r^2} \cdot \mu_{0,0}(r \cdot S)\right| = O\left(r^{-\left(\frac{15}{11} - \varepsilon\right)}\right) .$$

The same upper error bound holds for the estimates

$$\frac{1}{r} \cdot \frac{\mu_{1,0}(r \cdot S)}{\mu_{0,0}(r \cdot S)} \quad \text{and} \quad \frac{1}{r} \cdot \frac{\mu_{0,1}(r \cdot S)}{\mu_{0,0}(r \cdot S)}$$

of the coordinates

$$\frac{m_{1,0}(S)}{m_{0,0}(S)} \quad \text{and} \quad \frac{m_{0,1}(S)}{m_{0,0}(S)}$$

of the center of gravity, respectively. For the estimate of the orientation only sets S with $\mathcal{M}_{2,0}(S) \neq \mathcal{M}_{0,2}(S)$ are relevant. Then S's orientation $\mathcal{D}(S)$ can be recovered within an worst-case error of $\mathcal{O}(r^{-\frac{15}{11}+\varepsilon})$, by using the estimate

$$\tan(2 \cdot \mathcal{D}(S)) \approx \frac{2 \cdot \bar{\mu}_{1,1}(r \cdot S)}{\bar{\mu}_{2,0}(r \cdot S) - \bar{\mu}_{0,2}(r \cdot S)} .$$

The elongation $\mathcal{E}$ of a 3-smooth convex set S can be estimated by

$$\Theta(r \cdot S) = \frac{t_1(r \cdot S) + \sqrt{t_2(r \cdot S)}}{t_1(r \cdot S) - \sqrt{t_2(r \cdot S)}} ,$$

for

$$t_1(r \cdot S) = \overline{\mu}_{2,0}(r \cdot S) + \overline{\mu}_{0,2}(r \cdot S)$$

and

$$t_2(r \cdot S) = 4 \cdot (\overline{\mu}_{1,1}(r \cdot S))^2 + (\overline{\mu}_{2,0}(r \cdot S) - \overline{\mu}_{0,2}(r \cdot S))^2 \ ,$$

for a planar set S. The error in the approximation $\mathcal{E}(S) \approx \Theta(r \cdot S)$ has an upper error bound in $O\left(r^{-\frac{15}{11}+\varepsilon}\right)$.

Of course, Theorems 3, 4, and 5 may also be used to derive error bounds for features defined by moments of higher order than just up to order two.

4 Length of Curves in 2D and 3D

There are provable convergent length estimators in 2D and 3D, where linear multigrid convergence has been shown for planar convex curves. A superlinear convergence $O(r^{-1.5})$ of asymptotic length estimation has been achieved in [ ] just for the case of digitized straight lines, and there are superlinear length estimators in 2D if sampling of curves is assumed instead of digitization.

4.1 Polygonal Approximations of Curves in 2D

Boundaries of digitized planar sets, or digitized planar curves, can be approximated by a polygon, using a *digital straight segment* (DSS) procedure to segment the boundary into a sequence of maximal-length DSSs. The resulting polygon depends on the starting point and the orientation of the scan. Besides this DSS-based approach to the approximation of digital curves by polygons, there are other possible approaches using minimum-length polygons; see [, ,]. We review all three methods with respect to multigrid convergence.

Given a connected region S in the Euclidean plane and a grid resolution r, the *r-frontier* $\partial\mathbf{G}_r(S)$ of S is uniquely determined. Note that an r-frontier may consists of several non-connected curves even in the case of a bounded convex set S. A set S is *r-compact* iff there is a number $r_S > 0$ such that $\partial\mathbf{G}_r(S)$ is just one (connected) curve, for any $r \geq r_0$.

Theorem 6. (Kovalevsky/Fuchs 1992, Klette/Zunic 2000) *Let S be a convex, r-compact polygonal set in $\Re^2$. Then there exists a grid resolution r_0 such that for all $r \geq r_0$, any DSS approximation of the r-frontier $\partial\mathbf{G}_r(S)$ is a connected polygon with perimeter p_r satisfying the inequality*

$$|\mathcal{L}(\partial(S)) - p_r| \leq \tfrac{2\pi}{r}\left(\varepsilon_{DSS}(r) + \frac{1}{\sqrt{2}}\right) .$$

This theorem and its proof can be found in [ ]. The proof is based to a large extent on material given in [ ]. The value of r_0 depends on the given set, and $\varepsilon_{DSS}(r) \geq 0$ is an algorithm-dependent approximation threshold specifying the maximum Hausdorff-Chebyshev distance (generalizing the Euclidean distance between points to a distance between sets of points) between the r-frontier

$\partial \mathbf{G}_r(S)$ and the constructed (not uniquely specified !) DSS approximation polygon. Assuming $\varepsilon_{DSS}(r) = 1/r$, it follows that the upper error bound for DSS approximations is characterized by[1]

$$\frac{2\pi}{r^2} + \frac{2\pi}{r \cdot \sqrt{2}} \approx \frac{4.5}{r} \quad \text{if} \quad r \gg 1 \quad (\text{i.e. } r \text{ is large}) \,.$$

The grid resolution $1/r$ is assumed in the chord property in [ ], where a DSS is defined to be a finite 8-path. In the case of using cell complexes it is appropriate to consider a finite 4-path as a DSS iff its main diagonal width is less than $\sqrt{2}$, see [,].

A second approach, see [ ], is based on Jordan digitization of sets S in the Euclidean plane. The difference set $\mathbf{O}_r(S) \setminus \mathbf{I}_r(S)$ can be transformed into a subset such that the Hausdorff-Chebyshev distance (generated by the d_∞ metric) between its inner and outer boundary is exactly $1/r$, i.e. the grid constant. The perimeter of S can be estimated by the length of a minimum-length polygon (MLP) contained in this subset, and circumscribing the inner boundary of this subset, which is homeomorphic to an annulus. The subset can be described by a sequence of r-squares, where any r-square has exactly two edge-neighboring r-squares in the sequence. Such a sequence is called a *one-dimensional grid continuum* (1D-GC). Such 1D-GCs are treated in the theory of 2D cell complexes in the plane. This specifies an alternative approach (GC-MLP in short) to the approximation of digital curves; it has been experimentally compared with the DSS method in [ ].

For the case of GC-MLP approximations there are several convergence theorems in [ ], showing that the perimeter of the GC-MLP approximation is a convergent estimator of the perimeter of a bounded, convex, smooth or polygonal set in the Euclidean plane. The following theorem is basically a quotation from [ ]; it specifies the asymptotic constant for GC-MLP perimeter estimates.

Theorem 7. (Sloboda/Zaťko/Stoer 1998) *Let γ be a (closed) convex curve in the Euclidean plane which is contained in a 1D-GC of r-squares, for $r \geq 1$. Then the GC-MLP approximation of this 1D-GC is a connected polygonal curve with length l_r satisfying the inequality*

$$l_r \leq \mathcal{L}(\gamma) < l_r + \frac{8}{r} \,.$$

Finally we sketch a third method, see [,], which is also based on minimum-length polygon calculation. Assume an r-frontier of S which can be represented in the form $P = (v_0, v_1, \ldots, v_{n-1})$ where vertices are clockwise ordered and the interior of S lies to the right. For each vertex of P we define forward and backward shifts: The *forward shift* $f(v_i)$ of v_i is the point on the edge (v_i, v_{i+1}) at the distance δ from v_i. The *backward shift* $b(v_i)$ is that on the edge (v_{i-1}, v_i) at the distance δ from v_i.

---

[1] Let $\kappa(r) = 2\pi/r^2 + 2\pi/r \cdot \sqrt{2}$. Then it follows that $\kappa(r) \to \pi\sqrt{2}$ as $r \to \infty$.

In the approximation scheme as detailed below we replace an edge (v_i, v_{i+1}) by a line segment $(v_i, f(v_{i+1}))$ interconnecting v_i and the forward shift of v_{i+1}, which is referred to as the *forward approximating segment* and denoted by $L_f(v_i)$. The *backward approximating segment* $(v_i, b(v_{i-1}))$ is defined similarly and denoted by $L_b(v_i)$. Now we have three sets of edges, original edges of the r-frontier, forward and backward approximating segments. Let $0 < \delta \leq 0.5/r$. Based on these edges we define a connected region $A_r^\delta(S)$, which is homeomorphic to the annulus, as follows:

Given a polygonal circuit P describing an r-frontier in clockwise orientation, by reversing P we obtain a polygonal circuit Q in counterclockwise order. In the initialization step of our approximation procedure we consider P and Q as the *external* and *internal* bounding polygons of a polygon P_B homeomorphic to the annulus. It follows that this initial polygon P_B has area contents zero, and as a set of points it coincides with $\partial \mathbf{G}_r(S)$.

Now we 'move' the external polygon P 'away' from $\mathbf{G}_r(S)$, and the internal polygon Q 'into' $\mathbf{G}_r(S)$ as specified below. This process will expand P_B step by step into a final polygon which contains $\partial \mathbf{G}_r(S)$, and where the Hausdorff-Chebyshev distance between P and Q becomes non-zero. For this purpose, we add forward and backward approximating segments to P and Q in order to increase the area contents of the polygon P_B.

To be precise, for any forward or backward approximating segment $L_f(v_i)$ or $L_b(v_i)$ we first remove the part lying in the interior of the current polygon P_B and updating the polygon P_B by adding the remaining part of the segment as a new boundary edge. The direction of the edge is determined so that the interior of P_B lies to the right of it. The resulting polygon P_B^δ is referred to as the *approximating sausage* of the r-frontier and denoted by $A_r^\delta(S)$. The width of such an approximating sausage depends on the value of δ. An *AS-MLP curve* for approximating the boundary of S is defined as being a shortest closed curve $\gamma_r^\delta(S)$ lying entirely in the interior of the approximating sausage $A_r^\delta(S)$, and encircling the internal boundary of $A_r^\delta(S)$. It follows that such an AS-MLP curve $\gamma_r^\delta(S)$ is uniquely defined, and that it is a polygonal curve defined by finitely many straight segments. Note that this curve depends upon the choice of the approximation constant δ.

Theorem 8. (Asano/Kawamura/Klette/Obokkata 2000) *Let S be a bounded, r-compact convex polygonal set. Then, there exists a grid resolution r_0 such that for all $r \geq r_0$ it holds that any AS-MLP approximation of the r-frontier $\partial \mathbf{G}_r(S)$, with $0 < \delta \leq .5/r$, is a connected polygon with a perimeter l_r and*

$$|\mathcal{L}(\partial S) - l_r| \leq (4\sqrt{2} + 8 * 0.0234)/r = 5.844/r. \tag{1}$$

These three Theorems 6, 7 and 8 show that the DSS error bound of $4.5/r$ is smaller than the AS-MLP bound $5.844/r$, and the AS-MLP is smaller than the GC-MLP bound $8/r$. Further theoretical and experimental measures may be used for performance comparisons such as *effectiveness* defined by the product of error and number of generated line segments, or the time efficiency of implemented algorithms. With respect to asymptotic time complexity, a linear-time algorithm is known for any of these three linear approximation (i.e. polygonalization) methods [,   ].

4.2   Higher-Order Approximation for Sampled 2D Curves

Higher-order approximations of curves with the purpose of length estimations have been studied in [,   ]: for $k \geq 1$, estimate the length $\mathcal{L}(\gamma)$ of a C^k regular parametric curve $\gamma : [0,1] \to \Re^n$ from $(m+1)$-tuples $Q_m = (q_0, q_1, \ldots, q_m)$ of points $q_i = \gamma(t_i)$ positioned on the curve γ. The parameters t_i's are not assumed to be given. Of course, sampling (see both examples in the Introduction) is a different situation compared to digitization. An increase in grid resolution r defining a scale for two dimensions in the plane, corresponds to an increase in the number m of sampling points defining a one-dimensional scale on the curve.

Some assumptions about the distribution of the t_i's are needed to make the sampling problem solvable [   ]. The problem is the easiest when the t_i's are chosen in a perfectly uniform manner, namely $t_i = \frac{i}{m}$. In such a case it seems natural to estimate γ by a curve $\widetilde{\gamma}$ that is piecewise polynomial of degree $a \geq 1$. Then we prove

Theorem 9. (Noakes/Kozera/Klette 2001) *Let γ be C^{s+2} and let t_i's be sampled perfectly uniformly. Then $\mathcal{L}(\widetilde{\gamma}) = \mathcal{L}(\gamma) + O(\frac{1}{m^{s+s_0}})$, where s_0 is 1 or 2 according as s is odd or even.*

It is known [   ] that Lagrange estimates of length based on a uniform grid do not always converge to $\mathcal{L}(\gamma)$ when the unknown t_i's are non-uniform. In [   ] it is shown that there are some approximately uniform samplings of t_i's for which those estimates are well-behaved. More precisely

Definition 4. *For $\varepsilon \geq 0$, the t_i's are (ε, k)-uniformly sampled if there is a C^k reparameterization $\phi : [0,1] \to [0,1]$, for $k \geq 1$, such that $t_i = \phi(\frac{i}{m}) + O(\frac{1}{m^{1+\varepsilon}})$.*

Lagrange estimates of length can behave badly for $(0, k)$-uniform sampling, see [   ], but for $0 < \varepsilon \leq 1$ the following theorem holds [   ], using piecewise Lagrange interpolants $\widetilde{\gamma}$.

Theorem 10. (Noakes/Kozera/Klette 2001) *Let the t_i's be sampled (ε, k)-uniformly where $0 < \varepsilon \leq 1$ and $k \geq 4$. Then, for piecewise-quadratic Lagrange interpolants $\widetilde{\gamma}$, determined by a sampled $(m+1)$-tuple Q_m and based on a uniform grid, $\mathcal{L}(\widetilde{\gamma}) = \mathcal{L}(\gamma) + O(\frac{1}{m^{4\varepsilon}})$.*

Whereas Theorem 9 permits length estimates of arbitrary accuracy (for s sufficiently large), Theorem 10 refers only to piecewise-quadratic estimates (i.e. $s = 2$), and accuracy is limited accordingly. However, even in the latter case it holds that the quartic convergence speed[2] is three magnitudes faster then the linear convergence speed discussed for DSS, GC-MLP and AS-MLP polygonalizations (with $s = 1$). These sampling-based results encourage further research on higher-order approximations for digitized curves.

---

[2]   Note that m specifies an increase in the order of $\sqrt{m}$ only with respect to a scale for two dimensions in the plane, i.e. the quartic convergence speed in m may be compared with a quadratic convergence speed in r.

4.3   A Polygonal Approximation of Curves in 3D

Consider the length estimation problem for rectifiable curves γ in the three-dimensional Euclidean space. We assume curves γ which lead to simple cube curves for the digitization model $J_r^+(\gamma)$.

A *cube-curve* is a sequence $g = (f_0, c_0, f_1, c_1, ..., f_n, c_n)$ of r-faces f_i and r-cubes c_i in $\Re^3$, for $0 \le i \le n$, such that r-faces f_i and f_{i+1} are sides of r-cube c_i, for $0 \le i \le n$ and $f_{n+1} = f_0$. Such a cube-curve is *simple* iff $n \ge 4$, and for any two r-cubes c_i, c_k in g with $|i - k| \ge 2 \pmod{n+1}$ it holds that if $c_i \cap c_k \ne \emptyset$ then either $|i - k| = 2 \pmod{n+1}$ and $c_i \cap c_k$ is an r-edge, or $|i - k| = 3 \pmod{n+1}$ and $c_i \cap c_k$ is an r-vertex. A *tube* **g** is the union of all r-cubes contained in a cube-curve g. Such a tube is a polyhedrally-bounded compact set in $\Re^3$, and it is homeomorphic with a torus in case of a simple cube-curve.

A curve is *complete in* **g** iff it has a non-empty intersection with any r-cube contained in g.

Definition 5. *A minimum-length polygon (MLP) of a simple cube-curve g is a shortest polygonal simple curve σ which is contained and complete in tube* **g**.

Following [ ], the *length* of a simple cube-curve g is defined to be the length $\mathcal{L}(\sigma)$ of an MLP σ of g. Theorem 7 states that this length estimation approach is multigrid convergent to the true value in case of planar convex curves γ as specified in this theorem.

An algorithm for approximating such an MLP in a simple cube-curve has been specified in [ ]. It is based on the following theorem [ ]. An edge contained in a tube **g** is *critical* iff this edge is the intersection of three cubes contained in the cube-curve g.

Theorem 11. (Buelow/Klette 2000) *Let g be a simple cube-curve. Critical edges are the only possible locations of vertices of a shortest polygonal simple curve contained and complete in tube* **g**.

The algorithmic solution in [ ] provides a polygonal approximation of desired MLP's, and thus a length measurement method for simple cube-curves in 3D space. The algorithm possesses a measured time complexity in $O(n)$. However, two open problems remain at this stage: the time complexity might be provable always in $O(n)$, and the convergence might be provable always towards the MLP. For details of the algorithm see [ ].

5   Surface Area of Regular Solids

A '3D object' can be modeled by a regular solid, which is defined to be a simply-connected compact set having a measurable surface area [ ]. Algorithms for multigrid-convergent surface area estimation are still a research topic. Obviously, increasing the grid resolution in a digitization of a regular solid in the form of a 3D cell complex [ ], and measuring the area of the resulting isothetic surface, does not result in convergence to the true value. This might be compared with

Fig. 3. Three Euclidean sets digitized for increasing grid resolution and approximated by marching-cube polyhedrizations

the fact that 4-path length is not related to the length of a digitized curve in 2D. Marching-cube based polyhedrizations, see Fig. 3 do not support multigrid-convergent surface area estimations toward the true value [19]. This might be compared with the fact that 8-path length (with weighting factor $\sqrt{2}$ for diagonal steps) is not related to the length of a digitized curve in 2D.

Polyhedrization is a common goal of segmenting the surface of a digitized regular solid, normally given in the form of boundary points of a 3D grid point set (e.g. using 3D Gauss digitization) or in the form of a two-dimensional grid continuum (2D-GC) defined by a difference between the inner and outer Jordan digitizations.

5.1   Experimentally Measured Convergence

Expanding the ideas of DSS approximations into 3D leads to a *digital plane segment* (DPS) approach for achieving multigrid-convergent measurement of surface area: the boundary of a Gauss- or Jordan-digitized set is segmented into maximum-size DPSs, and surface areas of related polyhedral faces are added to form a final estimate.

[32] introduced *arithmetic geometry* which allows characterizations of hyperplanes in n-dimensional spaces. [2] proposed a general definition that linked planes and topology, introducing $|a| + |b| + |c|$ thick planes. These planes were further specified and used in [11]. For a generalization to n dimensions see [3]. Digital plane segments can be defined within arithmetic geometry as follows: r-cubes have eight directed diagonals. The *main diagonal* of a Euclidean plane is those directed diagonal (out of these eight) that has the largest dot product (inner product) with the normal of the plane. Note that in general there may be

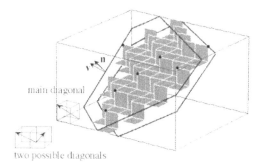

Fig. 4. Illustration of the main diagonal of a DPS

more than one main diagonal for a Euclidean plane; if so, we can choose any of them as the main diagonal. The distance between two parallel Euclidean planes in the main diagonal direction is called the *main diagonal distance* between these two planes.

Now consider a finite set of faces of r-cubes in 3D space. A Euclidean plane is called a *supporting plane* of this set if it is incident with at least three non-collinear vertices of the set of faces, and all the faces of the set are in only one of the (closed) halfspaces defined by the plane. Note that any non-empty finite set of faces has at least one supporting plane. Any supporting plane defines a *tangential plane*, which is the nearest parallel plane to the supporting plane such that all faces of the given set are within the closed slice defined by the supporting and tangential planes. Note that a tangential plane may be a supporting plane as well. Figure 4 gives a rough sketch of such a set of faces, where **n** denotes the normal to the two parallel planes, and **v** is the main diagonal.

Definition 6. *A finite, edge-connected set of faces in 3D space is a* digital planar segment *(DPS) iff it has a supporting plane such that the main diagonal distance between this plane and its corresponding tangential plane is less than* $\sqrt{3}/r$ *(i.e. the length of a diagonal of an r-cube).*

Such a supporting plane is called *effective* for the given set of grid faces. Let **v** be a vector in a main diagonal direction with a length of $\sqrt{3}/r$, let **n** be the normal vector to a pair of parallel planes, and let $d = \mathbf{n} \cdot \mathbf{p}_0$ be the equation of one of these planes. According to our definition of a DPS, all the vertices **p** of the faces of a DPS must satisfy the following inequality:

$$0 \leq \mathbf{n} \cdot \mathbf{p} - d < \mathbf{n} \cdot \mathbf{v}$$

Let $\mathbf{n} = (a, b, c)$. Then this inequality becomes

$$0 \leq ax + by + cz - d < |a| + |b| + |c| \ ,$$

i.e. an DPS is an edge-connected subset of faces in a *standard plane* [11]. A *simply-connected DPS* is such that the union of its faces is topologically equivalent (in Euclidean space) to the unit disk.

The general DPS recognition problem can be stated as follows: Given n vertices $\{\mathbf{p}_1, \mathbf{p}_2, \ldots, \mathbf{p}_n\}$, does there exist a DPS such that each vertex satisfies the inequality system

$$0 \leq \mathbf{n} \cdot \mathbf{p}_i - d < \mathbf{n} \cdot \mathbf{v}, \quad i = 1, \ldots, n,$$

[ ] suggests a method of turning this into a linear inequality system, by eliminating the unknown d as follows:

$$\mathbf{n} \cdot \mathbf{p}_i - \mathbf{n} \cdot \mathbf{p}_j < \mathbf{n} \cdot \mathbf{v}, \quad i, j = 1, \ldots, n,$$

This system of n^2 inequalities can be solved in various ways. [,] use a Fourier elimination algorithm. However, this algorithm is not time-efficient even for very small cell complexes. In fact, in [ ] a more advanced elimination technique than Fourier-Motzkin was proposed to eliminate unknowns from systems of inequalities. This technique eliminates all variables at once, whereas the Fourier-Motzkin technique eliminates one variable at a time. Eliminating all variables at once leads to an $O(n^4)$ algorithm for recognizing a DPS, which is faster than the algorithm sketched in [ ]. [ ] included results for hyperplanes of arbitrary dimension. Note that two different definitions are actually used to define digital planes, depending on what kind of connectivity relation is required:

$$0 \leq ax + by + cz - d < |a| + |b| + |c|$$

or

$$0 \leq ax + by + cz - d < max(|a|, |b|, |c|) .$$

The second definition was used in [ ], but the results obtained there for the elimination technique are equally valid for the first definition.

An incremental algorithm for DPS recognition has been proposed in [ ], based on updating lists of effective supporting planes. This algorithm can be used for segmenting boundaries of digitized 3D sets into maximal-size DPSs, see Fig. 5 for two examples.

Actually, any DPS recognition algorithm could be used for segmenting a surface of a 3D cell complex into maximal-size DPSs. However, the starting point and the search strategy during the process of 'growing' a DPS are critical for the behavior of such an algorithm. Also, after obtaining maximal-size DPSs, it is not straightforward to derive a polyhedron from the resulting segmentation of the surface.

Analytical surface area calculation of an ellipsoid, with all three semi-axes a, b, c allowed to be different, is a complicated task. If two semi-axes coincide, i.e. in the case of an ellipsoid of revolution, the surface area can be analytically specified in terms of standard functions. The surface area formula in the general case is based on standard elliptic integrals. Example 2 in [ ], reporting recent work by G.Tee, specifies an analytical method of computing the surface area of a general ellipsoid. This area can be used in experimental studies as *ground truth* to evaluate the performance of DPS algorithms in surface area estimation. Figure 6 shows the error in the estimated value relative to the true value for an

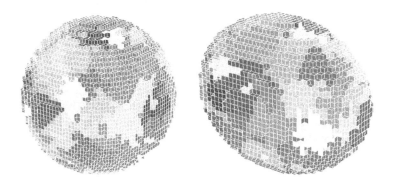

Fig. 5. Agglomeration of faces of a sphere and an ellipsoid into DPSs

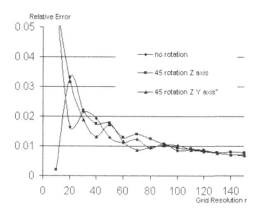

Fig. 6. Relative errors in surface area estimation for an ellipsoid in three orientations for increasing grid resolution. Figure 5 illustrates resolution $r = 40$

ellipsoid in three different orientations, using a search depth of 10 in region growing (breadth-first search). In general these DPS-based estimates behave 'better' than those based on convex hull (for digitized convex sets) or on marching-cube algorithms. Convex hull and marching-cube methods lead to relative errors of 3.22% and 10.80% for $r = 100$, respectively, while the DPS error is less than 0.8%. The DPS method shows a good tendency to converge, but theoretical work needs to be done to prove this. Altogether, there are working algorithms which appear to provide multigrid-convergent surface area estimations, but there is no related theorem stating this property for some type of 3D sets.

5.2   Multigrid Convergence of Estimated Surface Area

Recently there is actually progress on proving multigrid-convergent behavior of surface area estimation, but so far without an algorithmic solution for the proposed method! [,] introduce the *relative convex hull* $\mathrm{CH}_Q(P)$ of a polyhedral solid P which is completely contained in the interior of another polyhedral solid Q. If the convex hull $\mathrm{CH}(P)$ is contained in Q then $\mathrm{CH}_Q(P) = \mathrm{CH}(P)$; otherwise $\mathrm{CH}_Q(P) \subseteq Q$ is a 'shrunk version' of the convex hull. To be precise, let $\overline{\mathbf{pq}}$ be the (real) straight line segment from point $\mathbf{p}$ to point $\mathbf{q}$ in $\Re^3$, and introduce the following definition[ ]:

Definition 7. *A set* $A \subseteq Q \subseteq \Re^3$ *is* Q-*convex iff for all* $\mathbf{p}, \mathbf{q} \in A$ *such that* $\overline{\mathbf{pq}} \subseteq Q$ *we have* $\overline{\mathbf{pq}} \subseteq A$. *Let* $P \subseteq Q$. *The* relative convex hull $CH_Q(P)$ *of* P *with respect to* Q *is the intersection of all* Q-*convex sets containing* P.

For a set $S \subseteq \Re^3$ we defined the inner and outer Jordan digitizations $\mathbf{J}_r^-(S)$ and $\mathbf{J}_r^+(S)$ for grid resolution $r \geq 1$. If S is a regular solid with a defined surface area, let $\mathcal{A}(S)$ be its surface area in the Minkowski sense [ ].

Theorem 12. (Sloboda/Zaťko 2000) *Let* $S \subset \Re^3$ *be a compact set bounded by a smooth closed Jordan surface* ϑS. *Then*

$$\lim_{r \to \infty} s\left(CH_{\mathbf{J}_r^+(S)}\left(\mathbf{J}_r^-(S)\right)\right) = \mathcal{A}(S) \ .$$

This theorem, from [ ], specifies a method of multigrid convergence which still requires research on algorithmic implementation, theoretical and experimental convergence speed, and performance evaluation in comparison with other methods such as the DPS segmentation method sketched above.

6   Conclusions

Euclidean geometry specifies the ground truth, the correct moment, length or surface area prior to digitization. The concept of multigrid convergence may provide a general methodology for evaluating and comparing different approaches. The measurement of quantitative properties is certainly a main topic in digital geometry. [ ] is one of the early publications in this area, and [ ] is one of the more recent ones, both focusing on length estimates. Probability-theoretical aspects of digitization errors [,] have only been studied for a few elementary figures and simple geometric problems; further studies should provide answers to open problems such as those listed in [ ].

There is still no solution with respect to multigrid convergence for surface area estimation which combines a convergence theorem and an algorithmic implementation. The study of non-polygonal approximations of digitized curves with respect to improvements in convergence speed appears as another important open problem.

For all multigrid-convergence problems, it is important to determine what optimum convergence speed $\kappa(r)$ is actually possible (for example, see open problem defined by Theorem 2). A test set of six curves has been specified in [ ] for evaluations of curve length estimations, and general ellipsoids are proposed in [   ] for surface area performance evaluations. Evaluation measures might be, e.g., absolute error, efficiency (error times number of generated segments), and computing time. A classification of properties $\mathcal{L}, \mathcal{A}, \mathcal{M}_{p,q}, \mathcal{V}, \mathcal{E}, \mathcal{D}, \ldots$ with respect to families of sets, optimum convergence speed, and optimum algorithmic time complexity might be a long-term project.

Acknowledgments

This chapter informs about work which has been done partially in collaboration with colleagues as specified by citations, and the author thanks all these colleagues for the excitement in joint work. Thanks also to the reviewers for valuable comments.

References

1. T. A. Anderson and C. E. Kim. Representation of digital line segments and their preimages. *Computer Vision, Graphics, Image Processing*, 30:279–288, 1985. 327
2. E. Andres. Discrete circles, rings and spheres. *Computers & Graphics*, 18:695–706, 1994. 331
3. E. Andres, R. Acharya, and C. Sibata. Discrete analytical hyperplanes. *Graphical Models and Image Processing*, 59:302–309, 1997. 331
4. E. Artzy, G. Frieder, and G. T. Herman. The theory, design, implementation and evaluation of a three-dimensional surface detection algorithm. *Computer Vision, Graphics, Image Processing*, 15:1–24, 1981. 330
5. T. Asano, Y. Kawamura, R. Klette, and K. Obokkata. A new approximation scheme for digital objects and curve lengthy estimations. Internat. Conf. *IVCNZ'00*, 27-29 November, Hamilton (2000) 26–31. 326, 327
6. T. Asano, Y. Kawamura, R. Klette, and K. Obokkata. Minimum-Length Polygons in Approximation Sausages. In: Proc. *Visual Form 2001*, Capri May 2001, LNCS 2059, 103–112, 2001. 326, 327, 329
7. J. Bresenham. An incremental algorithm for digital plotting. In *ACM Natl. Conf.*, 1963. 321
8. T. Buelow and R. Klette: Rubber band algorithm for estimating the length of digitized space-curves. In: Proc. *ICPR*, Barcelona, September 2000, IEEE, Vol. III, 551-555. 330
9. M. Chleík and F. Sloboda. Approximation of surfaces by minimal surfaces with obstacles. Technical report, Institute of Control Theory and Robotics, Slovak Academy of Sciences, Bratislava, 2000. 335
10. L. Dorst and A. W. M. Smeulders. Discrete straight line segments: parameters, primitives and properties. In: R. Melter, P. Bhattacharya, A. Rosenfeld (eds): *Ser. Contemp. Maths.*, Amer. Math. Soc. 119:45–62, 1991. 326
11. J. Françon, J.-M. Schramm, and M. Tajine. Recognizing arithmetic straight lines and planes. In *Proc. DGCI,* LNCS 1176, pages 141–150. Springer, Berlin, 1996. 331, 332, 333
12. H. Freeman. Techniques for the digital computer analysis of chain-encoded arbitrary plane curves. In *Proc. Natl. Elect. Conf.*, volume 17, pages 421–432, 1961. 321

13. R. M. Haralick and L. G. Shapiro. *Computer and Robot Vision, Volume II.* Addison-Wesley, Reading, Massachusetts, 1993. 319
14. M. Hu. Visual pattern recognition by moment invariants. *IRE Trans. Inf. Theory* **8** (1962) 179–187. 323
15. M. N. Huxley. Exponential sums and lattice points. *Proc. London Math. Soc.*, 60:471–502, 1990. 323
16. R. Jain, R. Kasturi, and B. G. Schunck. *Machine Vision*, McGraw-Hill, New York, 1995. 324
17. C. Jordan. Remarques sur les intégrales définies. *Journal de Mathématiques* (4e série), 8:69–99, 1892. 321
18. Y. Kenmochi and R. Klette. Surface area estimation for digitized regular solids. In *Proc. Vision Geometry IX,* SPIE 4117, pages 100–111, 2000. 319, 322, 330, 333, 336
19. R. Klette. Measurement of object surface area. Proc. *Computer Assisted Radiology,* Tokyo (1998) 147–152. 331
20. R. Klette. Cell complexes through time. In *Proc. Vision Geometry IX,* SPIE 4117, pages 134–145, 2000. 322
21. R. Klette and T. Buelow. Critical edges in simple cube-curves. In: Proc. *DGCIÔ2000,* Uppsala December 2000, LNCS, 467-478. 330
22. R. Klette and H.-J. Sun. Digital Planar Segment Based Polyhedrization for Surface Area Estimation. In: Proc. *Visual Form 2001,* Capri May 2001, LNCS 2059, 356–366, 2001. 333
23. R. Klette and B. Yip. The length of digital curves. *Machine Graphics & Vision*, 9:673–703, 2000 (extended version of: R. Klette, V. V. Kovalevsky, and B. Yip. Length estimation of digital curves. In *Proc. Vision Geometry VIII,* SPIE 3811, pages 117–129). 327, 329, 335
24. R. Klette and J. Žunić. Interactions between number theory and image analysis. In *Proc. Vision Geometry IX,* SPIE 4117, pages 210–221, 2000. 323
25. R. Klette and J. Žunić. Multigrid convergence of calculated features in image analysis. *J. Mathematical Imaging Vision,* 13:173–191, 2000. 321, 323, 325, 326, 335
26. V. Kovalevsky and S. Fuchs. Theoretical and experimental analysis of the accuracy of perimeter estimates. In W. Förstner and S. Ruwiedel, editors, *Robust Computer Vision,* pages 218–242. Wichmann, Karlsruhe, 1992. 326
27. Z. Kulpa. Area and perimeter measurements of blobs in discrete binary pictures. *Computer Graphics Image Processing,* 6:434–454, 1977. 319, 335
28. E. Landau. *Ausgewählte Abhandlungen zur Gitterpunktlehre.* Deutscher Verlag der Wissenschaften, Berlin, 1962. 323
29. H. Minkowski. *Geometrie der Zahlen.* Teubner, Leipzig, 1910. 322, 335
30. L. Noakes, R. Kozera, and R. Klette. Length estimation for curves with different samplings. (in this book). 329
31. L. Noakes and R. Kozera. More-or-less uniform sampling and lengths of curves. Submitted. 329
32. J.-P. Reveillès. Géométrie discrète, calcul en nombres entiers et algorithmique. Thèse d'état, Univ. Louis Pasteur, Strasbourg, 1991. 327, 331
33. A. Rosenfeld. Digital straight line segments. *IEEE Trans. Computers,* 23:1264–1269, 1974. 327
34. W. Scherrer. Ein Satz über Gitter und Volumen. *Mathematische Annalen,* 86:99–107, 1922. 322
35. H. A. Schwarz: Sur une définition erronée de l'aire d'une surface courbe, *Ges. math. Abhandl.* **2** (1890) 309–311. 319, 320

36. J. Serra. *Image Analysis and Mathematical Morphology*. Academic Press, New York, 1982. 322
37. F. Sloboda and B. Zaťko. On polyhedral form for surface representation. Technical report, Institute of Control Theory and Robotics, Slovak Academy of Sciences, Bratislava, 2000. 335
38. F. Sloboda, B. Zaťko, and J. Stoer. On approximation of planar one-dimensional continua. In R. Klette, A. Rosenfeld, and F. Sloboda, editors, *Advances in Digital and Computational Geometry*, pages 113–160. Springer, Singapore, 1998. 326, 327, 330
39. B. L. van der Waerden. *Geometry and Algebra in Ancient Civilizations*. Springer, Berlin, 1983. 319
40. P. Veelaert. Digital planarity of rectangular surface segments. *IEEE Trans. PAMI*, 16:647–652, 1994. 333
41. K. Voss. Digitalisierungseffekte in der automatischen Bildverarbeitung. *EIK*, 11:469–477, 1975. 335
42. K. Voss and H. Süsse. *Adaptive Modelle und Invarianten für zweidimensionale Bilder*, Shaker, Aachen, 1995. 324
43. A. M. Vossepoel and A. W. M. Smeulders. Vector code probability and metrication error in the representation of straight lines of finite length. *Computer Graphics Image Processing*, 20:347–364, 1982. 335

Length Estimation for Curves with Different Samplings

Lyle Noakes[1], Ryszard Kozera[2], and Reinhard Klette[3]

[1] The University of Western Australia, Department of Mathematics and Statistics
35 Stirling Highway, Crawley WA 6009, Australia
[2] The University of Western Australia
Department of Computer Science and Sofware Engineering
35 Stirling Highway, Crawley WA 6009, Australia
[3] The University of Auckland, Centre for Image Technology and Robotics
Tamaki Campus, Building 731, Auckland, New Zealand

Abstract. This paper[*] looks at the problem of approximating the length of the unknown parametric curve $\gamma : [0, 1] \to \mathbb{R}^n$ from points $q_i = \gamma(t_i)$, where the parameters t_i are not given. When the t_i are uniformly distributed Lagrange interpolation by piecewise polynomials provides efficient length estimates, but in other cases this method can behave very badly [ ]. In the present paper we apply this simple algorithm when the t_i are sampled in what we call an ε-*uniform* fashion, where $0 \le \varepsilon \le 1$. Convergence of length estimates using Lagrange interpolants is not as rapid as for uniform sampling, but better than for some of the examples of [ ]. As a side-issue we also consider the task of approximating γ up to parameterization, and numerical experiments are carried out to investigate sharpness of our theoretical results. The results may be of interest in computer vision, computer graphics, approximation and complexity theory, digital and computational geometry, and digital image analysis.

1 Introduction

Recent research in digital and computational geometry and image analysis concerns estimation of lengths of digitized curves; indeed the analysis of digitized curves in $\mathbb{R}^2$ or $\mathbb{R}^3$ is one of the most intensively studied subjects in image data analysis. This paper contributes to this topic by showing that there are possible improvements in convergence speed compared to all known methods in digital geometry, however, based on sampling of curves (as common in approximation theory) compared to digitization (as common in digital geometry).

A digitized curve is the result of a process (such as contour tracing or 2D thinning extraction) which maps a curve γ (such as the boundary of a region) onto a computer-representable curve. An analytical description of $\gamma : [0, 1] \to \mathbb{R}^2$

---

[*] This research was performed at the University of Western Australia, while the third author was visiting under the UWA Gledden Visiting Fellowship scheme.[1,2] Additional support was received under an Australian Research Council Small Grant[1] and under an Alexander von Humboldt Research Fellowship.[1b]

G. Bertrand et al. (Eds.): Digital and Image Geometry, LNCS 2243, pp. 339–351, 2001.
© Springer-Verlag Berlin Heidelberg 2001

is not given, and numerical measurements of points on γ are corrupted by a process of *digitization*: γ is digitized within an orthogonal grid of points $(\frac{i}{m}, \frac{j}{m})$, where i, j are permitted to range over integer values, and m is a fixed positive integer called *the grid resolution*.

Depending on the digitization model [ ], γ is mapped onto a digital curve and approximated by a polygonal curve $\hat{\gamma}_m$ whose length is an estimator for $d(\gamma)$. This is a standard approach for approximating a digital curve with respect to geometric analysis tasks. However, different smooth approximations (e.g. snake model) are used in image analysis as well, where the convergence analysis with respect to geometric figures is omitted. Approximating polygons $\hat{\gamma}_m$ based on local configurations of digital curves do not ensure multigrid length convergence, but global approximation techniques yield *linearly* convergent estimates, namely $d(\gamma) - d(\hat{\gamma}_m) = O(\frac{1}{m})$ [ ], [   ], [   ] or [   ]. Recently, experimentally based results reported in [ ] and [   ] confirm a similar rate of convergence for $\gamma \subset \mathbb{R}^3$. In the special case of discrete straight line segments in $\mathbb{R}^2$ a stronger result is proved, for example, [ ], where $O(\frac{1}{m^{1.5}})$ order of asymptotic length estimates are given. In Theorems 1 and 2 presented in this paper the convergence is of order at least $O(\frac{1}{m^{r+1}})$ and $O(\frac{1}{m^{4\varepsilon}})$ when $0 < \varepsilon \le 1$, respectively.

For $k \ge 1$, consider the problem of estimating the length $d(\gamma)$ of a C^k regular parametric curve $\gamma : [0, 1] \to \mathbb{R}^n$ from $m + 1$-tuples $Q = (q_0, q_1, \ldots, q_m)$ of points $q_i = \gamma(t_i)$ on the curve γ. The parameters t_i are not assumed to be given, but some assumptions are needed to make our problem solvable. For example, if none of the t_i lie in $(0, \frac{1}{2})$ the task becomes intractable. The problem is easiest when the t_i are chosen in a perfectly uniform manner, namely $t_i = \frac{i}{m}$ (e.g. see also [   ] or [   ]). In such a case it seems natural to estimate γ by a curve $\tilde{\gamma}$ that is piecewise polynomial of degree $r \ge 1$. We prove first in this paper:

Theorem 1. *Let γ be C^{r+2} and let t_i be sampled perfectly uniformly. Then there exists piecewise-r-degree polynomial $\tilde{\gamma}$, determined by Q such that $d(\tilde{\gamma}) = d(\gamma) + O(\frac{1}{m^{r+p}})$, where p is 1 or 2 according as r is odd or even.*

As usual, $O(g(m))$ means a quantity whose absolute value is bounded by some multiple of $g(m)$ as $m \to \infty$. We are principally concerned with non-uniform sampling. More precisely

Definition 1. *For $0 \le \varepsilon \le 1$, the t_i's are said to be ε-uniformly sampled when there is an order-preserving C^k reparameterization $\phi : [0, 1] \to [0, 1]$ such that $t_i = \phi(\frac{i}{m}) + O(\frac{1}{m^{1+\varepsilon}})$.*

Note that ε-uniform sampling arises from two types of perturbations of uniform sampling: first via a diffeomorphism $\phi : [0, 1] \to [0, 1]$ combined subsequently with added extra distortion term $O(\frac{1}{m^{1+\varepsilon}})$. In particular, for $\phi = id$ and $\varepsilon = 0$ $(\varepsilon = 1)$ the perturbation is *linear* i.e. of uniform sampling order *(quadratic)*, which constitutes asymptotically a big (small) distortion of a uniform partition of $[0, 1]$. The extension of Definition 1 for $\varepsilon > 1$ could also be considered. This case represents, however, a very small perturbation of uniform sampling (up to a ϕ-shift) which seems to be of less interest in applications.

Lagrange estimates of length can behave badly for 0-uniform sampling (the more elaborate algorithm of [ ] is needed for this case), but for $0 < \varepsilon \leq 1$ we prove the following, using piecewise-quadratic Lagrange interpolants $\mathcal{Q}^i$ (see Section 4).

Theorem 2. *Let the t_i be sampled ε-uniformly, where $0 < \varepsilon \leq 1$, and suppose that $k \geq 4$. Then, there is a function** $\widetilde{\gamma}$, determined by $\mathcal{Q}$, such that $d(\widetilde{\gamma}) = d(\gamma) + O(\frac{1}{m^{4\varepsilon}})$.*

Whereas Theorem 1 permits length estimates of arbitrary accuracy (for r sufficiently large), Theorem 2 refers only to piecewise-quadratic estimates, and accuracy is limited accordingly. The interest in this (the main result of the present paper) lies in the non-uniform distribution of the unknown parameters t_i's. The proofs of Theorems 1 and 2 also permit uniform estimates of γ up to reparameterization. Note that the construction of the piecewise-r-degree polynomial interpolant P_r^j including $\mathcal{Q}^i$ (see Sections 3 and 4) requires neither the explicit knowledge of γ nor of the parameters t_i (each P_r^j is constructed over a uniform local grid in $s \in [0, r]$; for $\mathcal{Q}^i$ over uniform grid in $[0, 2]$). The latter are used merely to compare $d(\gamma)$ with $d(\widetilde{\gamma})$ and $\widetilde{\gamma}$ with γ, respectively. More specifically, in order to prove Theorems 1 and 2 both *global* and *local* t- and s-parameterizations shall be used. On the other hand, the explicit construction of the interpolant P_j^r (or $\mathcal{Q}^i$) approximating γ (and thus $d(\gamma)$) resorts exclusively to the local parameterization.

For these results $\mathcal{Q}$ arises from uniform or ε-uniform samplings as opposed to digitization. So strict comparisons cannot be made. Our results seem relevant to digital and image geometry nonetheless for the following reasons. They provide comparisons with the interpolation and indicate potential problems which might arise in digitization based on non-uniform distribution of t_i. Moreover, they show that using piecewise Lagrange polynomial approach to estimate length of a digiztized curve $\hat{\gamma}$ may not always be appropriate. Finally, as a special case we provide upper bounds for optimal rates of convergence when piecewise polynomials are applied to the digitized curves. Related work can also be found in [, , , , ,]. There is also some interesting work on complexity [, , ,].

2   Sampling and Curves

We are going to discuss different ways of forming ordered samples $0 = t_0 < t_1 < t_2 < \ldots < t_m = 1$ of variable size $m + 1$ from the interval*** $[0, 1]$. The simplest procedure is *uniform sampling*, where $t_i = \frac{i}{m}$ (where $0 \leq i \leq m$). Uniform sampling is not invariant with respect to *reparameterizations*, namely order-preserving C^k diffeomorphisms $\phi : [0, 1] \to [0, 1]$, where $k \geq 1$. A small perturbation of uniform sampling is no longer uniform, but may approach uniformity in some asymptotic sense, at least after some suitable reparameterization.

** See section 4 for details.
*** In the present context there is no real gain in generality from considering other intervals $[0, T]$.

The ε-uniform sampling in Definition 1 of the previous Section is a possible example of such perturbation. Note that ϕ and the asymptotic constants are chosen independently of $m \geq 1$, and that ε-uniform implies δ-uniform for $0 \leq \delta < \varepsilon$. Uniform sampling is ε-uniform for any $0 \leq \varepsilon \leq 1$. At the other extreme are examples, where sampling increments $t_i - t_{i-1}$ are neither large nor small, considering m, and yet sampling is not ε-uniform for any $0 < \varepsilon \leq 1$:

Example 1. Set t_i to be $\frac{i}{m}$ or $\frac{2i-1}{2m}$ according as i is even or odd. Then $(1/2m) \leq t_i - t_{i-1} \leq (3/2m)$ for all $1 \leq i \leq m$ and all $m \geq 1$. Thus sampling is 0-uniform. To see that sampling is not ε-uniform for $\varepsilon > 0$ assume the opposite. Then, for some C^1 reparameterization $\phi : [0,1] \rightarrow [0,1]$, $t_{i+1} - t_i = \frac{1}{2m} = \phi\left(\frac{i+1}{m}\right) - \phi\left(\frac{i}{m}\right) + O(\frac{1}{m^{1+\varepsilon}})$ and $t_{i+2} - t_{i+1} = \frac{3}{2m} = \phi\left(\frac{i+2}{m}\right) - \phi\left(\frac{i+1}{m}\right) + O(\frac{1}{m^{1+\varepsilon}})$. By the Mean Value Theorem

$$\frac{1}{2m} = \frac{\phi'(\xi_{1i}^{(m)})}{m} + O(\frac{1}{m^{1+\varepsilon}}), \qquad \frac{3}{2m} = \frac{\phi'(\xi_{2i}^{(m)})}{m} + O(\frac{1}{m^{1+\varepsilon}}), \tag{1}$$

for some $\xi_{1i}^{(m)} \in (\frac{i}{m}, \frac{i+1}{m})$ and $\xi_{2i}^{(m)} \in (\frac{i+1}{m}, \frac{i+2}{m})$. Fixing i and increasing m, $\phi'(\xi_{1i}^{(m)}) \rightarrow \phi'(0)$ and $\phi'(\xi_{2i}^{(m)}) \rightarrow \phi'(0)$. On the other hand, by (1), $\phi'(\xi_{1i}^{(m)}) \rightarrow 1/2$ and $\phi'(\xi_{2i}^{(m)}) \rightarrow 3/2$: a contradiction.

Let $\| \cdot \|$ be the Euclidean norm in $\mathbb{R}^n$, where $n \geq 1$, with $< \cdot, \cdot >$ the corresponding inner product. The *length* $d(\gamma)$ of a C^k parametric curve ($k \geq 1$) $\gamma : [0,1] \rightarrow \mathbb{R}^n$ is defined as $d(\gamma) = \int_0^1 \|\dot{\gamma}(t)\| dt$, where $\dot{\gamma}(t) \in \mathbb{R}^n$ is the derivative of γ at $t \in [0,1]$. The curve γ is said to be *regular* when $\dot{\gamma}(t) \neq \mathbf{0}$, for all $t \in [0,1]$. A *reparameterization* of γ is a parametric curve of the form $\gamma \circ \psi : [0,1] \rightarrow \mathbb{R}^n$, where $\psi : [0,1] \rightarrow [0,1]$ is an order-preserving C^k diffeomorphism. Clearly, $\gamma \circ \psi$ has the same image and length as γ. Let γ be regular: then so is any reparameterization $\gamma \circ \psi$. The curve γ is *parameterized proportionally to arc-length* when $\|\dot{\gamma}(t)\|$ is constant for $t \in [0,1]$. We want to estimate $d(\gamma)$ from ordered $m+1$-tuples $\mathcal{Q} = (q_0, q_1, q_2, \ldots, q_m) \in (\mathbb{R}^n)^{m+1}$, where $q_i = \gamma(t_i)$, whose parameter values $t_i \in [0,1]$ are not known but sampled in some reasonably regular way: sampling might be ε-uniform for some $0 \leq \varepsilon \leq 1$. ε-uniform sampling is invariant with respect to C^k reparameterizations $\psi : [0,1] \rightarrow [0,1]$. So suppose, without loss of generality, that γ is parameterized proportionally to arc-length.

We close this section with Figure 1 indicating why arbitrary sampling and piecewise-quadratic Lagrange interpolation (see Section 3) in most cases gives poor estimates for $d(\gamma)$ (and indeed for γ). In Figure 1 only the uniform data yields reasonable approximations. In the next sections we show that some kinds of non-uniform sampling also give good approximations.

3   Uniform Sampling

We first consider length estimates of γ in the easier case, where the t_i's are sampled perfectly uniformly: $t_i = \frac{i}{m}$ (with $0 \leq i \leq m$). Suppose $k = r+2$, where $r \geq$

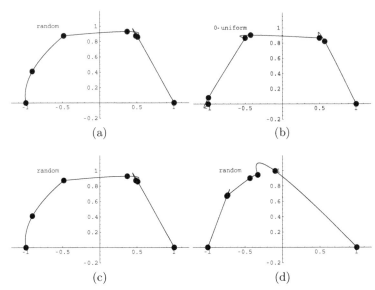

Fig. 1. Absolute errors $E = |\pi - d(\widetilde{\gamma})|$ for a unit semicircle approximated with the piecewise-quadratic interpolant $\widetilde{\gamma}$: (**a**) For perfectly uniform sampling $E = 0.00362662$. (**b**) For 0-uniform sampling (where $t_i = \frac{i}{6}$ for i even and $t_i = \frac{i}{6} - \frac{1}{12}$ for i odd; $0 \le i \le 6$) $E = 0.323189$. (**c**) For some random sampling $E = 0.22992$. (**d**) For another random sampling $E = 0.15394$

1, and (without loss of generality) that m is a multiple of r. Then Q gives $\frac{m}{r}$ $r+1$-tuples of the form $(q_0, q_1, \ldots, q_r)$, $(q_r, q_{r+1}, \ldots, q_{2r})$, $\ldots,(q_{m-r}, q_{m-r+1}, \ldots, q_m)$. The j-th $r + 1$-tuple is interpolated by the r-degree Lagrange polynomial $P_r^j :$ $[0, r] \to \mathbb{R}^n$, here $1 \le j \le \frac{m}{r}$: $P_r^j(0) = q_{(j-1)r}$, $P_r^j(1) = q_{(j-1)r+1}, \ldots, P_r^j(r) = q_{jr}$. Note that P_r^j is defined in terms of a local variable $s \in [0, r]$. Recall Lemma 2.1 of Part 1 of [ ]:

Lemma 1. *Let* $f : [a, b] \to \mathbb{R}^n$ *be* C^l, *where* $l \ge 1$ *and assume that* $f(t_0) = \mathbf{0}$, *for some* $t_0 \in (a, b)$. *Then there exists a* C^{l-1} *function* $g : [a, b] \to \mathbb{R}^n$ *such that* $f(t) = (t - t_0)h(t)$.

The proof of Lemma 1 shows that $g = O(\frac{df}{dt})$. If f has multiple zeros $t_0 < t_1 < \ldots < t_k$ then $k + 1$ applications of Lemma 1 give

$$f(t) = (t - t_0)(t - t_1)(t - t_2) \ldots (t - t_k)h(t), \tag{2}$$

where h is $C^{l-(k+1)}$ and $h = O(\frac{d^{k+1}f}{dt^{k+1}})$.

Assuming that γ is C^{r+2} (i.e. $k = r + 2$) we are now going to prove Theorem 1, where estimation of $d(\gamma)$ is based on piecewise-r-degree polynomial interpolation. For each j-th r-tuple consider the interpolating polynomial P_r^j. Let $\psi : [t_{(j-1)r}, t_{jr}] \to [0, r]$ be the affine mapping given by $\psi(t_{(j-1)r}) = 0$

and $\psi(t_{jr}) = r$, namely $\psi(t) = mt - (j-1)r$. Thus $\dot{\psi}(t)$ is identically m (a diffeomorphism). Note that since both intervals $[t_{(j-1)r}, t_{jr}]$ and $[0, r]$ are *uniformly sampled*, ψ maps the t_i's to the corresponding grid points in $[0, r]$. Define $\tilde{\gamma}_j = P_r^j \circ \psi : [t_{(j-1)r}, t_{jr}] \to \mathbb{R}^n$. Then as ψ is affine, $\tilde{\gamma}_j$ is a polynomial of degree at most r. Note that $f = \tilde{\gamma}_j - \gamma : [t_{(j-1)r}, t_{jr}] \to \mathbb{R}^n$ is C^{r+2} and satisfies $f(t_{(j-1)r}) = f(t_{(j-1)r+1}) = \cdots = f(t_{jr}) = 0$. By (2)

$$f(t) = (t - t_{(j-1)r})(t - t_{(j-1)r+1}) \cdots (t - t_{jr})h(t), \tag{3}$$

where $h : [t_{(j-1)r}, t_{jr}] \to \mathbb{R}^n$ by Lemma 1 is C^1. Still by proof of Lemma 1

$$h(t) = O(\frac{d^{r+1}f}{dt^{r+1}}) = O(\frac{d^{r+1}\gamma}{dt^{r+1}}) = O(1), \tag{4}$$

because $deg(\tilde{\gamma}_j) \leq r$ and $\frac{d^{r+1}\gamma}{dt^{r+1}}$ is $O(1)$. Thus by (3) and (4)

$$f(t) = O(\frac{1}{m^{r+1}}), \tag{5}$$

for $t \in [t_{(r-1)j}, t_{rj}]$. Differentiating function h (defined as a $r+1$-multiple integral of $f^{(r+1)}$ over the compact cube $[0, 1]^{r+1}$; see proof of Lemma 1) yields

$$\dot{h}(t) = O(\frac{d^{r+2}f}{dt^{r+2}}) = O(\frac{d^{r+2}\gamma}{dt^{r+2}}) = O(1), \tag{6}$$

as $deg(\tilde{\gamma}_j) \leq r$. By (3) and (6) $\dot{f} = O(\frac{1}{m^r})$ and hence for $t \in [t_{(j-1)r}, t_{jr}]$

$$\dot{\gamma}(t) - \dot{\tilde{\gamma}}_j(t) = \dot{f}(t) = O(\frac{1}{m^r}). \tag{7}$$

Let $V_{\dot{\gamma}}^{\perp}(t)$ be the orthogonal complement of the line spanned by $\dot{\gamma}(t)$. Since $\|\dot{\gamma}(t)\| = d(\gamma)$,

$$\dot{\tilde{\gamma}}_j(t) = \frac{<\dot{\tilde{\gamma}}_j(t), \dot{\gamma}(t)>}{d(\gamma)^2}\dot{\gamma}(t) + v(t), \tag{8}$$

where $v(t)$ is the orthogonal projection of $\dot{\tilde{\gamma}}_j(t)$ onto $V_{\dot{\gamma}}^{\perp}(t)$. Since $\dot{\tilde{\gamma}}_j(t) = \dot{f}(t) + \dot{\gamma}(t)$, we have $\dot{\tilde{\gamma}}_j(t) = (1 + \frac{<\dot{f}(t), \dot{\gamma}(t)>}{d(\gamma)^2})\dot{\gamma}(t) + v(t)$. Furthermore, by (7), $v = O(\frac{1}{m^r})$. Since $< \dot{\gamma}(t), v(t) >= 0$, by the Binomial Theorem the norm $\|\dot{\tilde{\gamma}}_j(t)\| =$

$$\|\dot{\gamma}(t)\|\sqrt{1 + 2\frac{<\dot{f}(t), \dot{\gamma}(t)>}{d(\gamma)^2} + O(\frac{1}{m^{2r}})} = \|\dot{\gamma}(t)\|(1 + \frac{<\dot{f}(t), \dot{\gamma}(t)>}{d(\gamma)^2}) \tag{9}$$

up to the $O(\frac{1}{m^{2r}})$ term; note that by (7) $|2\frac{<\dot{f}(t), \dot{\gamma}(t)>}{d(\gamma)^2} + O(\frac{1}{m^{2r}})| < 1$ holds asymptotically. Integrating by parts, $\int_{t_{(j-1)r}}^{t_{jr}} (\|\dot{\tilde{\gamma}}_j(t)\| - \|\dot{\gamma}(t)\|) dt =$

$$\int_{t_{(j-1)r}}^{t_{jr}} \frac{<\dot{f}(t), \dot{\gamma}(t)>}{d(\gamma)} dt + O(\frac{1}{m^{2r+1}}) = -\int_{t_{(j-1)r}}^{t_{jr}} \frac{<f(t), \ddot{\gamma}(t)>}{d(\gamma)} dt + O(\frac{1}{m^{2r+1}}).$$

Since γ is compact and at least C^3 by (4) and $h = O(1)$ we have $< h(t), \ddot{\gamma}(t) >= O(1)$ and $< h(t), \gamma^{(3)}(t) >= O(1)$. Similarly, by (6) we have $< \dot{h}(t), \ddot{\gamma}(t) >= O(1)$. Hence, by (3) and Taylor's Theorem applied to $r(t) =< h(t), \ddot{\gamma}(t) >$ at $t = t_{(j-1)r}$, we get $< f(t), \ddot{\gamma}(t) >= (t - t_{(j-1)r})(t - t_{(j-1)r+1}) \ldots (t - t_{jr})(a + O(\frac{1}{m}))$, where a is constant in t and $O(1)$. Since sampling is uniform the integral $\int_{t_{(j-1)r}}^{t_{jr}} (t - t_{(j-1)r})(t - t_{(j-1)r+1}) \ldots (t - t_{jr}) dt$ vanishes when r is even. So $\frac{1}{d(\gamma)} \int_{t_{(j-1)r}}^{t_{jr}} < f(t), \ddot{\gamma}(t) > dt$ is either $O(\frac{1}{m^{r+2}})$ or $O(\frac{1}{m^{r+3}})$, according as r is odd or even. As $2r + 1 \geq r + 3$ (for $r \geq 2$) and $2r + 1 \geq r + 2$ (for $r \geq 1$),

$$\int_{t_{(j-1)r}}^{t_{jr}} (\|\ddot{\tilde{\gamma}}_j(t)\| - \|\dot{\gamma}(t)\|) \, dt = \begin{cases} O(\frac{1}{m^{r+2}}) & \text{if } r \geq 1 \text{ is odd} \\ O(\frac{1}{m^{r+3}}) & \text{if } r \geq 2 \text{ is even.} \end{cases}$$

Take $\tilde{\gamma}$ to be a track-sum of the $\tilde{\gamma}_j$, i.e. $d(\tilde{\gamma}) = \Sigma_{j=0}^{\frac{m}{r}-1} d(\tilde{\gamma}_j) = d(\gamma) + O(\frac{1}{m^{r+p}})$, where p is 1 or 2 according as r is odd or even. This proves Theorem 1.

Notice that, by (5), perfectly uniform sampling permits estimates of γ with uniform $O(\frac{1}{m^{r+1}})$ error. Next we consider non-uniform samplings, for which piecewise-quadratic interpolation gives good length estimates.

4 ε-Uniform Sampling

Let $k = 4$, so that $\gamma : [0,1] \to \mathbb{R}^n$ and its reparameterizations are at least C^4. Fix $0 < \varepsilon \leq 1$, and let the t_i's be sampled ε-uniformly. We are going to prove Theorem 2. Without loss of generality m is even. For each triple (q_i, q_{i+1}, q_{i+2}), where $0 \leq i \leq m - 2$, let $Q^i : [0, 2] \to \mathbb{R}^n$ be the quadratic curve (expressed in local parameter $s \in [0, 2]$) satisfying $Q^i(0) = q_i$, $Q^i(1) = q_{i+1}$, and $Q^i(2) = q_{i+2}$. Write $Q^i(s) = q_i + a_1 s + a_2 s^2$, where $s \in [0, 2]$. Then

$$a_0 = q_i, \quad a_1 = \frac{4q_{i+1} - 3q_i - q_{i+2}}{2} \quad \text{and} \quad a_2 = \frac{q_{i+2} - 2q_{i+1} + q_i}{2}. \tag{10}$$

By Taylor's Theorem $\gamma(t_q) = \gamma(t_i) + \dot{\gamma}(t_i)(t_q - t_i) + (1/2)\ddot{\gamma}(\xi_q))(t_q - t_i)^2$, for either $q = i + 1$ or $q = i + 2$ and some $t_i < \xi_q < t_q$. Combining the latter with $\gamma(t_i) = q_i$, $\gamma(t_{i+1}) = q_{i+1}$, $\gamma(t_{i+2}) = q_{i+2}$ and substituting into (10) yields

$$a_2 = (1/2)\ddot{\gamma}(t_i)(t_{i+2} - 2t_{i+1} + t_i) + O(\frac{1}{m^2}). \tag{11}$$

Because sampling is ε-uniform the Mean Value Theorem gives

$$t_q - t_i = \phi'(\eta_q)\frac{1}{m} + O(\frac{1}{m^{1+\varepsilon}}), \tag{12}$$

for either $q = i + 1$ or $q = i + 2$ and some $t_i < \eta_q < t_q$. Thus by (11) and (12)

$$a_2 = \frac{t_{i+2} - 2t_{i+1} + t_i}{2}\ddot{\gamma}(t_i) + O(\frac{1}{m^2}). \tag{13}$$

Furthermore

$$t_{i+2} - 2t_{i+1} + t_i = \phi(\frac{i+2}{m}) - \phi(\frac{i+1}{m}) - (\phi(\frac{i+1}{m}) - \phi(\frac{i}{m})) + O(\frac{1}{m^{1+\varepsilon}}) \tag{14}$$

because sampling is ε-uniform. By Taylor's Theorem the following holds

$$\phi(\frac{i+1}{m}) = \phi(\frac{i}{m}) + \dot\phi(\frac{i}{m})\frac{1}{m} \quad \text{and} \quad \phi(\frac{i+2}{m}) = \phi(\frac{i}{m}) + \dot\phi(\frac{i}{m})\frac{2}{m}, \tag{15}$$

up to a $O(\frac{1}{m^2})$ term. Substituting (15) into (14) and taking into account ε-uniformity renders

$$t_{i+2} - 2t_{i+1} + t_i = O(\frac{1}{m^{1+\varepsilon}}). \tag{16}$$

The latter combined with (13) and ε-uniform sampling yields

$$a_2 = O(\frac{1}{m^{1+\varepsilon}}) + O(\frac{1}{m^2}) = O(\frac{1}{m^{1+\varepsilon}}). \tag{17}$$

A similar argument results in

$$a_1 = \frac{4t_{i+1} - 3t_i - t_{i+2}}{2}\dot\gamma(t_i) + O(\frac{1}{m^2}) = O(\frac{1}{m}). \tag{18}$$

From (17) and (18),

$$\frac{dQ^i}{ds} = a_1 + 2sa_2 = O(\frac{1}{m}) \quad \text{and} \quad \frac{d^2Q^i}{ds^2} = 2a_2 = O(\frac{1}{m^{1+\varepsilon}}), \tag{19}$$

where $s \in [0, 2]$. Let $\psi : [t_i, t_{i+2}] \to [0, 2]$ be the quadratic $\psi(t) = b_0 + b_1 t + b_2 t^2$ satisfying $\psi(t_i) = 0$, $\psi(t_{i+1}) = 1$, and $\psi(t_{i+2}) = 2$ (although ψ depends on i we suppress this in the notation). Inspection reveals $b_1 = (t_{i+1} - t_i)^{-1} - b_2(t_{i+1} + t_i)$, and $b_2 = ((t_{i+1} - t_i) - (t_{i+2} - t_{i+1}))[(t_{i+1} - t_i)(t_{i+2} - t_{i+1})(t_{i+2} - t_i)]^{-1}$. Furthermore, as before, by ε-uniformity $(t_{i+1} - t_i) - (t_{i+2} - t_{i+1}) = O(\frac{1}{m^{1+\varepsilon}})$, and $m^3(t_{i+1} - t_i)(t_{i+2} - t_{i+1})(t_{i+2} - t_i) = O(1)$, where the right-hand side of the latter is bounded away from 0 (as ϕ is a diffeomorphism defined over a compact set $[0, 1]$). Hence,

$$b_2 = O(m^{2-\varepsilon}) \quad \text{and} \quad \ddot\psi(t) = 2b_2 = O(m^{2-\varepsilon}). \tag{20}$$

As easily verified $(t_{i+1} - t_i)^{-1} = O(m)$. Hence, coupling $b_1 = (t_{i+1} - t_i)^{-1} - b_2(t_{i+1} + t_i)$ with (20) yields

$$\dot\psi(t) = b_1 + 2b_2 t = O(m) + b_2(2t - (t_{i+1} + t_i)) = O(m), \tag{21}$$

as sampling is ε-uniform and $2t - (t_{i+1} + t_i) = O(\frac{1}{m})$, for $t \in [t_i, t_{i+2}]$. In particular, ψ is a diffeomorphism for m large. Define $\tilde\gamma_i = Q^i \circ \psi : [t_i, t_{i+2}] \to \mathbb{R}^n$. Then $\tilde\gamma_i$ is polynomial of degree at most 4. Its derivatives of order $2 \leq p \leq 4$, are $O(m^{(p-1)(1-\varepsilon)})$. Indeed, by (19), (20), (21), $deg(\psi) \leq 2$ and $deg(Q^i) \leq 2$

$$\ddot{\tilde\gamma}_i = Q^{i''}\dot\psi^2 + Q^{i'}\ddot\psi = O(\frac{1}{m^{1+\varepsilon}})O(m^2) + O(\frac{1}{m})O(m^{2-\varepsilon}) = O(m^{1-\varepsilon}), \tag{22}$$

$$\tilde\gamma_i^{(3)} = 3Q^{i''}\dot\psi\ddot\psi = O(\frac{1}{m^{1+\varepsilon}})O(m)O(m^{2-\varepsilon}) = O(m^{2-2\varepsilon}), \tag{23}$$

$$\tilde\gamma_i^{(4)} = 3Q^{i''}\ddot\psi^2 = O(\frac{1}{m^{1+\varepsilon}})O(m^{4-2\varepsilon}) = O(m^{3-3\varepsilon}). \tag{24}$$

Then $f = \tilde{\gamma}_i - \gamma : [t_i, t_{i+2}] \to \mathbb{R}^n$ is C^4 and satisfies $f(t_i) = f(t_{i+1}) = f(t_{i+2}) = 0$. By (22), (23), (24) and ε-uniformity we have

$$\frac{d^2 f}{dt^2} = O(m^{1-\varepsilon}), \quad \frac{d^3 f}{dt^3} = O(m^{2(1-\varepsilon)}), \quad \frac{d^4 f}{dt^4} = O(m^{3(1-\varepsilon)}). \tag{25}$$

Use Lemma 1 to write $f(t) = (t - t_i)(t - t_{i+1})(t - t_{i+2})h(t)$, where $h : [t_i, t_{i+2}] \to \mathbb{R}^n$ is C^1, respectively. Then again by Lemma 1 and (25) we have $h = O(\frac{d^3 f}{dt^3}) = O(m^{2(1-\varepsilon)})$. Furthermore (6) coupled with (25) renders $\dot{h} = O(m^{3(1-\varepsilon)})$. The latter combined with the ε-uniformity yields

$$\dot{f} = O(\frac{1}{m^{2\varepsilon}}) \quad \text{and} \quad f = O(\frac{1}{m^{1+2\varepsilon}}). \tag{26}$$

As in the proof of Theorem 1 define $V_{\dot{\gamma}}^{\perp}(t)$ to be the orthogonal complement to the space spanned by $\dot{\gamma}(t)$. Then expand $\dot{\tilde{\gamma}}_j(t)$ according to (8), where $v(t)$ is the orthogonal projection of $\dot{\tilde{\gamma}}_j(t)$ onto $V_{\dot{\gamma}}^{\perp}(t)$. Similarly to (9), by using (26) we arrive at $v = O(\frac{1}{m^{2\varepsilon}})$ and thus

$$\|\dot{\tilde{\gamma}}_j(t)\| = \|\dot{\gamma}(t)\|(1 + \frac{<\dot{f}(t), \dot{\gamma}(t)>}{d(\gamma)^2}) + O(\frac{1}{m^{4\varepsilon}}), \tag{27}$$

for which we use $\varepsilon \in (0, 1]$. Consequently, by (9), (26), and (27) the integral $\int_{t_i}^{t_{i+2}} (\|\dot{\tilde{\gamma}}_i(t)\| - \|\dot{\gamma}(t)\|) \, dt =$

$$\int_{t_i}^{t_{i+2}} \frac{<\dot{f}(t), \dot{\gamma}(t)>}{d(\gamma)} dt + O(\frac{1}{m^{1+4\varepsilon}}) = -\int_{t_i}^{t_{i+2}} \frac{<f(t), \ddot{\gamma}(t)>}{d(\gamma)} dt$$

up to $O(\frac{1}{m^{1+4\varepsilon}})$. Now $<f(t), \ddot{\gamma}(t)> = (t - t_i)(t - t_{i+1})(t - t_{i+2})r(t)$, where $r(t) = <h(t), \ddot{\gamma}(t)>$. Taylor's Theorem applied to r at t_i yields $r(t) = r(t_i) + (t - t_i)\dot{r}(\xi)$, for some $t_i < \xi < t_{i+2}$. Similarly to the argument used for uniform sampling, $r(t_i) = O(m^{2(1-\varepsilon)})$ and $\dot{r} = O(m^{3(1-\varepsilon)})$. Consequently, by (16) $\int_{t_i}^{t_{i+2}} (t - t_i)(t - t_{i+1})(t - t_{i+2})dt = \frac{1}{12}(t_i - t_{i+2})^3(t_{i+2} - 2t_{i+1} + t_i) = O(\frac{1}{m^{4+\varepsilon}})$ and hence the integral $\int_{t_i}^{t_{i+2}} \frac{<f(t), \ddot{\gamma}(t)>}{d(\gamma)}) dt =$

$$\frac{r(t_i)}{d(\gamma)} \int_{t_i}^{t_{i+2}} (t - t_i)(t - t_{i+1})(t - t_{i+2})dt + O(\frac{1}{m^5})O(m^{3(1-\varepsilon)})$$

is $O(\frac{1}{m^{2+3\varepsilon}})$. So again by ε-uniformity $\int_{t_i}^{t_{i+2}} (\|\dot{\tilde{\gamma}}_i(t)\| - \|\dot{\gamma}(t)\|) \, dt = O(\frac{1}{m^{2+3\varepsilon}}) + O(\frac{1}{m^{1+4\varepsilon}}) = O(\frac{1}{m^{1+4\varepsilon}})$, and we finally arrive at $d(\tilde{\gamma}) = \Sigma_{j=0}^{\frac{m}{2}-1} d(Q^{2j}) = d(\gamma) + O(\frac{1}{m^{4\varepsilon}})$. This proves Theorem 2.

Notice that ε-uniform sampling, permits (by (26)) estimates of γ with uniform $O(\frac{1}{m^{1+2\varepsilon}})$ error. Moreover, by taking $\varepsilon = 1$ in Theorem 2 we obtain a stronger statement (as ϕ-perturbation of uniform sampling is still allowed; see Definition 1) than Theorem 1 when $r = 2$. Note also that Theorem 2 has nothing to say about the case $\varepsilon = 0$. This is dealt with in [ ] using a different approach.

5   Experiments

Next we test the sharpness of the theoretical results in Theorems 1 and 2 with some numerical experiments. Our test curves are a semicircle and cubic $\gamma_s, \gamma_c : [0,1] \to \mathbb{R}^2$, given by $\gamma_s(t) = (\cos(\pi(1-t)), \sin(\pi(1-t)))$ and $\gamma_c(t) = (\pi t, (\frac{\pi t+1}{\pi+1})^3)$. Of course $d(\gamma_s) = \pi$, and numerical integration gives $d(\gamma_c) = 3.3452$. Experiments were performed with Mathematica.

5.1   Uniform Sampling

We first discuss convergence of length estimates for piecewise polynomial approximations and perfectly uniform sampling. Experiments were conducted for both test curves, with $r = 1, 2, 3, 4$ for which Theorem 1 asserts errors that are $O(\frac{1}{m^2})$, $O(\frac{1}{m^4})$, $O(\frac{1}{m^4})$, and $O(\frac{1}{m^6})$, respectively. For each $r = 1, 2, 3, 4$ the minimum and maximum number of interpolation points were $(min_1, max_1) = (min_2, max_2) = (7, 101)$, $(min_3, max_3) = (7, 100)$, and $(min_4, max_4) = (9, 101)$. Let $\tilde{\gamma}_{r,m_r}$ represent a piecewise-r-degree polynomial interpolating m_r points. In each row of Tables 1 and 2 (for a fixed $1 \le r \le 4$) we list only some specific values obtained from the set of absolute errors $E_{m_r}(\gamma) = |d(\gamma) - d(\tilde{\gamma}_{r,m_r})|$ (here m_r indexes $\tilde{\gamma}_{r,m_r}$, where $min_r \le m_r \le max_r$ and $m_r = rn + 1$), namely: $E_{min_r}^{max_r}(\gamma) = \max_{min_r \le m_r \le max_r} E_{m_r}(\gamma)$ and $E_{max_r}(\gamma)$. Moreover, for each r, in searching for the estimate of convergence rate $O(\frac{1}{m^\alpha})$ (where $m + 1 = m_r$ is a number of interpolation points) a linear regression is carried out on pairs of points $(\log(m_r - 1), -\log(E_{m_r}(\gamma)))$, where $min_r \le i_r \le max_r$ and $i_r = rn + 1$. Here are the results. Both Tables 1 and 2 suggest that (in these cases at least) the statements in Theorem 1 are sharp. (The last two rows of Table 2 are somewhat irrelevant in that Lagrange interpolation returns, for $r \ge 3$, the same curve γ_c, up to machine precision.)

5.2 ε-Uniform Sampling

A full report on experiments with piecewise-quadratic Lagrange interpolation with ε-uniform sampling is given in [ ]. We experimented with $\varepsilon_1 = 1$, $\varepsilon_{1/2} = 1/2$, $\varepsilon_{1/3} = 1/3$, and for $l = 1, 2, 3$, with diffeomorphisms $\phi_l : [0,1] \to [0,1]$,

Table 1. Results for length estimation $d(\gamma_s)$ of semicircle γ_s.

r	min_r	max_r	α (where $E_{m_r} \propto m^{-\alpha}$)	$E_{min_r}^{max_r}(\gamma_s)$	E_{max_r}
1	7	101	1.99943	0.0357641	1.29191×10^{-4}
2	7	101	3.98485	0.0036266	5.09874×10^{-8}
3	7	100	3.97964	0.0026509	3.98087×10^{-8}
4	9	101	5.98218	0.0001136	3.19167×10^{-11}

Table 2. Results for length estimation $d(\gamma_s)$ of cubic curve γ_c.

r	min_r	max_r	α (where $E_{m_r} \propto m^{-\alpha}$)	$E_{min_r}^{max_r}(\gamma_c)$	E_{max_r}
1	7	101	2.00006	0.0357641	5.18348×10^{-6}
2	7	201	4.09546	0.0036266	1.22657×10^{-12}
3	7	100	n/a[a]	5.90639×10^{-14} [a]	4.44089×10^{-16} [a]
4	9	101	n/a[a]	2.73115×10^{-13} [a]	4.44089×10^{-16} [a]

[a] not applicable (see above).

given by $\phi_1(t) = t$, $\phi_2(t) = \frac{1}{\pi+1}t(\pi t + 1)$, and $\phi_3(t) = \frac{\exp(\pi t)-1}{\exp(\pi)-1}$. These functions are used to define first ε-uniform *random sampling*

$$t_i = \phi_l(\frac{i}{m}) + (Random[\,] - 0.5)\frac{1}{m^{1+\varepsilon}}, \tag{28}$$

where $Random[\,]$ takes the pseudorandom values from the interval $[0,1]$ and $0 \leq i \leq m$. In addition, we experimented with two other families of *skew-symmetric* ε-uniform samplings with ϕ_1 and $0 \leq i \leq m$:

$$(i)\ t_i = \frac{i}{m} + \frac{(-1)^{i+1}}{2m^{1+\varepsilon}} \qquad (ii)\ t_i = \begin{cases} \frac{i}{m} & \text{if } i \text{ even}, \\ \frac{i}{m} + \frac{1}{2m^{1+\varepsilon}} & \text{if } i = 4k+1, \\ \frac{i}{m} - \frac{1}{2m^{1+\varepsilon}} & \text{if } i = 4k+3. \end{cases} \tag{29}$$

In all cases $t_0 = 0, t_1 = 1$. Piecewise-quadratic interpolation was implemented for both kinds of sampling, with m even running from $m = 6$ up to $m = 100$ and to $m = 200$, respectively. These experiments with γ_s and γ_c showed faster convergence than proved in Theorem 2 for sampling (28). However, the statement of Theorem 2 appears to be sharp for the samplings (29): the observed rates of convergence nearly coincide with those asserted by the theorem: $\alpha_1 = 4$, $\alpha_{1/2} = 2$ and $\alpha_{1/3} = 4/3$. Note also that for 0-uniform sampling (29)(i), and for semicircle γ_s and cubic curve γ_c, a piecewise-quadratic Lagrange polynomial interpolant does not provide good estimates of $d(\gamma_s)$ and $d(\gamma_c)$, respectively (see Figure 2).

6   Conclusion

The problem of estimating $d(\gamma)$ of a C^k curve seems rather straightforward when the parameter values $t_i \in [0,1]$ are given, for example when sampling is uniform. This paper examines a class of samplings for which the same simple methods give length estimates converging to the true value $d(\gamma)$, including investigation of convergence rates. Our results appear to be sharp for the class of samplings studied in this paper. Piecewise Lagrange interpolation does not work well for 0-uniform samplings (more elaborate methods for dealing with these are given in [ ]) and so the class of ε-uniform samplings is of special interest where

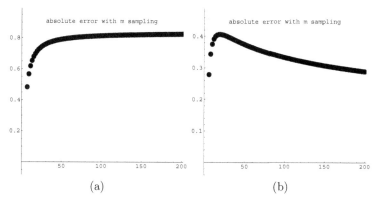

(a)   (b)

Fig. 2. Absolute errors plotted for 0-uniform skew sampling $(29)(i)$ (where m_2 is even and $6 \leq m_2 \leq 200$): (**a**) $E_{m_2}(\gamma_s)$ against m_2. (**b**) $E_{m_2}(\gamma_c)$ against m_2

$0 < \varepsilon \leq 1$. In general, the relationship between convergence of length estimates and uniform convergence to the image of γ seems not quite straightforward. Because the methods used herein are relatively simple, they are also widely applicable. Unlike the situation in [ ] there is no convexity requirement on γ, and there is no need to restrict attention to planar curves.

References

1. Asano T., Kawamura Y., Klette R., Obokkata K. (2000) A new approximation scheme for digital objects and curve length estimation. In: Cree M. J., Steyn-Ross A. (Eds) Proceedings of Image and Vision Computing Conference, Hamilton, New Zealand. November 27-November 29, 2000. Department of Physics and Electronic Engineering, University of Waikato Press, 26–31   340
2. Barsky B. A., DeRose T. D. (1989) Geometric continuity of parametric curves: three equivalent characterizations. IEEE. Comp. Graph. Appl. **9**, (6):60–68   341
3. Boehm W., Farin G., Kahmann J. (1984) A survey of curve and surface methods in CAGD. Comput. Aid. Geom. Des. **1**, 1–60   341
4. Bülow T., Klette R. (2000) Rubber band algorithm for estimating the length of digitized space-curves. In: Sneliu A., Villanva V. V., Vanrell M., Alquézar R., Crowley J., Shirai Y. (Eds) Proceedings of 15th International Conference on Pattern Recognition, Barcelona, Spain. September 3-September 8, 2000. IEEE, Vol. III, 551-555   340
5. Dąbrowska D., Kowalski M. A. (1998) Approximating band- and energy-limited signals in the presence of noise. J. Complexity **14**, 557–570   341
6. Dorst L., Smeulders A. W. M. (1991) Discrete straight line segments: parameters, primitives and properties. In: Melter R., Bhattacharya P., Rosenfeld A. (Eds) Ser. Contemp. Maths, Vol. 119. Amer. Math. Soc., 45–62   340
7. Epstein M. P. (1976) On the influence of parametrization in parametric interpolation. SIAM. J. Numer. Anal. **13**, (2):261–268   341

8. Hoschek J. (1988) Intrinsic parametrization for approximation. Comput. Aid. Geom. Des. **5**, 27–31 341

9. Klette R. (1998) Approximation and representation of 3D objects. In: Klette R., Rosenfeld A., Sloboda F. (Eds) Advances in Digital and Computational Geometry. Springer, Singapore, 161–194 340

10. Klette R., Bülow T. (2000) Critical edges in simple cube-curves. In: Borgefors G., Nyström I., Sanniti di Baja G. (Eds) Proceedings of 9th Discrete Geometry for Computer Imagery Conference, Uppsala, Sweden, December 13-December 15, 2000. Springer LNCS Vol. 1953, 467–478 340

11. Klette R., Kovalevsky V., Yip B. (1999) On the length estimation of digital curves. In: Latecki L. J., Melter R. A., Mount D. M., Wu A. Y. (Eds) Proceedings of SPIE Conference, Vision Geometry VIII, Denver, USA, July 19-July 20, 1999. The International Society for Optical Engineering, 3811:52–63 340

12. Klette R., Yip. B. (2000) The length of digital curves. Machine Graphics and Vision **9**, 673–703 340

13. Milnor J. (1963) Morse Theory. Annals of Math. Studies 51, Princeton University Press 343

14. Moran P. A. P. (1966) Measuring the length of a curve. Biometrika **53**, (3-4):359–364 340

15. Noakes L., Kozera, R. More-or-less-uniform sampling and lengths of curves. Quart. Appl. Maths. In press 339, 341, 347, 349, 350

16. Noakes L., Kozera R., Klette R. (2001) Length estimation for curves with ε-uniform sampling. In: Skarbek W. (Ed.) Proceedings of 9th International Conference on Computer Analysis of Images and Patterns, Warsaw, Poland, September 5-September 7, 2001. Springer LNCS Vol. 2124. In press 348

17. Piegl L., Tiller W. (1997) *The NURBS Book*. Springer, Berlin 341

18. Plaskota L. (1996) Noisy Information and Computational Complexity. Cambridge Uni. Press, Cambridge 341

19. Sederberg T. W., Zhao J., Zundel A. K. (1989) Approximate parametrization of algebraic curves. In: Strasser W., Seidel H. P. (Eds) Theory and Practice in Geometric Modelling. Springer, Berlin, 33–54 341

20. Sloboda F., Zaťko B., Stör J. (1998) On approximation of planar one-dimensional continua. In: Klette R., Rosenfeld A., Sloboda F. (Eds) Advances in Digital and Computational Geometry. Springer, Singapore, 113–160 340

21. Steinhaus H. (1930) Praxis der Rektifikation und zum Längenbegriff (In German). Akad. Wiss. Leipzig, Berlin **82**, 120–130 340

22. Traub J. F., Werschulz A. G. (1998) Complexity and Information. Cambridge Uni. Press, Cambridge 341

23. Werschulz A. G., Woźniakowski H. What is the complexity of surface integration? J. Complexity. In press 341

The 2-D Leap-Frog:
Integrability, Noise, and Digitization

Lyle Noakes[1] and Ryszard Kozera[2]

[1] The University of Western Australia, Department of Mathematics and Statistics
35 Stirling Highway, Crawley WA 6009, Australia
[2] Department of Computer Science and Software Engineering
35 Stirling Highway, Crawley WA 6009, Australia

Abstract. The 1-D Leap-Frog Algorithm [ ] is an iterative scheme for solving a class of nonlinear optimization problems. In the present paper we adapt Leap-Frog to solve an optimization problem in computer vision. The vision problem in the present paper is to recover (as far as possible) an integrable vector field (over an orthogonal grid) from a field corrupted by noise or the effects of digitization of camera images. More generally, we are dealing with integration of discrete vector fields, where every vector represents a surface normal at a grid position within a regular orthogonal grid of size $N \times N$. Our 2-D extension of Leap-Frog is a scheme which we prove converges linearly to the optimal estimate. 1-D Leap-Frog [ ] can deal with nonlinear problems such as are encountered in computer vision. In the present paper we exploit Leap-Frog's capacity to handle large number of variables for a linear problem (in this situation Leap-Frog becomes an extension of Gauss-Seidel), and we offer a geometrical proof of convergence for the case of photometric stereo (see e.g. [,]), where data is corrupted by noise or digitization. In the present paper, noise enters in an especially simple way, as Gaussian noise added to gradient estimates. So this as a first step towards more realistic (and demanding) applications, where Leap-Frog's capacity to deal with nonlinearities is needed. The present paper also offers an alternative to other methods in photometric stereo [ ], [ ], [ ], and [ ]. The performance of 2-D Leap-Frog was demonstrated in [ ] without proof of convergence: established methods are faster, but without Leap-Frog's capacity for generalization to nonlinear problems.

1 Introduction

A monochrome picture of a smooth object typically exhibits brightness variation or *shading*. *Shape-from-shading* is the problem of determining the visible part of the object from the picture. As shown in [ ] (see also [ ]), this corresponds to solving a nonlinear first-order partial differential equation. Specifically, one seeks a function u, representing surface depth in the direction of the z-axis,

---

[1] This research was supported by an Australian Research Council Small Grant^{a,b} and an Alexander von Humboldt Research Fellowship.^{b}

G. Bertrand et al. (Eds.): Digital and Image Geometry, LNCS 2243, pp. 352–364, 2001.

satisfying an *image irradiance equation*. For a *a Lambertian surface* (a perfect light diffuser) with constant albedo, illuminated from the light-source direction (p_1, p_2, p_3), the image irradiance equation is

$$\frac{p_1 u_x(x, y) + p_2 u_y(x, y) - p_3}{\sqrt{p_1^2 + p_2^2 + p_3^2} \sqrt{u_x^2(x, y) + u_y^2(x, y) + 1}} = E(x, y) \tag{1}$$

over an image $\Omega \subset \mathbb{R}^2$. Here E is an image intensity formed by orthographic projection onto the plane of the image Ω. The unknown surface S is the graph of the function u, which is to be determined up to translations $u \mapsto u + c$. A single image shape-from-shading problem (with no additional constraints) is, in general, *ill-posed* in the sense that not enough data is given for the problem (1) to be uniquely solved. For background we refer to [ ], [ ] or [ ].

In *multiple light-source shape from shading* the situation is simpler in some respects, namely a Lambertian surface is generically determined up to translation by multiple images, at least over the intersection of their respective domains (see [, subsection 10.16] or [,]). Shape reconstruction, as in multiple light-source photometric stereo, can be decomposed into: *gradient computation* (an algebraic step) and *gradient integration* (an analytic step). What complicates the problem, especially when images are subject to noise or digitization, is that the tableaux of computed values v^1, v^2 of u_x, u_y usually do not correspond to a function u. To see why not suppose that u is C^2. Then $u_{xy}(x, y) = u_{yx}(x, y)$. It follows that a necessary condition for $\boldsymbol{v} = (v^1, v^2)$ to correspond to a C^2 function is $v_y^1 = v_x^2$. When Ω is simply-connected this *integrability condition* is also sufficient for the computed values of v^1, v^2 to correspond to an image. From now on take Ω to be the unit square $[0, 1] \times [0, 1]$, and consider the problem of correcting (v^1, v^2) to an integrable field before passing to gradient integration. This second step is easy, and then u is determined up to translation.

A natural way to correct $\boldsymbol{v}$ is to choose the closest integrable vector field, depending on what is meant by "close" (here we use the standard definition). This, if attempted directly, results in a large-scale linear algebra, which is intractible by elementary methods. By contrast, each step of 2-D Leap-Frog is a small-scale optimization problem, and the method consists of blending these elementary tasks by means of an iterative scheme.

Known techniques for handling the problem include [, subsections 11.7-11.8] and [ ] which minimizes different best fit functionals. One of Horn's variants minimizes the following functional $\int_\Omega ((u_x(x, y) - v^1(x, y))^2 + (u_y(x, y) - v^2(x, y))^2) d\Omega$ whose Euler-Lagrange equation is a Poisson equation $\triangle u = f$. A *natural boundary condition* makes the corresponding problem well-posed. Fast methods of solving Poisson's equation exploiting the linearity of $\triangle u = f$ can be then be used [ ] or [ ]. A different minimization scheme was introduced in [ ], using projection to the closest function $\widetilde{u}$ expanded as a Fourier series. Another alternative is the *Lawn-Mowing Algorithm* [ ], which gives suboptimal solutions.

Theorem 1 of the present paper asserts linear convergence of 2-D Leap-Frog to the optimal solution estimate $(\hat{v}^1, \hat{v}^2)$. There is no claim that our method competes for the speed with established methods based on Poisson's equation, but our geometrical proof seems interesting in its own right, and the construction is capable of considerable generalization [ ], [ ]. Our proof also seems more accessible than known algebraic proofs for classical iterative schemes for large linear systems. In the present linear setting 2-D Leap-Frog can be seen as a subclass of such methods, but 1-D Leap-Frog solves nonlinear optimization problems in a geometrical setting. So extensions of 2-D Leap-Frog to the nonlinear case are likely to be useful for nonlinear optimization problems arising in computer vision. In fact a number of further nonlinear applications are planned, to demonstrate the generality of this technique.

Although we focus here on rectifying nonintegrable vector fields in the context of noisy or digitized photometric stereo images, similar techniques can also be applied to digitized or noisy optical flow or in physics where the unknown potenial u gives rise to the observable vector fields.

2   The 2-D Leap-Frog Global Optimizer

For $n > m$ let $L : \mathbb{R}^n \to \mathbb{R}^m$ be a linear transformation of maximal rank m. Given $L(v) \neq u$, for some $v \in \mathbb{R}^n$ and $u \in \mathbb{R}^m$, consider the task of finding the closest $\hat{v} \in \mathbb{R}^n$ to $v \in \mathbb{R}^n$ such that $L(\hat{v}) = u$. It is well-known that $\|v - \hat{v}\|$ is minimized by the orthogonal projection of v on the affine subspace $A_L = v_p + Ker(L) \subset \mathbb{R}^n$, where $Ker(L) = span\{f_1, f_2, \ldots, f_k\}$ is the kernel of L and v_p is a particular solution to $L(x) = u$ (here $k = n - m$). In order to actually find $\hat{v}$ we need to compute $\hat{\alpha}_i$ using the following $k \times k$-linear system: $\sum_{i=1}^{k} \hat{\alpha}_i < f_i | f_j >=< v - v_s | f_j >$, in principle solvable by the Gaussian elimination. When k is large and the matrix of L has special properties, Gauss-Seidel or, more generally, multiplicative Schwarz (see e.g. [ ]) may be more appropriate. Alterntively one can use the *pseudoinverse* $L_{ps} = L^T(LL^T)^{-1} : \mathbb{R}^m \to \mathbb{R}^n$ for which

$$\hat{v} = v - L^T(LL^T)^{-1}(L(v) - u). \tag{2}$$

As with ordinary inverses it is not always convenient to use the pseudoinverse. Direct or iterative methods of solution may be more efficient. From this point of view, 2-D Leap-Frog is an iterative scheme, particularly well suited to photometric stereo, and with links to other problems of interest.

For many purposes the problem of approximating a noisy or digitized vector field by an integrable field can be reduced to linear algebra, by transforming the integrability condition into its discrete analogue. In more detail, for fixed $k < l \in \mathbb{N}$ divide the unit square Ω into 2^{2l} *grid cells* of the form $S_{i_g j_g} = [(i_g - 1)/2^l, i_g/2^l] \times [(j_g - 1)/2^l, j_g/2^l]$ (for $1 \leq i_g, j_g \leq 2^l$). Of course, grid resolution is here 2^l. From values of u on such a grid we could calculate *central-difference derivative approximations* with $\Delta x = \Delta y = 1/2^l$ $u_x[i, j] \approx \frac{u_{i+1}^j - u_i^j}{2^l}$ and $u_y[i, j] \approx \frac{u_i^{j+1} - u_i^j}{2^l}$, for each side of a subsquare in the x and y directions, accordingly whether the side is horizontal or vertical. Along each grid cell, the analytic integrability condition (modulo truncation error assumed here to be dominated by noise or digitization) translates into the *discrete analogue*

$$u_x[i, j+1] - u_x[i,j] = u_y[i+1,j] - u_y[i,j]. \qquad (3)$$

Assume that either a uniform Gaussian noise with mean zero is added to each $u_x[i,j]$ and each $u_y[i,j]$ independently, or alternatively that noise is generated by digitization (the specific value of standard deviation shall be set later; see the closing section). The problem now is to estimate (u_x, u_y) from $2^{l+1}(2^l+1)$ noise contaminated differences in such a way as to be closest to the discrete nonintegrable field $\boldsymbol{v} = (v^1, v^2)$. Accordingly, $\hat{\boldsymbol{v}} = (\hat{u}_x, \hat{u}_y)$ should minimize the sum of squared residuals

$$\sum_{1 \le i,j \le 2^l(2^l+1)} \left((\hat{u}_x[i,j] - v^1[i,j])^2 + (\hat{u}_y[i,j] - v^2[i,j])^2 \right). \qquad (4)$$

Such a solution is said to be *optimal*, and finding it leads to a large system of linear equations in many unknowns: the need for some sort of iterative scheme is quite apparent.

We now confine attention to a given S_{ij}^{kl}, with $1 \le i,j \le 2^{l+1-k} - 1$ (denoted also by S_t) having 2^{2k} grid cells, and review various optimization problems considered over S_t subject to nine different boundary conditions, namely: *top-right, left-top-right, left-top, top-right-bottom, top-right-bottom-left, bottom-left-top, right-bottom, right-bottom-left, and bottom-left.* These are referred to in the description of 2-D Leap-Frog in subsection 2.1 (see also Figure 1). Note that $1 \le i,j \le 2^{l+1-k} - 1$ as we consider here only half S_{ij}^{kl}-overlaps.

In doing so, we assign to nonintegrable vector field $\boldsymbol{v}_h = (v^1[i,j], v^2[i,j])$ (over each S_t) $2^{k+1}(2^k + 1)$ free variables $\boldsymbol{v}_z$ corresponding to the unknown corrected values of the closest integrable vector field $\hat{\boldsymbol{v}}_h$. More specifically, over each S_t, define $2^k(2^k + 1)$ free variables (corresponding to $v^1[i,j]$), namely: $x_1 = v^1[0,0], \ldots, x_{2^k} = v^1[2^k - 1, 0], x_{2^k+1} = v^1[0,1], \ldots, x_{2^{k+1}} = v^1[2^k - 1, 1]$, and $x_{2^{2k}+1} = v^1[0, 2^k], \ldots, x_{2^k(2^k+1)} = v^1[2^k - 1, 2^k]$. Analogously, we have $2^k(2^k + 1)$ variables (corresponding to $v^2[i,j]$), namely: $y_1 = v^2[0,0], \ldots, y_{2^k} = v^2[0, 2^k - 1], y_{2^k+1} = v^2[1,0], \ldots, y_{2^k+1} = v^2[1, 2^k - 1], y_{2^{2k}+1} = v^2[2^k, 0], \ldots, y_{2^k(2^k+1)} = v^2[2^k, 2^k - 1]$. Assuming temporarily, that $k = l$ (3) applied to each grid cell yields a homogeneous optimization system of 2^{2l} linear equations in $2^{l+1}(2^l + 1)$ unknowns $L_l^h(\boldsymbol{v}_z) = \boldsymbol{0}$, where $\boldsymbol{0} \in \mathbb{R}^{2l}$ and $\boldsymbol{v}_z \in \mathbb{R}^{2^{l+1}(2^l+1)}$ with $L_l^h(\boldsymbol{v}_h) \ne \boldsymbol{0}$ (note that for $l = k$ we have $\boldsymbol{v}^p = \boldsymbol{0}$ as defined previously in this section). Direct methods for solving such a global optimization problem (for which $L_l^h(\hat{\boldsymbol{v}}_h) = \boldsymbol{0}$) constitute, for $l \ge 7$, an unwieldly computational task. So further discussion focuses on local optimizations over each S_t (with $k < l$) and on melding local optima into a global optimum (the 2-D Leap-Frog). Note that as simple verification shows L_l^h has *maximal rank* i.e. $rank(L_l^h) = 2^{2l}$ and hence $dim(ker(L_l^h)) = 2^{l+1}(2^l + 1) - 2^{2l} = 2^l(2^l + 2)$. This is the dimension of the space of integrable vector fields.

As the analysis of the nine cases of different boundary constraints imposed on S_t for the 2-D Leap-Frog is similar we discuss here only *top-right* boundary conditions. Assume that 2^{k+1} boundary values are given, i.e. $x_{2^{2k}+1} = x^0_{2^{2k}+1}, x_{2^{2k}+2} = x^0_{2^{2k}+2}, \ldots, x_{2^k(2^k+1)} = x^0_{2^k(2^k+1)}$ (representing top boundary conditions for S_t) and $y_{2^{2k}+1} = y^0_{2^{2k}+1}, y_{2^{2k}+2} = y^0_{2^{2k}+2}, \ldots, y_{2^k(2^k+1)} = y^0_{2^k(2^k+1)}$ (representing right boundary conditions for S_t). Applying 2^{2k} integrability constraints (3) (along each grid cell) we arrive at an inhomogeneous system of 2^{2k} linear equations in $2^{k+1}(2^k+1) - 2^{k+1} = 2^{2k+1}$ unknowns. Using notation from section 2 this system can be treated as $L^{tr}_k(\boldsymbol{x}_{tr}) = \boldsymbol{u}_{tr}$ (for $L^{tr}_k(\boldsymbol{v}_{tr}) \neq \boldsymbol{u}_{tr}$), with $\boldsymbol{x}_{tr}, \boldsymbol{v}_{tr} \in \mathbb{R}^{2^{2k+1}}$, $\boldsymbol{u}_{tr} \in \mathbb{R}^{2^{2k}}$ and $L^{tr}_k : \mathbb{R}^{2^{2k+1}} \to \mathbb{R}^{2^{2k}}$ being a linear operator. Note that for $n = 2^{2k+1}$ and $m = 2^{2k}$ condition $n > m$ holds. It is straightforward to show that $Rank(L^{tr}_k) = m$. As $L^{tr}_k(\boldsymbol{v}_{tr}) \neq \boldsymbol{u}_{tr}$, the closest vector $\hat{\boldsymbol{v}}_{tr}$ satisfying $L^{tr}_k(\boldsymbol{x}_{tr}) = \boldsymbol{u}_{tr}$ can be found as in section 2. Note that as k increases the dimensions of both linear spaces $\mathbb{R}^m$ and $\mathbb{R}^n$ grow exponentially. The original problem (4) is reduced to a collection of computationally tractable problems, where l is replaced by $1 \leq k < l$. The operators $L^{ltr}_k, L^{lt}_k, L^{trb}_k, L^{trbl}_k, L^{blt}_k L^{rb}_k, L^{rbl}_k$, and L^{bl}_k are determined similarly.

2.1 2-D Leap-Frog

Suppose given a nonintegrable discrete vector field $\boldsymbol{v}$ over a rectangular grid in $\Omega = [0,1] \times [0,1]$ of grid size $1/2^l$. The optimization problem (4) is solvable in principle by the methods of section 2, but in practice this is unworkable by direct methods unless l is small. Certainly $l \geq 7$ makes diffficulties. So the following algorithm uses a sequence of optimizations where l is replaced by a smaller integer k.

Cover Ω by a familiy of overlapping subsquares $\mathcal{F}^{kl} = \{S^{kl}_{ij}\}_{1 \leq i,j \leq 2^{l-k+1}-1}$ each comprising 2^{2k} grid cells $S^l_{i_q j_q}$. Each S^{kl}_{ij} is defined as follows $S^{kl}_{ij} = [(i - 1)2^{k-l-1}, (i-1)2^{k-l-1} + 2^{k-l}] \times [(j-1)2^{k-l-1}, (j-1)2^{k-l-1} + 2^{k-l}]$, where $1 \leq i, j \leq 2^{l-k+1} - 1$. The case when $k = l - 1$ (and $l \geq 2$ is arbitrary) is shown in Figure 1. Let $\hat{\boldsymbol{v}}^l$ be any integrable vector field. Many choices are possible. For example we might use the methods introduced in [ ], [ ] or [ ] to approximate $\hat{\boldsymbol{v}}$. The easiest (and perhaps best) choice is $\hat{\boldsymbol{v}}^1 = \boldsymbol{0}$. We present Leap-Frog with half S^{kl}_{ij} overlaps, but Theorem 1 can be proved for overlaps between 1 and $2^k - 1$ pixels, using the same argument.

For $n = 1, 2, \ldots$ repeat the following sequence of steps until some halting condition is flagged (e.g. bounds imposed on the number of iterations, on estimated error, or on deficiency angle as in specified in section 4).

 – start with the left bottom subsquare S^{kl}_{11} and apply a least-square optimization with respect to[2] $\boldsymbol{v}$, with fixed top-right boundary constraints inherited from the values computed in the $n - 1$th iteration (for $k = l - 1$ see Figure 1 (a)).

---

[2] More precisely, adjust the variables in the snapshot to minimize the distance to the given nonintegrable field $\boldsymbol{v}$.

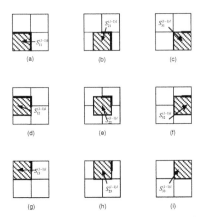

Fig. 1. Covering an image Ω by the family $\mathcal{F}^{kl} = \{S_{ij}^{(l-1)l}\}_{1\leq i,j\leq 3}$ of subsquares (here $k = l-1$). Each $S_{ij}^{(l-1)l}$ consists of $2^{2(l-1)}$ grid cells

- pass now to the second subsquare of the first row S_{21}^{kl} and optimize with fixed left-top-right boundary constraints. The left and left-half top boundary conditions are inherited from computed values over S_{11}^{kl} obtained in the nth iteration. The right and right-half top boundary conditions are inherited from the computed values in obtained $n - 1$th iteration (for $k = l - 1$ see Figure 1 (b)).
- continue until the last subsquare $S_{(2^{l-k+1}-1)1}^{kl}$ in the first row. Over this subsquare optimize with left-top boundary constraints fixed. The left and left-half top boundary conditions are given by the previously computed values over $S_{(2^{l-k+1}-2)1}^{kl}$ in nth iteration. The right-half top boundary conditions are given by the values obtained in the $n - 1$th iteration (for $k = l - 1$ see Figure 1 (c)). Thus the first row of nth iteration is completed.
- next pass to the second row. Start with the S_{12}^{kl} subsquare and optimize with fixed top-right-bottom boundary constraints. The bottom and bottom-half right boundary constraints are inherited from nth iteration values computed over subsquares S_{11}^{kl}, S_{21}^{kl} and S_{31}^{kl}. The top and top-half right boundary conditions are inherited from values computed in the $n - 1$th iteration (for $k = l - 1$ see Figure 1 (d)).
- pass to the second subsquare S_{22}^{kl} over which we optimize with top-right-bottom-left boundary constraints (*a generic case*). The left-half top, the left, the bottom and the bottom-half right boundary conditions are inherited from the values of the nth iteration computed over S_{12}^{kl}, S_{31}^{kl} and S_{41}^{kl}. The remaining boundary conditions are inherited from computed values obtained in the $n - 1$th iteration (for $k = l - 1$ see Figure 1 (e); note that in case when $k = l - 1$ there is no dependence on S_{41}^{kl}-boundary).
- continue until the last subsquare $S_{(2^{l-k+1}-1)2}^{kl}$ in the second row is reached. Over this subsquare optimize with bottom-left-top boundary constraints

fixed. The left-half top, the left, and the bottom boundary conditions are inherited from the values computed in the nth iteration over $S^{kl}_{(2^{l-k+1}-2)2}$ and $S^{kl}_{(2^{l-k+1}-1)1}$. The right-half top boundary conditions are inherited from values computed in the $n-1$th iteration. Thus the second row of the nth iteration is completed (for $k = l-1$ see Figure 1 (f)).

- continue row by row (as specified in the previous step), until the last row is reached. Now optimize over $S^{kl}_{12^{l-k+1}-1}$ with fixed right and bottom boundary constraints. The bottom and the bottom-half right boundary conditions are inherited from values computed in the nth iteration over $S^{kl}_{12^{l-k+1}-2}$, $S^{kl}_{22^{l-k+1}-2}$, and $S^{kl}_{32^{l-k+1}-2}$. The top-half right boundary conditions are inherited from values computed in the $n-1$th iteration (for $k = l-1$ see Figure 1 (g)).

- pass to the second subsquare of the last row $S^{kl}_{22^{l-k+1}-1}$ over which we optimize with right-bottom-left boundary constraints fixed. The bottom-half right, the bottom and the left boundary conditions are inherited from values of the nth iteration computed over $S^{kl}_{12^{l-k+1}-1}$, $S^{kl}_{32^{l-k+1}-2}$ and $S^{kl}_{42^{l-k+1}-2}$. The top-half right boundary conditions are inherited from computed values obtained in the $n-1$th iteration (for $k = l-1$ see Figure 1 (h); note that when $k = l-1$ there is no dependence on $S^{kl}_{42^{l-k+1}-2}$-boundary).

- this process continues, up until the last subsquare $S^{kl}_{(2^{l-k+1}-1)(2^{l-k+1}-1)}$, in the last row is reached. Over this subsquare optimize with the bottom-left boundary constraints fixed and inherited from computed values over $S^{kl}_{(2^{l-k+1}-2)(2^{l-k+1}-1)}$ and $S^{kl}_{(2^{l-k+1}-1)(2^{l-k+1}-2)}$ in the nth iteration (for $k = l-1$ see Figure 1 (i)). This completes the nth iteration, and the resulting integrable vector field is labelled $\hat{v}^{n+1}$.

Remark 1. During the n_0th 2-D Leap-Frog iteration once the mth S_m snapshot is reached we update free variables according to one of 9 different boundary constraints. Let $\hat{v}^{n_0}_{m-1} \in \mathbb{R}^{2^{l+1}(2^l+1)}$ be the current estimate of $\hat{v} \in \mathbb{R}^{2^{l+1}(2^l+1)}$ obtained upon completion of local minimization over S_{m-1} within n_0th iteration. Let $w^{n_0}_m$ represent currently updated and fixed boundary values of S_m (all or some of them) taken from $\hat{v}^{n_0}_{m-1}$ according to one of the nine admissible rules. As previously the integrability condition over S_m reads $L_m(X) = u^{n_0}_m$ (there are 9 different operators L_m). Assume $v_{\pi(m)}$ defines the vector which instead of free variables (over a snapshot S_m) inherits the corresponding values from v. Leap-Frog solves the local optimization problem by

$$\hat{v}^{n_0}_{\pi(m)} = v_{\pi(m)} - L^m_{ps}(L_m(v_{\pi(m)}) - u^{n_0}_m), \tag{5}$$

where $L^m_{ps} = L^T_m(L_m L^T_m)^{-1}$, and corrects a generic local misfit $L_m(v_{\pi(m)}) \neq u^{n_0}_m$ to $L_m(\hat{v}^{n_0}_{\pi(m)}) = u^{n_0}_m$, minimizing $\|\hat{v}^{n_0}_{\pi(m)} - v_{\pi(m)}\|$ over all integrable vector fields over S_m (subject to relevant boundary constraints). For $m > 1$, Leap-Frog updates $\hat{v}^{n_0}_{m-1}$ to $\hat{v}^{n_0}_m$ by replacing variables in $\hat{v}^{n_0}_{m-1}$ (corresponding to free variables in S_m) with $\hat{v}^{n_0}_{\pi(m)}$. Alternatively, 2-D Leap-Frog, can be viewed as successive orthogonal projections of $z^{n_0}_m$ ($z^{n_0}_m$ arises by replacing entries in $\hat{v}^{n_0}_{m-1}$ with those in

v coinciding with free variables in S_m) on some affine subspaces $A_m^{n_0} \subset Ker(L_l^h)$ of integrable vector fields in $\mathbb{R}^{2^{l+1}(2^l+1)}$.

3   Convergence of 2-D Leap-Frog

Here is our main result:

Theorem 1. *Let $\hat{v}$ be the optimal integrable vector field for the problem* (4) *described in section* 2 *and $\hat{v}^n$ its estimate upon completion of nth iteration of 2-D Leap-Frog. Then for some constant $C \in [0,1)$ and for each $n \in \mathbb{N}$*

$$\|\hat{v}^{n+1} - \hat{v}\| \le C\|\hat{v}^n - \hat{v}\|. \tag{6}$$

Preliminaries: To prove Theorem 1 (but not to implement 2-D Leap-Frog) it is necessary to change the way we look at the $\hat{v}^n$. Rather than regard these as vectors in $\mathbb{R}^{2^{l+1}(2^l+1)}$, it is better to view the $\hat{v}^n$ as abstract vectors or *tableaux*: a *tableau* is an array of real numbers of the form

$$v_T = \begin{pmatrix} & x_{2^{2l}+1} & & \cdots & & \cdots & & x_{2^l(2^l+1)} & \\ y_{2^l} & \diamond & y_{2^{2l}} & \cdots & & y_{2^{2l}} & \diamond & y_{2^l(2^l+1)} \\ & x_{2^{2l-1}+1} & & \cdots & & \cdots & & x_{2^{2l}} & \\ \vdots & \vdots & \vdots & & \vdots & & \vdots & \vdots \\ & x_{2^l+1} & & \cdots & & \cdots & & x_{2^{2l}} & \\ y_1 & \diamond & y_{2^l+1} & \cdots & y_{2^{2l-1}+1} & \diamond & y_{2^{2l}+1} \\ & x_1 & & \cdots & & \cdots & & x_{2^l} & \end{pmatrix}. \tag{7}$$

Even without the $\diamond$ symbols, a tableau is not a matrix (neither rows nor columns have the same number of entries). A $\diamond$ symbol in a tableau is called a *nucleus* and its adjacent real entries are its *neighbours*. Given l, the set of all tableaux is denoted by V_T.

For $v_{1T}, v_{2T} \in V_T$ define $v_{1T} +_T v_{2T} \overset{\text{def}}{=} v_T$ by $x_s^{v_T} = x_s^{v_{1T}} + x_s^{v_{2T}}$ and $y_s^{v_T} = y_s^{v_{1T}} + y_s^{v_{2T}}$ (where $1 \le s \le 2^l(2^l+1)$). For $v_T \in V_T$ and $\mu \in \mathbb{R}$, define $w_T \overset{\text{def}}{=} \mu \cdot_T v_T$ by $x_s^{w_T} = \mu x_s^{v_T}$ and $y_s^{w_T} = \mu y_s^{v_T}$ (here similarly $1 \le s \le 2^l(2^l + 1)$). With respect to these operations of vector addition and scalar multiplication, $(V_T, +_T, \cdot_T)$ is a vector space, equipped with a natural isomorphism $\mathcal{I}$ to $\mathbb{R}^{2^{l+1}(2^l+1)}$ i.e. $\mathcal{I}(v_T) = (x_1, x_2, \ldots, x_{2^l(2^l+1)}, y_1, y_2, \ldots, y_{2^l(2^l+1)})$. Define an inner product $< \cdot \,|\, \cdot >_T$ on V_T by multiplying corresponding real entries in two tableaux and then summing. Denote the associated norm by $\|\cdot\|_T$. Then $\mathcal{I}$ is an isometry to Euclidean space $\mathbb{R}^{2^{l+1}(2^l+1)}$. A tableau (7) is said to be *integrable* when, for each nucleus, the *weighted sum* of its four neighbours vanishes. More specifically, for an integrable tableau:

$$x_{(m-1)2^l+n} + y_{n2^l+m} - x_{m2^l+n} - y_{(n-1)2^l+m} = 0, \tag{8}$$

where $1 \le m, n \le 2^l$. The integrable tableaux comprise a vector subspace $V_T^{\oplus}$ of the vector space V_T of all tableaux. (8) coupled with $dim(ker(L_l^h)) = 2^l(2^l + 2)$

yields $dim(V_T^{\oplus}) = 2^l(2^l + 2)$. We shall prove now that one interation of 2-D Leap-Frog can be viewed as an affine mapping $A_T : V_T^{\oplus} \to V_T^{\oplus}$. Indeed

Lemma 1. *There is an affine transformation $A_T : V_T^{\oplus} \to V_T^{\oplus}$ such that*

$$\hat{\boldsymbol{v}}_T^{n+1} = A_T \hat{\boldsymbol{v}}_T^n, \tag{9}$$

where $\hat{\boldsymbol{v}}_T^k = \mathcal{I}(\hat{\boldsymbol{v}}^k)$ (for $k \in \mathbb{N}$) denotes the representation (via $\mathcal{I}$) of $\hat{\boldsymbol{v}}^k$ in the space V_T. The linear part of A_T (i.e. A_T^L) is constant over each iteration.

Proof: With half-snapshot overlaps, one iteration of 2-D Leap-Frog consists of $(2^{l+1-k} - 1)^2$ snapshot optimizers. From formula (2) we infer that each snapshot optimizer defines an affine map $V_T^{\oplus} \to V_T^{\oplus}$. Indeed (2) says $\boldsymbol{v}$ is an orthogonal projection of $\boldsymbol{v}_h$ (see section 2). The term $\boldsymbol{u}$ in (2) is linear in the integrable vector field. So the right-hand side of (5) is affine on integrable vector fields (depending on the position of the snapshot). Thus on each iteration 2-D Leap-Frog defines an affine transformation (as a composite of N_0 affine mappings) $A(\hat{\boldsymbol{v}}^n) = A^L(\hat{\boldsymbol{v}}^n) + \hat{\boldsymbol{w}}^n$ from the subspace of integrable vector fields to itself; A^L forms its linear component. Note that from (5) it is clear that only the linear component for each iteration remains unchanged. Hence the transformation $A_T(\hat{\boldsymbol{v}}_T^n) = \mathcal{I}(A(\mathcal{I}^{-1}(\hat{\boldsymbol{v}}_T^n))) = A_T^L(\hat{\boldsymbol{v}}_T^n) + \hat{\boldsymbol{w}}_T^n$, where $A_T^L = \mathcal{I} \circ A^L \circ \mathcal{I}^{-1}$ and $\mathcal{I}(\hat{\boldsymbol{w}}^n) = \hat{\boldsymbol{w}}_T^n$, yields (9).

We shall now choose a special basis in the space $V_T^{\oplus}$. In doing so let W_M be the vector space of real $(2^l + 1) \times (2^l + 1)$ matrices. Think of the entries $\{a_{ij}\}$ of such a matrix as values of the function u calculated at grid points labelled (i, j), where $1 \le i, j \le 2^l + 1$. Define a linear transformation $\Phi : W_M \to V_T$ by mapping from function values to the corresponding discretised vector field, namely $X = (a_{ij})_{1 \le i,j \le n}$ (where $n = 2^l + 1$) and

$$
Y = \begin{pmatrix}
& a_{12} - a_{11} & & \cdots & \\
a_{11} - a_{21} & \diamond & a_{12} - a_{22} & \cdots & a_{1n} - a_{2n} \\
& a_{22} - a_{21} & & \cdots & \\
\vdots & \vdots & \vdots & & \vdots \\
& a_{(n-1)2} - a_{(n-1)1} & & & \\
a_{(n-1)1} - a_{n1} & \diamond & a_{(n-1)2} - a_{n2} & \cdots & a_{(n-1)n} - a_{nn} \\
& a_{n2} - a_{n1} & & \cdots &
\end{pmatrix}.
$$

Notice that Φ maps into the space of integrable tableaux, namely $\Phi(W_M) \subset V_T^{\oplus} \subset V_T$ (see (8)). As easily verified the kernel of Φ is spanned by the constant matrix whose real entries are everywhere $a_{ij} = 1$ (for $1 \le i, j \le n$). Consequently Φ has rank $(2^l + 1)^2 - 1 = 2^l(2^l + 2)$. As showed before this is the dimension of $V_T^{\oplus}$. So the image under Φ of any $(2^l + 1)^2 - 1$ linearly independent matrices from W_M (which do not belong to $ker(\Phi)$) constitutes a basis of $V_T^{\oplus}$. In particular, define $\mathcal{B}_T^{\oplus} = \{h_{11}^{\oplus}, h_{12}^{\oplus}, \dots, h_{1n}^{\oplus}, h_{21}^{\oplus}, h_{22}^{\oplus}, \dots, h_{2n}^{\oplus}, \dots, h_{n1}^{\oplus}, h_{n2}^{\oplus}, \dots, h_{n(n-1)}^{\oplus}\}$, where $h_{11}^{\oplus} = \Phi(X_{11})$ (matrix X_{11} has $a_{n1} = 1$ and zeros elsewhere), $h_{12}^{\oplus} = \Phi(X_{12})$ (matrix X_{12} has $a_{n2} = 1$ and zeros elsewhere),..., $h_{1n}^{\oplus} = \Phi(X_{1n})$ (matrix X_{1n} has $a_{nn} = 1$ and zeros elsewhere), $h_{21}^{\oplus} = \Phi(X_{21})$ (matrix X_{21} has $a_{(n-1)1} = 1$

and zeros elsewhere), $h_{22}^{\oplus} = \Phi(X_{22})$ (martix X_{22} has $a_{(n-1)2} = 1$ and zeros else-where),..., $h_{2n}^{\oplus} = \Phi(X_{2n})$ (martix X_{2n} has $a_{(n-1)n} = 1$ and zeros elsewhere),..., $h_{n1}^{\oplus} = \Phi(X_{n1})$ (matrix X_{n1} has $a_{11} = 1$ and zeros elsewhere), $h_{n2}^{\oplus} = \Phi(X_{n2})$ (ma-trix X_{n2} has $a_{12} = 1$ and zeros elsewhere),..., and finally $h_{n(n-1)}^{\oplus} = \Phi(X_{n(n-1)})$ (matrix $X_{n(n-1)}$ has $a_{1(n-1)} = 1$ and zeros elsewhere). Then $\mathcal{B}_T^{\oplus}$ is a basis of the space $V_T^{\oplus}$ of integrable tableaux. Let $\boldsymbol{v}_T \in V_T$ be a nonintegrable tableau corre-sponding to the noisy or digitized data for problem (4), and define for $\hat{\boldsymbol{x}}_T \in V_T^{\oplus}$ the function $f : \mathbb{R}^s \to \mathbb{R}$ (here $s = 2^l(2^l + 2)$), called *performance index* such that, for $n = 2^l + 1$, we have

$$f(x) = \| \sum_{\substack{1 \le i,j \le n \\ (i,j) \neq (n,n)}} x_{ij} h_{ij}^{\oplus} - \boldsymbol{v}_T \|_T^2 = \bar{f}(\hat{\boldsymbol{x}}_T), \tag{10}$$

where $\| \cdot \|_T$ is computed from the inner product $< \cdot \, | \cdot >_T$ on V_T and $x_{ij} \in \mathbb{R}$. Then $x_o \in \mathbb{R}^{2^l(2^l+2)}$ is a *critical point* of f when, for $\hat{\boldsymbol{x}} = \mathcal{I}^{-1}(\hat{\boldsymbol{x}}_T)$ and $\boldsymbol{v} = \mathcal{I}^{-1}(\boldsymbol{v}_T)$,

$$\bar{f}(\hat{\boldsymbol{x}}_T) = \|\hat{\boldsymbol{x}}_T - \boldsymbol{v}_T\|_T^2 = \|\mathcal{I}^{-1}(\hat{\boldsymbol{x}}_T) - \mathcal{I}^{-1}(\boldsymbol{v}_T)\|^2 = \|\hat{\boldsymbol{x}} - \boldsymbol{v}\|^2 \tag{11}$$

attains its (global) minimum at $\hat{\boldsymbol{x}}_{To}$. Note that $\hat{\boldsymbol{x}}_{To}$ is an orthogonal projection $\hat{\boldsymbol{v}}_T$ of $\boldsymbol{v}_T$ on $V_T^{\oplus}$. The importance of $\mathcal{B}_T^{\oplus}$ for 2-D Leap-Frog is due to the following lemma.

Lemma 2. *The vector $\hat{\boldsymbol{x}}_T \in V_T^{\oplus}$ is not optimal if and only if 2-D Leap-Frog improves the performance index* (10).

Proof. No snapshot step of 2-D Leap-Frog increases f. To see this, note that once a given snapshot S_{ij}^{kl} is reached, 2-D Leap-Frog improves generically internal values of S_{ij}^{kl} while keeping the others fixed. In a rare situation if the boundary values coincide with those corresponding to $\hat{\boldsymbol{x}}_o$ (and thus $\hat{\boldsymbol{x}}_{To}$) no change is made. Hence (4) cannot be globally increased. Evidently 2-D Leap-Frog cannot improve the global optimum $\hat{\boldsymbol{x}}_o$. So 2-D Leap-Frog fixes $\hat{\boldsymbol{x}}_o$. On the other hand, if $\hat{\boldsymbol{x}}$ (and thus $\hat{\boldsymbol{x}}_T$) is not already optimal, the gradient of f at the corresponding point in $v \in \mathbb{R}^{2^l(2^l+2)}$ is nonzero. Let (a,b) (here $1 \le a, b \le n$ and $(a,b) \neq (n,n)$, $n = 2^l + 1$) be the first indices when $(\partial f/\partial x_{ab})(v) \neq 0$ and when reached by step (i_0, j_0) in 2-D Leap-Frog, where the corresponding basis element $h_{ab}^{\oplus}$ is in the interior of the snapshot $S_{i_0 j_0}^{kl}$ (the existence of $S_{i_0 j_0}^{kl}$ is justified in the closing section). Since f can be decreased by changing only one variable v_{ab}, optimizing the snapshot gives a better performance index, which feeds through to the global tableau.

We are ready now to complete the proof of Theorem 1. Note that combining Lemma 1 and (10) with $A_T(\hat{\boldsymbol{v}}_T) = \hat{\boldsymbol{v}}_T$ (see Lemma 2) yields $\|\hat{\boldsymbol{x}}^{n+1} - \hat{\boldsymbol{v}}\| = \|\hat{\boldsymbol{x}}_T^{n+1} - \hat{\boldsymbol{v}}_T\|_T = \|A_T(\hat{\boldsymbol{x}}_T^n) - \hat{\boldsymbol{v}}_T\|_T =$

$$\|A_T^L(\hat{\boldsymbol{x}}_T^n - \hat{\boldsymbol{v}}_T)\|_T \le \|A_T^L\|_T \|\hat{\boldsymbol{x}}_T^n - \hat{\boldsymbol{v}}_T\|_T = C\|\hat{\boldsymbol{x}}^n - \hat{\boldsymbol{v}}\|,$$

where $C = \|A_T^L\|_T = sup_{\|\boldsymbol{x}_T\|=1}\|A_T^L(\boldsymbol{x}_T)\|_T$. It suffices to show $C < 1$ (it is essential here that the A_T^L is the same over each 2-D Leap-Frog iteration). By

Lemma 2 if $\hat{\boldsymbol{x}} \neq \hat{\boldsymbol{v}}$ then $\|A_T(\hat{\boldsymbol{x}}_T) - \boldsymbol{v}_T\|_T < \|\hat{\boldsymbol{x}}_T - \boldsymbol{v}_T\|_T$, and by Pythagoras' Theorem

$$\|A_T(\hat{\boldsymbol{x}}_T) - \hat{\boldsymbol{v}}_T\|_T < \|\hat{\boldsymbol{x}}_T - \hat{\boldsymbol{v}}_T\|_T. \tag{12}$$

Let now $S_T(\hat{\boldsymbol{v}}_T, 1)$ be the unit sphere in the space $V_T^{\oplus}$ of integrable tableaux centered on the optimum $\hat{\boldsymbol{v}}_T$. Defining $g : S_T(\hat{\boldsymbol{v}}_T, 1) \to \mathbb{R}$ as $g(\hat{\boldsymbol{x}}_{TS}) = \|A_T^L(\hat{\boldsymbol{x}}_{TS} - \hat{\boldsymbol{v}}_T)\|_T$ and taking into account (12) we get for $\hat{\boldsymbol{x}}_{TS} \in S_T(\hat{\boldsymbol{v}}_T, 1)$ (of course here $\hat{\boldsymbol{x}}_{TS} \neq \hat{\boldsymbol{v}}_T$) $g(\hat{\boldsymbol{x}}_{TS}) = \|A_T(\hat{\boldsymbol{x}}_{TS}) - \hat{\boldsymbol{v}}_T\|_T < \|\hat{\boldsymbol{x}}_{TS} - \hat{\boldsymbol{v}}_T\|_T = 1$. Note that g is defined on a compact set $S_T(\hat{\boldsymbol{v}}_T, 1)$ (as $dim(V_T^{\oplus}) = 2^l(2^l + 2) < \infty$) and thus attains its supremum for some $\hat{\boldsymbol{x}}^*_{TS} \in S_T(\hat{\boldsymbol{v}}_T, 1)$ which with $g(\hat{\boldsymbol{x}}_{TS}) < 1$ yields $\sigma = sup\{g(\hat{\boldsymbol{x}}_{TS}) : \hat{\boldsymbol{x}}_{TS} \in S_T(\hat{\boldsymbol{v}}_T, 1)\} = g(\hat{\boldsymbol{x}}^*_{TS}) < 1$. The latter combined with $\sigma = \|A_T^L\|_T$ guarantees $C < 1$. Theorem 1 is proved.

Of course, by (6) we infer $\|\boldsymbol{v}^{n+1} - \hat{\boldsymbol{v}}\| \leq C^n\|\boldsymbol{v}^1 - \hat{\boldsymbol{v}}\|$ and thus as $C < 1$ is independent from n the convergence $\boldsymbol{v}^n \to \hat{\boldsymbol{v}}$ holds.

4   Experimentation and Conclusions

We briefly describe the performance of 2-D Leap-Frog. Tests were run in Mathematica.[3] We took $k = 4$ and $l = 7$: each subsquare S_{ij}^{kl} consists of 16×16 pixel resolution and an image Ω consists of 128×128 pixels. In generating $\boldsymbol{v}$, Gaussian noise was added, with mean zero and standard deviation 0.04, to the integrable gradient field $\boldsymbol{v}_{grad} = (u_x, u_y)$ obtained from the function $u_i : [0,1] \times [0,1] \to \mathbb{R}$ (for $i = 1, 2$) defined as $u_1(x, y) = x^2 + 3xy + 2y^2$ (*quadratic polynomial*) and $u_2(x, y) = \cos(20((x - 0.5)^2 + 2(y - 0.3)^2))$ (*wavy function*). Alternatively digitization of camera images can be performed, prior to computation of $\boldsymbol{v}$. 2-D Leap-Frog was then applied with an *a priori* iteration bound set to $n_0 = 21$. Lawn-Mowing [ ] was used in the first iteration and then followed by $n_0 - 1 = 20$ 2-D Leap-Frog iterations. For evaluation we introduce some auxiliary notions. Given the estimate $\hat{\boldsymbol{v}}^{n_0}$ of $\hat{\boldsymbol{v}}$ obtained after 21 iterations, define $e = \|\boldsymbol{v} - \boldsymbol{v}_{grad}\|^2$, $c = \|\boldsymbol{v} - \hat{\boldsymbol{v}}^{n_0}\|^2$, $d = \|\boldsymbol{v}_{grad} - \hat{\boldsymbol{v}}^{n_0}\|^2$. Pythagoras' Theorem applied to the triangle $\Delta(\boldsymbol{v}, \hat{\boldsymbol{v}}^{n_0}, \boldsymbol{v}_{grad})$ yields the cosine of the angle between sides $\overline{\boldsymbol{v}\hat{\boldsymbol{v}}^{n_0}}$ and $\overline{\boldsymbol{v}_{grad}\hat{\boldsymbol{v}}^{n_0}}$ as: $\cos(\alpha) = \frac{d+c-e}{2\sqrt{dc}}$. Convergence of $\hat{\boldsymbol{v}}^n$ to $\hat{\boldsymbol{v}}$ amounts to $\lim_{n \to \infty} \cos(\alpha_n) = 0$. Denoting $\beta_n = \frac{\pi}{2} - arccos(\alpha_n)$ as the *angle deficiency*, expressed in radians, upon completion of nth iteration of the 2-D Leap-Frog (β_n should be close to zero) the following experimental results (each five-multiple iteration onwards) were obtained[4]:

We close this section with some observations.

– clearly k cannot be too big. Thus n_0 iterations of the 2-D Leap-Frog take approximately (for large l, $k < k_{max}$ fixed and $l - k > 2$) $\lambda = n_0(2^{l+1-k} - 1)^2/2^{2(l-k)} \approx 4n_0$ times longer than Lawn-Mowing [ ]. If however, l is small

---

[3] Mathematica runs very slowly and further experimentation is needed with C in order to make comparisons with other methods.

[4] The (very high quality) ideal and reconstructed surfaces (with and without noise rectification) can be viewed at `http://www.cs.uwa.edu.au/~ryszard/leap/pics/`.

Table 1. Performance of 2-D Leap-Frog for u_1 and u_2

Iteration number	$n = 1^a$	$n = 2$	$n = 6$	$n = 11$	$n = 16$	$n_0 = 21$
β_n^1 for u_1	0.04256	0.00298	0.00050	0.00028	0.00018	0.00013
β_n^2 for u_2	0.03786	0.00286	0.00081	0.00033	0.00018	0.00011

^{a} First iteration was Lawn-Mowing, but could be zero.

and k is comparable with l i.e. when $l - k = 1$ or $l - k = 2$ then $\lambda = 9/4$ or $\lambda = 49/16$, respectively. In general, $\lambda = 4 - 2^{k-l+2} + 2^{2(k-l)}$.

- the removable noise or digitization errors are those from the orthogonal complement $V_T^{\oplus\perp}$ to the space of integrable tableaux. The unremovable proportion is approximately equal to $(dim(V_T^{\oplus}))/dim(V_T)) = (2^l(2^l+2))/(2^{l+1}(2^l+1)) = \frac{1}{2}(1 + \frac{1}{2^{l+1}}) \approx 0.5$ (for l large).
- 2-D Leap-Frog is related to classical algebraic methods in numerical analysis, as well as more recent work in nonlinear optimization [ ]. In the special case $k = 1$ 2-D Leap-Frog is Gauss-Seidel for a system with respect to our canonical basis. Alternatively Gauss-Seidel could be applied directly to the larger system (2) with respect to the standard Euclidean basis. For $k > 1$ is multiplicative Schwarz (see e.g. [ ]). See also [ ] for other methods for sparse linear systems.
- in the proof of Theorem 1 a single coefficient variation x_{ab} associated with $h_{ab}^{\oplus}$ changes only 4 points (or 3 or 2 if the corresponding snapshot $S_{i_0 j_0}^{kl}$ forms an Ω-boundary snapshot) positioned inside $S_{i_0 j_0}^{kl}$ (as $k \geq 1$, a minimal snapshot size consists of 2x2 pixel block). This is the reason for the use of our canonical basis $\mathcal{B}_T^{\oplus}$. Examining this argument, we see that the size of the half snapshot overlap can be varied between 1 and $2^k - 1$.
- 1-D Leap-Frog solves a global problem, is iterative, converges regardless of the initial guess, works within the space of feasibile objects, has each step relatively small-scale, is amenable to the parallelism, and deals with nonlinearities. 2-D Leap-Frog has these features also, except that in the present paper there is no need to deal with nonlinearities.

References

1. Benzi M., Nabben R., Szyld D. (2000) Algebraic theory of multiplicative Schwarz methods. Tech. Report 00210 Dep. Math., Temple Uni., Pliladelphia, USA, http://www.math.temple.edu/~szyld/. Numerische Mathematik. In press 354, 363
2. Buzbee B. L., Golub G. H., Nielson C. W. (1970) On direct methods for solving Poisson's equations. SIAM J. Numer. Anal. **7**, (4):627–656 353
3. Frankot R. T. Chellappa R. (1988) A method of enforcing integrability in shape from shading algorithms. IEEE **10**, (4):439–451 352, 353, 356

4. Hackbush W. (1994) Iterative Solution of Large Sparse Systems of Equations. Springer, New York, Heidelberg, Berlin 363
5. Horn, B. K. P. (1986) Robot Vision. McGraw-Hill, New York Cambridge, MA 352, 353
6. Horn B. K. P. (1990) Height and gradient from shading. Int. J. Comp. Vision **5**, (1):37–75 352, 353, 356
7. Horn B. K. P., Brooks M. J. (1989) Shape from Shading. MIT Press, Cambridge, MA 352, 353
8. Kaya C. Y., Noakes L. (1998) A Leap-Frog Algorithm and optimal control: theoretical aspects. In: Caccetta L., Teo K. L., Siew P. F., Leung Y. H., Jennings L. S., Rehbock V. (Eds) Proc. 2nd Int. Con. Optim. Tech. Appl., Perth, Australia, July 1- July 3 1998. Curtin Uni. of Technology, 843–850 354
9. Klette R., Schlüns K. R., Koschan A. (1998) Computer Vision - Three Dimensional Data from Images. Springer, Singapore 353
10. Kozera, R. (1991) Existence and uniqueness in photometric stereo. Appl. Math. Comput. **44**, (1):1–104 352, 353
11. Kozera R. (1992) On shape recovery from two shading patterns. Int. J. Patt. Rec. Art. Int. **6**, (4):673–698 352, 353
12. Noakes L. (1999) A global algorithm for geodesics. J. Math. Australian Soc. Series A. **64**, 37–50 352, 354, 363
13. Noakes L., Kozera R., Klette R. (1999) The Lawn-Mowing Algorithm for noisy gradient vector fields. In: Latecki L. J., Melter R. A., Mount D. M., Wu A. Y. (Eds) Proceedings of SPIE Conference, Vision Geometry VIII, Denver, USA, July 19-July 20 1999. The International Society for Optical Engineering, 3811:305–316 352, 353, 356, 362
14. Noakes L., Kozera R. (1999) A 2-D Leap-Frog Algorithm for optimal surface reconstruction. In: Latecki L. J., Melter R. A., Mount D. M., Wu A. Y. (Eds) Proceedings of SPIE Conference, Vision Geometry VIII, Denver, USA, July 19-July 20 1999. The International Society for Optical Engineering, 3811:317–328 352
15. Simchony T., Chellappa R., Shao M. (1990) Direct analytical methods for solving Poisson equations in computer vision problems. IEEE Trans. Patttern Rec. Machine Intell. **12**, (5):435–446 352, 353

On Approximation of Jordan Surfaces in 3D

Fridrich Sloboda and Bedrich Zaťko

Institute of Control Theory and Robotics, Slovak Academy of Sciences
Dubravska 9, 842 37 Bratislava, Slovak Republic
utrrzatk@savba.sk

Abstract. A polyhedral approximation of closed Jordan surfaces is described. The approximation is based on the notion of a relative convex hull in a polyhedrally bounded compact set obtained by gridding technique and can be applied also to approximation of surfaces of functions.

1 Introduction

Measures are of importance in mathematics in two different ways. Measures can be used for estimating the size of sets, and measures can be used to define integrals. Borel, Lebesgue, Peano and Jordan associated to each set of points A an outer measure m^*A and an inner measure m_*A. A set of points A is called measurable if $m^*A = m_*A$. Caratheodory [ ] defined a p-dimensional measure in a q-dimensional space for p, q integer. Hausdorff [ ] generalized the Caratheodory's p-dimensional integer measure for p real and so defined also fractal dimensions. For example the fractal dimension of the Cantor's dust is $\frac{\log 2}{\log 3}$. Hausdorff measures have played a dominant role in the development of surface area theory and, in particular, in the theory of minimal surfaces [, , ,].

Let A be an arbitrary set of points in q-dimensional Euclidean space, $q > 1$. Let $U_1, U_2, \ldots$ be a sequence of countable sets of points (closed or open) of the property

a) $A \subseteq \bigcup_k U_k$
b) $d(U_k) < \epsilon$, $k = 1, 2, \ldots$, where $d(.)$ denotes the diameter of U_k.

Let us denote

$$\inf \sum_{k=1}^{\infty} d_k^{(p)} = L_\epsilon^{(p)}, \quad p = 1, 2, \ldots,$$

for all U_k, $k = 1, 2, \ldots$, of the property a) and b), where $d_k^{(p)}$ denotes the p-dimensional diameter of U_k. Then

$$\lim_{\epsilon \to 0} L_\epsilon^{(p)} A = {}^*L^{(p)} A$$

is called the outer p-dimensional Caratheodory measure. If $p = 1$ the measure is called linear, if $p = 2$ it represents the 2-dimensional area (surface area) of A. If U_k are balls with diameter ϵ then the 2-dimensional diameter of U_k equals $\frac{1}{4}\pi\epsilon^2$. In the next we will use the following notation

$$ {}^*L^{(2)} A = H^2(A),$$

G. Bertrand et al. (Eds.): Digital and Image Geometry, LNCS 2243, pp. 365–386, 2001.
© Springer-Verlag Berlin Heidelberg 2001

where $H^2(A)$ denotes the 2-dimensional Hausdorff measure of A, i.e., the surface area.

The first topological definition of the surface area was given by Minkowski in 1901 [ ]. Let $S \subset R^3$ be a two-dimensional continuum homeomorphic to the unit sphere. The surface area of S, $s(S)$, is according to H. Minkowski defined as

$$s(S) = \lim_{\delta \to 0} \frac{V(M_{\delta,S})}{2\delta},$$

provided that the limit exists, where $V(.)$ denotes the volume of $(.)$, and for $\delta > 0$, $, y \in S$

$$M_{\delta,S} = \bigcup U_{\delta,y},$$

$$U_{\delta,y} := \{x \in R^3 \mid \mathrm{dist}_2(y,x) \leq \delta\}.$$

Since 1918 a number of other approaches to surface area definitions have been published. Unlike the curve length definitions, which are all equivalent, the surface area definitions are not equivalent [ ]. They are equivalent only for a subclass of all surfaces, for example for sufficient smooth surfaces. The most general surface area definition is the Caratheodory–Hausdorff definition.

The 2-dimensional Caratheodory outer measure defines the surface area but it does not allow to visualize the surface itself. The most efficient way how to visualize the surface itself is to find its polyhedral approximation. A two-dimensional continuum in R^3 homeomorphic to the unit sphere is called a closed Jordan surface, and a two-dimensional continuum in R^3 homeomorphic to the unit square is called a Jordan surface. An efficient polyhedral approximation of closed Jordan surfaces which with arbitrary accuracy approximates the surface area of a given closed Jordan surface is still an open problem. The first attempt to a polyhedral approximation of closed Jordan surfaces was made by C. Jordan by the end of the 19-th century. This approach was based on a triangulation method. But only few years later Peano and Schwarz showed that this approach is not functional even for a cylinder.

In this paper a new approach to polyhedral approximation of closed Jordan surfaces is described. It is based on the notion of a relative convex hull related to a polyhedrally bounded compact set obtained by gridding. The notion of a relative convex hull represents a generalization of the notion of a convex hull. The relative convex hull related to a polyhedrally bounded compact set represents the smoothest polyhedral approximation of closed Jordan surfaces and can be applied also to approximation of surfaces of functions. It enables an efficient representation and visualization of surfaces and has applications in computer aided geometric design, computer graphics and image processing.

2   Polyhedron

Polyhedron in R^3 is a simply connected compact set bounded by a surface homeomorphic to the unit sphere which is represented by union of convex faces. A face is a simply connected compact set bounded by a simple closed convex polygonal curve and consists of vertices and edges.

According to Euler for a polyhedron it holds

$$v - e + f = 2,$$

where v, e, f denotes the number of vertices, edges and faces, respectively.

There are three types of vertices of a polyhedron: convex, concave and saddle vertices. Let $v \in R^3$ be a point and let $\pi \subset R^3$ be a plane, such that $v \in \pi$. Let for $\delta > 0$

$$M(v, \pi, \delta) := \{x \in \pi \subset R^3 \mid \mathrm{dist}_2(x, v) \le \delta\}.$$

Definition 1. *A vertex v of a polyhedron P_S is called a convex vertex of P_S if there exists a plane π and $\delta > 0$, such that*

$$M(v, \pi, \delta) \setminus \{v\} \subset {}^c P_S,$$

where ${}^c P_S$ denotes the complement of P_S.

Definition 2. *A vertex v of a polyhedron P_S is called a concave vertex of P_S if there exists a plane π and $\delta > 0$, such that*

$$M(v, \pi, \delta) \setminus \{v\} \subset P_S^o,$$

where P_S^o denotes the interior of P_S.

Definition 3. *A vertex v of a polyhedron P_S is called a saddle vertex of P_S if there does not exist a plane π and $\delta > 0$, such that*

$$M(v, \pi, \delta) \setminus \{v\} \subset {}^c P_S, \quad or \quad M(v, \pi, \delta) \setminus \{v\} \subset P_S^o.$$

According to the Gauss-Bonnet theorem the global curvature of a surface homeomorphic to the unit sphere equals 4π. In the case of a polyhedron the curvature is related to the vertices of the polyhedron only. With each vertex of a polyhedron is associated an angle. This angle is defined as the sum of angles between neighbouring edges related to a vertex in radians. With each vertex of a polyhedron is associated a curvature which is defined as $2\pi - \alpha$, where α is the angle related to the vertex. A convex vertex of a polyhedron not necessarily possesses a positive curvature and a polyhedron which possesses convex vertices only is not necessarily convex, see Fig. 1.

For further considerations we are introducing the following

Definition 4. *Let e be an edge of a polyhedron P_S and let y be a point of e, $y \in e$. We say that e is convex if there exists a straightline $l : y \in l$ and $\delta > 0$ that*

$$M_1(y, l, \delta) \setminus \{y\} \subset {}^c P_S,$$

where

$$M_1(y, l, \delta) := \{x \in l \subset R^3 \mid dist_2(x, y) \le \delta\}$$

and ${}^c P_S$ denotes the complement of P_S.

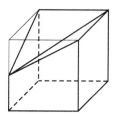

Fig. 1. A polyhedron which possesses convex vertices only and itself is not convex

Similarly also a concave edge is defined. Now it holds that a polyhedron which consists of convex edges only is convex.

In the following we will consider isothetic polyhedrons. Isothetic polyhedrons have the property that they possess only faces which are parallel with xy-, xz-, zy-plane, respectively.

For an isothetic polyhedron it holds that its convex, concave vertices possess positive curvature which equals $\pi/2$ and there are only two types of saddle vertices and these possess $-\pi/2$, $-\pi$ curvature, respectively.

3   Relative Convex Hull

In the following the notion of a relative convex hull will be introduced and its properties are summarized.

Definition 5. *Let $U \subseteq R^n$ be an arbitrary set. A set $C \subseteq U$ is said to be U-convex iff for every $x, y \in C$ such that $\overline{xy} \subseteq U$ it holds that $\overline{xy} \subseteq C$.*

$A \subseteq U$ is U-convex iff for every convex set $C \subseteq U$ the set $C \cap A$ is convex. Any convex $A \subseteq U$ is U-convex. If U is convex then $A \subseteq U$ is U-convex iff A is convex. U itself is U-convex. If $A \subseteq B \subseteq U$ and A is U-convex then A is also B-convex.

Let $A \subset B \subset U \subset R^3$, and $\partial A, \partial B, \partial U$ be homeomorphic to the unit sphere. Let A, B be U–convex and $H^2(\partial U) < \infty$. Then

$$H^2(\partial A) < H^2(\partial B) < H^2(U).$$

If $U = R^3$ and $H^2(\partial B) < \infty$ then A, B are convex and

$$H^2(\partial A) < H^2(\partial B) < \infty,$$

which follows from the theory of convexity [,].

Definition 6. *Let $V \subseteq U \subseteq R^n$ be given. The intersection of all U-convex sets containing V will be termed U-convex hull of V and denoted by $conv_U(V)$.*

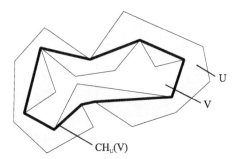

Fig. 2. V, $U \subset R^2$ and $CH_U(V)$

For the relative convex hull $conv_U(V)$ it holds the following:

a) $V \subseteq conv_U(V) \subseteq U \cap conv(V)$,
b) $V = conv_U(V)$ holds iff V is U-convex,
c) $conv_U(V) = conv(V)$ iff $conv(V) \subseteq U$.

Let $V \subset U \subset R^3$, and $\partial V, \partial U$ be homeomorphic to the unit sphere. Further let $H^2(\partial U) < \infty$. Then

$$H^2(\partial conv_U(V)) = \min_{A \ \ U-\text{convex containing } V} \{H^2(\partial A)\},$$

which is the property of minimality of the relative convex hull. Because $U \subset R^3$ itself is U-convex by definition and $V \subset U$ it holds that

$$H^2(\partial conv_U(V)) < H^2(\partial U).$$

Let $V, U \subset R^3$ be polyhedrons such that $V \subset U^\circ$. Then $\partial conv_U(V)$ is a polyhedral surface. The vertices of $conv(V)$ belong to the vertices of $conv_U(V)$. But not necessarily all vertices of $conv_U(V)$ belong to the vertices of V, U, respectively. There might exist vertices of $conv_U(V)$ which are lying on an edge of V, U, respectively.

Let $V, U \subset R^2$ be polygons such that $V \subset U^\circ$ and let $A \subset R^2$ be a simply connected compact set. Then for $conv_U(V)$ it holds

a) $H^1(\partial conv_U(V)) = \min_{\partial A \subset U \backslash V^\circ, \ V \subseteq A} H^1(\partial A)$

b) $conv_U(V)$ is a polygon, so called minimum perimeter polygon,
    whereby $\partial conv_U(V) = CH_U(V)$ is called a stretched thread, see Fig. 2.

According to a) $\partial conv_U(V)$ is the shortest Jordan curve in $G := U \backslash V^\circ$ encircling V. The properties of the shortest path in a polygonally bounded compact set are described in [ ].

4   Polyhedral Approximation

In the following we will consider a polyhedral approximation of surfaces homeomorphic to the unit sphere in R^3. The approximation is based on the method of
shrinking and on the notion of the relative convex hull. The method of shrinking
could be also called a method of compression and it is related to a well-known
property of nested set of continua [ ]:

Lemma 1: Let

$$C_1 \supset C_2 \supset \ldots \supset C_i \supset \ldots$$

be a set of continua in R^3. Then

$$C = \bigcap_i C_i, \quad i = 1, 2, \ldots$$

is a continuum, i.e., a connected compact set.

Suppose that $C = S \subset R^3$ is a two-dimensional continuum homeomorphic to
the unit sphere with $H^2(S) < \infty$ and

$$C_i = P_2^{(i)} \setminus P_1^{(i)o}, \quad i = 1, 2, \ldots,$$

where $P_2^{(i)}$, $P_1^{(i)}$ are polyhedrons such that $P_1^{(i)} \subset P_2^{(i)o}$. The approximation
of the surface area of S homeomorphic to unit sphere in R^3 is related to the
following

Theorem 1: Let $M \subset R^3$ and let $\partial M = S$ be a two-dimensional continuum
homeomorphic to the unit sphere with $H^2(S) < \infty$. Let

$$G_0 \supseteq G_1 \supseteq G_2 \supseteq \ldots \supseteq G_i \supseteq \ldots \supseteq S, \quad S = \bigcap_{i=1}^{\infty} G_i,$$

where $G_i := P_2^{(i)} \setminus P_1^{(i)o}$; $P_2^{(i)}$, $P_1^{(i)}$ are polyhedrons such that

$$a)\ P_1^{(i)} \subset P_2^{(i)o}, \quad i = 0, 1, 2, \ldots$$

$$b)\ \lim_i \ \sup H^2(\partial P_2^{(i)}) < \infty.$$

Then

$$\lim_{i \to \infty} H^2(CH_{P_2^{(i)}}(P_1^{(i)})) = H^2(S),$$

where $CH_{P_2^{(i)}}(P_1^{(i)}) = \partial conv_{P_2^{(i)}}(P_1^{(i)})$.

Comment 1: For the relative convex hull boundary $CH_{P_2^{(i)}}(P_1^{(i)})$ it holds that

$$a)\ H^2(CH_{P_2^{(i)}}(P_1^{(i)})) \leq H^2(S) \text{ or } \ H^2(CH_{P_2^{(i)}}(P_1^{(i)})) \geq H^2(S)$$

$$b)\ H^2(S_{min}^{(i)}) \leq H^2(CH_{P_2^{(i)}}(P_1^{(i)}) < \limsup_i H^2(\partial P_2^{(i)}),$$

where $S_{\min}^{(i)}$ denotes the minimal area surface homeomorphic to the unit sphere in $G_i := P_2^{(i)} \setminus P_1^{(i)o}$ containing $P_1^{(i)}$. For $S_{\min}^{(i)}$ it holds [ ]

$$\lim_i H^2(S_{\min}^{(i)}) = H^2(S).$$

The relative convex hull boundary $CH_{P_2^{(i)}}(P_1^{(i)})$ can be called as stretched elastic surface.

Comment 2: Theorem 1 for the two-dimensional case is proved in [ ].

5   Method of Gridding

The method of gridding enables to realize the above described general approach to surface area approximation of surfaces in R^3, to represent and to visualize them. Regular grids represent a theoretical tool of set theoretic topology and mathematical analysis. The grid was introduced by C. Jordan and G. Peano by the end of the 19-the century in order to define measurable sets. In three dimensional space the most important grid is the orthogonal one. More formally, for $p = 0, 1, 2, \ldots$, and for each triple (w_1, w_2, w_3) of integer numbers let

$$N_{(w_1,w_2,w_3)}^p := \{x \in R^3 \mid w_i 2^{-p} \le x_i \le (w_i + 1) 2^{-p}, \ i = 1, 2, 3\}$$

represents the topological unit of an orthogonal grid. Let $M \subset R^3$ be a compact set bounded by a smooth closed Jordan surface S, $\partial M = S$. Let us denote

$$^-M_p := \{N_{(w_1,w_2,w_3)}^p \mid N_{(w_1,w_2,w_3)}^p \subset M^o\}$$

$$^+M_p := \{N_{(w_1,w_2,w_3)}^p \mid N_{(w_1,w_2,w_3)}^p \cap M \ne \emptyset\}.$$

$N_{(w_1,w_2,w_3)}^p$ elements which belong to ^-M_p will be called inner elements of M and elements for which $N_{(w_1,w_2,w_3)}^p \cap \partial M \ne \emptyset$ will be called boundary elements of M, see Fig. 3.

M is Jordan measurable if

$$I(M) = \inf_p I(^+M_p) = \sup_p I(^-M_p),$$

where $I(.)$ denotes the volume of $(.)$. If M is Jordan measurable it does not imply that $H^2(\partial M) < \infty$. We say that ^+M_p, ^-M_p are related to a proper grid if for some $p_0 \ge 0$

a) ^+M_p, ^-M_p for $p = p_0, p_0 + 1, \ldots$, are simply connected compact sets

b) for $N_{(w_1,w_2,w_3)}^p \in {}^+M_p \setminus {}^-M_p^o$, $p = p_0, p_0 + 1, \ldots$, $\partial N_{(w_1,w_2,w_3)}^p \cap S$ is a simple closed Jordan curve or a single point.

^+M_p, ^-M_p related to a proper grid possess a proper polyhedral representation of their boundaries, i.e., ^+M_p, ^-M_p with proper polyhedral representation are polyhedrons, so that they do not possess collinear vertices, see Fig. 4.

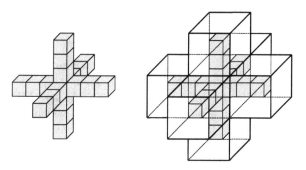

Fig. 3. ^-M_p and ^+M_p related to an $M \subset R^3$

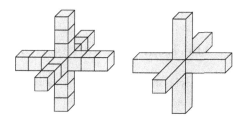

Fig. 4. ^-M_p and its proper polyhedral representation

Let us denote by

$$G_p := {}^+M_p \setminus {}^-M_p^o, \quad p = p_0, p_0 + 1, \ldots.$$

According to the definition of ^+M_p, ^-M_p, respectively, it holds

$$a) \; {}^-M_p \subset {}^+M_p^o$$

$$b) \; G_p \supseteq G_{p+1} \supseteq S$$

$$c) \; S = \bigcap_{p=p_0}^{\infty} G_p.$$

For $S = \partial M \subset R^3$ smooth $H^2(S) < \infty$ and $\lim_p \sup H^2(\partial^+ M_p) < \infty$, so that Theorem 1 applies. Fig. 7 shows ^-M_p related to an $M \subset R^3$, ^+M_p, $CH(^-M_p)$ and $CH_{+M_p}(^-M_p)$. Fig. 8 shows ^-M_p related to an $M \subset R^3$, $CH(^-M_p)$ and $CH_{+M_p}(^-M_p)$.

Comment 3: If $M \subset R^3$ is convex then

$$a) \; \partial conv_{+M_p}(^-M_p) = CH_{+M_p}(^-M_p) = CH(^-M_p)$$

$$b) \; H^2(CH(^-M_p)) \leq H^2(CH(^-M_{p+1}))$$

$$c) \ \lim_{p \to \infty} H^2(CH(^-M_p)) = H^2(S),$$

for $p = p_0, p_0 + 1, \ldots$, which follows from the theory of convexity [,].

Comment 4: Let $S : z = f(x,y)$, $H^2(S) < \infty$, where $f > 0$ is defined on a square D related to a regular orthogonal grid. Then ^-M_p related to a proper grid is a simply connected compact set, $^-M_p \subset {}^+M_p$, and $CH_{+M_p}(^-M_p)$ is a polyhedral surface homeomorphic to the unit sphere which approximates also the surface $S : z = f(x,y)$ for $(x,y) \in D$, see Fig. 9 and Fig. 22. There is an one to one map between a digital image and its 3D grid model. The gray-level related to a pixel represents the number of voxels in the corresponding column of its 3D model denoted by ^-M_p and vice versa, see Fig. 22. ^+M_p is obtained by bordering of ^-M_p by voxel elements so that $^-M_p \subset {}^+M_p^o$ and the corresponding $CH_{+M_p}(^-M_p)$ represents also a polyhedral approximation and representation of the surface related to a digital image.

6 A Method for the Relative Convex Hull Calculation

Let $^+M_p, {}^-M_p \subset R^3$ be polyhedrons defined by union of $N^p_{(w_1,w_2,w_3)}$ elements such that $^-M_p \subset {}^+M_p$. A set theoretic method for the relative convex hull $conv_{+M_p}(^-M_p)$ calculation is defined as follows

Step (1): Set $A_0 := {}^-M_p$ and define

$$A_k := \bigcup_{x,y \subset A_{k-1}, \ \overline{xy} \subset {}^+M_p} \overline{xy}, \quad k = 1, 2, \ldots.$$

Step (2): The relative convex hull $conv_{+M_p}(^-M_p)$ is defined as

$$conv_{+M_p}(^-M_p) = \bigcup_{k=0}^{s} A_k,$$

whereby $^-M_p = A_0 \subset A_1 \subset A_2 \subset \ldots \subset A_s = A_{s+1}$, i.e., the relative convex hull is defined by finite number of steps.

7 Implicit Surfaces

Implicit surfaces become interesting in computer aided geometric design and computer graphics. The regular orthogonal grids represent the most popular tools used by methods for representation and visualization of implicit surfaces [ ]. Besides regular orthogonal grids also adaptive grids are used. A commonly used method for representation of implicit surfaces is based on so called polygonization techniques [ ].

Let

$$F(q) = 0, \quad q \in R^3$$

be a smooth two-dimensional continuum homeomorphic to the unit sphere, the boundary of a compact set $M \subset R^3$, $\partial M \equiv F(p) = 0$. A point $q \in R^3$, for which sign $F(q) = -1$, sign $F(q) = +1$ is called inner, outer point of M, $\partial M \equiv F(q) = 0$, respectively.

Suppose that for $p = p_0, p_0 + 1, \ldots,$ $^{+}M_p$, $^{-}M_p$ related to $M \subset R^3$, $\partial M \equiv F(q) = 0$, defined in the previous section, are related to a proper grid. For the bounary elements

$$N^p_{(w_1, w_2, w_3)} \in {}^{+}M_p \setminus {}^{-}M^o_p$$

for which $\partial N^p_{(w_1, w_2, w_3)} \cap (F(q) = 0)$ is

a) a simple closed Jordan curve which does not cut an edge of $N^p_{(w_1, w_2, w_3)}$ or

b) a point which is not a vertex of $N^p_{(w_1, w_2, w_3)}$

it holds that all their vertices, v_i, $i = 1, 2, \ldots, 8$, have the same sign $F(v_i) \neq 0$, see Fig. 5.

However, the relative convex hull boundary $CH_{+M_p}(^{-}M_p)$ cannot intersect these boundary $N^p_{(w_1, w_2, w_3)}$ elements, so that they need not be considered for the $CH_{+M_p}(^{-}M_p)$ calculation. For all other boundary elemetns $N^p_{(w_1, w_2, w_3)} \in {}^{+}M_p \setminus {}^{-}M^o_p$ it holds that not all their vertices, v_i, $i = 1, 2, \ldots, 8$, have the same sign $F(v_i)$, so that the set $G_p := {}^{+}M_p \setminus {}^{-}M^o_p$ in which the boundary of the relative convex hull lies is well defined. $CH_{+M_p}(^{-}M_p)$ calculation does not require numerical root finding to locate polygon vertices on edges of each boundary $N^p_{(w_1, w_2, w_3)}$ element as the polygonization methods do require [ ].

In case of approximation of planar implicit curves homeomorphic to the unit circle on orthogonal grids the boundary of the relative convex hull is a polygonal curve whose vertices belong to convex, concave vertices of $^{-}M_p$, $^{+}M_p$, respectively, so that they belong to the grid points which are represented by integer coordinates only. The corresponding algorithm for $CH_{+M_p}(^{-}M_p)$ calculation is based on integer arithmetic and has linear time complexity [ ], see Fig. 6 which shows the approximation of $F(x, y) \equiv (x^2 + y^2)^3 - 9(x^4 + y^4) = 0$ by $CH_{+M_p}(^{-}M_p)$ related to 512x512 grid point resolution.

In case of approximation of convex implicit surfaces on orthogonal grids the boundary of the relative convex hull $CH_{+M_p}(^{-}M_p) = CH(^{-}M_p)$, so that it

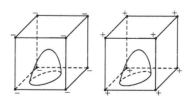

Fig. 5. Boundary elements whose all vertices v_i, $i = 1, 2, \ldots, 8$, have the same sign $F(v_i)$

is again represented by grid point vertices with integer coordinates only, see Fig. 10 – Fig. 17. The polygonization methods do require numerical root finding procedure whereby the resulting polyhedron is not necessarily convex [ ].

Fig. 18 and Fig. 20 show ^-M_p related to non-convex implicit surfaces and Fig. 19 and Fig. 21 show the corresponding $CH(^-M_p)$. Fig. 22 shows ^-M_p related to a surface of a function and Fig. 23 shows the corresponding $CH(^-M_p)$.

8 Conclusion

A new approach to polyhedral approximation of closed Jordan surfaces is described. The approximation is based on the notion of a relative convex hull in a polyhedrally bounded compact set obtained by gridding technique. The relative convex hull enables an efficient representation and visualization of surfaces and has significant applications in computer aided geometric design, computer graphics and image processing.

References

1. J. Bloomenthal, Surface tiling, in Introduction to implicit surfaces, ed. J. Bloomenthal, Morgan Kaufmann Publisher, San Francisco, (1997), 126–165   373, 374, 375

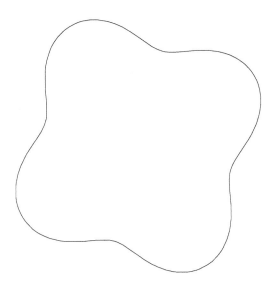

Fig. 6. The boundary of the relative convex hull, which approximates $F(x, y) \equiv (x^2 + y^2)^3 - 9(x^4 + y^4) = 0$ on a 512x512 orthogonal grid, possesses 153 grid point vertices and provides 0.1 % length error

2. C. Caratheodory, Gesammelte mathematische Schriften, C. H. Becksche Verlags-buchhandlung, München, (1956), 247–277 365
3. M. Chlebik and F. Sloboda, Approximation of surfaces by minimal surfaces with obstacles, Tech. report of the Institute of Control Theory and Robotics, Bratislava, (2000), 1-11 371
4. E. Giusti, Minimal surfaces and functions of bounded variation, Birkhäuser, (1994) 365
5. H. Federer, Geometric measure theory, Springer–Verlag, New York, (1969) 365
6. H. Hadwiger, Altes und Neues über konvexe Körper, Birkhäuser, (1955) 368, 373
7. F. Hausdorff, Dimension und äußeres Maß, Mathematische Annalen, (1919), 157–179 365
8. M. Kneser, Einige Bemerkungen über das Minkowskische Flächenmaß, Archiv der Mathematik, (1955), 382–390 366
9. H. Minkowski, Über die Begriffe Länge, Oberfläche und Volumen, Jahresbericht der Deutschen Math.–Vereinigung, (1901), 115–121 366, 368, 373
10. A. S.Parchomenko, Was ist eine Kurve, VEB Deutscher Verlag der Wissenschaften, Berlin (1957) (original Moscow (1954)). 370
11. C. A. Rogers, Hausdorff measures, Cambridge Univ. Press, (1970) 365
12. F. Sloboda, B. Zaťko and J. Stoer: On approximation of planar one-dimensional continua, Advances in Digital and Computational Geometry, eds. R. Klette, A. Rosenfeld and F. Sloboda, Springer Verlag, Singapore, (1998), 113–160. 369, 374
13. F.Sloboda and J.Stoer, On Piecewiese Linear Approximation of Planar Jordan Curves, J.Comp. Applied Mathematics 55 (1994) 369–383. 371

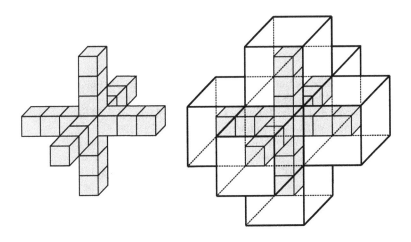

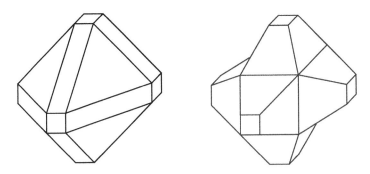

Fig. 7. ^-M_p, ^+M_p, $CH(^-M_p)$ and $CH_{+M_p}(^-M_p)$

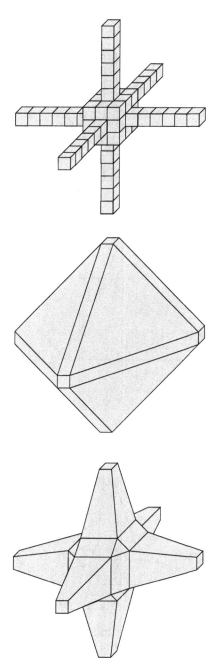

Fig. 8. ^-M_p, $CH(^-M_p)$ and $CH_{+M_p}(^-M_p)$, whereby ^+M_p is defined by bordering of ^-M_p by $N^p_{(w_1,w_2,w_3)}$ elements so that $^-M_p \subset {}^+M^o_p$

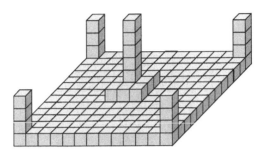

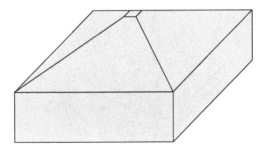

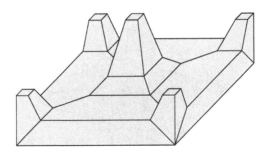

Fig. 9. ^-M_p related to a surface of a function, $CH(^-M_p)$ and $CH_{+M_p}(^-M_p)$, whereby ^+M_p is defined by bordering of ^-M_p by $N^p_{(w_1,w_2,w_3)}$ elements so that $^-M_p \subset {}^+M_p^o$

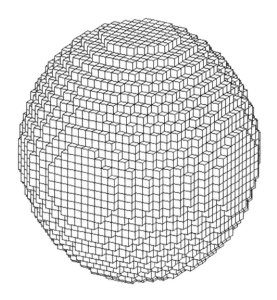

Fig. 10. ^-M_p related to $F(x, y, z) \equiv x^2 + y^2 + z^2 - 1 = 0$

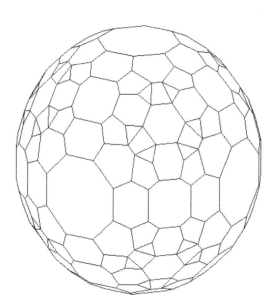

Fig. 11. $CH_{+M_p}(^-M_p) = CH(^-M_p)$ related to the previous figure, whereby ^+M_p is defined by bordering of ^-M_p by $N^p_{(w_1, w_2, w_3)}$ elements so that $^-M_p \subset {}^+M_p^o$

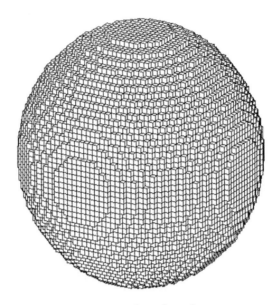

Fig. 12. $^{-}M_p$ related to $F(x, y, z) \equiv x^2 + y^2 + z^2 - 1 = 0$ which corresponds to a higher grid point resolution

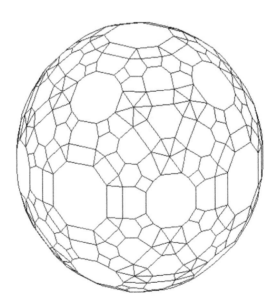

Fig. 13. $CH_{+M_p}(^{-}M_p) = CH(^{-}M_p)$ related to the previous figure, whereby $^{+}M_p$ is defined by bordering of $^{-}M_p$ by $N^p_{(w_1, w_2, w_3)}$ elements so that $^{-}M_p \subset {}^{+}M_p^o$

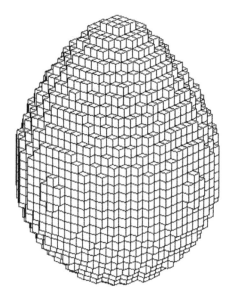

Fig. 14. ^-M_p related to $F(x,y,z) \equiv (x^2 + y^2 + (z+0.3)^2)^2 - 0.18(-x^2 - 5.0y^2 + z^2) + +0.3^4 - 0.31^4 = 0$

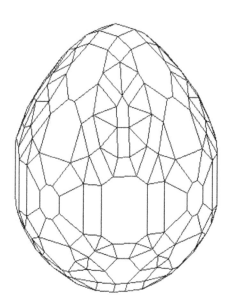

Fig. 15. $CH_{+M_p}(^-M_p) = CH(^-M_p)$ related to the previous figure, whereby ^+M_p is defined by bordering of ^-M_p by $N^p_{(w_1,w_2,w_3)}$ elements so that $^-M_p \subset {}^+M_p^o$

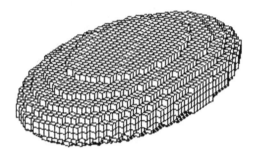

Fig. 16. ^-M_p related to $F(x, y, z) \equiv x^2 + 0.5^{-2}y^2 + 0.2^{-2}z^2 - 1 = 0$

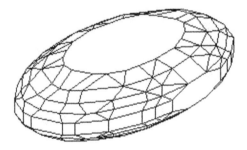

Fig. 17. $CH_{+M_p}(^-M_p) = CH(^-M_p)$ related to the previous figure, whereby ^+M_p is defined by bordering of ^-M_p by $N^p_{(w_1,w_2,w_3)}$ elements so that $^-M_p \subset {}^+M_p^o$

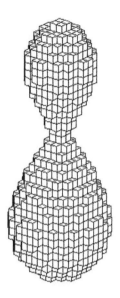

Fig. 18. ^-M_p related to $F(x, y, z) \equiv (x^2 + y^2 + (z + 0.06)^2)^2 - 0.7688(-x^2 - y^2 + z^2)+ +0.62^4 - 0.63^4 = 0$

Fig. 19. $CH(^-M_p)$ related to the previous figure, whereby ^+M_p is defined by bordering of ^-M_p by $N^p_{(w_1, w_2, w_3)}$ elements so that $^-M_p \subset {}^+M_p^o$

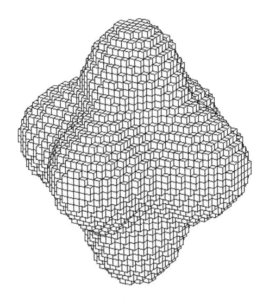

Fig. 20. ^-M_p related to $F(x, y, z) \equiv (x^2 + y^2 + z^2)^3 - (x^4 + y^4 + z^4) = 0$

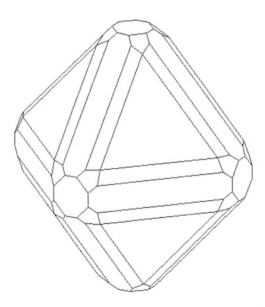

Fig. 21. $CH(^-M_p)$ related to the previous figure, whereby ^+M_p is defined by bordering of ^-M_p by $N^p_{(w_1, w_2, w_3)}$ elements so that $^-M_p \subset {^+M_p^o}$

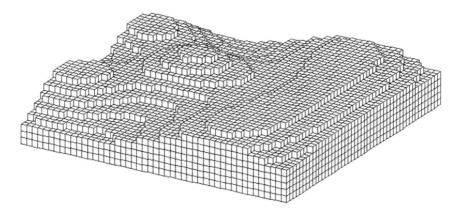

Fig. 22. ^-M_p related to a surface of a function

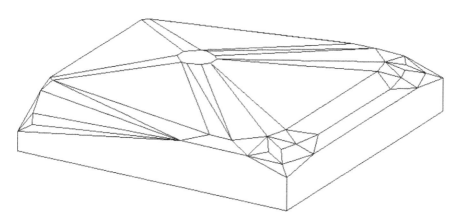

Fig. 23. $CH(^-M_p)$ related to the previous figure, whereby ^+M_p is defined by bordering of ^-M_p by $N^p_{(w_1,w_2,w_3)}$ elements so that $^-M_p \subset {}^+M_p^o$

Part V

Shape Similarity and Simplification

Similarity Measure Computation of Convex Polyhedra Revisited

Jos B.T.M. Roerdink and Henk Bekker

Institute for Mathematics and Computing Science, University of Groningen
P.O. Box 800, 9700 AV Groningen, The Netherlands
{roe,bekker}@cs.rug.nl

Abstract. We study the computation of rotation-invariant similarity measures of convex polyhedra, based on Minkowski's theory of mixed volumes. To compute the similarity measure, a (mixed) volume functional has to be minimized over a number of critical orientations of these polyhedra. These critical orientations are those relative configurations where faces and edges of the two polyhedra are as much as possible parallel. Two types of critical orientations exist for two polyhedra A and B. Type-1 critical orientations are those relative orientations where a face of B is parallel to a face of A, and an edge of B is parallel to a face of A, or vice versa. Type-2 critical orientations correspond to the case that three edges of A are parallel to three faces of B, or vice versa. It has been conjectured that to perform minimization of the volume functional, it is sufficient to consider Type-1 critical orientations only. Here we present experimental proof showing this conjecture to be false.

1 Introduction

Shape comparison is one of the fundamental problems of machine vision. Shape similarity may be quantified by introducing a similarity measure. The requirement of invariance under some set of shape transformations in general leads to complicated optimization problems. Therefore, one often studies shape classes and transformation sets for which a compromise between generality and efficiency can be found.

Recently, a new approach to similarity measure computation of convex polyhedra has been developed based on Minkowski addition [,]. These similarity measures are based upon the Minkowski inequality and its descendants, and the central operation is the minimization of (mixed) volume functionals. An attractive property of this family of similarity measures is that they are invariant under translations and possibly under scaling, rotation, and reflection. The method may be used in any-dimensional space, but we will concentrate on the 3D case.

For computing a rotation-invariant similarity measure of two convex polyhedra, a (mixed) volume functional has to be evaluated over a number of special relative orientations of these polyhedra, the so-called critical orientations. These critical orientations are those relative configurations where faces and edges of

G. Bertrand et al. (Eds.): Digital and Image Geometry, LNCS 2243, pp. 389–399, 2001.
© Springer-Verlag Berlin Heidelberg 2001

the two polyhedra are parallel as much as possible. (Two faces are called parallel when they have the same outward normal.) Given two polyhedra A and B, the set of critical orientations can be divided in two classes, denoted by Type 1 and 2, respectively. Type 1 occurs when a face of B is parallel to a face of A, and an edge of B is parallel to a face of A, or vice versa; Type 2 occurs when three edges of A are parallel to three faces of B, or vice versa. It was proved in [ ] that (i) for a given rotation axis, it is sufficient to compute the (mixed) volume functionals only for a finite number of critical angles, thus generalizing a result for the 2D case [ ]; and (ii) the number of rotation axes to be checked is finite. The second result is trivial for Type-1 orientations, but the proof for Type 2 is more involved, and only establishes finiteness of the number of axes to be checked, without giving an explicit upper bound on the number of axes. Such an explicit upper bound for the number of Type-2 critical orientations was given in [ ], where it was shown that the problem can be reduced to solving an algebraic equation of degree 8, which has to be solved numerically. So, given three edges of A and three faces of B, the number of critical orientation axes is at most 8.

Experiments on mixed-volume minimization were reported by Tuzikov and Sheynin in [ ]. Only Type-1 critical orientations were taken into account, and the authors conjectured that to find the global minimum of the mixed-volume functional it is sufficient to consider Type-1 critical orientations only. In this paper we reconsider this issue, and present experimental proof showing this conjecture to be false. Experiments were carried out with randomly generated pairs of tetrahedra, and the minimum of the mixed volume functional was computed by taking into account either Type-1 or Type-2 critical orientations. As a result, we have found that the minimum value of the mixed volume functional for Type-2 minimization can be larger as well as smaller than that of Type-1 minimization.

The paper is organized in the following way. In Section 2 we define Minkowski addition of convex polyhedra and their slope diagram representation, and introduce a rotation-invariant similarity measure based on inequalities for the (mixed) volume. In Section 3 similarity measure computation by minimization of a mixed-volume functional is considered, and the main results from the literature are summarized. In Section 4 we give experimental results on minimization of the mixed-volume functional, by taking into account Type-1 and Type-2 critical orientations, respectively. Conclusions are summarized in Section 5.

2   Preliminaries

In this section the Minkowski sum, mixed volumes, a similarity measure based on the Minkowski like inequality, and the slope diagram representation of convex polyhedra are introduced. The compact convex subsets of $\mathbb{R}^3$ are denoted by $\mathcal{C} = \mathcal{C}(\mathbb{R}^3)$. Two shapes A and B are said to be *equivalent* if they differ only by translation; we denote this as $A \equiv B$.

2.1   Minkowski Sum and Mixed Volumes

The Minkowski sum of two sets $A, B \subseteq \mathbb{R}^3$ is defined as

$$A \oplus B = \{a + b | a \in A, b \in B\}. \tag{1}$$

It is well known [ ] that every convex set A is uniquely determined by its *support function*, given by:

$$h(A, u) = \sup\{\langle a, u \rangle \mid a \in A\}, \qquad u \in S^2.$$

Here $\langle a, u \rangle$ is the inner product of vectors a and u, and S^2 denotes the unit sphere in $\mathbb{R}^3$. Also [ ]:

$$h(A \oplus B, u) = h(A, u) + h(B, u), \qquad u \in S^2, \tag{2}$$

for $A, B \in \mathcal{C}$.

Denote by $V(A)$ the volume of the set $A \subset \mathbb{R}^3$. Given convex sets $A, B \subset \mathbb{R}^3$ and $\alpha, \beta \geq 0$, the following holds:

$$V(\alpha A \oplus \beta B) = \alpha^3 V(A) + 3\alpha^2 \beta V(A, A, B) + 3\alpha\beta^2 V(A, B, B) + \beta^3 V(B). \tag{3}$$

Here $V(A, A, B)$ and $V(A, B, B)$ are called *mixed volumes*.

The Minkowski inequality for convex sets $A, B \in \mathcal{C}(\mathbb{R}^3)$ reads [ ]

$$V(A, A, B)^3 \geq V(A)^2 V(B), \tag{4}$$

where equality holds if and only if $B \equiv \lambda A$ for some $\lambda > 0$.

2.2   Similarity Measure

Using the Minkowski inequality (4), a similarity measure σ may be defined as follows:

$$\sigma(A, B) = \sup_{R \in \mathcal{R}} \frac{V(B)^{\frac{2}{3}} V(A)^{\frac{1}{3}}}{V(B, B, R(A))} = \sup_{R \in \mathcal{R}} \frac{V(B)^{\frac{2}{3}} V(A)^{\frac{1}{3}}}{V(A, R(B), R(B))} \tag{5}$$

where $\mathcal{R}$ denotes the set of all spatial rotations, and where $R(B)$ denotes a rotation of B by $R \in \mathcal{R}$. The second equality follows from the fact that $V(B, B, R(A)) = V(R(A), B, B) = V(A, R^{-1}(B), R^{-1}(B))$. Obviously, $0 \leq \sigma(A, B) \leq 1$, where $\sigma(A, B) = 1$ when $B \equiv \lambda R(A)$ for some rotation R and some $\lambda > 0$. The similarity measure σ is invariant under rotations and scalar multiplications. It is not symmetric in its arguments. Symmetric versions may be defined in various ways; an example is the measure $\sigma'(A, B) = \frac{1}{2}(\sigma(A, B) + \sigma(B, A))$.

To find the maximum in (5), the mixed volume $V(A, R(B), R(B))$ has to be minimized over all orientations of A.

If B is a convex polyhedron with faces F_i and corresponding outward unit normal vectors u_i, $i = 1, \ldots, k$, then [ ]

$$V(A, B, B) = V(B, B, A) = \frac{1}{3} \sum_{i=1}^{k} h(A, u_i) S(F_i), \tag{6}$$

where $S(F_i)$ is the area of the face F_i of B and $h(A, u_i)$ is the value of the support function of A for the normal vector u_i.

2.3 Slope Diagram Representation

Denote face i of polyhedron A by $F_i(A)$, edge j by $E_j(A)$, and vertex k by $V_k(A)$. The slope diagram representation (SDR) of polyhedron A, denoted by SDR(A), is a unit sphere covered with spherical polygons. A vertex of A is represented by the interior of a polygon on SDR(A), an edge by a spherical arc on SDR(A), and a face by a vertex of some polygon on SDR(A). To be more precise:

- *Face representation.* $F_i(A)$ is represented on the sphere by a point SDR($F_i(A)$), located at the intersection of the outward unit normal vector u_i on $F_i(A)$ with the unit sphere.
- *Edge representation.* An edge $E_j(A)$ is represented by the arc of the great circle connecting the two points corresponding to the two adjacent faces of $E_j(A)$.
- *Vertex representation.* A vertex $V_k(A)$ is represented by the interior of the polygon bounded by the arcs corresponding to the edges of A meeting at $V_k(A)$.

In Fig. 1 an example of a polyhedron and its SDR is given.

It is easily verified that the faces $F_i(A)$ and $F_j(B)$ are parallel (that is, have the same outward normal) when SDR($F_i(A)$) coincides with SDR($F_j(B)$). Also, an edge $E_i(A)$ is parallel to $F_j(B)$ when SDR($F_j(B)$) lies on SDR($E_i(A)$). Therefore, the maximum in (5) is obtained when points of SDR(A) coincide with points or edges of SDR(B).

3 Similarity Measure Computation

In this section, we consider the problem of computing the similarity measure (5).

Let ℓ be an axis passing through the coordinate origin and $r_{\ell,\alpha}$ be the rotation in $\mathbb{R}^3$ about ℓ by an angle α in a counter-clockwise direction. The problem to be considered is the minimization of the functional $V(A, r_{\ell,\alpha}(B), r_{\ell,\alpha}(B))$. Given a fixed axis ℓ and angle α, (6) can be used to compute this functional.

While rotating the slope diagram of polyhedron A, situations arise when spherical points of the rotated SDR of A intersect spherical arcs or points of the SDR of B. Such relative configurations of A w.r.t. B are *critical* in the sense that they are candidates for (local) minima of the objective functional to be minimized. For more precise definitions we refer to [ ].

3.1 Fixed Rotation Axis

Let ℓ be a fixed rotation axis. The ℓ-critical angles of B with respect to A for mixed volume $V(A, r_{\ell,\alpha}(B), r_{\ell,\alpha}(B))$ are the angles $\{\alpha'_j\}$, $0 \le \alpha'_1 < \alpha'_2 < \ldots < \alpha'_N < 2\pi$, for which spherical points of the rotated slope diagram SDR($r_{\ell,\alpha'_j}(B)$) intersect spherical points or arcs of SDR(A). The following result was proved in [ ].

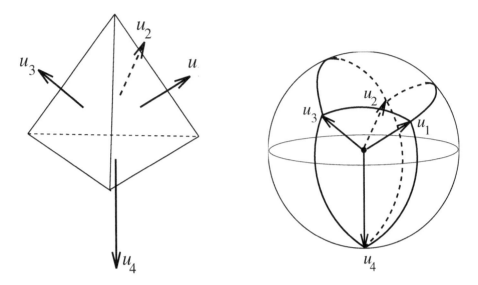

Fig. 1. (a): A tetrahedron with unit normal vectors on its faces. (b): Its slope diagram representation

Proposition 1. *Given an axis of rotation ℓ, the mixed volume of the convex polyhedra A and B, i.e. $V(A, r_{\ell,\alpha}(B), r_{\ell,\alpha}(B))$, is a function of α which is piecewise concave on $[0, 2\pi)$, i.e., concave on every interval $(\alpha'_k, \alpha'_{k+1})$, for $k = 1, 2, \ldots, N$ and $\alpha'_{N+1} = \alpha'_1$.*

This result implies that in order to minimize the mixed volume for any *fixed* rotation axis ℓ, it is sufficient to compute it for all ℓ-critical angles (which are clearly finite in number), and take the minimum of the values thus obtained.

3.2 Minimization over All Rotation Axes

An extensive analysis in [ ] showed that two types of critical orientations have to be considered for obtaining the global minimum of the mixed volume $V(A, R(B), R(B))$:

Type 1 A face of A is parallel to a face of B, and an edge of A is parallel to a face of B.

Type 2 Three edges of A are parallel to three faces of B.

To find the orientations of Type 1 is trivial. When a face $F_j(B)$ is parallel to a face $F_i(A)$, B has only one degree of freedom left, being a rotation around an axis through the origin and the spherical point $\mathrm{SDR}(F_j(B))$. Using the slope diagram representations of A and B, it is easy to find those rotations of B around this axis that make the slope diagram representations of faces of B coincide with the slope diagram representations of edges of A. The problem can be restated as solving a quadratic equation in one variable [ ].

To find the orientations of Type 2 means looking for those orientations of B where three points on SDR(B) (representing three faces of B) lie on three spherical arcs of SDR(A). (Notice that no more than three points have to be checked, since a rotation is uniquely determined by three parameters.) This problem can be reformulated as follows: given two triples of 3D vectors a, b, c and k, l, m, find the rotation R that transforms the vectors a, b, c such that $R(a)$ is perpendicular to k, $R(b)$ is perpendicular to l, $R(c)$ is perpendicular to m. That is, the following system of equations has to be solved for R:

$$\langle k, R(a) \rangle = 0$$
$$\langle l, R(b) \rangle = 0$$
$$\langle m, R(c) \rangle = 0$$

Using the computer-algebra program MAPLE©, this system was reduced in [ ] to the solution of an algebraic equation in one variable of degree 8, whose coefficients are lengthy expressions in the elements of the vectors a, b, c and k, l, m. This equation can be solved numerically using Laguerre's method [ ].

In [ ] the minimization of the mixed volume was carried out by taking into account Type-1 critical orientations only. The authors conjectured that this is actually sufficient to find the global minimum.

In the next section we report on a number of experiments we did in order to verify this conjecture.

4   Experimental Results

In this section, we give results for minimization of the mixed volume $V(A, R(B), R(B))$ of convex polyhedra A and B when R runs over the set of all spatial rotations. Both polyhedra were chosen to be tetrahedra, whose edge sizes varied randomly. For each pair A and B, we performed minimization in two ways, depending on the set of rotations taken into account:

Type-1 minimization All critical rotations are of Type 1: a face of A is parallel to a face of B, and an edge of A is parallel to a face of B.
Type-2 minimization All critical rotations are of Type 2: three edges of A are parallel to three faces of B.

To verify the conjecture, we checked for each pair of tetrahedra A and B whether the result for Type-2 minimization was larger than that of Type-1 minimization.

Remark In fact, in our implementation we use two routines, one which performs Type-1 minimization, and another one which minimizes over the combined set of both Type-1 and Type-2 critical rotations. If we find that the minimum in the second case is smaller than that in the first case, then we know that the conjecture is false. Also, instead of fixing A and rotating B, we may as well fix B and rotate A in view of the identity $V(A, R(B), R(B)) = V(B, B, R^{-1}(A))$.

In the experiments, we found cases among the randomly generated pairs of tetra-hedra for which the result for Type-1 minimization was actually larger than that of Type-2 minimization, although the differences were often small. An example where the difference is substantial is shown in in Fig. 2. In this case the mixed volume for Type-1 minimization equals 1.81213e+006, and that for Type-2 mini-mization equals 1.59156e+006, which is significantly smaller. The corresponding tetrahedra are shown in Fig. 2.

It is interesting to look at the Minkowski sum of $B \oplus R^*(A)$, with R^* the rotation which realizes the minimum of the mixed volume for Type-1 and Type-2 minimization, respectively, see Fig. 3. The corresponding slope diagrams are shown in Fig. 4, with the spherical arcs of A shown in bold. From the pictures, one can verify that indeed for Type-1 minimization, a spherical point of B coincides with a spherical point of A, and another spherical point of B is on a spherical arc of A, whereas for Type-2 minimization, three spherical points of B are on spherical arcs of A.

5 Conclusion

We have studied the computation of a rotation-invariant similarity measure of convex polyhedra A and B, involving the minimization of a mixed-volume func-tional $V(A, R(B), R(B))$ with R running over the set of critical rotations. Two types of critical orientations were distinguished: for Type-1 critical orientations a face of A is parallel to a face of B, and an edge of A is parallel to a face of B; for Type-2 critical orientations three edges of A are parallel to three faces of B. We performed experiments with randomly generated tetrahedra, and computed the minimum of the volume functional by taking into account either Type-1 or Type-2 critical orientations. We found that the result for Type-2 minimization can be larger as well as smaller than that of Type-1 minimization. Therefore, in contrast to what has been conjectured in [ ], one has in general to take both Type-1 and Type-2 critical orientations into account to compute the global min-imum of the mixed volume.

References

1. Bekker, H., and Roerdink, J. B. T. M. Calculating critical orientations of polyhedra for similarity measure evaluation. In *Proc. 2nd Annual IASTED International Conference on Computer Graphics and Imaging, Palm Springs, California USA, Oct. 25-27* (1999), pp. 106–111. 390, 394
2. Heijmans, H. J. A. M., and Tuzikov, A. Similarity and symmetry measures for convex shapes using Minkowski addition. *IEEE Trans. Patt. Anal. Mach. Intell.* 20, 9 (1998), 980–993. 389, 390
3. Press, W. H., Flannery, B. P., Teukolsky, S. A., and Vetterling, W. T. *Numerical Recipes, the Art of Scientific Computing.* Cambr. Univ. Press, New York, 1986. 394
4. Schneider, R. *Convex Bodies. The Brunn-Minkowski Theory.* Cambridge Univer-sity Press, 1993. 391

5. Tuzikov, A. V., Roerdink, J. B. T. M., and Heijmans, H. J. A. M. Similarity measures for convex polyhedra based on Minkowski addition. *Pattern Recognition 33*, 6 (2000), 979–995. 389, 390, 392, 393

6. Tuzikov, A. V., and Sheynin, S. A. Minkowski sum volume minimization for convex polyhedra. In *Mathematical Morphology and its Applications to Image and Signal Processing*, J. Goutsias, L. Vincent, and D. S. Bloomberg, Eds. Kluwer Acad. Publ., Dordrecht, 2000, pp. 33–40. 390, 394, 395

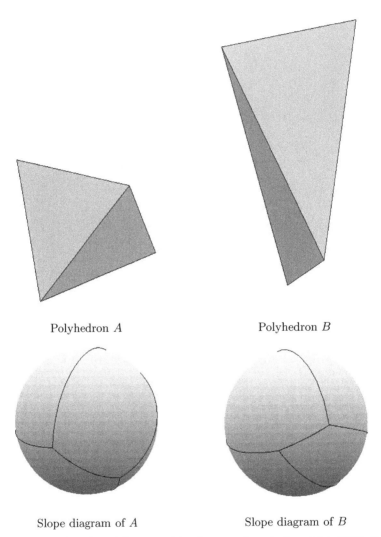

Polyhedron A Polyhedron B

Slope diagram of A Slope diagram of B

Fig. 2. Polyhedra A and B as used in the experiment, for which $V(B, B, R_2^*(A))$, with R_2^* the rotation which realizes the minimum of Type 2, is smaller than $V(B, B, R_1^*(A))$, with R_1^* the rotation which realizes the minimum of Type 1

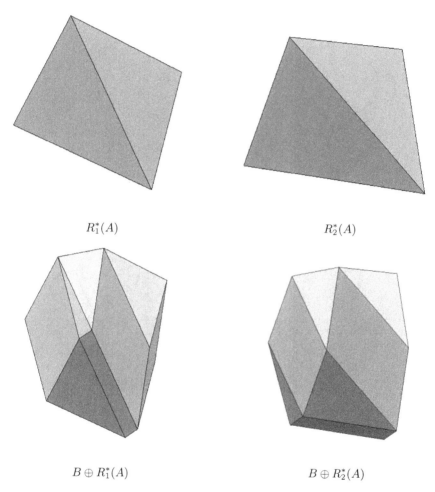

$R_1^*(A)$ $R_2^*(A)$

$B \oplus R_1^*(A)$ $B \oplus R_2^*(A)$

Fig. 3. Top row: polyhedron A in the rotated configuration which minimizes mixed volume according to Type 1 ($R_1^*(A)$) and Type 2 ($R_2^*(A)$). Bottom row: Minkowski sums of polyhedron B and rotated polyhedron $R_1^*(A)$, c.q. $R_2^*(A)$

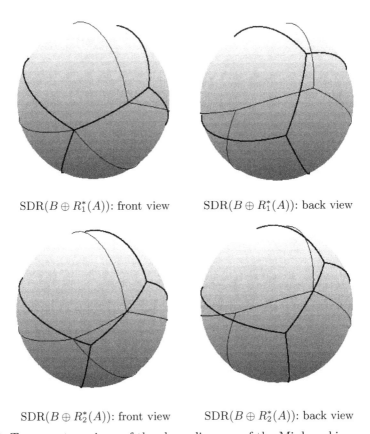

SDR($B \oplus R_1^*(A)$): front view   SDR($B \oplus R_1^*(A)$): back view

SDR($B \oplus R_2^*(A)$): front view   SDR($B \oplus R_2^*(A)$): back view

Fig. 4. Top row: two views of the slope diagram of the Minkowski sum of polyhedron B and rotated polyhedron $R_1^*(A)$. Bottom row: two views of the slope diagram of the Minkowski sum of polyhedron B and rotated polyhedron $R_2^*(A)$. Here R_1^* and R_2^* are the rotations which realize the minimum of Type 1 and Type 2, respectively. Bold curve segments indicate spherical arcs of polyhedron A

Reversible Surface Skeletons of 3D Objects by Iterative Thinning of Distance Transforms

Stina Svensson

Centre for Image Analysis, Swedish University of Agricultural Sciences
Lägerhyddvägen 17, SE-75237 Uppsala, Sweden
stina@cb.uu.se

Abstract. We describe an approach to compute the surface skeleton of a 3D object using distance information. It is based on iterative thinning of the distance transform of the object. The surface skeleton of an object should be topologically equivalent to the original object, centred within the object, a thin subset of the object, and such that the original object can be recovered from the surface skeleton. We emphasize the last property, i.e., the importance of having a reversible surface skeleton, and present a general framework for distance based skeletonization algorithms. Resulting surface skeletons for the D^6 and the D^{26} distance cases are shown.

1   Introduction

The surface skeleton of a 3D object provides a representation of the object which is simpler and contains a smaller amount of data than the original object, without loss of the information necessary to perform shape analysis. This representation can be used directly for shape analysis or can be further reduced to a curve skeleton. An overview of surface skeletonization algorithms closely related to the algorithms presented here is given in Section 4.

To be a useful representation, a surface skeleton should satisfy some properties. It should be *topologically equivalent* to the original object, *centred* within the object, and a *thin subset* of the object. Another important property, not often considered among existing skeletonization algorithms for 3D objects, is that the surface skeleton should be *reversible*, i.e., that the original object can be recovered from the surface skeleton. This can be achieved by using distance information. If each voxel in the surface skeleton is labelled with the distance to its closest voxel in the original background, according to the chosen distance, and the set of centres of maximal balls is included in the surface skeleton, then the original object can be recovered, e.g., by applying the reverse distance transformation. The set of centres of maximal balls can easily be detected on the distance transform. By including this set, we also guarantee that the surface skeleton will be centred within the original object with respect to the chosen distance.

The surface skeleton can be detected in several different ways. Most often, iterative thinning of the object is used. In this case, the border of the object is

G. Bertrand et al. (Eds.): Digital and Image Geometry, LNCS 2243, pp. 400–411, 2001.
© Springer-Verlag Berlin Heidelberg 2001

iteratively removed, keeping voxels necessary for topology and shape preservation, until the surface skeleton remains. This approach can be combined with distance information, i.e., the distance transform is computed and then iterated removal is directly performed distance layer after distance layer. In fact, on the distance transform, we have an easy identification of the border, since the successive borders are proper subsets of the successive layers. If we use the set of centres of maximal balls as anchor-points, i.e., non-removable voxels, during the thinning, a reversible surface skeleton is obtained.

We start from previous work, [,], and present a novel way of detecting a reversible surface skeleton using iterative thinning of the distance transform of the object. Surface skeletons centred within the original object with respect to two different distances, D^6 and D^{26}, are considered (Section 3). The algorithms are presented in a general way, aimed towards a skeletonization algorithm which can, with small changes, be computed by iterative thinning of distance transforms based on any distance. With respect to the algorithms introduced in [,], the algorithms presented here have a considerably smaller computational cost; they are framed into a common scheme which allows the user to select the preferred distance directly when the input object is provided; and the most important difference is that the algorithms better preserve shape information, which is needed if the skeleton is used for applications. A brief presentation of other skeletonization algorithms can be found in Section 4 and a discussion about future work in Section 5.

2   Definitions

The images are, in our case, 3D images and consist of object and background, i.e., are binary volume images. Each voxel can have three types of neighbours: face, edge, and point neighbours. We assume that the object is 26-connected and, thus, the background is 6-connected.

The algorithms presented in this paper are based on the use of distance information from the background to voxels in the object. The distance is computed by propagating distance information locally and the distance increases from the border towards the inside of the object. This is done in one forward and one backward scan over the image. The result is a distance transform (DT) in which each voxel in the object is labelled with the distance to its closest voxel in the background. The labels in the DT depend on the chosen distance. Here, we will use the simple distances D^6 and D^{26} for which the distance between two voxels is equal to the number of permitted steps in a minimal path between the two voxels. For D^6, only steps in face directions are possible, while for D^{26}, steps in face, edge, and point directions are possible. We will denote the corresponding distance transforms DT^6 and DT^{26}, respectively. Information about DTs for 3D images and on distances providing better approximations to the Euclidean distance can be found in [ ].

In a similar way as for the DT, we can compute the reverse DT. Also in this case, the distance is computed by propagating distance information locally but the distance decreases starting from a number of voxels with distance labels. This is done in one forward and one backward scan.

Each distance label in the DT of an object can be interpreted as the radius of a ball (in the underlying distance) fully enclosed in the object. A voxel is said to be a *centre of a maximal ball* (CMB) if its corresponding ball is not completely covered by any other ball. The union of the maximal balls coincides with the object.

The set of CMBs can for D^6 and D^{26} be detected as voxels having no neighbouring voxels with higher label. Again, for D^6, only face neighbours are considered. An object can be represented by its set of CMBs, since the object can be recovered from the set of CMBs by applying the reverse distance transformation.

A voxel v is a *simple voxel* if its removal does not alter the topology, i.e., the removal of v does not change the number of object components, the number of cavities, or the number of tunnels in a $3 \times 3 \times 3$ set of voxels centred on v, [,]. To verify whether the number of object components changes, the number of 26-connected components (N^{26}) in the 26-neighbourhood of v, which consists of all face, edge, and point neighbours of v, can be counted. Object voxels with $N^{26} \neq 1$ should be ascribed to the skeleton. To preserve the number of cavities and tunnels, a subset of the 26-neighbourhood is used, which consists of all face and edge neighbours of v, the 18-neighbourhood of v. The number of 6-connected background components in the 18-neighbourhood that are 6-adjacent to v ($\overline{N}_f^{18}$) is counted. Voxels with $\overline{N}_f^{18} \neq 1$ should be ascribed to the skeleton.

Note that since a simple voxel is defined by $N^{26} = 1$ and $\overline{N}_f^{18} = 1$, a simple voxel has a face neighbour in the background. In fact, a voxel with no face neighbours in the background has $\overline{N}_f^{18} = 0$. The set of voxels with $\overline{N}_f^{18} > 0$ are denoted as the *border* of the object.

As a running example for this paper, we will use a Euclidean ball. This might seem to be a trivial object, but this is not the case since we use distances (D^6 and D^{26}) different from the Euclidean distance. The object together with its CMBs for DT^6 and DT^{26} are shown in Fig. 1. We remind that the D^6 and the D^{26} balls are shaped as octahedra and boxes, respectively, see Fig. 2.

3 Skeletonization by Iterative Thinning of the Distance Transform

To compute the skeleton of a 2D planar pattern, it is possible to use an algorithm based on iterative thinning guided by distance information. The set of centres of maximal discs together with pixels needed for preserving the topology are kept during the thinning, see, e.g., [ ]. If we generalize the algorithm for 2D objects to deal with 3D objects, a possible algorithm would be:

Fig. 1. The running example, a Euclidean ball with radius 28 voxels, with its CMBs for DT^6 and DT^{26}

Fig. 2. The D^6 ball with radius 28 voxels, left, and the D^{26} ball with the same radius, right

1. Compute the DT of the object.
2. Detect the set of centres of maximal balls on the DT.
3. For each voxel with distance label k and a face neighbour in the background:
   (a) mark simultaneously the simple voxels that are not CMBs.
   (b) remove sequentially marked voxels that are still simple when they are visited.

Step 3 might need several iterations, depending on the underlying distance, until every voxel with distance label k is either a skeletal voxel or is removed. This is because the voxels with distance label k do not necessarily have a face neighbour with label $k-1$ on the DT, unless D^6 is used. Hence, some voxels with distance label k can not be considered for removal directly after the voxels with label $k-1$ have been processed. Those can be considered only after a face neighbour with distance label k has been removed.

When applying the skeletonization algorithm described above to our running example, we obtain the D^6 and the D^{26} skeletons shown in Fig. 3. They are topologically equivalent to the original object, centred within the object with respect to the underlying distance, and fully reversible. However, visually they are not what would be considered as good surface representations of the original object, but are rather "combs" of curves. For shape analysis purposes, this is not a convenient representation. We aim to have a surface representation centred within the object instead of a set including "combs" of curves. Thus, for 3D objects, skeletonization is more complicated, since also the gaps between the curves need to be filled, or rather not to be created. What is needed is a modification of the above algorithm to add a suitable condition for *surface preservation*. In detail, Step 3 becomes:

3.' For each voxel with distance label k and a face neighbour in the background:
   (a) mark simultaneously the simple voxels that are not CMBs.
   (b) remove sequentially marked voxels that are still simple and satisfy $Condition_{rm}$.

The condition for removal, $Condition_{rm}$, differs for different distances. $Condition_{rm}$ together with the resulting surface skeletons extracted from DT^6 and DT^{26} will be described in the following Sections 3.1 and 3.2.

3.1 D^6 Surface Skeletons

In an algorithm for detecting a surface skeleton of a 3D object based on iterative thinning of DT^6, we use the following condition (see Fig. 4(a))

Condition$_{rm}^{D^6}$ There exists a pair of opposite face neighbours of a voxel v such that one is a voxel not included in the skeleton being a background voxel and the other is an internal voxel, i.e., a voxel with higher distance label.

This is equivalent to assigning a voxel to the skeleton if it has no pair of opposite face neighbours such that one has lower distance label without being included in the skeleton and the other has higher distance label. $Condition_{rm}^{D^6}$ prevents removal of CMBs, since voxels having only face neighbours with lower or equal distance labels, i.e., CMBs, do not satisfy $Condition_{rm}^{D^6}$ and, hence, are not removed. Besides the CMBs also voxels in *saddle configurations* are prevented from

Fig. 3. D^6 and D^{26} skeletons topologically correct but not preserving surfaces

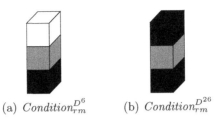

(a) $Condition_{rm}^{D^6}$ (b) $Condition_{rm}^{D^{26}}$

Fig. 4. Voxels involved in the $Conditions_{rm}$ used during voxel removal/marking for (a) D^6 and (b) D^{26} surface skeletonization. White, grey, and black denote background voxels, border voxels, and internal voxels, respectively

removal. These are voxels with pairs of opposite face neighbours both with labels higher or equal, or lower or equal than the voxel itself. Moreover, some voxels needed for topology preservation are prevented from removal by $Condition_{rm}^{D^6}$, without computing the more computationally heavy $\overline{N}_f^{18}$.

We start with computing DT^6 of the object. Although CMBs would be automatically kept, as they do not satisfy $Condition_{rm}^{D^6}$, we prefer to identify them separately. This will allow us to have within the same framework both this algorithm and the one based on D^{26}, but for which a similar condition, meaning also keeping CMBs, does not hold. On DT^6, the CMBs are detected. The labels in DT^6 can be interpreted as iteration numbers for the iterative thinning. Only border voxels need to be considered for removal. During each iteration, the border consists of all voxels having the current distance label together with voxels already detected as non-removable, i.e., with skeletal voxels with lower distance labels.

Each iteration is done in two scans. During the first scan, simple voxels that are not CMBs are marked. This is done simultaneously for all voxels in the border. During the second scan, marked voxels, still being simple voxels and fulfilling $Condition_{rm}^{D^6}$ are removed. This is done sequentially to prevent altering the topology. When all successive borders have been processed, all non-skeletal voxels have been removed and the voxels constituting the surface skeleton remain. The resulting D^6 surface skeleton of the running example is shown in Fig. 5, left.

3.2 D^{26} Surface Skeletons

For DT^{26}, the iterative thinning is slightly more complicated. The voxels with distance label k are not guaranteed to have a face neighbour with distance label $k-1$ since the D^{26} distance is used. This implies the use of more than two scans in each iteration. In fact, Step 3' has to be repeated three times to give all voxels with distance label k the possibility either to become a skeletal voxel or to be removed. Moreover, $Condition_{rm}^{D^6}$ is not suitable, since it is specifically tailored for DT^6.

A suitable $Condition_{rm}^{D^{26}}$ for a voxel v is (see Fig. 4(b))

Fig. 5. D^6 and D^{26} surface skeletons

Condition$_{rm}^{D^{26}}$ No pair of opposite face neighbours of v exists where the voxels in the pair are either internal voxels or border voxels that are not marked for removal.

This is equivalent to saying that v is prevented from being removed if it has a pair of opposite face neighbours each of which has either higher distance label, or the same distance label without being marked for removal.

Also in this case we start by computing the DT, DT^{26}, of the object. On DT^{26}, the CMBs are detected. For each iteration k in the thinning process, only voxels with distance label k and a face neighbour in the background are considered for removal. In the same way as for detecting the D^6 surface skeleton, we first mark suitable voxels simultaneously and, among the marked voxels, remove voxels sequentially. During the marking step, simple voxels that are not CMBs are marked. During the removal step, marked voxels, still being simple voxels and fulfilling $Condition_{rm}^{D^{26}}$, are removed. The resulting D^{26} surface skeleton of the running example is shown in Fig. 5, right.

4   Related Work

The number of papers dealing with reversible skeletonization for 3D objects or at least with skeletonization guided by the DT is increasing. This is not surprising since the number of applications where 3D images are used is increasing. Skeletonization is an efficient way of reducing the amount of data that needs to be analyzed. This is especially important for 3D images since the amount of data is large. Moreover, by reversible skeletonization, (little or) no information is lost. Most papers on reversible skeletonization published so far deal with iterative thinning in some way guided by the DT.

In [ ], Borgefors *et al.* presented an algorithm for iterative thinning where the border for each iteration is detected by identification of voxels having $\overline{N}_f^{18} = 1$. This can easily be changed to be guided by DT^6, so obtaining a fully reversible, distance guided skeletonization algorithm. However, a number of unnecessary voxels are added on the surface compared to the D^6 surface skeleton presented in this paper, especially on the edge or in the regions close to the edge of the surfaces in the surface skeleton. The unnecessary voxels added on the surface cause that the analysis of the skeleton becomes more difficult, e.g., the curve skeleton computed from the surface skeleton will have more (unnecessary) branches.

An algorithm for which the iterative thinning is distance guided was presented by Pudney in [ ]. There the weighted $\langle 3, 4, 5 \rangle$ DT for which the distance is given by a minimal path where the weight for a step in face direction is 3, in edge direction 4, and in point direction 5, [ ], is used. Two alternatives to skeletonization algorithms are given. One of these includes the set of CMBs in the skeleton. It is, however, not clear whether the skeleton is a surface skeleton or "combs" of curves. The latter is most possibly the case. Even though nothing is mentioned about reversibility in [ ], the original object can be recovered from the skeleton, since it includes the set of CMBs.

In 1995, Toriwaki and Saito introduced an algorithm where thinning is guided by the Euclidean DT of the object to be able to have a rotation invariant result [ ]. The result is a skeleton which is topologically equivalent to the original object but produces a lot of unnecessary branches. This has recently been combined with the approach introduced for 2D images by Ragnemalm in 1993, [ ], to have a less branchy skeleton. The resulting algorithm is called anchor-point thinning and was presented during the winterschool "Digital and Image Geometry", [,]. It was among the last works by the late Dr. Saito. The algorithm is computationally heavy, but produces good and stable results, and aimed to be a tool for specific applications, e.g., virtual endoscopy systems, where the object is a structure with cylindrical shape.

Another algorithm for which the Euclidean DT is used was presented by Malandain and Fernández-Vidal in [ ]. Two "skeletons" are combined, one giving the desired level of detail, i.e., giving a more robust skeleton, and the other preserving the topology of the original object. The final skeleton is obtained by topological reconstruction of the skeleton giving the desired level of detail from the topologically correct skeleton and allow almost full recovery of the original object (depending on the chosen level of details).

The idea of iterative thinning of a DT to preserve topology has been used also by Bouix and Siddiqi in [ ]. They combine previous work, [ ], not guaranteed to preserve topology with iterative thinning guided by the Euclidean DT. The resulting surface skeletons allow almost full recovery and are rotation invariant.

In [ ], a fully reversible skeletonization algorithm resulting in a D^6 surface skeleton was presented, which does not use iterative thinning of the DT. Instead skeletal voxels are detected *directly* on DT^6 in a number of scans, generally smaller than that required by iterative thinning of the DT, especially for thick objects. The number of scans is independent of the thickness of the object, which is not the case when iterative thinning is used.

5   Discussion

In this paper, we have presented a general framework for fully reversible surface skeletonization. The algorithm is simple and reconstruction of the original object is possible in an easy way, e.g., by using the reverse distance transformation. We have presented algorithms for the D^6 and the D^{26} distances. Work still remains to obtain a fully reversible surface skeletonization algorithm that can be based on any distance.

For the algorithms presented here, some minor improvements can be done. $Condition_{rm}$ used during the removal part of the iterative thinning is not optimal, as can be seen in Fig. 5. Still some problems remain when building the surfaces, resulting in rather jagged border of the surface, especially for the D^{26} surface skeleton.

Surface skeletons for a number of different objects are shown in Fig. 6. All objects are rather small ($64 \times 64 \times 64$ images) and created by merging different simple geometrical shapes such as balls and cylinders. The intention is to be able to show how the algorithms perform in detail. In Fig. 7, a slightly more complex and larger object is shown together with its D^6 and D^{26} surface skeletons.

We are aware that D^6 and D^{26} are rough approximations of the Euclidean distance. Hence, the surface skeletons are not rotation invariant. Future work will be to find similar algorithms based on, e.g., weighted distances such as $\langle 3, 4, 5 \rangle$. Such algorithms would result in surface skeletons that are almost rotation invariant.

When the Euclidean DT is used, [, ,], the computational cost is higher than if a weighted DT, is used. It is also more computationally heavy to guarantee full reversibility and to actually reconstruct the original object.

Note that as long as iterative thinning is used, the number of iterations needed is proportional to the thickness of the original object. By using an approach similar to the one introduced in [ ], this could be avoided. More investigation should be done to develop such algorithms for other distances than DT^6. One step in that direction is to first develop iterative algorithms for other distances and then use the experiences from that work.

Acknowledgements

Prof. Gunilla Borgefors and Dr. Ingela Nyström, both Centre for Image Analysis, Uppsala, Sweden, and Dr. Gabriella Sanniti di Baja, Istituto di Cibernetica, CNR, Arco Felice (Naples), Italy, are gratefully acknowledged for their valuable scientific support.

References

1. S. Banjo, T. Saito, and J.-i. Toriwaki. An improvement of three dimensional axis thinning method - quality controllable thinning concerning spike branches and shrinkage of branches. Technical Report PRMU2000-21 (2000.5.), Technical Report of IEICE, 2000. 407

2. G. Bertrand and G. Malandain. A new characterization of three-dimensional simple points. *Pattern Recognition Letters*, 15:169–175, Feb. 1994. 402

3. G. Borgefors. On digital distance transforms in three dimensions. *Computer Vision and Image Understanding*, 64(3):368–376, 1996. 401, 407

4. G. Borgefors, I. Nyström, and G. Sanniti di Baja. Computing skeletons in three dimensions. *Pattern Recognition*, 32(7):1225–1236, July 1999. 406

5. S. Bouix and K. Siddiqi. Divergence-based medial surfaces. In *Proceedings of European Conference on Computer Vision*, Dublin, Ireland, 2000. 407, 408

6. G. Malandain and S. Fernández-Vidal. Euclidean skeletons. *Image and Vision Computing*, 16:317–327, 1998. 407, 408

7. C. Pudney. Distance-ordered homotopic thinning: A skeletonization algorithm for 3D digital images. *Computer Vision and Image Understanding*, 72(3):404–413, Dec. 1998. 407

8. I. Ragnemalm. Rotation invariant skeletonization by thinning using anchorpoints. In K. A. Høgda, B. Braathen, and K. Heia, editors, *Scandinavian Conference on Image Analysis (SCIA'93)*, pages 1015–1022. Norwegian Society for Image Processing and Pattern Recognition, 1993. 407

9. P. K. Saha and B. B. Chaudhuri. 3D digital topology under binary transformation with applications. *Computer Vision and Image Understanding*, 63(3):418–429, May 1996. 402

10. T. Saito and K. Mori. Algorithms for 3D distance transformation and skeletonization and its applications for medical image processing. Private communication with Dr. Mori and Reported by Dr. Mori during Winterschool "Digital and Image Geometry", Schloß Dagstuhl, Wadern, Germany, December 18 - December 22, 2000. 407, 408

11. T. Saito and J.-i. Toriwaki. Reverse distance transformation and skeletons based upon the Euclidean metric for n-dimensional digital binary pictures. *IEICE Transactions on Information and Systems*, E77-D(9):1005–1016, Sept. 1994. Special Issue on 3D Image Processing. 407

12. G. Sanniti di Baja and S. Svensson. Surface skeletons detected on the D^6 distance transform. In F. J. Ferri, J. M. Iñetsa, A. Amin, and P. Pudil, editors, *Proceedings of S+SSPR 2000: Advances in Pattern Recognition*, volume 1876 of *Lecture Notes in Computer Science*, pages 387–396, Alicante, Spain, 2000. Springer-Verlag, Berlin Heidelberg. 407, 408

13. K. Siddiqi, S. Bouix, A. Tannenbaum, and S. W. Zucker. The Hamilton-Jacobi skeleton. In *Proceedings of International Conference on Computer Vision (ICCV'99)*, Corfu, Greece, Sept. 1999. 407

14. S. Svensson. Detecting a D^6 surface skeleton using iterative thinning. In J. Bigün and K. Malmqvist, editors, *Proceedings Symposium on Image Analysis (SSAB'00)*, pages 37–40, Halmstad, Sweden, 2000. Halmstad University. 401

15. S. Svensson, G. Borgefors, and I. Nyström. On reversible skeletonization using anchor-points from distance transforms. *Journal on Visual Communication and Image Representation*, 10(4):379–397, 1999. 402

16. S. Svensson, I. Nyström, and G. Borgefors. Fully reversible skeletonization for volume images based on anchor-points from the D^{26} distance transform. In B. K. Ersbøll and P. Johansen, editors, *Proceedings of the 11th Scandinavian Conference on Image Analysis (SCIA'99)*, pages 601–608, Kangerlussuaq, Greenland, 1999. The Pattern Recognition Society of Denmark. 401

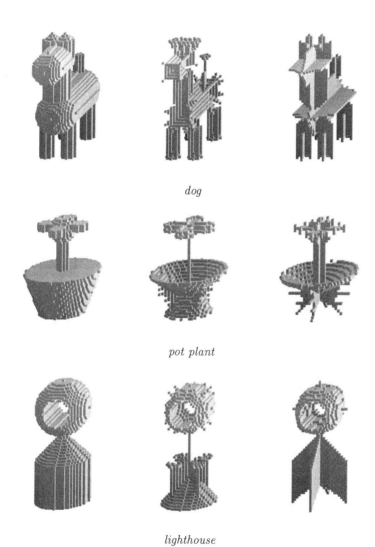

dog

pot plant

lighthouse

Fig. 6. From left to right, original object, D^6 surface skeleton and D^{26} surface skeleton

Fig. 7. Above, the original object, *horse*. Below, D^6 and D^{26} surface skeletons

Distance Transformation and Skeletonization of 3D Pictures and Their Applications to Medical Images

Jun-ichiro Toriwaki and Kensaku Mori

Graduate School of Engineering, Nagoya University
Furo-cho, Chikusa-ku, Nagoya 464-8603, Japan
{toriwaki,mori}@nuie.nagoyap-u.ac.jp
http://www.toriwaki.nuie.nagoya-u.ac.jp

Abstract. In this article, we present the Euclidean distance transformation (EDT) of three-dimensional (3D) binary pictures and applications, with stress on Dr. Toyofumi Saito's work in collaboration with one of the authors. First, EDT, skeletons, and reverse EDT are defined, and then the algorithms to perform them are presented. Next, a sequential-type 3D thinning algorithm that preserves the topology of an input object is introduced. This algorithm is derived by combining an ordinary two-dimensional thinning algorithm with EDT. Finally, applications to medical image processing are presented such as automated path finding in virtual endoscopy, as well as analysis of 3D pathological sample images. *Dr. Toyofumi Saito, an associate professor who was one of the most active researchers in the field of image processing in Japan and a young leader of the author's laboratory, and who was initially scheduled to be the first author of this paper, passed away on 26 October 2000. We have lost a most reliable and most promising colleague, an experienced supervisor, and a very sincere friend. He is deeply missed by all in his family, his friends, and his colleagues. On behalf of the laboratory the authors wish to dedicate this short note in memory of Dr. Toyofumi Saito and to express our sincere condolences to his loved ones.*

1 Introduction

Distance transformation (DT) is the transformation of a binary picture where the value of a 1-pixel is replaced by the distance to the nearest 0-pixel. DT is a very useful tool in binary image processing for shape feature extraction and thinning. Various studies have been reported concerning algorithms and applications of distance transformation since the early years of image processing [,].

From the viewpoint of parallel type and sequential type algorithms, the main issues are extension to 3D images and consideration on different distance metrics. Saito's most important contributions to these topics were in studies on DT using a Euclidean metric and the applications of DT to thinning for two- and three-dimensional picture.

G. Bertrand et al. (Eds.): Digital and Image Geometry, LNCS 2243, pp. 412–129, 2001.
© Springer-Verlag Berlin Heidelberg 2001

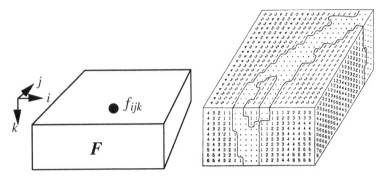

Fig. 1. Three-dimensional (3D) digital picture and 3D distance picture. The numbers in the right object show 26-neighbour distances

In this short note, we introduce an outline of results concerning Euclidean distance transformation (EDT), obtained by Dr. Toyofumi Saito in collaboration with J. Toriwaki of Nagoya University in Japan. First, we introduce an efficient EDT algorithm, which gives strict Euclidean distances. Second, we define what a skeleton is and present the reverse EDT briefly. Thirdly, we discuss a few applications including an improved thinning algorithm for 3D objects. The last topic is about a new thinning algorithm that has the ability to adjust the degree of occurrence of spurious short branches and degeneration, which was reported in the last paper of Dr. Saito submitted about a month before his death [ ].

2   Distance Transformation

2.1   Definition

Let us consider a 3D digital picture $\mathbf{F} = \{f_{ijk}\}$ where f_{ijk} is the density (gray value) at a voxel (i, j, k). The picture $\mathbf{F}$ is called a binary picture if f_{ijk} takes only densities 0 and 1. We call a set of 0-voxels, $\mathbf{S_0}$, the background and a set of 1-voxels, $\mathbf{S_1}$, an object, respectively (Fig. 1).

Then, the distance transformation (DT) of binary picture $\mathbf{F}$ is defined as follows:

Definition 1. $DT : \mathbf{F} = \{f_{ijk}\} \rightarrow \mathbf{G} = \{g_{ijk}\}$

$$g_{ijk} = \begin{cases} \min\{dist((i,j,k),(p,q,r)); (p,q,r) \in \mathbf{S_0}\}, \, if \, (i,j,k) \in \mathbf{S_1}, \\ 0, \qquad\qquad\qquad\qquad\qquad\qquad\qquad if \, (i,j,k) \in \mathbf{S_0}, \end{cases} \quad (1)$$

where $dist(a, b) =$ distance between two voxels a and b. We call the picture $\mathbf{G} = \{g_{ijk}\}$ the distance picture.

A transformation in which $dist^2(a, b)$ is calculated instead of $dist(a, b)$ above is called a squared DT.

Intuitively, DT is a transformation that gives the distance of each voxel to its nearest 0-voxel. Various distance metrics are used as $dist(a, b)$. For 3D digital pictures, the 6-neighbour distance, the 18-neighbour distance, and the 26-neighbour distance are popular, and the Euclidean distance is familiar by the analogy of the continuous space and the real world. Weighted DT has also been studied for 3D pictures [, ,]. For simplicity of description, we write DT using the above distance measures as the 6-n DT, 18-n DT, 26-n DT, and Euclidean DT (EDT), respectively.

2.2   Algorithm

Algorithms to calculate DT for a given 3D binary picture have been studied since the 1970's [, , ,], and efficient algorithms have been found for 6-n, 18-n, and 26-n DT, including both parallel type and sequential type algorithms. However, only a very few have been found for EDT.

Strategies for an algorithm to obtain EDT are classified as follows:

1. Record the differences of coordinate values in three axes and perform an exhaustive search for voxels in a suitable neighbourhood to find the voxel minimizing the Euclidean distance [ ].
2. Use vector propagation using masks of neighbourhood voxels whose sizes often are larger than 3 x 3 x 3 voxels [, ,]. Both parallel and sequential type algorithms are known [, ,].
3. Calculate the square of EDT instead of the exact EDT value [, ,].

The first case is the straightforward implementation of distance minimization and easy to understand. The computational load and memory requirement rapidly increase for large images and especially when working in higher dimensions.

The second method is easily implemented both in parallel type and sequential type algorithms. Basically, it gives approximations to the Euclidean distance, and the accuracy of an approximation is good with masks of suitable shapes and a computation of sufficient iterations.

The last procedure is to minimize $[dist((i, j, k), (p, q, r))^2]$ in Eq. (1) with the use of an Euclidean metric. This is the simplest procedure and is executed most effectively by computer among the three methods. The square roots of all obtained values must be calculated in order to know the exact distance values. Actually, this is not always necessary in most applications. The relations among the distance values often provide enough information for practical problems using DT. The extraction of a local maximum (or minimum) and equidistance voxel detection are typical examples. In this paper, we treat only EDT.

2.3   Reverse Distance Transformation

Reverse distance transformation (Rev. DT) is a transformation to restore a DT picture from a finite set of voxels and the distance values on them. We define it as follows.

[**Definition 2**] Given a set of voxels $\mathbf{S} = \{\mathbf{s}_1, \mathbf{s}_2, \ldots, \mathbf{s}_m\}$, where $\mathbf{s}_m = (i_m, j_m, k_m)$ is a voxel with a density value $f_{i_m j_m k_m}$, we consider the processing that obtains the set of all voxels $\mathbf{P}$ such that

$$\mathbf{P} = \{(i, j, k); \exists m, (i - i_m)^2 + (j - j_m)^2 + (k - k_m)^2 < f_{i_m, j_m, k_m},$$
$$\mathbf{s}_m = (i_m, j_m, k_m) \in \mathbf{S}\}.$$

If $f_{i_m j_m k_m}$ is the squared EDT value at a voxel $\mathbf{s}_m$, the above processing is called the reverse squared EDT.

The above equation means that for an arbitrary voxel $q = (i, j, k)$ in the set $\mathbf{P}$, there exists at least one voxel $\mathbf{s}_m$ on which the squared EDT value is larger than the squared distance between q and $\mathbf{s}_m$. This also means that the 0-voxel nearest to $\mathbf{s}_m$ is further from $\mathbf{s}_m$ than q. Therefore, the voxel q is inside an object in an original input picture.

Rev. DT is effectively performed by an algorithm similar to EDT as shown in [ ]. The details of the algorithm are skipped here. Reverse EDT is also presented in [,]. Its strategy is to put the balls of the radii given by the distance values on skeleton voxels with small numbers of scans after selecting the minimum set of necessary balls [,] or to extend skeleton voxels by an algorithm similar to EDT [ ]. The comparison of efficiency level has not been reported to date. How the voxel set $\mathbf{S}$ (=skeleton) can be obtained is presented briefly in Section 3.

3   Skeleton and Skeletonization

3.1   Definition of a Skeleton

The words "skeleton" and "skeletonization (= the process to extract a skeleton)" have different meanings, and they are not always defined uniquely.

A skeleton in relation to DT seems to be considered in different ways as follows.

1. The minimum set of voxels such that all input objects are restored exactly using their locations and the distance values on them. The term of "minimum" here means the smallest number of voxels of the set. Practically, the detection of the strictly minimal set is not easy, and a suboptimal set is utilized. The resultant skeleton is not unique; this is so even if strict minimization can be achieved.
2. The set of local maxima of DT $\mathbf{G} = \{g_{ijk}\}$. The size and the shape of the neighbourhood used for the determination of the local maxima vary corresponding to the kind of connectivity.

3. The center line or center surface of an input object. In this case, the preservation of topological features of the original pictures is strongly required.
4. A set of center voxels of balls included in an input object and touching the border of the object at more than one voxel.

None of these exactly coincides with the others. Obviously, the first is the subset of the fourth. We think the third case should be called thinning (the word skeletonization might be acceptable, but confusing) and excluded in discussions concerning DT.

In the case of EDT, the problem is more complicated. The definition of the term "skeleton" has not been discussed systematically except in [ ]. In this paper, we adopt the first case as the skeleton, and discuss its properties.

Rev. EDT provides a method to restore at least parts of an object from a suitable set of voxels with the EDT values on them. Then, the problem is how to find the minimum set from which the original object can be restored exactly. One available algorithm is the iterative testing of the coverage relation among balls and objects as presented in [,]. A strict minimization is still difficult and we have not succeeded yet in finding a better algorithm for this. Presently, we define skeletons of EDT as follows.

[Definition 3] (skeleton type 1) Given the squared EDT $\mathbf{G} = \{g_{ijk}\}$ of a 3D picture $\mathbf{F} = \{f_{ijk}\}$, the set of voxels $\mathbf{S}$ defined below is called the skeleton (type 1) of picture $\mathbf{F}$ and the EDT $\mathbf{G}$.

$$\mathbf{S} = \{(i, j, k); \exists(x, y, z), (i-x)^2 + (j-y)^2 + (k-z)^2 < g_{ijk},$$
$$and \max_{(u,v,w)} \{g_{uvw}; (x-u)^2 - (y-v)^2 - (z-w)^2 < g_{ijk}\} = g_{ijk}\}$$

This means that the skeleton is a set of centers of balls with a radius equal to an EDT value having at least one voxel as its dominating area. The dominating area of a ball C means a set of voxels covered by only the ball C itself.

We have proposed another definition of the term "skeleton" slightly different from the above in [ ].

[Definition 4] (skeleton type 2) Given the squared EDT $\mathbf{G} = \{g_{ijk}\}$ of a 3D picture $\mathbf{F} = \{f_{ijk}\}$, the skeleton of $\mathbf{G}$ (and $\mathbf{F}$) is defined as the set of voxels $\mathbf{S}'$ given below.

$$\mathbf{S}' = \{(i, j, k); \exists(x, y, z), (i-x)^2 + (j-y)^2 + (k-z)^2 < g_{ijk},$$
$$and \max_{(u,v,w)} \{g_{uvw} - (x-u)^2 - (y-v)^2 - (z-w)^2\} =$$
$$g_{ijk} - (x-i)^2 - (y-j)^2 - (z-k)^2\}$$

The underlying idea is as follows. Let us assume that a voxel (i, j, k) is the only one skeleton voxel in a given object and its squared distance value is g_{ijk}.

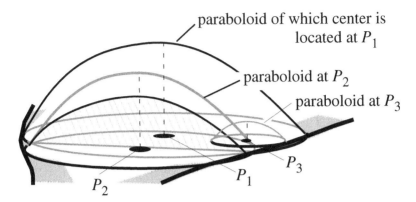

Fig. 2. Illustration of the skeleton type 2 for the two dimensional case (Pixels P_1 and P_2 are included in the skeletons. P_3 is not included)

Then, a squared distance value at a voxel (x, y, z) in this object will be given by $g_{xyz} = g_{ijk} - (i - x)^2 - (j - y)^2 - (k - z)^2$. The skeleton in this case is a set of centers of elliptic paraboloids, which are not covered perfectly by any other elliptic paraboloid. The elliptic paraboloid here is the one having the maximum value at its center voxel and the value is equal to the squared distance value of the center voxel (Fig. 2).

Examples are shown in Fig. 3. For the details of these definitions including intuitive interpretations and examples see [ ]. Note that the skeletons defined here do not preserve the topologies of the input pictures.

3.2 Restoration

Let us assume that a set of voxels and the distance values on them are given. If the distance value at a voxel P is d_p, then no 0-voxel exists within the distance d_p from P. In other words, a ball with a radius d_p and its center at P is included in the input object. Then, a single voxel P is extended (dilated) in all directions in the 3D space within the distance of d_p from P without going outside the object. The result is a part of the original object.

If a suitable set of voxels is given as seeds associated with the values of the degree of dilation, then the original object can be recovered as the sum of the sets obtained by the dilation. The dilation procedure is also implemented by a simple algorithm similar to EDT as shown in [ ]. The skeleton defined above is used as the most suitable seed set.

3.3 Skeleton Extraction and Restoration

To extract the skeleton, all local DT maxima are first extracted. Then, the next step is critical where the set of local maxima has to be reduced without loss of the restoration property.

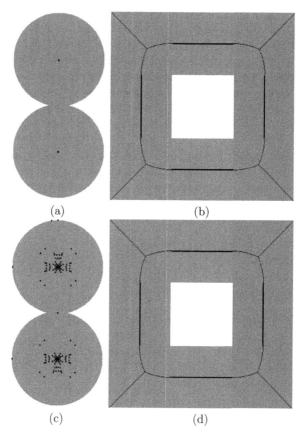

(a)   (b)

(c)   (d)

Fig. 3. Examples of skeletons based upon Euclidean distance transformation [ ]. (a,b) Type 1 (Definition 3) (c,d) Type 2 (Definition 4)

As mentioned above, the restoration is performed by Rev. EDT from the skeleton given by Definitions 3 and 4. This algorithm is easily implemented by an ordinary computer. It is obvious from the definitions that the input object is exactly restored by the above procedures from the skeleton. The following items are known concerning skeletons.

1. Skeletons by Definitions 3 (Type 1) and 4 (Type 2) do not coincide with each other. The number of skeleton voxels is usually larger for type 2 than for type 1.
2. The Rev. squared EDT shown in [ ] does not restore the EDT values, although the shape of the input object is recovered. If we need to recover the EDT values, the best way currently available is to recalculate the squared EDT from the restored object. Since both the squared EDT and Rev. squared EDT are fast, the sequence of the processing ≪ skeleton → Rev. squared EDT → squared EDT ≫ is not very time-consuming.

3. The set of local maxima detected by the 6- neighbourhood or the 26- neighbourhood does not guarantee the exact restoration of the original object. One algorithm [ ] is available to extract a skeleton satisfying all restoration requirements by simple local operations with a look up table. The algorithm to extract the type 2 skeleton is similar to reverse squared EDT and extended to higher dimensions straightforwardly [ ]. Computation time of the extraction of skeleton 2 was smaller than skeleton 1 in our limited size of experiments [ ]. More systematic study of efficiency is expected in the future.

4. The minimization of the number of skeleton voxels satisfying Definitions 3 and 4 is difficult except by an exhaustive search. After extracting a set of candidate skeleton voxels, the procedure to reduce skeleton voxels is available as shown in [,]. More efficient algorithm has not been found yet although a few suboptimal methods can be found in [ ].

4   Applications

The information obtained by using DT includes:

(a) the distance value between two objects or from a point to objects on a picture,
(b) shape information about the equi-distance surface,
(c) skeleton,
(d) how information is propagated.

The use of EDT instead of the popular 6n-DT or 26n-DT improves the results in (a), (b), and (c) significantly. We show several examples in this section.

The advantages of EDT over other types of DT are first that it gives a real Euclidean distance value used in the real world, and second that it is homogeneous in all directions. The development of the algorithms mentioned in the previous section has greatly contributed toward making use of these advantages in practical image processing.

4.1   Thinning

Thinning (which is often called skeletonization) is the processing to extract the center line (line thinning) or to decide the center surface (surface thinning) of a thick 3D object [, ,]. The connectivity of an input picture is kept unchanged in the result of a case of thinning. This requires that the resulting line or surface object be located at the center (or near the center) of the input object.

One of popular approaches among algorithms thus far developed is as described below [, , , , ,]

1. Test the 1-voxels at the edge of an object in terms of deletability, and replace them by 0-voxels (deleted) if deletable. There are two possible strategies.
   (a) Delete a deletable 1-voxel at the moment it is detected (sequential type).

(b) Examine all edge voxels and give a mark to a deletable voxel. Then, delete all of the voxels given marks simultaneously (parallel type).

2. Apply the above process to only the upper edges first. After processing all of the upper edge voxel, process the left edges. Apply the procedure to the lower edges, the right edges, the near side edges, and the far side edges sequentially. Typically, use six iterations for 3D pictures (subcycle). However, subcycles are not always used in some of the thinning process.
3. Iterate (1) and (2) until no deletable voxel can be found (main cycle).

A subcycle is needed to proceed with the deletion of 1-voxels at an equal rate from all directions and by doing so, to locate thinned results at or near the center of an object.

There are many variations in the details, but most existing algorithms employ this type of subcyle - main cycle structure. One problem common to this structure is that the results are too sensitive to the rotation (or the direction) of the input object. For instance, the thinned result of a 45-degree rotated object is often severely deformed from the rotation after thinning at the same angle. This problem is greatly improved by utilizing EDT in the above procedure.

The algorithm is modified as follows [   ].

The squared EDT of a given input picture is used as the input picture to the revised version of the algorithm. (The word "squared" is neglected for simplicity of description.)

1' Find the minimum of the EDT. In the first step of the revised algorithm only the voxels with the minimum EDT value are processed just as the edge voxels in (1).
2' Apply process (2) to the voxels with the minimum EDT value.
3' After completing (2'), process those voxels with the second smallest EDT value in the same way as in (2').
4' Iterate the whole of these for all values of EDT, starting from the voxels with the smallest value of EDT to the voxels with the largest ones.

Experimental results are shown in Figs. 4, 5 and 6. Although Ref. [ ] is often referred as one of the classic papers on 3D thinning, different methods have appeared recently [, , ,]. More systematic comparative studies can now be expected. Fig.6(b) was obtained by only deleting the subset of deletable voxels (belonging to three-dimensional simplexes), deriving Fig. 6(c) by deleting all deletable voxels while keeping the topology unchanged. For a more detailed evaluation of the algorithm, it is necessary to compare the algorithm with more recently published methods such as [ ].

The computation time is not affected much by the use of EDT here because the EDT is performed only once during the thinning process and its weight is not so heavy compared to the remaining parts of the algorithm.

This algorithm can be combined with anchor point thinning [ ], and by doing so, we can derive a thinning method with a function to adjust the degree of occurrence of spurious branches, which is in one of the latest papers by Saito

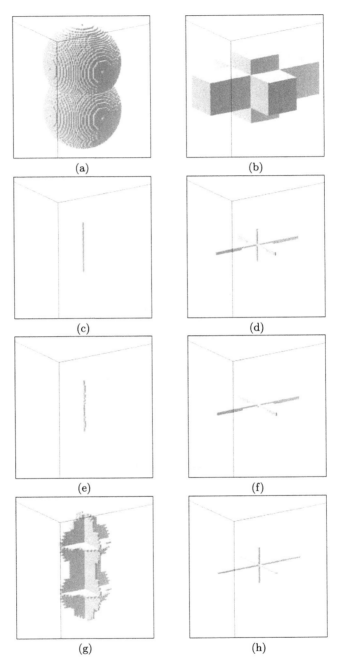

Fig. 4. Experimental results of thinning by [,]. (a) Input object (connected balls), (b) Input object (cross), (c,d) Algorithm by Saito et al. [,], (e,f) Algorithm by [ ], (g,h) Algorithm by [ ]

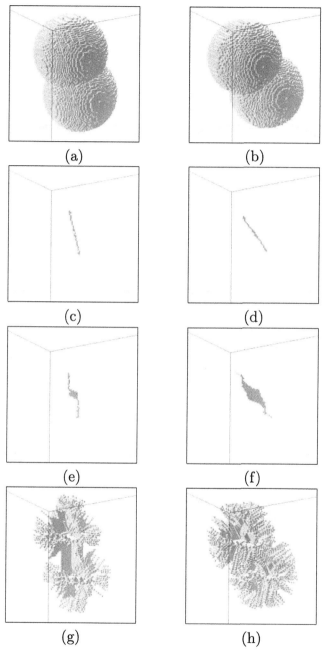

Fig. 5. Experimental results of thinning rotated pictures (a) Input object rotated by 15 deg. and (b) 35 deg., (c,d) Result of thinning by [,   ], (e,f) Result of thinning by [ ], (g,h) Result of thinning by [   ]

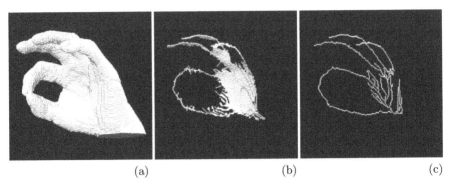

(a)   (b)   (c)

Fig. 6. Thinning experiments (Hand) (a) Input picture (Hand) (b) Result of thinning (surface thinning) (c) Result of thinning (axis thinning)

and his colleagues [24]. If we want to guarantee the preservation of some voxels after the thinning, we can give a kind of mark to each of these voxels before starting the thinning regardless of deletability. We call such voxels here anchor points. They are never deleted during the thinning process. Accordingly, the resultant thinning lines or surfaces always pass these anchor points. The method consists of the following steps:

a. Skeleton extraction
b. Skeleton selection
c. Anchor point thinning with EDT
d. Post-thinning

The first step (a) is presently performed according to the direct implementation of Definition 3 or 4. The step (b) is the key step of this procedure [1,24]. Let us consider two skeleton voxels P_1 and P_2 with EDT values g_1 and g_2, respectively (Fig. 7). Assuming $g_1 > g_2$, the skeleton voxel P_2 is excluded from the anchor point set if both of the following conditions hold:

[**Condition i**] A ball with a radius r_2 and its center at a voxel P_2 is covered by an ellipsoid E specified as follows:
center = P_1, radii=$(\alpha \cdot g_1, g_1, g_1)$, where α is a parameter, and the direction of the largest radius coincides with the line connecting P_1 and P_2.
[**Condition ii**] $g_1 + g_2 \geq d$, where d is the distance between P_1 and P_2.

Condition i means that skeleton voxels with small DT values near the skeletion voxels with large DT values are eliminated. **Condition ii** requires that skeleton voxels P_1 and P_2 belong to the same object. The parameter α can be adjusted in each applications. For $\alpha = 1.0$, no skeleton point is eliminated. For larger α, skeletons with larger EDT values are excluded. In addition, the occurrence of spurious branches is greatly suppressed, while the possibility of degeneration increases in the thinning result.

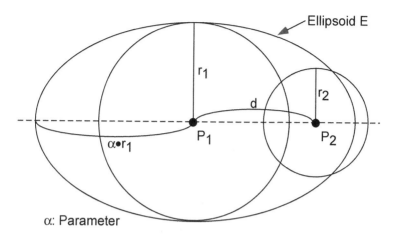

Fig. 7. Illustration of the condition to trim a skeleton point. See text

The generation of spurious short branches and the occurrence of degeneration or the disappearance of correct short branches are in a trade-off relationship. The application determines which branches are false and which are correct. It is difficult to give a general mathematical definition of the "correctness" of branches. In individual applications, however, it is possible to estimate the relations among them as a function of the parameter α like the ROC curve (receiver operating characteristic curve) in statistical decision-making by executing a preliminary experiment beforehand. We can select a suitable operation point on this curve. An example is shown in Fig. 8.

4.2   Path Finding in Virtual Endoscopy

With the recent progress in imaging technology in the medical field, a three-dimensional image of the human body can now be taken by a computer much more easily than before. We call an image a virtualized or virtual human body (VHB) [ ]. On VHB we can freely navigate the part of the human body we want to examine. We call this navigation diagnosis. In particular, flying through a pipe structured organ or an organ with a cavity inside (e.g. bronchus, vessel, colon, or stomach) can be considered as a kind of simulation of endoscopy, and called a virtual (or virtualized) endoscopy system (VES) [,] (Fig. 9).

In a VES for screening use, the automated selection of the path for fly-through can often reduce a doctor's load. Accordingly, we tried to apply the thinning algorithm to the bronchus and colon extracted from a 3D CT image by segmentation techniques in order to use the center line as the path of the viewpoint for an automated VES. Examples are shown in Fig. 10 [ ]. Results were presently accepted as reasonable by medical doctors, and validation by large number of clinical cases remain for future study.

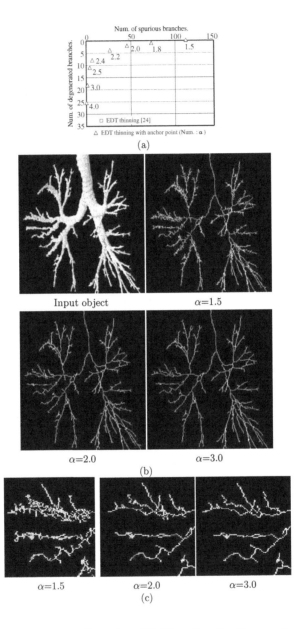

(a)

Input object $\alpha=1.5$

$\alpha=2.0$ $\alpha=3.0$

(b)

$\alpha=1.5$ $\alpha=2.0$ $\alpha=3.0$

(c)

Fig. 8. Control of spurious branches (a) Variation in the number of spurious branches and degeneration with the parameter α. The horizontal axis = the number of spurious branches, and the vertical axis = the numbers of degenerated branches. The numerals in the graph show the values of the parameter α. A black square at the bottom of the graph shows a result of thinning with EDT by [ ], (b) Examples. Input object and results for $\alpha=$ 1.5, 2.0, 3.0, (c) Magnified views of a part of branches for each α

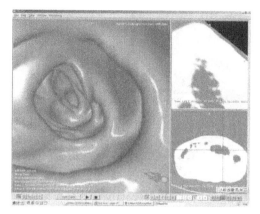

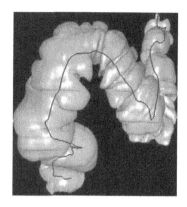

Fig. 9. Virtualized endoscopy system developed by the author's group [8]

Fig. 10. An example of the automated determination of a path with the virtualized endoscopy system shown in Fig. 9 [9]

4.3 Analysis of 3D Pathological Images

The third example is an analysis of a 3D microscope image of human liver. ¿From cross sections of a pathological sample of the liver, we extracted portal vein and hepatic vein borders, and reconstructed the 3D micro structure of capillary blood vessels in the liver. We then analyzed the shape features of 3D patterns or arrangement patterns of vessels by using 3D EDT and a 3D digital Voronoi diagram [18]. As one example, we could confirm the constancy of the distances to the portal and hepatic veins from an arbitrary point in the liver, and could also show characteristics of the relationship between the veins and necromatic zones. Figure 11 shows experimental results.

5 Conclusion

In this article, we introduced research work concerning distance transformation and applications of DT performed by Dr. Toyofumi Saito and his colleagues at Nagoya University. First, a Euclidean distance transformation algorithm was presented briefly, and then applications of the algorithm to thinning and medical image processing were shown. Here we explained how the 3D thinning was improved by utilizing EDT and presented a method to control the occurrence of spurious short branches.

Acknowledgements

The authors thank colleagues for their useful suggestions and discussions. Parts of this research were supported by the Grant-In-Aid for Scientific Research from the Ministry of Education, the Grant-In-Aid for Scientific Research from Japan

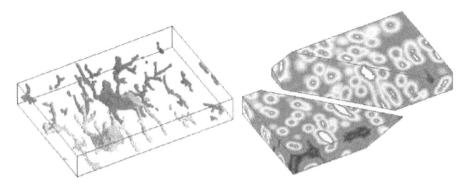

Fig. 11. Analysis of 3D pathological images. (a) 3D reconstruction of a hepatic vein in the liver and (b) Distance to the nearest hepatic vein

Society for Promotion of Science, and the Grant-In-Aid for Cancer Research from the Ministry of Health and Welfare, of the Japanese Government.

References

1. S. Banjo, T. Saito, and J. Toriwaki: An improvement of three dimensional axis thinning method - quality controllable thinning concerning spike branches and shrinkage of branches -, Technical Report of IEICE*, PRMU2000-21 (2000.5) (* IEICE = The Institute of Electronics, Information And Communication Engineers, Japan) 423
2. G. Borgefors: Distance transformation in arbitrary dimensions, Computer Vision, Graphics, And Image Processing, 27, pp.321-345 (1984) 414
3. G. Borgefors: On Digital Distance Transforms in Three Dimensions, Computer Vision and Image Understanding, 64, 3, pp.368-376 (1996) 414
4. G. Borgefors, I. Ragnemalm, and G. Sanniti di Baja: The Euclidean distance transform; finding the local maxima and reconstructing the shape, Proc. of 7th SCIA (Scandinavian Conf. on Image Analysis), pp.974-981 (1991) 414, 415
5. G. Borgefors and I. Nystrom: Efficient Shape Representation by Minimizing the Set of Centres of Maximal Discs/Balls, Pattern Recognition Letters, 18, pp.465-472 (1997) 415, 416, 419
6. P. E. Danielsson: Euclidean distance mapping, Computer Graphics And Image Processing, 14, pp.227-248 (1980) 414
7. Y. Kawase, S. Yokoi, and J. Toriwaki: On a sequential thinning algorithm for three-dimensional digitized picture, IEICE Trans., Japan, J68-D, 4, pp.481-488 (1985) 419, 421, 422
8. K. Mori, A. Urano, J. Hasegawa, J. Toriwaki, H. Anno, and K. Katada: Virtualized endoscope system - an application of virtual reality technology to diagnostic aid, IEICE Trans. Inf. and Syst., Japan, E79-D, 6, pp.809-819 (1996.6) 424, 426
9. K. Mori, Y, Hayashi, Y. Suenaga, J. Toriwaki, J. Hasegawa, and K. Katada: A Method for Detecting Unobserved Regions in Virtual Endoscopy System, Proc. of SPIE, 4321, pp.134-145 (2000) 424, 426
10. N. Nikolaridis: 3-D Image Processing Algorithm, John Wiley & Sons, New York, USA (2001) 419

11. F. Nilsson and P. E. Danielsson: Finding the Minimal Set of Maximum Disks for Binary Objects, Graphical Models and Image Processing, 59, 1, pp.55-60 (1997) 416, 419

12. K. Palágyi and A. Kuba: A Parallel 3D 12-Subiteration Thinning Algorithm, Graphical Models and Image Processing, 61, pp.199-221 (1999) 419, 420

13. C. Pudney: Distance-Ordered Homotopic Thinning: A Skeletonization Algorithm for 3D Digital Images, Computer Vision and Image Understanding, 72, 3, pp.404-413 (1998) 419, 420

14. I. Ragnemalm: The Euclidean distance transform, Linkoping Studies in Science And Technology, Dissertations No.304, Dep. of Electrical Eng., Linkoping Univ., Linkoping, Sweden, 1993 414, 419, 420

15. P. Rogalla, J. Terwisscha van Scheltinga, and B. Hamm eds.: Virtual Endoscopy and Related 3D Techniques, Springer, Heidelberg, Germany (2000) 424

16. A. Rosenfeld and A. C. Kak: Digital Picture Processing, Academic Press, 1976 412

17. P. K. Saha and B. B. Chaudhuri: Detection of 3-D Simple Points for Topology Preserving Transformations with Application to Thinning, IEEE Transactions on Pattern Analysis and Machine Intelligence, 16, 10, pp.1028-1032 (1994) 419, 420

18. T. Saito and J. Toriwaki: Algorithms of three dimensional Euclidean distance transformation and extended digital Voronoi diagram, and analysis of human liver section images, The Journal of the Institute of Image Electronics Engineers of Japan, 21, 5, pp.468-474 (1992) 414, 426

19. T. Saito and J. Toriwaki: Fast algorithms for n-dimensional Euclidean distance transformation, Proc. of the 8th SCIA, pp.747-754 (1993.5) 414

20. T. Saito and J. Toriwaki : New algorithms for n-dimensional Euclidean distance transformation, Pattern Recognition, 27, 11, pp.1551-1565 (1994) 414

21. T. Saito and J. Toriwaki : Reverse distance transformation and skeletons based upon the Euclidean metric for n-dimensional digital binary pictures, IEICE Trans. on Inf. and Syst., Japan, E77-D, 9, pp.1005-1016 (1994.9) 415, 416, 417, 418, 419

22. T. Saito and J. Toriwaki: A sequential thinning algorithm for three dimensional digital pictures using the Euclidean distance transformation, Proc. 9th SCIA, pp.507-516 (1995.6) 419, 420, 421, 422

23. T. Saito, K. Mori, and J. Toriwaki: A sequential thinning algorithm for three dimensional digital pictures using the Euclidean distance transformation and its properties, IEICE Trans. on Inf. and Syst., Japan, J79-D-II, 10, pp.1675-1685 (1996.10) 419, 421, 422, 425

24. T. Saito, S. Banjo, and J. Toriwaki: An improvement of three dimensional thinning method using a skeleton based on the Euclidean distance transformation - a method to control spurious branches, IEICE Trans. on Inf. and Syst., Japan, J84-D-II, 8, pp.1628-1635 (2001.8) 413, 423

25. J. Toriwaki, M. Okada, and T. Saito: Distance transformation and skeletons for shape feature analysis, in C. Arcelli, L. P. Cordella, and G. Sanniti di Baja eds.: Visual Form : Analysis and Recognition (Proc. of the International Workshop on Visual Form, 1991.5.27-30), pp.547-563, Plenum Press, New York, 1992 412, 414

26. J. Toriwaki and K. Mori: Recent progress in biomedical image processing - Virtualized human body and computer-aided surgery, IEICE Trans. on Inf. and Syst., Japan, E-82D, 3, pp.611-628 (1999,3) 424

27. Y. F. Tsao and K. S. Fu: A parallel thinning algorithm for 3-D pictures, Computer Graphics And Image Processing, 17, pp.315-331 (1981) 419, 420, 421, 422

28. H. Yamada: Complete Euclidean distance transformation by parallel operation, Proc. of 7th ICPR, pp.69-71 (1984) 414

29. Y. Zhou, A. Kaufman, and A. W. Toga: Three-Dimensional Skeleton and Center-line Generation Based on an Approximate Minimum Distance Field, The Visual Computer, 14, 7, pp.303-314 (1998)   420

About the Limiting Behaviour of Iterated Robust Morphological Operators

Johan Van Horebeek[1] and Ernesto Tapia-Rodriguez[1,2]

[1] Centro de Investigación en Matemáticas
Ap. Postal 402, 36000 Guanajuato, Gto., Mexico
horebeek@cimat.mx
[2] Freie Universität Berlin, Institut für Informatik
Tacostr. 9, D-14195 Berlin, Germany
tapia@inf.fu-berlin.de

Abstract. We study the limiting behaviour of the repeated application of a discrete image operator that belongs to a recently introduced class of morphological operators. To this purpose, we describe the implied iterative process by a discrete dynamical system.
The convergence is derived for binary and gray scale images.

1 Introduction

Mathematical morphology is an important branch of non-linear signal processing. It has its roots in discrete geometry and set theory, using as a starting point the probing of an image by a small(er) structural element. We refer to [ ], [ ] and [ ] for an overview.

This geometric foundation often facilitates an intuitive interpretation of morphological operators in terms of geometric concepts and is also extremely useful for the derivation of many fundamental properties.

Despite the vast amount of literature about mathematical morphology, the limiting behaviour of the repeated applications of a morphological operator has not received a lot of attention. Partly, this is because in many cases it is a trivial problem: a large class of operators are idempotent (e.g., openings, closings) or decreasing (increasing) functions with a natural lower (resp. upper) bound (e.g., erosions, dilations, openings by reconstruction). This papers aims to make a contribution for non-trivial cases.

The organization is as follows: in section 2, we describe the class of morphological operators, previously introduced in [ ], and for which we prove the convergence in section 3. Section 4 explains how the results can be extended to other morphological operators.

In what follows, X will denote a discrete image defined on a finite rectangular grid; X_x returns the gray level value at position x. In the case of a binary image, in order to simplify the notation, X denotes also the set of black pixels, depending on the context. The cardinality of a set S is given by $|S|$; the translation of S by t, $\{s + t, s \in S\}$ is denoted by $S_{(t)}$ and the symmetric set of S with respect to the origin, by $\tilde{S}$.

G. Bertrand et al. (Eds.): Digital and Image Geometry, LNCS 2243, pp. 429–438, 2001.
© Springer-Verlag Berlin Heidelberg 2001

2 Robust Morphological Operators

2.1 The Binary Case

Our starting point is the classical definition of an opening by a symmetric set S (called a *structural element*): a pixel x belongs to the opening of a binary image X by the structural element S iff x belongs to at least *one* translation of S *totally* included in X.

A natural generalization is to weaken the total inclusion requirement to several partial ones, i.e., we require that x belongs to *at least* r_2 translations of S, which have *at least* r_1 pixels in common with the image X. The above can be formalized in the following way [ ], [ ].

Definition 1. *For a given structural element S, we define the morphological operator $R_S^{r_1,r_2}(\cdot)$ as:*

$$R_S^{r_1,r_2}(X) = \rho_{r_2,\tilde{S}}(\rho_{r_1,S}(X)),\tag{1}$$

where $\rho_{r,S}(\cdot)$ denotes the $r-$rank operator [1]:

$$\rho_{r,S}(X) = \left\{x : |S_{(x)} \cap X| \geq r\right\}.\tag{2}$$

One observes that $R_S^{1,|S|}(\cdot)$, $R_S^{|S|,1}(\cdot)$, $R_S^{|S|,|S|}(\cdot)$ and $R_S^{1,1}(\cdot)$ correspond respectively to a closing $(X \bullet S)$, opening $(X \circ S)$, a double erosion and dilation.

2.2 The General Case

One way to extend Def. 1 for binary images to gray scale images is through the use of the set of *slices* $X^{-1}(u) = \{x, X_x \geq u\}$.

Definition 2. *If X is a gray scale image and S a subset of pixels, we define the morphological operator $R_S^{r_1,r_2}(\cdot)$ as:*

$$R_S^{r_1,r_2}(X)_x = \max\{u : x \in \rho_{r_2,\tilde{S}}(\rho_{r_1,S}(X^{-1}(u)))\}.\tag{3}$$

Using the geometric interpretation of the operator for the binary case, one obtains:

Proposition 1. *The value of $R_S^{r_1,r_2}(X)$ at x corresponds to the largest possible value u such that x is contained in at least r_2 translations of S, that have at least r_1 pixels in common with $X^{-1}(u)$.*

---

[1] Also known in the literature as order statistic filter, percentile filter or Ξ filter.

2.3   Motivation

A motivation to introduce the above class of operators was given in [ ]. As the result of $X \circ S$ and $X \bullet S$ is a basic step in the calculation of the *Pattern Spectrum* and is used in many composed operators like *Top-hats*, it is important to have at one's disposal a good estimator for $X \circ S$ and $X \bullet S$, when only $N(X)$, a noisy version of X, is available.

In [ ] it was shown that if $N(X)$ is the result of limited ($< 5\%$) pepper and salt noise distortion on X, $R_S^{r_1, r_2}(N(X))$ is a robust approximation for $X \circ S$ and $X \bullet S$ (using appropriate values of r_1 and r_2).

The above should be contrasted with the construction of the *Rank-Max filters* [ ] or, equivalently, *Regulated Openings* [ ], where one embeds the classical morphological openings in a larger (parametrized) family, conserving the properties of an algebraic opening. Although less sensitive to noise than the morphological openings, several experiments showed that $R_S^{r_1, r_2}(\cdot)$ outperforms them.

Of course, a drawback is the absence of the idempotency property. In the next section, we study the limit behaviour under repeated applications. To this end, we define for any value of r_1 and r_2:

$$T^1(X) = R_S^{r_1, r_2}(X), \quad T^t(X) = R_S^{r_1, r_2}(T^{t-1}(X)).$$

3   Limiting Behaviour

3.1   The Binary Case

Proposition 2. *For any X, $T^t(X)$ will converge to a fixed image when t tends to infinity.*

Proof. It is illustrative to describe the dynamical behaviour of $T^t(X)$ by means of a cellular automaton. To this purpose, consider an image X of size $n \times m$ as one long vector of size nm with elements belonging to $\{0, 1\}$. We define the state space of the automaton as a vector Y^t with two components, $Y = (Y_1^t, Y_2^t)$, each of size nm. The dynamics of Y^t is described by:

$$Y^t = (\rho_{r_2, \tilde{S}}(Y_2^{t-1}), \rho_{r_1, S}(Y_1^{t-1})) \quad Y^0 = (X, X) \text{ with } X \text{ a given image.} \quad (4)$$

Consequently, Y_1^{2t} corresponds to $T^t(X)$, the application of t times $R_S^{r_1, r_2}(\cdot)$. We construct:

$$M_S = \begin{pmatrix} 0 & N_{\tilde{S}} \\ N_S & 0 \end{pmatrix} \quad (5)$$

where N_S is the $nm \times nm$ matrix:

$$(N_S)_{x,y} = \begin{cases} 1 \text{ if } y \in S_{(x)} \\ 0 \text{ otherwise.} \end{cases}$$

As $y \in S_{(x)}$ iff $x \in \tilde{S}_{(y)}$, M_S is a symmetric matrix. Next, define the vector $\theta = (r_2, \cdots, r_2, r_1, \cdots, r_1)$ where r_1 and r_2 are repeated nm times.

All together, because of (4), one obtains:

$$Y^t = \mathbb{1}(M_S Y^{t-1} - \theta), \qquad (6)$$

where $\mathbb{1}(\cdot)$ is the vectorized threshold function (i.e. $(y_1, \cdots, y_n) = \mathbb{1}((x_1, \cdots, x_n))$ iff $y_i = I(x_i \geq 0)$ and $I(\cdot)$ denotes the Indicator function).

As there are only a finite number of elements in our state space, we can always find a θ_2 in such a way that:

$$M_S Y^{t-1} - \theta \geq 0 \text{ iff } M_S Y^{t-1} - \theta_2 > 0$$

and

$$M_S Y^{t-1} - \theta < 0 \text{ iff } M_S Y^{t-1} - \theta_2 < 0.$$

In the sequel we suppose that $\theta = \theta_2$ to guarantee that:

$$M_S Y^{t-1} - \theta \neq 0. \qquad (7)$$

Following the derivation in [ ] - which for the sake of completeness we include here - a Liapunov function is introduced, associating an energy with every state Y:

$$E(Y^t) = -\langle Y^t, M_S \cdot Y^{t-1} \rangle + \langle \theta, Y^t + Y^{t-1} \rangle,$$

where $\langle .,. \rangle$ denotes the usual inner product.

Using the fact that M_S is a symmetric matrix,

$$\Delta E := E(Y^t) - E(Y^{t-1}) = -\sum_x \{(Y_x^t - Y_x^{t-2})(\sum_y (M_S)_{x,y} Y_y^{t-1} - \theta_x)\}. \qquad (8)$$

By construction each term in the outer sum in (8) is positive, hence ΔE is always non-positive. As the state space is finite, from a certain moment on, E remains constant[ ]. Using once more that each term in the outer sum in (8) is positive, together with (7), $\Delta E = 0$ implies that there exists a given t_0, such that for $t > t_0$:

$$Y^{2t} = Y^{2t_0},$$

i.e., one obtains convergence of $T^t(\cdot)$ to a fixed image.   □

As an illustration, in Fig. 1 the fixed points are shown under repeated application of $R_S^{r_1, r_2}(\cdot)$ for different parameter values.

The following property describes the fixed points (images) in the particular case where S equals a discrete circle of radius 1 in the city-block metric.

Proposition 3. *If S is a circle of radius 1 in the city-block metric,*

1. *the limit points of $R_S^{2,3}(\cdot)$ are of the form $\cup_l R_l$, where R_l are rectangles with a minimum (Euclidean) distance of $\sqrt{2}$ between them, i.e.:*

$$min_{r_k \in R_k, r_l \in R_l} d(r_k, r_l) \geq \sqrt{2}, \text{ for } l \neq k; \qquad (9)$$

2. *the limit points of $R_S^{2,2}(\cdot)$ are of the form $\cup_l R_l$, where R_l are rectangles with a minimum (Euclidean) distance of two between them, i.e.:*

$$min_{r_k \in R_k, r_l \in R_l} d(r_k, r_l) \geq 2, \; for \; l \neq k.$$

Proof.

1. It is sufficient to show that the configurations of 2×2 pixels, shown in Fig. 2(a), can not form part of a fixed point(image). These configurations represent the only possible situations that can not be written as the union of rectangles satisfying (9).

Suppose the opposite holds; independent of the values of the neighbouring points, the application of $R_S^{2,3}(\cdot)$ to any of the configurations of Fig.2(a), will give rise to the configuration shown in Fig.2(b) causing a contradiction with the fact that Fig. 2(a) forms part of a limiting point.

In Fig. 3 a limit point is shown that illustrates that the lower bound of the distance between the rectangles can be attained.

2. In this case, we obtain a classical 2−threshold cellular automaton for which Prop. 4.5 of [ ] can be applied.

$\square$

3.2   The General Case

In order to extend Prop. 2. to gray scale images, we make use of the concept of an *umbra*, defined as [ ]:

$$Um(X) = \{(x, u) : u \leq X_x\}. \tag{10}$$

If we abbreviate $\{(a, x), \; a \in A\}$ by (A, x) with A an arbitrary set, an umbra of an image is equivalent to:

$$Um(X) = \cup_u (X^{-1}(u), u). \tag{11}$$

We denote by $Um^{-1}(\cdot)$ the operator that maps $Um(X)$ into X.

Using the above, we can associate with any order preserving set function $f(\cdot)$ (i.e., $A \subset B$ implies $f(A) \subset f(B)$) an operator $\mathcal{F}(\cdot)$ on X in the following way:

$$\mathcal{F}(X) = Um^{-1}(\cup_u (f(X^{-1}(u)), u)).$$

The following lemma shows how the result of a composition of certain functions acting on an image, is related to the calculation of a composition of corresponding functions acting on each slice of the image.

Lemma 1. *Suppose that $f(\cdot)$ and $g(\cdot)$ are two order preserving set functions. If we define $\mathcal{F}(\cdot)$ and $\mathcal{G}(\cdot)$ by:*

$$\mathcal{F}(X) = Um^{-1}(\cup_u (f(X^{-1}(u)), u)), \;\; \mathcal{G}(X) = Um^{-1}(\cup_u (g(X^{-1}(u)), u)),$$

then

$$\mathcal{F}(\mathcal{G}(X)) = Um^{-1}(\cup_u (f(g(X^{-1}(u))), u)).$$

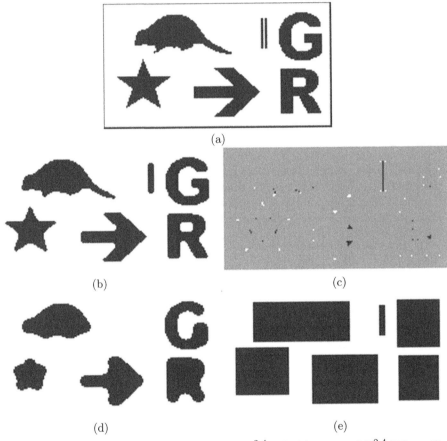

Fig. 1. (a): the given image X; (b): limit of $R^{2,4}_{S_1}(X)$; (c): limit of $R^{2,4}_{S_1}(X) - X$ (black denotes a positive difference, white a negative one); (d): limit of $R^{4,10}_{S_2}(X)$; (e): limit of $R^{2,3}_{S_1}(X)$ and S_1, S_2 denotes a circle of radius 1 resp. 2

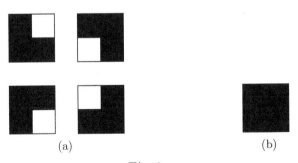

Fig. 2.

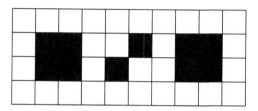

Fig. 3. A limit point of $R_S^{2,3}(\cdot)$, composed of four rectangles, the two in the middle are separated by a distance of $\sqrt{2}$

Proposition 4. *For any structural element, S, repeated applications of $R_S^{r_1,r_2}(\cdot)$ will converge to a fixed image.*

Proof. If $f(\cdot)$ is an order preserving set function and Z a gray scale image, we have the following equivalence, which is easy to demonstrate:

$$Um(Z) = \cup_u(f(X^{-1}(u)), u) \quad \text{iff} \quad \forall x : Z_x = \max\{u : x \in f(X^{-1}(u))\}. \quad (12)$$

Hence, Def. 2 is equivalent to:

$$Um(R_S^{r_1,r_2}(X)) = \cup_u(\rho_{r_2,\tilde{S}}(\rho_{r_1,S}(X^{-1}(u))), u). \quad (13)$$

Using Lemma 1., if we choose $f(\cdot) = g(\cdot) = R_S^{r_1,r_2}(\cdot)$, then repeated applications of $R_S^{r_1,r_2}(\cdot)$ is equivalent to repeatedly applying this operator to each slice and taking the union (or equivalent the supremum) at the end. For each slice, Prop. 2 guarantees convergence. □

An example is shown in Fig. 4 for different values of $R_S^{r_1,r_2}(\cdot)$. As one observes in Fig. 4(d), the limit point is not sensitive to the presence of low salt and pepper noise in the original image.

Although for all the above examples convergence was reached in less than ten iterations, a considerable speed-up (25%) of the calculation time was obtained using a *moving-histogram* algorithm to calculate $\rho_{r,S}(X)$, instead of its naive direct implementation. The basic idea of the *moving-histogram* algorithm [ ] is to update the histogram of the gray values of the pixels covered while sliding the structural element over the image. This is usually a much smaller set than all pixel values, the update step is extremely efficient and the value of $\rho_{r,S}(X)$ can be easily obtained.

In the particular case of binary images, other alternatives are available. We refer to [ ] for a recent overview.

4   Extensions to Soft Morphological Filters

The above construction is not only useful for the class of operators introduced in section 2. As an example, we consider the following class of soft morphological

(a)   (b)

(c)   (d)

Fig. 4. (a): the given image X; (b): limit of $R_{S_1}^{2,4}(X)$; (c): limit of $R_{S_2}^{3,7}(X)$; (d): limit of $R_{S_2}^{3,7}(N(X))$ where $N(X)$ denotes X corrupted by 1% salt and pepper noise and S_1, S_2 denote a circle of radius 1 resp. a 3×3 rectangle

filters (several definitions appear in the literature; we follow the one of [ ]). For the sake of simplicity, we restrict ourselves to soft morphological openings for binary images.

Definition 3. *[ ] Given two symmetric sets $S^{\mathcal{H}}$, $S^{\mathcal{S}}$, $S^{\mathcal{H}} \subset S^{\mathcal{S}}$ and $r \leq |S^{\mathcal{S}} \setminus S^{\mathcal{H}}|$, define the soft morphological opening, $X \circ [S^{\mathcal{H}}, S^{\mathcal{S}}, r]$, as:*

$$X \circ [S^{\mathcal{H}}, S^{\mathcal{S}}, r] = (X \ominus [S^{\mathcal{H}}, S^{\mathcal{S}}, r]) \oplus [S^{\mathcal{H}}, S^{\mathcal{S}}, r], \tag{14}$$

where $(X \ominus [S^{\mathcal{H}}, S^{\mathcal{S}}, r])_x$, $(X \oplus [S^{\mathcal{H}}, S^{\mathcal{S}}, r])_x$ denote the rth smallest, respectively, rth largest value of the multiset^2 $(r \diamond (S^{\mathcal{H}}_{(x)} \cap X)) \cup ((S^{\mathcal{S}} \setminus S^{\mathcal{H}})_{(x)} \cap X)$, and where $\diamond$ denotes de repetition operator ($r \diamond S$ equals r times the multiset S).

In a similar way, a soft morphological closing is defined.

Proposition 5. *The repeated application of a soft morphological opening (closing) will converge to a fixed image.*

Proof. Only minor changes in the proof of Prop. 2 have to be done. The dynamics can be described by a dynamical system similar to (6). In (5), we should now define:

$$(N_S)_{x,y} = \begin{cases} r \text{ if } y \in S^{\mathcal{H}}_{(x)} \\ 1 \text{ if } y \in S^{\mathcal{S}}_{(x)} \setminus S^{\mathcal{H}}_{(x)} \\ 0 \text{ otherwise.} \end{cases}$$

As N_S is also symmetrical, the same reasoning of the proof of Prop. 2 applies and the same conclusions hold.

□

5   Conclusions

In this paper we have proven that the repeated application of a certain class of discrete image operators converge to a fixed image. Although only some modest but promising results concerning the characterization of those limiting points were included, a lot of work still has to be done.

Some open questions are: For which parameter values of r_1, r_2 do we obtain a union of convex sets as the limit image? When do we obtain a non-trivial limit (i.e., an image that is not completely black or white)? How do we obtain an tight upper and lower bound of this image? How do we obtain a good upper bound of the convergence speed?

Acknowledgement

The authors like to thank a referee and I. Terol for a variety of suggestions that helped in the revision of the paper.

---

^{2} A multiset is a collection where repetition of the objects is allowed.

References

1. Agam, G., Dinstein, I.: Regulated Morphological Operators. Pattern Recognition **32** (1999) 947–971 431
2. Goles, E., Martínez, S.: Neural and Automata Networks: Dynamical Behavior and Applications. Kluwer Academic Publishers (1990) 432, 433
3. Heijmans, H.: Morphological Image Operators. Academic Press (1994) 429
4. Koskinen, L., Astola, J., Neuvo, Y.: Soft Morphological Filters. SPIE Image Algebra and Morphological Image Processing II (1991) 262–270 437
5. Ronse, C.: Erosion of Narrow Image Features by Combination of Local Low Rank and Max Filters. Proceedings 2nd Int. Conf. on Image Processing and its Applications, London, (1986) 77–81 431
6. Serra, J.: Image Analysis and Mathematical Morphology, Volume I. Academic Press (1982) 429
7. Soille, P.: Morphological Image Analysis : Principles and Applications. Springer Verlag (1999) 429
8. Sternberg, S. R.: Parallel Architecture for Image Processing. In Proc. Third International IEEE Compsac, Chicago (1979) 433
9. Tapia-Rodriguez, E.: A Study of Robust Mathematical Morphology for the Analysis of Discrete Images (in spanish). Master Thesis, CIMAT (1998) 430
10. Van Droogenbroeck, M., Talbot, H.: Fast Computation of Morphological Operations with Arbitrary Structuring Elements. Pattern Recognition Letters **17** (1996) 1451-1460 435
11. Van Horebeek, J., Tapia-Rodriguez, E.: The Approximation of Openings and Closings in the Presence of Noise. Signal Processing (to appear) 429, 430, 431
12. Zmuda, M., Tamburino, A.: Efficient Algorithms for the Soft Morphological Operators. IEEE-PAMI **18** (1996) 1142–1147 435

Collinearity and Weak Collinearity
in the Digital Plane

Peter Veelaert

Hogent
Schoonmeersstraat 52, 9000 Ghent, Belgium
`Peter.Veelaert@hogent.be`

Abstract. We discuss the optimal grouping of collinear line segments, where we use a digital form of collinearity as the grouping criterion. We also introduce weak collinearity as a new concept to facilitate the design of an efficient line grouping algorithm. This algorithm is based on the use of simplicial elimination orderings, as known in graph theory.

1 Introduction

Extracting groups of lines according to some geometric criterion such as collinearity or parallelism is an important problem in perceptual grouping. In this paper we discuss optimal grouping with respect to digital collinearity, a property defined within the framework of geometry in the digital plane. To increase the efficiency of the grouping process, we also introduce the concept of weak collinearity as an approximate form of digital collinearity. Using weak collinearity, we propose one particular grouping algorithm, which is based on simplicial vertex elimination, an idea borrowed from graph theory.

In computer vision the partitioning of objects into groups according to some similarity criterion is viewed as a clustering problem. We can make the distinction between vectorial based clustering and relation based clustering [ ]. In vectorial based clustering, features are represented as vectors in an n-dimensional feature space, e.g., a feature may consist of the slope, length, and midpoint of a line segment. The distance between feature vectors is used as a measure of similarity. Alternatively, relation based approaches introduce the strength of a relation as a similarity measure. For example, Lowe as well as Nacken introduce a metric that measures how far two line segments are from being collinear [,]. The relation based approach has the advantage of greater flexibility as for the choice of an appropriate function that measures the similarity between features. A recurring problem, however, is that such a function does not necessarily satisfy the triangle inequality, a desirable condition when the criterion used for grouping is based on a transitive property, as is the case for collinearity [ ].

In this work we propose a relational approach that is still strongly linked with the feature vectors. Instead of using distances between feature vectors we will embed each feature vector into a region called an uncertainty region, and we will use the intersection properties of the uncertainty regions to determine

G. Bertrand et al. (Eds.): Digital and Image Geometry, LNCS 2243, pp. 439–153, 2001.
© Springer-Verlag Berlin Heidelberg 2001

which vector pairs represent collinear line pairs. Although we loose some of the flexibility of a purely relational approach, and the simplicity of purely vectorial approach, there are some important advantages:

- We have a coherent framework in which we can define geometric properties such as digital parallelism, collinearity and concurrency; in each case, line grouping can be formulated as a combinatorial optimization problem.
- Line grouping is hierarchical process, where lines are combined into groups, and groups of lines into higher level structures; by using uncertainty regions hierarchical grouping becomes a natural process;
- One of the main conclusions of this work is that the failure of transitivity can be related to other issues such as the non-perfectness of graphs; the use of uncertainty regions better explains this relation.

The digital versions of collinearity, parallelism and concurrency have been studied before [ ]. In previous work on line grouping we focused on the grouping of parallel lines. Grouping parallel lines is greatly facilitated by the fact that parallelism involves only one parameter, i.e. the slope of a line. In fact, one can show that parallelism satisfies a one dimensional Helly type property [ ], and that the parallel relationships present in an image can be represented by an interval graph. For interval graphs we have efficient combinatorial optimization algorithms, leading to fast algorithms to extract the geometric structure. For example, once the interval graph has been constructed for a collection of N lines, there are algorithms of $O(N \log N)$ time complexity for extracting the largest set of parallel lines, as well as for the optimal partitioning into parallel subgroups, or for finding the largest set of non-parallel lines.

Collinearity on the other hand involves at least two line parameters, slope and height, and, not surprisingly, it is more difficult to solve geometric optimization problems for collinearity than for parallelism. There are two causes for this difficulty. First, digital collinearity satisfies a two-dimensional Helly type property [ ], and as a result the appropriate combinatorial structure for representing collinearity is a hypergraph. Second, under certain assumptions, we can replace collinearity by the simpler property of "weak collinearity", whose relationships can be represented in an ordinary graph. However, a weak collinearity graph is the intersection graph of a collection of two-dimensional convex sets, and by nature this kind of intersection graph is more difficult to process than the intersection graphs of one-dimensional intervals, which represent parallelism.

In this paper, to illustrate the notion of weak collinearity, we restrict ourselves to one particular geometric problem, i.e., partitioning a collection of lines into a minimum number of (weakly) collinear groups. This problem is trivial in Euclidean geometry, but not in the digital plane. First, we reformulate the partitioning problem as a well posed, but NP-complete optimization problem in graph theory. Next, we propose a heuristic algorithm that often finds an optimal solution. Although the proposed algorithm fails occasionally, one of its advantages is that we know when the solution is optimal. In fact, the circumstances of failure can be related to the occurrence of certain line configurations in the digital plane.

Finally note that in this work we use the "discretization by dilation" scheme, which has been formalized by Heijmans and Toet [ ] . One of the advantages of using the discretization by dilation scheme is that it leads to a number of well defined combinatorial optimization problems [,   ,   ]. This advantage will also be exploited in this work.

Section 2 gives a brief overview of properties in the digital plane. In Section 3 we introduce the notion of weak collinearity, and we briefly consider the major optimization problems with respect to collinearity as well as the major obstacles that arise when designing efficient algorithms to solve these problems. In Section 4 we propose a simplicial grouping algorithm to solve the problem of partitioning a set of lines into a minimum number of collinear groups. Finally, in Sections 5 and 6 we give experimental results, and we conclude the work.

2   Geometric Properties in the Digital Plane

In this section we recall some results obtained in previous work [,   ,   ]. Since image pixels lie on a rectangular grid, we assume that all pixels are part of the digital plane $\mathbb{Z}^2$, although this is not strictly necessary for what follows. A digital set S is a subset of the digital plane.

We model uncertainty by an *uncertainty region* that we associate with each grid point. The discretization process that coincides naturally with this notion of uncertainty is the *discretization by dilation* scheme developed by Heijmans and Toet [ ]. Let U denote a set in $\mathbb{R}^2$, let A be a second subset of $\mathbb{R}^2$, called the structuring element, and let A_p be the translate of A by p. Then the discretization by dilation of U consists of all points $p \in \mathbb{Z}^2$ for which $A_p \cap U$ is non-empty [ ].

Furthermore, to simplify the exposition in this paper we shall restrict ourselves to one particular form of discretization by dilation, that is, we use a simple variant of grid-intersect discretization. To model the uncertainty of its position, for each digital point $p = (x, y)$, we introduce as translate of the structuring element the vertical line segment $C_p(\tau)$, which comprises all points $(x, b) \in \mathbb{R}^2$ that satisfy $y - \tau/2 \le b < y + \tau/2$. Here τ is a positive real number, called the *acceptable thickness*. To avoid confusion between a vertical "segment" and a digital straight line "segment" we shall call a vertical segment an uncertainty segment. Up to a certain multiplication factor, the notion of acceptable thickness coincides with Réveillès's notion of arithmetical thickness of a discrete straight line, also discussed by Andres [,]. To simplify what follows even further, we shall only discuss the extraction of straight lines of the form $y = \alpha x + \beta$, where the slope α satisfies $-1 < \alpha < 1$. We assume that each set S contains at least two points with distinct x-coordinates.

Although the foregoing restrictions do not change our results in a fundamental way, they allow us to discard several special cases discussed more thoroughly in previous work [ ], where we show, for example, that for lines whose slope lies between -1 and 1, vertical uncertainty segments are sufficient to model position uncertainty.

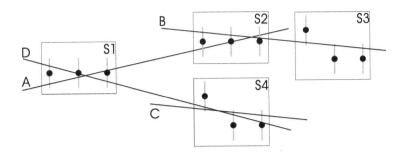

Fig. 1. Digital collinearity and parallelism

Definition 1. A digital set S is called *digitally straight* if there is a continuous straight line that cuts all uncertainty segments $C_p(\tau)$, $p \in S$.

A finite digitally straight set will also be called a digital straight line segment.

Definition 2. Let $S_1, \ldots, S_n$ a finite collection of finite digitally straight sets. We define the following digital geometric properties:

- The sets $\{S_i : i = 1, \ldots, n\}$ are called *digitally collinear* if there exists a common Euclidean straight line A that cuts the uncertainty segments of all the sets S_i;
- The sets in $\{S_i : i = 1, \ldots, n\}$ are called *digitally parallel* if there exist n Euclidean straight lines $A_1, \ldots, A_n$ that are parallel and such that for $i = 1, \ldots, n$ the line A_i cuts the uncertainty segments of the set S_i;

Thus the sets S_i are digitally collinear when their union is digitally straight. From now on, we shall often call two digital sets simply "collinear," when it is clear we mean digitally collinear. Digital collinearity and parallelism are illustrated in Figure 1. S_1 and S_2 are collinear since their uncertainty segments are cut by the common Euclidean line A. Likewise, S_2 and S_3 are collinear because of the common line transversal B. Moreover, S_2 is digitally parallel with S_4, since $B||C$ in the Euclidean sense. Finally note that although S_1 is collinear with S_2, S_2 with S_3, as well as S_1 with S_4, S_1 is not collinear with S_3 because there is no common line cutting all their uncertainty segments. For the same reason is S_4 not collinear with S_2, nor with S_3. Evidently, in the digital plane collinearity and parallelism are not transitive. In [ ] the following result is derived as a Helly-type theorem:

Theorem 1. *Let $\mathcal{S} = \{S_1, \ldots, S_n\}$ a finite collection of finite digital straight line segments, with $n > 2$. Then the collection $\mathcal{S}$ is digitally collinear if each subcollection containing three sets of $\mathcal{S}$ is digitally collinear.*

We shall need Theorem 1 when we reformulate optimal grouping as a combinatorial optimization problem.

Definition 3. Let S be a finite digital set that contains at least two points with distinct x-coordinates, and let τ be a chosen acceptable thickness with $0 < \tau$. Then the *domain* of S, denoted as $\mathrm{dom}_x(S; \tau)$, is the set of all parameter points $(\alpha, \beta) \in \mathbb{R}^2$ that satisfy the following system of inequalities:

$$-\tau/2 < \alpha x_i + \beta - y_i \leq \tau/2, \quad (x_i, y_i) \in S. \tag{1}$$

In other words, the domain contains the parameters of all the Euclidean lines that cut the uncertainty segments of S. It follows almost immediately that a collection of digital line segments is digitally collinear if and only if their domains have a common non-empty intersection [,].

As an example, consider the collection of digital sets representing lines in Fig. 2(a), whose domains are shown in Fig. 2(c), for an acceptable thickness $\tau = 1$. A domain of a digital set with N points is a convex bounded set, and can be computed in $O(N \log N)$ time as an intersection of $2N$ halfplanes [ ]. When we let τ vary, a domain gets larger for increasing values of τ. If the acceptable thickness is too small, however, the domain of a set will be empty. For a given set S, let $D \subset \mathbb{R}$ be the set of acceptable thicknesses for which the domain of S is non-empty. The *thickness* of S is defined as the infimum of D [ ].

We also need some basic notions of graph theory (see, e.g., [ ]). Let G be a simple graph, i.e, a graph that contains no loops or double edges. An *independent set* is a set of vertices that does not contain any pair of adjacent vertices. A *coloring* assigns an index to each vertex so that two adjacent vertices always receive distinct indices. An *optimal coloring* uses a minimum number of indices, i.e., the *chromatic number* of the graph. A graph is *perfect* if each of its induced subgraphs has a chromatic number equal to the size of its largest complete subgraph. A graph is *chordal* or *triangulated* if the only cycles that occur as induced subgraphs are triangles.

3   Collinearity and Weak Collinearity

Given a set of digital line segments, we shall represent collinear relationships in two different ways.

Definition 4. Let C be a collection of digital straight line segments. In the *weak collinearity graph* $G(V, E)$ each line segment in C is represented by a vertex. Two vertices in $G(V, E)$ are joined by an edge if the two corresponding line segments are digitally collinear. The *collinearity hypergraph* $H(V, E)$ also has one vertex for each line segment in C. The edge set E consists of all 2-membered and 3-membered subcollections that are digitally collinear. The *complement of the collinearity hypergraph* is defined as the hypergraph $H'(V, E)$ with one vertex for each line segment in C, and whose edge set E of $H'(V, E)$ consists of all 2-membered and 3-membered subcollections that are not digitally collinear.

A collection of digital line segments is called *weakly collinear* if the corresponding vertices induce a complete subgraph in the weak collinearity graph

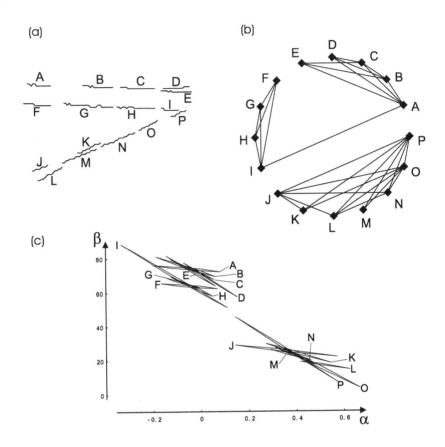

Fig. 2. (a) A set of lines, (b) the weak collinearity graph, and (c) the domains

$G(V, E)$. Figure 2(b) shows the weak collinearity graph of the lines in Figure 2(a). Two vertices in Figure 2(b) are adjacent if the corresponding domains in Figure 2(c) have a non-empty intersection. The collection of line segments A, B, C, and D is weakly collinear, since they induce a complete subgraph in Figure 2(b).

We shall concentrate on optimal grouping, i.e., to find the minimally sized collection of collinear groups. Due to Theorem 1 this problem can be solved with the collinearity hypergraph or its complement. To be precise, optimal grouping corresponds to an optimal coloring of the complement of the collinearity hypergraph. The edges of the complement $H'(V, E)$ are formed by all non-collinear pairs and triples. In a hypergraph coloring no edge is monochromatic. Thus, each color corresponds to a group that contains no non-collinear pairs or triples, or in other words, in each group all pairs and triples are collinear. By Theorem 1 it follows that the entire group is collinear.

Optimal grouping with the collinearity hypergraph is a difficult combinatorial problem. In this work our goal is to solve the optimal grouping problem with the weak collinearity graph, or at least to find a suboptimal solution. To appreciate the introduction of weak collinearity we note that weak collinearity suffices to solve at least one geometric optimization problem: to find the maximum set in which no pair is collinear. To solve this problem, it suffices to look for the maximum independent set of the weak collinearity graph. Also other optimization problems benefit from the introduction of weak collinearity, but we shall not consider them here.

As for optimal grouping, the introduction of weak collinearity is not without cost, however. Although a collinear collection of digital line segments is always weakly collinear, a collection that is weakly collinear may not be collinear. In addition, although weak collinearity simplifies problems, a weak collinearity graph may still be difficult to handle. In particular, a weak collinearity graph may not be chordal, or it may even fail to be perfect. In fact, if we are lucky it may happen that for a given collection of line segments all weakly collinear subcollections are also collinear, which means that we can use the weak collinearity graph to solve the optimal grouping problem. However, even in this case, if the weak collinearity graph is not perfect, then the grouping problem is still NP-complete [ ]. On the other hand, we may have a collection of line segments in which some triples are weakly collinear, but not collinear. Nonetheless, the weak collinearity graph can be still be chordal, so that we have an efficient algorithm to solve the grouping problem for weakly collinear groups.

Although the above difficulties seem to be different, closer examination shows, however, that each of the above anomalies arises from a similar cause, i.e., the existence of a cyclic configuration of the domains. We start with an example without anomalies. Figure 3(a) shows a collection of four line segments and their domains. In this case, the domains have a common non-empty intersection, the entire set is collinear, and collinearity coincides in a trivial way with weak collinearity. By contrast, Figure 3(b) shows three line segments and their domains. Each pair of domains has a non-empty intersection. The three domains do not have a common intersection, however. That is, although the three line segments are weakly collinear, they are not collinear. The difference between collinearity and weak collinearity reflects itself in a cyclic configuration of three domains that circle a hole in the parameter space. Figure 3(c) shows four line segments and their domains. In this case we have a cyclic chain of four domains that circle a hole. The weak collinearity graph is a chordless 4-cycle, and is therefore non-chordal. Finally, Figure 3(d) shows five lines. The weak collinearity graph of this set is a chordless 5-cycle, and the weak collinearity graph is not perfect. In fact, according to Berge's strong perfect graph conjecture a graph is perfect if and only if the graph neither has an induced subgraph that is an odd cycle of length at least five, nor the complement of such a cycle. In each of the cases (b), (c) and (d), an anomaly arises from the existence of a closed chain of domains that circle an empty hole.

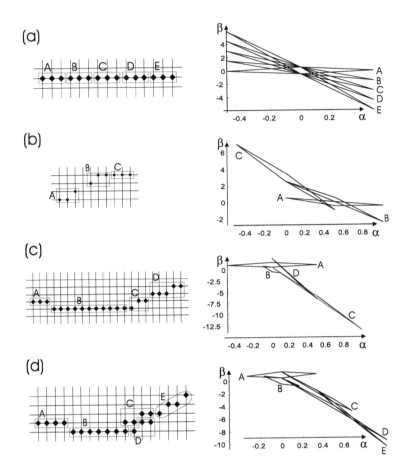

Fig. 3. (a) Normal case , (b) weak collinearity differs from collinearity, (c) non-chordal graph, (d) non-perfect graph

4   Simplicial Grouping

The problem of finding an optimal grouping of (weakly) collinear line segments is the same as finding a minimum clique covering for the weak collinearity graph, which in turn is equivalent with finding an optimal coloring of the complementary graph. Finding a minimum clique covering for a graph is NP-complete. Our goal is therefore to find a suboptimal algorithm, and we start with the following naive approach. First, we select one vertex at random, and then we look for the maximal clique that contains it. Next, after removing this maximal clique, we repeat the process for the smaller graph. There is no efficient procedure, however, for selecting a vertex whose maximal clique is also a maximum clique, because finding a clique of maximum size is in itself an NP-complete problem.

In fact, even finding the maximal clique that contains the selected vertex may be surprisingly difficult, when the degree of the vertex is large, since this problem is equivalent to finding the maximum clique of the graph induced by the neighborhood of the vertex.

Because the naive approach fails, we propose a more sophisticated method based on work by Gavril [ ]. Gavril has shown that we can find a minimum clique covering if instead of selecting vertices at random we restrict ourselves to simplicial vertices. The following definition is taken from West [ ].

Definition 5. A vertex of a graph is *simplicial* if its neighborhood (not including the vertex itself) in G induces a clique. A *simplicial elimination ordering* is an ordering $v_1, \ldots, v_n$ in which vertices can be deleted so that each vertex is a simplicial vertex of the remaining graph. A *simplicial construction ordering* is the reverse of a *simplicial elimination ordering*.

One can show that a graph is chordal if and only if it has a simplicial elimination ordering [ ]. Gavril has introduced the following minimum clique covering algorithm for chordal graphs. Let $v_1, \ldots, v_n$ be a simplicial elimination ordering, let V_i denote the neighborhood of v_i with v_i included, and let G_i denote the graph induced by all vertices that are in G but not in one of the neighborhoods $V_1, \ldots, V_{i-1}$. We construct a sequence of neighborhoods $Y_1, \ldots, Y_k$ as follows: $Y_1 = V_1$, and $Y_i = V_j$ where v_j is the first vertex in the ordering $v_1, \ldots, v_n$ that is still contained in G_i. Then $Y_1, \ldots, Y_k$ is a minimum clique covering.

Gavril's algorithm is intended for chordal graphs, but we shall extend the approach to graphs in general. We introduce the following algorithm, which constructs a sequence of cliques of G.

Simplicial Grouping Algorithm.
Input: a weak collinearity graph $G(V, E)$.
Output: a sequence T of disjoint cliques of G, and a sequence U of independent vertices.

1. Let $G' = G$, $i = 1$, and initialize T and U as empty sequences.
2. If G' has no simplicial vertices, then stop and return the current sequences T and U.
3. Select at random a simplicial vertex q_i of G', and let Q_i be the vertices of the clique formed by q_i and its neighbors.
4. Append Q_i to the sequence T, and append q_i to the sequence U.
5. Replace G' by the graph induced by all vertices of G' that are not in Q_i.
6. Go to the second step.

It is important to note that the sequence $U = \{q_1, q_2, \ldots\}$ does not have to form a simplicial elimination ordering. In a simplicial elimination ordering each vertex is or becomes simplicial when all preceding vertices have been deleted. In the simplicial grouping algorithm, a vertex is or becomes simplicial when all preceding vertices *and their neighbors* have been deleted.

The output of the simplicial grouping algorithm is a finite sequence of cliques $Q_1, \ldots, Q_m$. For some graphs, the vertices in this sequence will cover the entire graph. There are graphs, however, for which the algorithm terminates prematurely because no more simplicial vertices can be found. Now suppose a graph is not chordal, but that the simplicial grouping algorithm nonetheless succeeds in grouping all vertices in the graph G. Then we can show that the sequence $Q_1, \ldots, Q_m$ is in fact a minimum clique covering. We start with the following simple result, whose proof is almost immediate.

Lemma 1. *Let* $G(V, E)$ *be a weak collinearity graph, and let* $H(V, E')$ *be a second graph with the same vertex set, and with* $E' \subset E$. *Let* $\mathcal{P}$ *be a minimum clique covering of* $G(V, E)$. *If* $\mathcal{P}$ *is a clique covering for* $H(V, E')$, *then it is a minimum clique covering for* $H(V, E')$.

Proof. Clearly, any clique covering for H is a clique covering for G. If H can be covered with fewer cliques than the number of cliques in $\mathcal{P}$, then G also can be covered with fewer cliques and $\mathcal{P}$ would not be optimal.

Proposition 1. *The simplicial grouping algorithm yields a minimum clique covering* $\mathcal{P} = T$ *of the graph* G *if it succeeds in grouping all vertices of* G, *i.e., if all vertices occur in the sequence* T.

Proof. Let $Q_1, \ldots, Q_m$ be the output of the simplicial grouping algorithm covering all vertices in G. Let $U = \{q_1, \ldots, q_m\}$ be the simplicial vertices selected during the grouping process. Clearly, U is an independent set. Let $v_{m+1}, \ldots, v_n$ be the vertices of G that are not in U. Let K the complete graph with vertices $v_{m+1}, \ldots, v_n$, and let H be the union of the graphs K and G. We claim that H is chordal and that $q_1, \ldots, q_m, v_{m+1}, \ldots, v_n$ is a simplicial elimination ordering for H. In fact, no pair of vertices in U is adjacent, and each vertex in U is simplicial in H. It follows that H is a chordal graph, and that the collection C of all cliques of neighborhoods Q_i of the vertices q_i forms a minimum clique covering of H. By Lemma 1 it follows that C is also a minimum clique covering of G.

Hence, if it succeeds in grouping all vertices, the simplicial grouping algorithm produces an optimal clique covering, even for non-chordal graphs. Next, we show that for chordal graphs the algorithm will in fact be able to group all vertices, although the algorithm is not guided by a simplicial elimination ordering, as was noted previously.

Proposition 2. *Let* G *be a simple chordal graph, then the simplicial grouping algorithm is able to group all vertices of* G, *i.e., all vertices occur in* T.

Proof. Let $v_1, \ldots, v_n$ be any simplicial elimination ordering of G. If a vertex v_i is simplicial in a graph G, then it is simplicial in any induced subgraph of G. This follows from the simple fact that the neighbors of v_i still form a clique in any induced subgraph of G. Therefore, a simplicial elimination ordering is still a simplicial elimination ordering in any induced subgraph G' if we discard in the ordering $v_1, v_2, \ldots$ all vertices that are not in G'. In particular, assume that the simplicial grouping algorithm starts by removing the simplicial vertex v_i, $i \neq 1$ and its neighbors, then v_1 will still be a simplicial vertex in the new graph G'. As a result, after a finite number of steps v_1 will also be removed from the graph. Likewise, sooner or later any of the vertices v_j in the simplicial elimination ordering will become simplicial in the remaining graph.

From Propositions 1 and 2 it follows that for chordal graphs the simplicial grouping algorithm produces a minimum clique covering. However, since the simplicial vertices that are removed by the simplicial grouping algorithm can be randomly chosen from the set of simplicial vertices, the algorithm will produce different sequences T on different runs. We must therefore show that even when the algorithm terminates prematurely, it always removes the same set of vertices, although they may be classified in different groups. We need the following result.

Lemma 2. *Let G be a simple graph, and let u and v be two simplicial vertices of G. If u and v are adjacent then they belong to the same maximal clique.*

Proof. Since u is simplicial, u and its neighbors form a maximal clique, which is identical to the maximal clique formed by v and its neighbors.

Proposition 3. *Let G be a simple graph. Let T_1 and T_2 be the output sequences of the simplicial grouping algorithm after two different runs. Then the T_1 and T_2 cover the same subset of vertices of G, i.e., each vertex that occurs in T_1 also occurs in T_2, and vice versa.*

Proof. Suppose we run the grouping algorithm twice. Let $Y = \{\{y_1^1, \ldots, y_i^1\}, \{y_1^2, \ldots, y_j^2\}, \{y_1^3, \ldots, \}, \ldots\}$ be result of the first maximal grouping process, and let $V = \{\{v_1^1, \ldots, \}, \{v_1^2, \ldots\}, \ldots\}$ be the result of the second grouping process. Furthermore, without loss of generality, we assume that the groups have been ordered in such a way that in each group the vertex y_1^k is a simplicial vertex, that is, after the vertices of all previous groups have been removed from G.

It is sufficient to show that for any group $Y_k = \{y_1^k, \ldots\}$ in Y, each vertex in Y_k also belongs to some group in V. We first consider the group $Y_1 = \{y_1^1, \ldots, y_i^1\}$. Since this is the first group in Y, the vertex y_1^1 is simplicial in G and in any of its induced subgraphs that contain y_1^1. First, assume that y_1^1 has not been removed by V. This is impossible since it would mean that there is still a simplicial vertex in G, and it would follow that V is not maximal. Second, suppose y_1^1 has been removed by V. Then there is a group $V_l = \{v_1^l, \ldots\}$ containing y_1^1. By Lemma 2, V_l also contains all the vertices in Y_1, that have not yet been removed by the groups $V_1, \ldots, V_{l-1}$. In fact, v_1^l and y_1^1 are simplicial and adjacent to each other, and so they belong to the same clique.

Next, we consider Y_2. When we remove the vertices in Y_1 from G, the vertex y_1^2 becomes simplicial. Since all vertices of Y_1 have been removed by V, there must be a group V_j, $j \leq l$ in V such that either by removing V_j also y_1^2 and all its neighbors are removed, or that after removing V_j, the vertex y_1^2 becomes simplicial. By the same arguments as before, it follows that all vertices of Y_2 are removed by V. By induction it follows that the vertices in each group Y_k are removed by V.

Summarizing, the simplicial grouping algorithm will produce an optimal result for chordal graphs. For graphs that are not chordal, the algorithm yields an optimal result if it can group all vertices, which is often the case. Finally, for graphs for which the algorithm terminates prematurely, we know that it has removed as many vertices as possible, and although it is a random algorithm, running it twice will yield the same result. In fact, even when the algorithm terminates prematurely, we can derive upper and lower bounds for the size of a minimum clique covering. Let $\chi(\overline{G})$ denote the chromacity of the complement of G. Hence, $\chi(\overline{G})$ is the size of a minimum clique covering of G. Let G_1 and G_2 be two induced subgraphs of G, whose vertex sets partition the vertex set of G into two parts. Then $\chi(\overline{G}) \leq \chi(\overline{G_1}) + \chi(\overline{G_2})$. Furthermore, for any induced subgraph G_1 of G we have $\chi(\overline{G_1}) \leq \chi(\overline{G})$. Let H be the subgraph of G induced by all the vertices that can be removed by simplicial grouping. It is clear that H can be simplicially grouped and that this grouping is optimal. Let F denote be the subgraph of G induced by those vertices of G that are not in H. Then

$$\chi(\overline{H}) \leq \chi(\overline{G}) \leq \chi(\overline{H}) + \chi(\overline{F}).$$

Therefore, since the simplicial grouping algorithm is optimal for those vertices that have been removed, we know that if the algorithm is able to remove many vertices, i.e., when $\chi(\overline{F})$ is small, the result is almost optimal.

5   Experimental Results for Random Sets

Although we have shown that simplicial grouping gives optimal or suboptimal results, we must still examine how it performs in practice, i.e., how many times it yields optimal results, and if the results are not optimal, how close they are being optimal. There is an intuitive reason for using simplicial grouping: weakly collinear line segments tend to occur in clusters. In fact, a weak collinearity graph is not a completely random graph. If A is collinear with B, and B collinear with C, then the probability that A and C are also collinear is higher than in a random graph. Thus clustering works both ways: it increases the probability of cycles, but it also increases the probability of chords in cycles. The grouping algorithm performs well when most cycles have many chords.

The real performance of the grouping algorithms must be examined for real images. The random set of line segments in Figure 4(a) was generated as follows. We generated 250 random Euclidean line segments in 50x50 square. The slopes were uniformly distributed between -1 and 1. The length of the line segments was chosen such that after digitization we would have line segments of 8 pixels. After removing all line segments that were not collinear to at least one other line segment, 149 line segments were left, which are shown in Figure 4(a). Figure 4(b) shows the weak collinearity graph of this set, with 149 vertices and 163 edges.

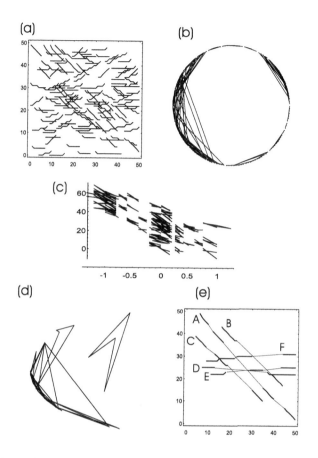

Fig. 4. (a) A random set of line segments, (b) the weak collinearity graph, (c) the domains, (d) what remains of the weak collinearity graph after simplicial grouping, and (e) the largest groups

Note that Figure 4(a) contains a large number of weakly collinear triples (35), but that only about one third (13) of these triples are collinear.

For line segments of 8 pixels, each digital segment can be represented as the digitization of a line segment with slope $k/7$, $k = -7, -6, \ldots, 7$. This explains the regular pattern of vertical strips shown by the domains in Figure 4(c). Figure 4(d) shows an induced subgraph of the graph in Figure 4(b). It contains 47 vertices that cannot be grouped by simplicial grouping. Figure 4(e) shows the largest groups that have been found by the grouping process. The groups A,B,C,D,E have been found by simplicial grouping of the weak collinearity graph. The group F was found by greedy grouping of the vertices that remained ungrouped after simplicial grouping, i.e., the vertices shown in Figure 4(d).

6   Concluding Remarks

A set of line segments is digitally collinear if their domains have a common non-empty intersection. Since the domains are two-dimensional convex sets satisfying a Helly-type property (Theorem 1), the optimal grouping problem for collinear line segments can be formulated as a combinatorial optimization problem. To solve this problem we have introduced the notion of weak collinearity. Weak collinearity enables us to replace the collinearity hypergraph by an ordinary graph that represents the weakly collinear relationships. With the weak collinearity graph we have described a grouping algorithm that often produces optimal or suboptimal results. Furthermore, we know precisely whether the outcome is optimal or not, and why the result may not be optimal. In fact, a suboptimal result is either due to the occurrence of chordless cycles in the weak collinearity graph, or to the occurrence of triangles that are weakly collinear but not collinear.

We did not discuss the combination of parallelism and collinearity. When we group lines according to parallelism as well as collinearity, a natural requirement is that collinear groups must be subgroups of parallel groups. Also in parallel grouping we use a simplicial grouping process [ ], but the precise relation between the two processes must still be further examined.

References

1. E. Andres, R. Acharya, and C. Sibata: Discrete analytical hyperplanes. *Graphical Models and Image Processing*, **59**, 302–309 (1997) 441
2. F. Gavril: Algorithms for minimum coloring, maximum clique, minimum covering by cliques and maximum independent set of a chordal graph. SIAM J. Comput. , **1**, 180–187 (1972) 447
3. M. Grötschel, L. Lovász, and A. Schrijver: Geometric Algorithms and Combinatorial Optimization. New York: Springer-Verlag, 2nd Ed. (1993) 445
4. H. J. A. M. Heijmans and A. Toet: Morphological sampling. Computer Vision, Graphics & Image Processing: Image Understanding **54**, 384–400 (1991) 441
5. D. Jacobs, D. Weinshall, and Y. Gdalyahu: Classification with nonmetric distances: image retrieval and class representation. *IEEE Trans. Pattern Anal. Machine Intell.*, **22**, 583–600, 2000. 439
6. D. Lowe: 3-d object recognition from single 2-d images. *Artificial Intelligence*, **31**, 355–395, 1987. 439
7. P. Nacken: A metric for line segments. *IEEE Trans. Pattern Anal. Machine Intell.*, **15**,1312–1318, 1993. 439
8. J.-P. Réveillès: Géométrie discrète, calcul en nombres entiers et algorithmique. Thèse de Doctorat d'Etat. Université Louis Pasteur, Strasbourg (1991) 441
9. P. Veelaert: Geometric constructions in the digital plane. J. Math. Imaging and Vision. **11**, 99–118 (1999) 440, 441, 442, 443
10. P. Veelaert: Algorithms that measure parallelism and concurrency of lines in digital images. Proceedings of SPIE's Conference on Vision Geometry VIII, (Denver), SPIE, 69–79 (1999) 441, 443

11. P. Veelaert: Line grouping based on uncertainty modeling of parallelism and collinearity. Proceedings of SPIE's Conference on Vision Geometry IX, (San Diego), SPIE, 36–45 (2000) 441
12. P. Veelaert: Parallel line grouping based on interval graphs. Proceedings of DGCI 2000, (Uppsala), Springer, 530–541 (2000) 441, 452
13. D. B. West: Introduction to Graph Theory. Upper Saddle River: Prentice Hall (1996) 443, 447

Author Index

Lecture Notes in Computer Science

For information about Vols. 1–2179
please contact your bookseller or Springer-Verlag